EAI/Springer Innovations in Communication and Computing

Series editor
Imrich Chlamtac, European Alliance for Innovation, Ghent, Belgium

Editor's Note

The impact of information technologies is creating a new world yet not fully understood. The extent and speed of economic, life style and social changes already perceived in everyday life is hard to estimate without understanding the technological driving forces behind it. This series presents contributed volumes featuring the latest research and development in the various information engineering technologies that play a key role in this process.

The range of topics, focusing primarily on communications and computing engineering include, but are not limited to, wireless networks; mobile communication; design and learning; gaming; interaction; e-health and pervasive healthcare; energy management; smart grids; internet of things; cognitive radio networks; computation; cloud computing; ubiquitous connectivity, and in mode general smart living, smart cities, Internet of Things and more. The series publishes a combination of expanded papers selected from hosted and sponsored European Alliance for Innovation (EAI) conferences that present cutting edge, global research as well as provide new perspectives on traditional related engineering fields. This content, complemented with open calls for contribution of book titles and individual chapters, together maintain Springer's and EAI's high standards of academic excellence. The audience for the books consists of researchers, industry professionals, advanced level students as well as practitioners in related fields of activity include information and communication specialists, security experts, economists, urban planners, doctors, and in general representatives in all those walks of life affected ad contributing to the information revolution.

About EAI

EAI is a grassroots member organization initiated through cooperation between businesses, public, private and government organizations to address the global challenges of Europe's future competitiveness and link the European Research community with its counterparts around the globe. EAI reaches out to hundreds of thousands of individual subscribers on all continents and collaborates with an institutional member base including Fortune 500 companies, government organizations, and educational institutions, provide a free research and innovation platform.

Through its open free membership model EAI promotes a new research and innovation culture based on collaboration, connectivity and recognition of excellence by community.

More information about this series at http://www.springer.com/series/15427

Anandakumar Haldorai • Arulmurugan Ramu
Sudha Mohanram • Mu-Yen Chen
Editors

2nd EAI International Conference on Big Data Innovation for Sustainable Cognitive Computing

BDCC 2019

Springer

EAI
RESEARCH MEETS INNOVATION

Editors
Anandakumar Haldorai
Department of Computer Science and Engineering
Sri Eshwar College of Engineering
Coimbatore, Tamil Nadu, India

Arulmurugan Ramu
Department of Computer Science and Engineering
Presidency University, Bengaluru
Karnataka, India

Sudha Mohanram
Electronics and Communication Engg
Sri Eshwar College of Engineering
Coimbatore, Tamil Nadu, India

Mu-Yen Chen
Dept of Info Mgmt, Sec 3
Natl Taichung Univ of Sci & Tech
Taichung, Taiwan

ISSN 2522-8595 ISSN 2522-8609 (electronic)
EAI/Springer Innovations in Communication and Computing
ISBN 978-3-030-47559-8 ISBN 978-3-030-47560-4 (eBook)
https://doi.org/10.1007/978-3-030-47560-4

This Springer imprint is published by the registered company Springer Nature Switzerland AG
The registered company address is: Gewerbestrasse 11, 6330 Cham, Switzerland

Fig. 1.31 Pollution in South Tangerang, Banten, Indonesia. (Source: Tom Fisk, Pexels)

Conclusions

This chapter has examined the basis of the multidisciplinary, interdisciplinary, and transdisciplinary science of biogeochemistry, including its history, composition, and approaches. The main focus of the chapter was the relationship of biogeochemistry with related sciences, an important factor for the development, and importance of the current science of biogeochemistry. The breadth and depth of this science identifies it as the backbone of many of the environmental and earth sciences. Considering the breadth of this subject, the chapter only covers the main topics that have been cited in the current literature. There is a tendency for biogeochemistry to diverge into close relations with some related sciences, such as the atmospheric, water, biological and earth sciences. The basic study topics of biogeochemistry were noted to be the chemical elements, especially carbon, nitrogen, phosphorous, potassium, oxygen, iron, calcium, selenium, sulfur, and mercury. Compounds including these elements included calcium carbonate, dolomite, carbon monoxide, carbon dioxide, calcium hydroxide, calcium oxide, ammonium, nitrite, nitrate, nitrous oxide, and nitric oxide. The model system illustrative approach was followed to give the reader a clear insight into the intra- and interrelations within and outside biogeochemistry. The succeeding chapters will explore these ideas and possibilities in more detail.

References

Achterberg, E. P. (2011). Grand challenges in marine biogeochemistry. *Frontiers in Marine Science*. https://doi.org/10.3389/fmars.2014.00007

Appl, M. (1999). *Ammonia: Principles and industrial practice*. WILEY-VCH Verlag GmbH.

Arrhenius, S. (1896). On the influence of carbonic acid in the air upon the temperature of the ground. *Philosophical Magazine Series, 5*(41), 273–276.

Barbash, J. E. (2014). The geochemistry of pesticides. *Treatise on Geochemistry, 11*, 535–572. https://doi.org/10.1016/B978-0-08-095975-7.00915-3

Bashkin, V. N., & Howart, R. W. (2002). *Modern Biogeochemistry*. Kluwer Academic Publishers.

Bianchi, T. S. (2021). The evolution of biogeochemistry: Revisited. *Biogeochemistry, 154*, 141–181. https://doi.org/10.1007/s10533-020-00708-0

Bianchi, T. S., Blair, N., Burdige, D., Eglinton, T. I., & Galy, V. (2018). Centers of organic carbon burial at the land-ocean interface. *Organic Geochemistry, 115*, 138–155.

Bianchi, T. S., Anand, M., Bauch, C., Canfield, D., Meester, L., Fennel, K., Groffman, P., Pace, M., Saito, M., & Simpson, M. (2020). *Ideas and perspectives: Biogeochemistry – Its future role in interdisciplinary Frontiers*. https://doi.org/10.5194/bg-2020-395

Böcker, N., Grahl, M., Tota, A., Häussinger, P., Leitgeb, P., & Schmücker, B. (2013). Nitrogen. In *Ullmann's Encyclopedia of industrial chemistry*. https://doi.org/10.1002/14356007.a17_457.pub2

Bodnar, M., Konieczka, P., & Namiesnik, J. (2012). The properties, functions, and use of selenium compounds in living organisms. *Journal of Environmental Science and Health. Part C, Environmental Carcinogenesis & Ecotoxicology Reviews, 30*(3), 225–252. https://doi.org/10.1080/10590501.2012.705164

Boero, F., De Leo, F., Fraschetti, S., & Ingrosso, G. (2019). The cells of ecosystem functioning: Towards a holistic vision of marine space. *Advances in Marine Biology, 82*, 129–153. https://doi.org/10.1016/bs.amb.2019.03.001

Brasseur, P., Gruber, N., Barciela, R., Brander, K., Doron, M., El Moussaoui, A., Hobday, A. J., Huret, M., Kremeur, A.-S., Lehodey, P., Matear, R., Moulin, C., Murtugudde, R., Senina, I., & Svendsen, E. (2009). Integrating biogeochemistry and ecology into ocean data assimilation systems. *Oceanography, 22*(3), 206–215. https://doi.org/10.5670/oceanog.2009.80

Britannica, T. (2020). *Editors of Encyclopaedia 2020*. Biogeochemical cycle. Encyclopedia Britannica. https://www.britannica.com/science/biogeochemical-cycle.

Brusseau, M. L., Pepper, I. L., & Gerba, C. P. (2019). *Environmental and pollution science book* (3rd ed.). Academic Press. https://doi.org/10.1016/C2017-0-00480-9

Butcher, S. S., Charlson, R. J., Orians, G. H., & Wolfe, G. V. (Eds.). (1992). *Global biogeochemical cycles*. Academic Press (Harcourt Brace Jovanovich).

CABI Digital Library. (2023). *Organic phosphorus in the environment*. B.L. Turner, E. Frossard, D.S. Baldwin. Retrieved from https://www.cabidigitallibrary.org/doi/book/. https://doi.org/10.1079/9780851998220.0000.

Calvo, M. S., & Lamberg-Allardt, C. J. (2015). Phosphorus. *Advances in Nutrition (Bethesda, Md.), 6*(6), 860–862. https://doi.org/10.3945/an.115.008516

Campbell, M. (2015). *Vultures: Their evolution, ecology and conservation*. Taylor and Francis (CRC Press).

Campbell, M. (2017). *Biological conservation in the 21st century: A conservation biology of large wildlife*. Nova Science Publishers.

Campbell, M. (2018). *Geomatics and conservation biology*. Nova Science Publishers.

Campbell, M. (2020). *The physical geography of El Salvador: A geophysical and ecological approach*. Nova Science Publishers.

Campbell, M. (2022). *The great eagles: Their evolution, ecology and conservation*. Taylor and Francis (CRC Press).

College of Earth, Ocean, and Atmospheric Sciences (2022). *Ocean Ecology and Biogeochemistry*. Corvallis, OR: College of Earth, Ocean, and Atmospheric Sciences 2022.

Cutter, G. A. (2005). Biogeochemistry: Now and into the future. In N. Noffke (Ed.), *Geobiology: Objectives, concepts, perspectives* (pp. 191–198). Elsevier Science.

Denman, K. L., Brasseur, G., Chidthaisong, A., Ciais, P., Cox, P. M., Dickinson, R. E., Hauglustaine, D., Heinze, C., Holland, E., Jacob, D., Lohmann, U., Ramachandran, S., da Silva Dias, P. L., Wofsy, S. C., & Zhang, X. (2007). Couplings between changes in the climate system and biogeochemistry. In S. Solomon, D. Qin, M. Manning, Z. Chen, M. Marquis, K. B. Averyt, M. Tignor, & H. L. Miller (Eds.), *Climate change 2007: The physical science basis. Contribution of Working Group I to the fourth assessment report of the intergovernmental panel on climate change*. Cambridge University Press.

Department of Marine Sciences. (2022). *Biogeochemistry*. Retrieved from https://www.marsci.uga.edu/research/content/biogeochemistry

Devillanova, F. A., & Mont, W. D. (2007). *Handbook of chalcogen chemistry: New perspectives in sulfur, selenium, and tellurium*. Royal Society of Chemistry.

Dobrovolsky, V. V. (1994). *Biogeochemistry of the world's land*. CRC Press.

Falkowski, P. (2003). Biogeochemistry of primary production in the sea. *Treatise on Geochemistry, 8*, 185–213. https://doi.org/10.1016/B0-08-043751-6/08129-9

Fantle, M. S., & Tipper, E. T. (2014). Calcium isotopes in the global biogeochemical Ca cycle: Implications for development of a Ca isotope proxy. *Earth-Science Reviews, 129*, 148–177. https://doi.org/10.1016/j.earscirev.2013.10.004

Fox, P. M., Carrero, S., Anderson, C. L., Dewey, C. P., Keiluweit, M., Conrad, M., Naughton, H. R., Fendorf, S., Carroll, R. W., Dafflon, B., Malenda-Lawrence, H., Dwivedi, D., Gilbert, B., Christensen, J. N., Boye, K., Beutler, C. A., Brown, W. S., Newman, A. W., Versteeg, R., Williams, K. H., & Nico, P. S. (2022). Sulfur biogeochemical cycling and redox dynamics in a shale-dominated mountainous watershed. *Journal of Geophysical Research: Biogeosciences, 127*(6), e2021JG006769.

Gandhi, S. M., & Sarkar, B. C. (2016). *Essentials of mineral exploration and evaluation*. Elsevier. https://doi.org/10.1016/C2015-0-04648-2

Gardner, C. B., Berry Lyons, W., & Long, D. T. (2014). *Defining urban geochemistry*. International Association of Geochemistry: Urban Geochemistry Working Group; Columbus, Ohio, 5–6 August 2014. Retrieved from https://eos.org/science-updates/defining-urban-geochemistry

Gorham, E. (1991). Biogeochemistry: Its origins and development. *Biogeochemistry, 13*(3), 199–239.

Greenwood, N. N., & Earnshaw, A. (1997). *Chemistry of the elements* (2nd ed.). Butterworth-Heinemann.

Grenne, T., & Slack, J. F. (2019). Mineralogy and geochemistry of silicate, sulfide, and oxide iron formations in Norway: Evidence for fluctuating redox states of early Paleozoic marine basins. *Mineralium Deposita, 54*, 829–848. https://doi.org/10.1007/s00126-018-0840-2

Grimm, N. B., Faeth, S. H., Golubiewski, N. E., Redman, C. L., Wu, J. G., Bai, X. M., & Briggs, J. M. (2008). Global change and the ecology of cities. *Science, 319*(5864), 756–760.

Hartikainen, H. (2005). Biogeochemistry of selenium and its impact on food chain quality and human health. *Journal of Trace Elements in Medicine and Biology: Organ of the Society for Minerals and Trace Elements (GMS), 18*(4), 309–318. https://doi.org/10.1016/j.jtemb.2005.02.009

Hellweger, F. L. (2008). Biogeochemical Models. In *Encyclopedia of ecology* (pp. 386–396). https://doi.org/10.1016/B978-008045405-4.00177-4

Hoffmann, J. E., & King, M. G. (2010). Selenium and selenium compounds. In *Kirk-Othmer Encyclopedia of chemical technology*. https://doi.org/10.1002/0471238961.190512.0508150606.a01.pub3

House, J. E. (2008). *Inorganic chemistry*. Academic Press. https://doi.org/10.1039/JM9920201123

Hutton, J. (1795). *The history of the earth, 2 volumes*. Cadell, Junior, and Davies. Seen in reprint (1959). Wheldon and Wesley, Cadicote, Hertfordshire, England.

Jäger, H., Frohs, W., Collin, G., von Sturm, F., Vohler, O., & Nutsch, G. (2010). Carbon, 1. General. In *Ullmann's Encyclopedia of industrial chemistry*. https://doi.org/10.1002/14356007.a05_095.pub2

Kabata-Pendias, A., & Pendias, H. (1999). *Biogeochemia pierwiastków śladowych (Polish)* (2nd ed., pp. 344–354). PWN Warszawa.

Kappler, A., Bryce, C., Mansor, M., Lueder, U., Byrne, J. M., & Swanner, E. D. (2021). An evolving view on biogeochemical cycling of iron. *Nature Reviews. Microbiology, 19*(6), 360–374. https://doi.org/10.1038/s41579-020-00502-7

Karl, D. M., & Schlesinger, W. H. (2014). Volume editors' introduction. In H. D. Holland & K. K. Turekian (Eds.), *Treatise on geochemistry* (Vol. 10, 2nd ed., pp. xxiii–xxvi). Elsevier Science. https://doi.org/10.1016/B978-0-08-095975-7.09983-6

Kaye, J. P., Groffman, P. N., Grimm, N. B., Baker, L. A., & Pouyat, R. V. (2006). A distinct urban biogeochemistry? *Trends in Ecology & Evolution, 21*(4), 192–199.

Kirschner, M. J., Alekseev, A., Dowy, S., Grahl, M., Jansson, L., Keil, P., Lauermann, G., Meilinger, M., Schmehl, W., Weckler, H., & Windmeier, C. (2017). Oxygen. In *Ullmann's Encyclopedia of industrial chemistry*. https://doi.org/10.1002/14356007.a18_329.pub2

Klein, C., & Hurlbut, C. S., Jr. (1985). *Manual of mineralogy (after James D. Dana)* (20th ed.). Wiley.

Kuypers, M., Marchant, H., & Kartal, B. (2018). The microbial nitrogen-cycling network. *Nature Reviews Microbiology, 16*, 263–276. https://doi.org/10.1038/nrmicro.2018.9

Kvakić, M., Tzagkarakis, G., Pellerin, S., Ciais, P., Goll, D., Mollier, A., & Ringeval, B. (2020). Carbon and phosphorus allocation in annual plants: An optimal functioning approach. *Frontiers in Plant Science, 11*, 149. https://doi.org/10.3389/fpls.2020.00149

Laane, R. W. P. M., & Middleburg, J. J. (2011). Biogeochemistry, an introduction. In E. Wolanski & D. McLusky (Eds.), *Treatise on estuarine and coastal science* (pp. 1–5). Academic Press.

Libes, S. M. (1992). *Introduction to marine biogeochemistry*. Academic Press.

Lide, D. R. (Ed.). (2005). *CRC handbook of chemistry and physics* (86th ed.). CRC Press.

Linderoth, S., & Mørup, S. (1992). Stability and magnetic properties of an iron-mercury alloy. *Journal of Physics: Condensed Matter, 4*, 8627–8634.

Malyuga, D. (1995). *Biogeochemical methods of prospecting*. Springer.

National Ocean Service. (2023). *What is the carbon cycle?* Retrieved from https://oceanservice.noaa.gov/facts/carbon-cycle.html#transcript

Nehb, W., & Vydra, K. (2006). Sulfur. In *Ullmann's Encyclopedia of industrial chemistry*. https://doi.org/10.1002/14356007.a25_507.pub2

Nogara, P. A., Madabeni, A., Bortoli, M., Teixeira Rocha, J. B., & Orian, L. (2021). Methylmercury can facilitate the formation of Dehydroalanine in Selenoenzymes: Insight from DFT molecular modeling. *Chemical Research in Toxicology, 34*(6), 1655–1663. https://doi.org/10.1021/acs.chemrestox.1c00073

Odum, E. P. (1971). *Fundamental of ecology*. W.B. Saunders.

Pickett, S. T. A., Cadenasso, M. L., Grove, J. M., Boone, C. G., Groffman, P. M., Irwin, E., Kaushal, S. S., Marshall, V., McGrath, B. P., Nilon, C. H., Pouyat, R. V., Szlavecz, K., Troy, A., & Warren, P. (2011). Urban ecological systems: Scientific foundations and a decade of progress. *Journal of Environmental Management, 92*(3), 331–362.

Rieuwerts, J. (2015). *The elements of environmental pollution*. Earthscan Routledge.

Rytuba, J. J. (2003). Mercury from mineral deposits and potential environmental impact. *Environmental Geology, 43*, 326–338. https://doi.org/10.1007/s00254-002-0629-5

Schlesinger, W. H. (1997). *Biogeochemistry: An analysis of global change* (2nd ed.). Academic Press.

Schlesinger, W. H. (2004). Better living through biogeochemistry. *Ecology, 85*(9), 2402–2407.

Schlesinger, W. H. (2021a). *Biogeochemistry*. Retrieved from https://www.oxfordbibliographies.com/view/document/obo-9780199830060/obo-9780199830060-0111.xml

Schlesinger, W. H. (2021b). Some thoughts on the biogeochemical cycling of potassium in terrestrial ecosystems. *Biogeochemistry, 154*, 427–432. https://doi.org/10.1007/s10533-020-00704-4

Schlesinger, W. H., & Bernhardt, E. S. (2020). *Biogeochemistry: An analysis of global change* (4th ed.). Academic Press.

Schmitz, O. J. (2020). Predators and rainfall control spatial biogeochemistry in a landscape of fear. *PNAS, 117*(39) https://www.pnas.org/doi/pdf/10.1073/pnas.2016449117

Schultz, H., Bauer, G., Schachl, E., Hagedorn, F., & Schmittinger, P. (2000). Potassium compounds. In *Ullmann's Encyclopedia of industrial chemistry*. https://doi.org/10.1002/14356007.a22_039

Schulze, E. D. (2015). Biogeochemistry: Historical and future perspectives. *Austin Journal of Earth Science, 2*(2), id1012.

Scripps Institution of Oceanography. (2022). *Earth Biogeochemistry*. Retrieved from https://scripps.ucsd.edu/research/topics/biogeochemistry

Sharma, V. K., McDonald, T. J., Sohn, M., Anquandah, G. A. K., Pettine, M., & Zboril, R. (2015). Biogeochemistry of selenium. A review. *Environmental Chemistry Letters, 13*, 49–58. https://doi.org/10.1007/s10311-014-0487-x

Singh, A. D., Khanna, K., Kour, J., Dhiman, S., Bhardwaj, T., Devi, K., Sharma, N., Kumar, P., Kapoor, N., Sharma, P., Arora, P., Sharma, A., & Bhardwaj, R. (2023). Critical review on biogeochemical dynamics of mercury (Hg) and its abatement strategies. *Chemosphere, 319*, 137917. https://doi.org/10.1016/j.chemosphere.2023.137917

Smart, L. E., & Moore, E. A. (2005). *Solid state chemistry: An introduction* (3rd ed.). CRC Press.

Steffen, W., Richardson, K., Rockström, J., Schellnhuber, H. J., Dube, O. P., Dutreuil, S., Lenton, T. M., & Lubchenco, J. (2020). The emergence and evolution of earth system science. *Nature Reviews Earth & Environment, 1*(1), 54–63.

Suess, E. (1875). *Die Entstehung der Alpen*. The Origin of the Alps. W. Braumüller.

Tartowski, S. L., & Howarth, R. W. (2013). Nitrogen, Nitrogen Cycle. In *Encyclopedia of biodiversity* (2nd ed., pp. 537–546). https://doi.org/10.1016/B978-0-12-384719-5.00098-8

Testa, J. M., & Kemp, W. M. (2011). Oxygen - dynamics and biogeochemical consequences. In E. Wolanski & D. S. McLusky (Eds.), *Treatise on Estuarine and coastal science* (pp. 163–199). Academic Press. https://doi.org/10.1016/B978-0-12-374711-2.00505-2

Thomas, I. M., & Weller, M. T. (1992). Synthesis, structure and thermal properties of phosphophyllite, Zn2Fe(PO4)2·4H2O. *Journal of Materials Chemistry, 2*, 1123–1126.

Turner, B. L., Frossard, E., & Baldwin, D. S. (2005). *Organic phosphorus in the environment*. CABI Publishing.

Vernadsky, V. I. (1926). *The biosphere*. Trans. Langmuir, D McMenamin. Springer.

Vernadsky, V. I. (1944). Problems in biogeochemistry. II. *Transactions of the Connecticut Academy of Arts and Sciences, 35*, 493–494.

Vernadsky, V. I. (1945). The biosphere and the noosphere. *Scientific American, 33*, 1–12.

Warf, B. (2010). *Biogeochemical cycles. Encyclopedia of geography*. SAGE Publications. https://doi.org/10.4135/9781412939591

Warren, J. (2000). Dolomite: Occurrence, evolution, and economically important associations. *Earth-Science Reviews, 52*(1–3), 1–81.

Wildermuth, E., Stark, H., Friedrich, G., Ebenhöch, F. L., Kühborth, B., & Silver, J. (2000). Iron compounds. In *Ullmann's Encyclopedia of industrial chemistry*. https://doi.org/10.1002/14356007.a14_591

Woods Hole Oceanographic Institution. (2022). *Biogeochemistry*. Retrieved from https://www.whoi.edu/know-your-ocean/ocean-topics/ocean-chemistry/ biogeochemistry/

Zavarzin, G. A. (2008). Microbial cycles. In *Encyclopedia of ecology* (Vol. 4, 2nd ed., pp. 129–134).

Chapter 2
Geochemistry as the Core of Biogeochemistry

Michael O'Neal Campbell

Abstract This chapter examines the science of geochemistry, as the core, background discipline of biogeochemistry. Biogeochemistry is based on geochemistry; in that, it is the framework for those aspects of biology that link to the chemical bases of earth materials, which are defined as within the study of geochemistry. Therefore, the justification for this chapter is that a fuller understanding of biogeochemistry requires a study of geochemistry and the possibilities for its linkage with the biological sciences. Geochemistry is the study of the interfaces of geology and chemistry, using the methodology of the chemical sciences to investigate the composition of earth materials and the occurrence and movements of chemical elements and compounds within the earth system. Geochemistry is consequently an extremely wide subject, as it examines all aspects of the chemistry of earth materials. However, few studies have explored the links between geochemistry and the biological sciences, as the basis for biogeochemical studies. This chapter examines the relevant topics of chemistry, the content of geochemistry, and the developments that have forged a closer link with biogeochemistry. The branches of chemistry, namely, analytical, inorganic, organic, physical chemistry, and biochemistry, are examined, as is their relationship with geochemistry. The branches of geochemistry include organic geochemistry, inorganic geochemistry, isotope geochemistry, aqueous geochemistry, cosmochemistry, trace-element geochemistry, igneous rock geochemistry, metamorphic rock geochemistry, photogeochemistry, and low-temperature or environmental geochemistry. The literary evidence indicates that developments in chemistry, biology, geology, and even archaeology and astronomy (the last two linked to isotope geochemistry and cosmochemistry) have benefited geochemistry as a discipline, and in combination, these have contributed to the advancement of biogeochemistry. Variable issues in biogeochemistry are principally the concern of some branches of geochemistry, such as carbon, inorganic and marine chemistry, and organic, inorganic, isotope, and aqueous geochemistry. This examination contributes to knowledge on the interfaces between the biological, chemical, and geological sciences.

M. O. Campbell (✉)
Simon Fraser University, Burnaby, BC, Canada
e-mail: evaz@torontomu.ca

M. O. Campbell, *Biogeochemistry and the Environment*,
https://doi.org/10.1007/978-3-031-47017-2_2

Introduction

Geochemistry, the science of the chemistry of the solid Earth, and its interactions with biological and biochemical topics are arguably core backgrounds to biogeochemistry; hence, a description of geochemistry is a necessary prelude to a detailed study of biogeochemistry (Campbell, 2020). Here, a short description of geochemical links to biogeochemistry is presented, followed by a detailed set of definitions of geochemistry. As geochemistry refers to the chemistry of the Earth, Hellweger (2008: 386) argues that "the field of biogeochemistry deals with the effect of biological organisms on the chemistry of the Earth. Since there are numerous living organisms, all of which affect the chemistry of their environment in multiple ways, biogeochemistry is a large subject area covering many processes." The noted authority Hallberg (2009: 435) adds that "biogeochemistry tries to combine biology and geochemistry to achieve a deeper understanding of the whole complex of chemical transformations occurring in nature, in the continuous interactions between its biotic and abiotic components." Biogeochemistry is argued to straddle three disciplines (Hallberg, 2009: 436), namely biology, geology, and chemistry, and as the "root of the term biogeochemistry is chemistry and above all chemistry related to geoscience: geochemistry," it is noted that "adding the prefix bio to geochemistry implies that biogeochemistry is concerned with the activities of living organisms in the migration, distribution, dissipation, and concentration of chemical elements" (Hallberg, 2009: 436).

Commenting on geochemistry's important, developing role in recent scientific development, White (2017) writes that "geochemistry has grown over the last 50 or 60 years to touch virtually every aspect of earth science. The contributions of geochemistry to this advance have been simply enormous. Much of this progress has come from innovation in analytical techniques and the ability to assess the nature of natural materials with evermore precision and on evermore finer scales." Bianchi (2021: 154) gives a copious account of the development and relevance of geochemistry to the development of a complex model of biological, geological, and chemical interactions (biogeochemistry) at the global and smaller scales, arguing that geology still played a role in the evolution of biogeochemistry in the twentieth and twenty-first centuries, including topics such as global sedimentary modeling, plate tectonics theory, carbon and climate change, and increasingly global chemical cycling (see also Gaillardet & Galy, 2008). Bianchi (2021, 155) further argues that geochemistry evolved out of the writing mostly of Clarke and Washington (1924) on the chemical composition of the Earth's crust, Goldschmidt (1954), and Strunz (1941). It is argued that "these important chemical developments now allow in part, for perspectives on how biological compounds interact with minerals in nature and how organic matter is chemically altered after being buried for millennia – we now see the emergence of yet another new discipline inherently linked with biogeochemistry, organic geochemistry" (Bianchi, 2021, 155; see also Kvenvolden, 2006).

The development of organic chemistry, and related topics in geochemistry also played a role in the development of biogeochemistry (Bianchi, 2021). This development acknowledged the importance of organic compounds in previously inorganic-focused analyses. Organic geochemistry redirected the focus to biochemical issues, such as botanical physiology and productivity, and climate change, and links to organic residues such as crude oil and other carbon compounds in the Earth's crust. Early writings in this regard included those of Treibs (1936) (considered by some as the father of organic geochemistry) and of Trask and Patnode (1942), which in some cases "introduced the broader geological linkages between bulk organic material in source sedimentary rock and source beds of petroleum" and examined links between organic geochemistry and biogeochemistry (Bianchi, 2021). These studies developed the notion that some fossils and fossil organic molecules can be used as chemical biomarkers to provide a window to the past ecosystems (Bianchi, 2021, 155; see also Peters et al., 2005; Kvenvolden, 2006; Gaines et al., 2009; Bianchi & Canuel, 2011; Greenwalt et al., 2013; Killops & Killops, 2013; Briggs & Summons, 2014; Whiteside & Grice, 2016). Bianchi (2021: 155) further elaborates that "the application of novel chemical biomarkers/proxies, both organic and inorganic, in paleoreconstruction has been particularly useful in understanding major biogeochemical shifts in Earth's 4-billion-year history, which are central in making predictions about future changes in global biogeochemical cycles" (see also Lenton & Daines, 2017).

Geochemistry is defined as the study of the Earth using the techniques of chemistry, which may also be used for studies linked to the relevant biological sciences, and hence biogeochemistry (Campbell, 2020). Geochemistry, as a complex discipline, has been defined and described by numerous authorities with sometimes variable foci, since its development in the late nineteenth century. These definitions point to geochemistry as principally a chemical science, which uses the techniques of the chemical sciences for the study, analyses, and understanding of Earth studies. Victor Goldschmidt (1888–1947) is the oft-cited "father" of geochemistry, especially due to his systematization of the "behavior of the chemical elements in an Earth Sciences context via his classification of them according to their preferred host phases on silicate Earth (Lithophile, siderophile, chalcophile and atmophile)" (Mather, 2013). Mather (2013) further notes that geochemistry is "a broad and fascinating subject both in terms of the topics it addresses, and the techniques employed." Lastovicka (2009) writes that "Goldschmidt's definition of this scientific discipline from the 1930s is reasonably applicable even now," which was "the distribution and amounts of the chemical elements in minerals, ores, soils, waters, and the atmosphere, and the circulation of the elements in nature, on the basis of the properties of their atoms and ions."

White (2017) argues that based on the name "geochemistry," this science is "a marriage between chemistry and geology, or more broadly geosciences or earth science, and it is arguably a subdiscipline of both. A better explanation of geochemistry is that it is applies chemistry and chemical principles to understanding the Earth and its cosmic environment and to using that understanding to better the human condition." Britannica (2014) defines geochemistry as the "scientific discipline that deals with the relative abundance, distribution, and migration of the Earth's chemical elements and their isotopes." Inamuddin et al. (2021: xiii) give a much broader definition: geochemistry is "a branch of earth science" and note that "since it is a field of study that uses the tools and principles of chemistry to explain the mechanisms in geologic environments, it often focuses on determining processes that control the abundance and composition of minerals and their distribution in the earth's crust..." and the subject also "plays a vital role in environmental soil and water systems in identifying and modulating environmental problems, and in studying the composition, structure, and processes of the Earth." Yale University (2022) gives a similar definition as "the branch of Earth Science that applies chemical principles to deepen an understanding of the Earth system and systems of other planets" and its practitioners "consider Earth composed of discrete spheres – rocks, fluids, gases and biology – that exchange matter and energy over a range of time scales."

Sarala (2015) writes that "geochemistry can be broadly defined as the science concerned with all geological studies involving chemical change. It includes the study of the distribution of elements in minerals, rocks, and soils along with the interaction between these earth materials." Mather (2013) points out that geochemistry "utilizes the tools and principles of chemistry to explain the mechanisms regulating the workings – past and present – of the Earth. The reach of geochemistry is broad and extends to the study of the formation of the Earth and the Solar System as well as topics such as the origins and evolution of the Earth's crust, oceans, and atmosphere." Lastovicka (2009) defines geochemistry as dealing with "the distribution and cycling of the chemical elements, and their isotopes, throughout nature. It is of great importance to understanding the Earth and planets, and their origin and nature" and as eventually evolving from "a substantially descriptive to a highly quantitative and predictive discipline."

Trueman et al. (2019: 271) look further at the geochemical research methods, defining geochemistry as the study of "the chemistry of natural earth materials and the chemical processes operating within and upon the Earth, both now and in the past," with geochemical analyses "carried out on any natural sample such as air, volcanic gas, water, dust, soil, sediment, rock, or biological hard tissues (especially ancient biological tissues) and also on anthropogenic materials such as industrial effluent and sewage sludge. Geochemical analyses, therefore, involve a wide range of materials and analytes of interest and may be performed for industrial, environmental, or academic reasons." They further argue that "all of the naturally occurring elements in the periodic table are important for one geochemical investigation or another." The application of geochemical research techniques may focus on the "analysis of the inorganic constituents of the materials most closely associated with geology – rocks and rock forming minerals" or may be broader, including "the chemical analysis of soils, waters, biological tissues, and organic geochemistry" (Trueman et al., 2019: 271).

Another set of definitions deals with the development and components of geochemistry as a discipline. Currently, geochemistry has developed into a well-established discipline, with numerous subdisciplines (Campbell, 2020). Mather (2013) determines the subdisciplines of geochemistry as distinguishable from each other variably, with considerable overlaps between them. The criteria for the division include "analytical geochemistry in contrast to theoretical studies, the technique being employed or the types of measurements an area is concerned with (e.g., isotope geochemistry); the type of sample (e.g., igneous geochemistry and organic and petroleum geochemistry); the class of scientific questions being addressed (e.g., cosmochemistry); the pressure-temperature regime that a certain subset of geochemists are engaged with (e.g., low-temperature geochemistry)." Geochemistry is also increasingly linked to other, related sciences (Campbell, 2020). For example, Lastovicka (2009) argues that geochemistry as an interdisciplinary subject links with the earth sciences of geology, geophysics, astronomy, planetary sciences and atmospheric science, and the more basic sciences of physics, chemistry, biology, and material sciences. Geochemical research methods also have important practical functions, including "the search for deposits of various raw materials like oil, gas, and various metallic ores" and "direct environmental applications" (Lastovicka, 2009). These complexities may intrude into the relationship between geochemistry and biogeochemistry, with both fields being extremely complex and variously linked to other disciplines (Campbell, 2020).

Lastovicka (2009) further notes that "the study of geochemistry is of great importance to the Earth and planetary sciences, on both purely scientific and more immediate practical grounds, because chemical processes are fundamental to understanding how the planetary bodies formed and evolved at all scales, from atomic to solar system." This vital background means that "geochemistry plays a central role in understanding a diverse set of scientific questions, such as the formation and differentiation of the Earth and planets, the origin and evolution of life, the controls on global climate and climate change, and the formation and management of natural resources" (Lastovicka, 2009). Scott (2014) concurs: "Geochemistry plays an essential role in our understanding of processes that produce economic concentrations of minerals whether by hydrothermal, magmatic, metamorphic, hydraulic (both surficial and subterranean) or weathering agents, or a combination of these. Geochemistry also contributes importantly to exploration."

The rest of this chapter continues the trend of the main thesis of this book, that an understanding of biogeochemistry requires an in-depth study of the associated sciences that contribute to the theoretical developments, methodologies and research technologies of biogeochemistry. The relevant developments in geochemistry include those within the fields of organic, inorganic, aqueous, trace-element and isotope geochemistry, which may link with biogeochemistry through chemical cycling, and some of the more solid earth subdisciplines such as igneous and metamorphic rock geochemistry. Cosmogeochemistry and photo-geochemistry are also important developing areas.

Links of Geochemistry with Biogeochemistry

The main link between geochemistry and biogeochemistry appears to be the extent of connections between geochemistry and biological topics, which add the "bio" prefix to the name of the science. Both disciplines retain similar structures. Indiana

University (2022) notes that "geochemistry and biogeochemistry are inherently multidisciplinary, positioned at the intersection of biology, geology, and chemistry, and typically combining field investigations and sampling with laboratory analyses and experimentation to yield empirical data that can constrain computational models." Recent, cutting-edge developments concerning the links between the two disciplines concern the addition of biological links to geochemistry, from which the foundations of biogeochemistry emerge.

Sahai et al. (2016: 389) note that the link between geochemistry and biogeochemistry is mostly concerned with two multidisciplinary subjects, but the addition of biological studies may include the origin of life to the Earth science perspective. Notably, "paradigm-changing discoveries about stellar and planetary evolution, the survival of organic molecules and microorganisms under extreme conditions, and geochemical environments on early Earth and other planets are sparking a synergistic dialogue between geoscientists, chemists, and biologists to understand how life originated" (Sahai et al., 2016: 389). It is argued that a principal approach is to "explain the non enzymatic synthesis of biologically relevant organic molecules under geologically plausible conditions," "overcome the rigid conceptual dichotomy of the "RNA world" versus the "metabolism-first" hypotheses" and "develop high-throughput analytical systems to sample the myriad possible combinations of environmental conditions to find those that could initiate life" (Sahai et al., 2016: 389). Hence, the links between geochemistry and biogeochemistry may include the study of developments with organic chemistry, and the organic molecules and other links with possible origins of life, this adding the "bio" to geochemistry. Sahai et al. (2016: 389) remark, concerning the issue of the journal *Elements* that it is vital to "highlight the roles of minerals and geochemical environments in the emergence of protocells, the cell-like entities that might have preceded the Last Universal Common Ancestor."

Bianchi (2021: 141) notes, concerning the development of biogeochemistry with its links with geochemistry, that "the development of organic geochemistry as a discipline, allowed for new roots to develop in the evolution of biogeochemistry through linkages between short and long-term carbon cycles." Mather (2013) adds points on the importance of organic geochemistry to the links with biogeochemistry, as "organic geochemistry studies the distribution, composition, and fate of organic matter in the geosphere on both bulk and molecular levels, combining aspects of geology, chemistry, and biology. This is often concerned with biogeochemical cycles, with the carbon cycle probably being one of the most fundamental, without which life could not exist." Sinha (2013) explores the area between organic geochemistry and biogeochemistry, as the former is "the study of the organic compounds found in geologic materials and meteorites, including those of problematic biological origin" and the latter is "the study of the behaviour of inorganic chemical elements in biological systems of geologic scope." The subject areas classified within biogeochemistry and organic geochemistry include ore deposit, petroleum and life origins, coal chemistry, the composition of primitive atmospheres, biogeochemical prospecting for mineral deposits, the chemistry of natural waters, and soil formation. As "almost all geologic processes that occur at Earth's surface are affected by biological activity" (Sinha, 2013), these links underscore the role of organic geochemistry and of geochemistry in general for the understanding and development of biogeochemistry.

Subfields of Geochemistry

Subfields of geochemistry are listed in Table 2.1. These are differentiated by their methodologies, and element, isotope, and regional focus (Langmuir, 1997; Dickin, 2005; Killops & Killops, 2013; Mather, 2013; Schlesinger & Bernhardt, 2013; Rass et al., 2014; Doane, 2017; White, 2020; Organic Geochemistry, 2022). Each of these subfields share a common interest: the distribution, change and formation of chemical elements and compounds on the Earth's surface (land, soil, water, rock) and sometimes that of other space bodies. Such chemical changes may be due to volcanism, deposition, erosion, heat, or even light. With such a broad range of factors for composition and change, each subfield may be related to other fields, such as geophysics and atmospheric physics, hydrology, biology, sedimentology, and stratigraphy, as well as the subfields of chemistry, including physical, organic, and inorganic chemistry (Housecroft & Sharpe, 2008; Atkins & de Paula, 2009; Burrows et al., 2009; White, 2020). Each of these subfields will be examined, including examples of some of the recent, cutting-edge research on related topics, based on published articles in refereed journals and critical book chapters.

Recent studies of geochemistry generally focus on ever more detailed examination and analyses of the composition of Earth surface materials, using increasingly

Table 2.1 Subfields of geochemistry

Subfield	Main research focus
Organic geochemistry	Role of organic compounds in geochemistry
Inorganic geochemistry	Role of inorganic and organometallic compounds (not carbon-based)
Isotope geochemistry	Element isotopes, measured with isotope ratio mass spectrometry
Aqueous geochemistry	Elements in natural waters (e.g., copper, sulfur, mercury)
Cosmochemistry	The chemical composition and formation of space bodies
Trace-element geochemistry	Elements too small to be included in a mineral's chemical formula
Igneous rock geochemistry	Igneous rock formation, including magmatic processes
Metamorphic rock geochemistry	Geochemical changes due to metamorphism
Photogeochemistry	Sunlight-induced chemical reactions on Earth surface materials
Low-temperature geochemistry	Chemical processes under normal Earth surface conditions, ecology

Sources: Langmuir (1997), Dickin (2005), Killops and Killops (2013), Mather (2013), Schlesinger and Bernhardt (2013), Rass et al. (2014), Doane (2017), White (2020), Organic Geochemistry (2022)

sophisticated methods. This can be seen by an examination of the articles published in the leading geochemically oriented journals. As noted by Lu et al. (2019: 682) "academic journals are important carriers of scientific activities, important academic communication media, and channels in the process of scientific development, and an important research unit in the structure and classification system of scientific knowledge." Leading journals include: Geochemistry, Geochemistry International, Organic Geochemistry, Applied Geochemistry, Results in Geochemistry, Geochimica et Cosmochimica Acta, Geochemical Journal, Geochemical Exploration, International Journal of Coal Geology, Lithos, Journal of Asian Earth Sciences: X, Journal of Palaeogeography, Journal of South American Earth Sciences, Engineering Geology, Geoscience Frontiers, Physics and Chemistry of the Earth, Parts A/B/C, and Marine and Petroleum Geology.

Others are Chemical Geology, Journal of Volcanology and Geothermal Research, Precambrian Research, Journal of Geodynamics, Oceanologia, Continental Shelf Research, Contributions to Mineralogy and Petrology, Journal of Geophysical Research, American Mineralogist, Bulletin of Volcanology, Precambrian Research, Earth and Planetary Science Letters, Journal of Volcanology and Geothermal Research, Geophysics, AAPG Bulletin, Solid Earth, Journal of Geosciences, Environmental Chemistry, Geological Journal, Journal of the Geological Society, Australian Journal of Earth Sciences, Tellus Series B: Chemical and Physical Meteorology, Nonlinear Processes in Geophysics, Applied Clay Science, Environmental Geochemistry, Physics and Chemistry of Minerals and Lithosphere. Tables 2.2, 2.3, 2.4, 2.5, 2.6, and 2.7 present the main research foci of these journals, topics ranging from geochemistry to related sciences.

Some recent, cutting-edge studies on the various branches of geochemistry are listed in the sections below. Geochemical research tends to advance with attention to greater detail and specialization, simultaneously with more studies linking geochemistry to developments in related or linked sciences (transdisciplinary, multidisciplinary, interdisciplinary, methodology), such as geophysics and related physics subdisciplines, branches of chemistry (including environmental chemistry and even pollution), space sciences, oceanography, biology and fossil studies, and computer sciences. In these blossoming relations, geochemistry is gradually spanning the entire range of the earth and environmental sciences, and with this development a possibly broader engagement with biogeochemistry (Geochemical Society, 2022; Geoscience Australia, 2022; Imperial College, 2022). Each section below comments on several such recent studies, mainly in the topic listed in the heading. These studies are taken from some of the journals listed above and indicate some of the directions of geochemistry.

Organic Geochemistry

Organic geochemistry is the subdiscipline of geochemistry that focuses on organic (carbon bearing) compounds found in geologic environments. Research methods are mostly focused on the study of organic, carbon-based compounds in earth

Table 2.2 Key research areas of leading geochemistry-focused journals

Journal	Key topics
Geochemistry	Cosmochemistry, meteoritics; igneous, metamorphic, and sedimentary petrology; volcanology, low- and high-temperature geochemistry, experimental–theoretical–field-related studies, Mineralogy – crystallography; environmental geosciences, Archaeometry. Source: Science Direct (2022i). These points are the main foci of the Journal Editors. Cited, "Geochemistry was founded as Chemie der Erde 1914 in Jena, and, hence, is one of the oldest journals for geochemistry-related topics."
Geochemistry International	Biogeochemistry, cosmochemistry; geochemistry of magmatic, metamorphic, hydrothermal, and sedimentary processes; Experimental isotope, and organic geochemistry; Chemistry of the environment, environmental geosciences, Archaeometry Source: Springer Nature (2022c).
Organic Geochemistry	Biogeochemistry, organic geology, sedimentology, organic-rich rocks and soils, organic petrography, environmental geochemistry, aqueous and sediment organic matter, chemical oceanography organic hydrology, molecular geochemistry, isotopic geochemistry, paleontology, organics in element geochemical cycling, petroleum, and natural gas Source: Science Direct (2022a). It is cited that "Organic Geochemistry serves as the only dedicated medium for the publication of peer-reviewed research on all phases of geochemistry in which organic compounds play a major role."
Applied Geochemistry	Environmental geochemistry and biogeochemical cycling, hydrogeochemistry and hydrogeology, geochemical dynamics across air-water-soil interfaces, contaminants processes, impacts and remediation, medical geochemistry and health, geochemistry in environmental disasters and sustainability, mineral and energy resources exploration and recovery Source: Science Direct (2022c). This journal focuses on any geochemistry topics (inorganic, organic and isotope geochemistry, geochemical processes, spatial and temporal monitoring case studies, theoretical and fundamental studies) "provided they meet the main criterion," which they have "have some practical applications or implications to human endeavour and wellbeing" (Science Direct, 2022d).
Results in Geochemistry	Biogeochemistry, cosmochemistry, Planetary geochemistry, Isotopic and elemental geochemistry, Organic and inorganic geochemistry; Hydrosphere, biosphere, lithosphere geochemical processes; Igneous, metamorphic, sedimentary petrology; Low- and high-temperature aqueous solutions; applied geochemistry; Paleoclimate Source: Science Direct (2022w). Additionally, the journal includes "papers dealing with experimental results and significant analytical advances in geochemistry."
Geochimica et Cosmochimica Acta	Organic geochemistry, isotope geochemistry, lunar science, planetary geochemistry, meteoritics and meteorite impacts, igneous and metamorphic petrology, chemical processes in the earth's surface (lithosphere etc.), chemistry of gases, solutions, glasses, crystalline solids Source: Science Direct (2022j). This includes "research papers in a wide range of subjects in terrestrial geochemistry, meteoritics, and planetary geochemistry" (Science Direct, 2022j).

(continued)

Table 2.2 (continued)

Journal	Key topics
Geochemical Journal	Organic Geochemistry, Isotope Geochemistry and Geochronology, Cosmochemistry, Mineral and Rock Chemistry, Volcanology and Hydrothermal Chemistry, Atmospheric Chemistry, Hydro- and Marine Chemistry, Environmental Chemistry Source: Geochemical Journal (2022). This is "an international journal devoted to original research papers in geochemistry and cosmochemistry" (Geochemical Journal, 2022).

Table 2.3 Key research areas of leading geochemistry-focused journals

Journal	Key topics
Journal of Geochemical Exploration	Development of mineral deposits, e.g., element transfer; Boundaries of biogeochemistry, metal, mineral accumulation; Quantification of natural and technogenic radioactivity; Anomalies, baseline references, compositional, geospatial analysis; Pollutant sources, transport in surfaces, solid, fluid, aerial matrices; Impacts of mining on the surface environment; Ecosystem pollution, health, deterministic, stochastic approaches. Source: Science Direct (2022n). This journal is "mostly dedicated to publication of original studies in exploration and environmental geochemistry and related topics" and focuses on methods in geochemistry (Science Direct, 2022n).
Chemical Geology	Biogeochemistry; Cosmochemistry; Chemistry of igneous, metamorphic, and sedimentary petrology; Low- and high-temperature aqueous solutions; Field, experimentally, or computationally based studies; Not remediation, applied geochemistry/economic geology studies; Significant analytical advances of wide interest; Broad international topics, rather than regional cases Source: Science Direct (2022e). This journal examines "geochemical studies of fundamental and broad significance to enhance the understanding of processes of the Earth and the solar system."
Lithos	Lithospheric Geological change; petrology, geochemistry and petrogenesis of igneous rocks; Petrology, geochemistry and petrogenesis of metamorphic rocks; Mineralogy related to petrology; Mineral physics related to petrogenetic problems; Related to the UN's sustainable development goals Source: Science Direct (2022s). The focus is on mineralogy, petrology, geochemistry, petrogenesis of igneous and metamorphic rocks.
Journal of Asian Earth Sciences	Geology and tectonics; Geophysics, geochemistry, petrology, volcanology; Landscape evolution; Ore genesis, hydrocarbon resources; Paleontology, paleoenvironment; Geological hazards, environmental geology Source: Science Direct (2022m). This journal publishes "research related to the full range of the Earth sciences of the Asian continent and its adjacent seas and oceans

(continued)

Table 2.3 (continued)

Journal	Key topics
Journal of Palaeogeography	Geochemistry and sedimentary environments; Stratigraphic paleogeography; Physiogeographic paleogeography; Paleoclimatology; Paleoseismology; Paleokarst (buried karst landforms); Basin analysis; Ichnology (tracks and traces); Lithofacies paleogeography and paleoecology; Historical environments, human impacts; Prediction, exploration, oil, gas, coal, etc. Source: Science Direct (2022p). The focus is sedimentology and paleogeography, with cases from China.
Engineering Geology	Geological and geotechnical engineering studies; Applied geomorphology and structural geology; Applied geophysics and geochemistry; Environmental geology and hydrogeology Soil and rock mechanics; Applied geotechnical engineering; Remote sensing techniques; Natural hazards; Land use planning; Engineering, environmental concerns, and safety Source: Science Direct (2022h). Engineering Geology. The focus is the integration of geological and engineering studies "that are of interest to engineering geologists, whether their initial training is in geology or civil/mining engineering" (Science Direct, 2022h).
Journal of South American Earth Sciences	Economic geology, metallogenesis, hydrocarbon genesis/reservoirs; Geophysics, geochemistry, volcanology, petrology; Tectonics, neo- and seismotectonics and geodynamic modeling; Geomorphology, geological hazards, environmental geology, climate; Stratigraphy, sedimentology, structure and basin evolution; Paleontology, paleoecology, paleoclimatology, Quaternary geology; Water/watershed dynamics, water cycle/climate change Source: Science Direct (2022q). This journal focuses on regional studies of South America.

Table 2.4 Key research areas of leading geochemistry-focused journals

Journal	Key topics
Geoscience Frontiers	Petrology/geochemistry of lithosphere and mantle dynamics; Astrogeology; Global tectonics (volcanology, seismology); Geophysics; Stratigraphy, paleontology, correlations and evolution; Environmental and engineering geology Source: Science Direct (2022k). The focus is on the "interdisciplinary fields of Earth and Planetary Sciences" (Science Direct, 2022k).
Physics and Chemistry of the Earth, Parts A/B/C	Geochemistry, tectonophysics, seismology, volcanology; Atmospheric sciences of planets, satellites; Cosmochemistry and exobiology; Solar, heliospheric, solar-planetary sciences, geophysics; Hydrology, Oceans and Atmosphere; Engineering/management, oceanography/oceanic chemistry; Lakes, rivers, atmospheric sciences, glaciology, chemistry Solid Earth and Geodesy; Geodesy, marine and environmental geosciences; Paleomagnetism rock magnetism, electromagnetism; Source: Science Direct (2022u). The journal's focus includes space, solid and water within physics, chemistry, and geological sciences.

(continued)

Table 2.4 (continued)

Journal	Key topics
Journal of Volcanology and Geothermal Research	Geochemistry of volcanoes, stratigraphy, rocks, lava, ash, degassing; Geophysics of volcanoes, rocks, silicate melts, magmas, heat flow; Geodesy and remote sensing; Hydrology/lakes, magmatic/hydrothermal fluids, gas, mineralization; Computational modeling of magma, lavas, gases, fluids, processes; Hazards, zonation methods, forecasting, assessment techniques Source: Science Direct (2022r). This journal focuses on "volcanic and geothermal processes and their impact on the environment and society" and "does not accept geothermal or other research papers that are not related to volcanism" (Science Direct, 2022r).
Marine and Petroleum Geology	Organic geochemistry (distribution, composition, of organic matter); Marine geology/Continental margins; Basin analysis, evaluation (formation, evolution of sedimentary basin); Sedimentary geology (or Sedimentology, study of sedimentary rock); Seismic stratigraphy (deep sedimentary rock, by seismic data); Structural geology/tectonics; Thermal models of basic evolution; Geophysical interpretation (geophysical methods, resource detection); Formation evaluation techniques; Reserve/resource estimation; Well logging Source: Science Direct (2022t). This journal publishes in "all areas covering marine and petroleum geology," thereby assisting in "environmentally sustainable exploration, exploitation, and utilization of natural resources of petroleum and gas hydrate" and in some cases promoting "science and engineering advancement in the form of using geological formations for safe storage of various gases to help achieve net-zero carbon goals…"
Contributions to Mineralogy and Petrology	Geochronology; Thermodynamic modeling of petrologic and geochemical processes; Experimental petrology; Igneous and metamorphic petrology; Mineralogy; Major and trace-element mineral chemistry; Major element and trace-element and isotope geochemistry Source: Springer Nature (2022b). This journal focuses on "igneous and metamorphic petrology, geochemistry and mineralogy," which "investigate petrologic, geochemical and mineralogical evidence recorded in rocks and minerals to explore these processes and their implications for the development of the Earth's crust, mantle and core through time" (Springer Nature, 2022b).
Precambrian Research	Cosmochemical, chemical, and biochemical evolution; Oceanic and Atmospheric evolution Paleobiology and fossil records; Geophysics of the early Earth and Precambrian terrains; Development of Precambrian lithosphere, mantle, processes; Precambrian mineral deposits Geochronology and isotope and elemental geochemistry Source: Science Direct (2022v). This journal publishes "all aspects of the early stages of the composition, structure and evolution of the Earth and its planetary neighbours… especially on process-oriented and comparative studies…" (Science Direct, 2022v).

materials, using geochemical techniques. Volkman (1998) defines organic geochemistry as "the study of the origins and fates of organic matter in the geosphere, and it encompasses studies of coals, oils, sediments and natural waters" and that "much of organic geochemistry can be classified into general themes" on organic

Table 2.5 Key research areas of leading geochemistry-focused journals

Journal	Key topics
International Journal of Coal Geology	Geology, geochemistry, mineralogy, petrology of coal, coal seams; Metamorphosis of coal and organic matter in rocks; Coal, coal seams, and organic matter-rich shales; Exploration and development of coal, carbon, petroleum, metals; Formation, transport, storage of carbon; energy gases in shales/coal; Critical metals in coal, coal-bearing strata, acid-mine drainage; Environmental coal impacts, acid-mine drainage, emissions; Coal fires and underground coal gasification; Computation for coal resources, big data, artificial intelligence; Analytical methods for geochemistry, mineralogy etc., modeling; Machine learning, automated reasoning/inference, reasoning, vision Source: Science Direct (2022l). This journal examines "fundamental and applied aspects of the geology, petrology, geochemistry and mineralogy of coal, oil/gas source rocks, and shales" and "aims to advance the environmentally sustainable exploration, exploitation, and utilization of these resources, as well as advancement of engineering and technology for effective resource management towards net-zero carbon goals."
Journal of Geodynamics	Plate tectonic processes: paleogeodynamics, modeling in tectonics; Mantle geodynamics: mantle convection, hot spots, plumes, zones; Geodynamics: crust, mantle, lithosphere, asthenosphere interactions; Seismic, gravity, magnetic, thermal processes, tectonic settings Stress field and deformation: geodetics of crustal movements; Seismic, electrical and magnetic anisotropy, seismicity; Crustal and mantle rheology: experiments, modeling, rock properties; Geodynamics of magma/lava flow, modeling, transport, igneous rocks; Geodynamics of topographic change: deep processes and topography; Geodynamics of ice sheets: melting, glacial isostasy, ice-seismicity; Geodynamic control on ore and hydrocarbon mineral deposits Source: Science Direct (2022o). This journal examines "the entire multitude of scientific methods aimed at understanding deep and shallow processes in various tectonic settings and acting at different time- and length-scales" and "publishes results based on a multitude of research disciplines, including geophysics, geodesy, geochemistry, geology, and based on field observations, laboratory experiments, and numerical simulations."
Continental Shelf Research	Physical sedimentology and geomorphology; Estuarine, coastal and shelf sea modeling and processes; Physical dynamics of ocean with shoreline features; Geochemistry of the coastal ocean (inorganic, organic); Coastal water and sediment quality, ecosystem health; Physical dynamics (e.g., waves) and biogeochemical cycles; Benthic-pelagic coupling (physical and biogeochemical); Benthic, phytoplankton and zooplankton ecology; Marine environment and anthropogenic effects Source: Science Direct (2022f). This journal focuses on the "the biological, chemical, geological and physical oceanography of the shallow marine environment, from coastal and estuarine waters out to the shelf break" and encourages "interdisciplinary process-oriented" and "innovative experimental studies" (Science Direct, 2022f).

(continued)

Table 2.5 (continued)

Journal	Key topics
Journal of Geophysical Research	Earth Surface, physical, chemical, biological processes, temporal, spatial; Biogeosciences, Earth system in time series, planetary studies; Oceans, physics, chemistry, biology and geology; Earth system. Earth systems, climate variability and change; Space Physics, aeronomy, magnetospheric, solar physics, cosmic rays; Planets, planetary geology, geophysics, geochemistry, atmospheres; Atmospheres, Source: American Geophysical Union (2022). This journal "publishes original scientific research on the physical, chemical, and biological processes that contribute to the understanding of the Earth, Sun, and solar system and all of their environments and components" (American Geophysical Union, 2022).
American Mineralogist	Biogeochemistry and biomineralogy; Mineralogy and crystallography; Mineral physics; High- and low-temperature geochemistry; Petrology; Geofluids; Relevant synthetic material Planetary evolution Source: American Mineralogist (2022). This journal provides "some of the most important advances in the Earth and Planetary Sciences with a mineralogical focus in the broadest sense" (American Mineralogist, 2022).

Table 2.6 Key research areas of leading geochemistry-focused journals

Journal	Key topics
Bulletin of Volcanology	Volcanic products; Volcanic eruptive behavior; Deeper structure of volcanoes; Evolution of magmatic systems; Geochemical, petrological, and geophysical techniques; Volcanic Hazards Source: Springer Nature (2022a). The main objective is to "improve understanding of the deeper structure of volcanoes, and the evolution of magmatic systems" (Springer Nature, 2022a).
Earth and Planetary Science Letters	Earth and Planetary Sciences (miscellaneous); Evolution and general properties of the Earth and planets; Geochemistry and Petrology; Geophysics; Physical and chemical processes Interior, surface, and atmospheric dynamics Source: Science Direct (2022g). This journal focuses on "physical and chemical processes, the evolution and general properties of the Earth and planets – from their deep interiors to their atmospheres" (Science Direct, 2022g).
AAPG Bulletin	Earth and Planetary Sciences (miscellaneous); Geochemistry and Petrology; Geology; Energy; Energy Engineering and Power Technology; Energy (miscellaneous); Fuel Technology Source: American Association of Petroleum Geologists (2022). This journal examines "geoscience and the associated technology of the energy industry" (American Association of Petroleum Geologists, 2022).
Solid Earth	Geochemistry, mineralogy, petrology, volcanology; Stratigraphy, sedimentology, and paleontology; Geodesy and gravity; Geodynamics: numerical, analogue modeling of processes; Geoelectrics and electromagnetics; Geomagnetism; Geomorphology, morphotectonics, and paleoseismology; Rock physics; Seismics and seismology; Rock deformation, structural geology, and tectonics Source: European Geosciences Union (2022b). This journal focuses on "multidisciplinary research on the composition, structure, dynamics of the Earth from the surface to the deep interior at all spatial and temporal scales" (European Geosciences Union, 2022b).

(continued)

Table 2.6 (continued)

Journal	Key topics
Environmental Chemistry	Biogeochemistry; Atmospheric chemistry; Marine and freshwater chemistry; Polar chemistry Fire chemistry; Soil and sediment chemistry; Chemical aspects of ecotoxicology; Climate change Source: Commonwealth Scientific and Industrial Research Organisation (CSIRO) (2022). This journal examines "the chemistry of the environment (air, water, earth, and biota), including the behaviour and impacts of contaminants and other anthropogenic disturbances," with primary focus on an "interdisciplinary approach, while advancing our understanding of the linkages between chemistry and physical or biological processes" (CSIRO, 2022).
Journal of the Geological Society	Interdisciplinary Earth Sciences; Fundamental geological processes; Regional studies Source: The Geological Society of London (2012). This is a broadly focused journal, publishing on the full range of the earth sciences.
Australian Journal of Earth Sciences	Earth and Planetary Sciences; Total earth science; Basin studies; Regional geophysical studies; Metallogeny (genesis, distribution of mineral deposits) Source: Taylor and Francis Online (2022a). This is a broad journal, covering the fields of the earth sciences (Taylor and Francis Online, 2022a).
Physics and Chemistry of Minerals	Atomic structures and crystalline states; General solid state spectroscopy; Experimental/theoretical analysis of chemical bonding; Physical properties; Thermal expansion, compressibility, elastic constants, atomic structure; Electron microscopy for physical and chemical studies; Mineral surfaces; Computational methods in mineralogy Source: Springer Nature (2022d). This journal covers "applications of modern techniques or new theories and models to interpret atomic structures and physical or chemical properties of minerals" (Springer Nature, 2022d).

Table 2.7 Key research areas of leading geochemistry-focused journals

Journal	Key topics
Lithosphere	Geochemistry and Petrology; Mineralogy and Crystallography; Petroleum Geology; Sedimentary Geology, Paleoclimate, Paleontology; Atmospheric Science; Geomorphology, Regional Geology, Past Environments; Structural Geology and Tectonics; Oceanography Space and Planetary Science; Applied Geophysics; Economic Geology; Engineering Geology and Natural Hazards Source: GeoScienceWorld (2022). This journal "covers research in all areas of earth, planetary, and environmental sciences, providing a unique publishing choice for authors in the geoscience community." The Journal emphasizes the topics in the above table, and "the intersections between them."

(continued)

Table 2.7 (continued)

Journal	Key topics
Journal of Geosciences	Major- and trace-element geochemistry, isotope geochemistry; Mineral chemistry and mineralogy; Experimental petrology and mineralogy; Theoretical models of igneous and metamorphic processes; Mineralizing processes and mineral deposits; Igneous and metamorphic petrology; Process-oriented studies of igneous, metamorphic complexes; Research in structural geology and tectonics; Dating igneous activity and metamorphic events Source: Czech Geological Society (2022). This journal examines "all aspects of the nature and origin of igneous and metamorphic rocks" (Czech Geological Society, 2022).
Geological Journal	Geochemistry and Geochronology, stable/unstable isotopes, elements; Metamorphic Geology, regional, fluids, crystallization paths, burial and uplift; Volcanic, Igneous Geology, igneous, magma genesis, composition, products; Sedimentology/Geomorphology, basin and facies analysis, sedimentation; Paleontology, macroevolution, biostratigraphy, taphonomy, ichnology; Quaternary Geology, climate/sea-level change, fossils, glaciations; Tectonics and Structural Geology, strain analysis, tectonics, deformation Source: John Wiley and Sons Ltd (2022). This journal "has a particular interest in publishing papers on regional case studies from any global locality which have conclusions of general interest. Such papers may emphasize aspects across the full spectrum of geological sciences" (John Wiley & Sons Ltd, 2022).
Applied Clay Science	Modification by biological microorganisms; Modification by chemical treatments with organic and inorganic molecules; Soil chemistry, agriculture, hydrocarbon and geothermal prospection; Interaction with water, with polar and apolar molecules; Thermal, structural, crystallographic, mineralogical properties of clay minerals; Genesis and deposits of clay minerals; Modification by thermal and physical treatments; Structural and textural modifications of clay minerals; Radioactive Wastes, Water Engineering, Heavy metal Depollution, Filtration; Clay minerals in formulations of pesticides, drugs, animal food production; Drilling fluids and muds in petroleum industry, Geotechnical applications; Ceramics, construction, iron ore and steel industry, paper, catalysis, cosmetics Source: Science Direct (2022b). For this journal, some of the topics mentioned are on the natural chemistry of clay, while others are more focused on industrial methods
Tellus Series B: Chemical and Physical Meteorology	Biogeochemical cycles of atmospheric constituents; Air/surface exchange processes; Long-range and global transport and dispersion; Atmospheric aerosols; Processes affecting atmospheric composition; Atmospheric chemical reactions; Cloud physics and chemistry, cloud radiative effects Source: Taylor and Francis Online (2022b). This journal focuses on "all aspects of atmospheric chemical cycling related to Earth science processes" (Taylor and Francis Online, 2022b).
Nonlinear Processes in Geophysics	Geoscience systems, disruption, and innovation; Related social and biological systems; Big data and artificial intelligence; New sensing analysis; Simulation technologies; Data-driven research; Mathematical physics Source: European Geosciences Union (2022a). This journal is committed to "breaking the deadlocks often faced by standard approaches in Earth and space sciences" and publishes "disruptive and innovative concepts and methodologies, as well as original applications of these to address the ubiquitous complexity in geoscience systems, and in interacting social and biological systems" (European Geosciences Union, 2022a).

matter. The Geochemical Society defines organic geochemistry as encompassing diverse research areas such as "biogeochemistry, aspects of climate change studies, petroleum geochemistry, aspects of archaeology, and studies of extraterrestrial organic matter" and documents that there is "potential for large, more-than incremental, advances which the diversity of organic geochemical approaches can collectively yield when applied to complex natural systems" (Geochemical Society, 2022). The definition of organic geochemistry is mirrored in the work of the Society's Organic Geochemistry Division, which is stated as encouraging and fostering "studies on the origin, nature, geochemical significance, and behavior during diagenesis and catagenesis of naturally occurring organic substances in the Earth, and also studies of extraterrestrial organic matter" (Geochemical Society, 2022). Utrecht University (2022) describes the work of organic geochemists as using a molecular approach to examine and study the origin and development of organic matter in the bio- and geosphere with the focus of research in that university being mostly on land-ocean organic carbon movement, paleoclimate reconstructions, proxy development, and ecosystem processes and the "the relationships between the occurrence, distribution, and/or isotopic composition of specific molecules, so-called biomarkers, and environmental parameters in modern systems."

Organic geochemistry may also examine evidence of environmental change in the Earth's crust. For example, Imperial College (2022) describe their Organic Geochemistry Research (ICOG) program as including research on the evidence for mass extinctions in rock chemistries, especially the big five mass extinctions at the end Ordovician, Late Devonian, end Permian, end Triassic, and end Cretaceous, examining the organic remains of the organisms that existed during these events, which "are entombed in rocks and can be extracted and analysed using organic geochemical methods. Interpreting these molecular fossils allows us to reconstruct the environments in which these organisms prevailed and thereby understand the causes and consequences of the extinction events" (Imperial College, 2022). Other research focuses on environmental change, using organic material such as plant spores, which contain pigments that protect their genetic materials from mutation by UV light. The analysis of the pigment contents of spores collected over long periods, it is possible to determine the trends in the amount of UV penetrating to the Earth's surface (Imperial College, 2022). Rock matrices deposited in layers at regular intervals, can also be examined, and the detailed study of these layers and the existing diagnostic molecules enables the abstraction of data that documents long term environmental change (Imperial College, 2022).

For practical results, organic geochemistry may also examine sources of petroleum in the Earth's crust, examining sources, compositional changes, and movement of such petroleum sources. Petroleum geochemistry has become one of the most important foci of organic geochemistry, with the main issues being the development of ever more sophisticated and detailed method for the detection of organic compounds and oil being strata in ground materials. Imperial College (2022), describes their Organic Geochemistry Research (ICOG) program as also including Petroleum Studies, arguing that "recently, the demand for oil is focusing scientific attention on unconventional hydrocarbon deposits. Imperial College Organic

Geochemistry activities involve research into both conventional and unconventional petroleum deposits." This is echoed by Mather (2013), who writes that "organic geochemistry studies the distribution, composition, and fate of organic matter in the geosphere on both bulk and molecular levels, combining aspects of geology, chemistry, and biology… Given the origins of fossil fuels such as crude oil, natural gas, or coal via the incorporation into sediments and burial of biomass from decayed organisms (followed by complex chemical reactions under the influence of temperature, catalysis, and, in shallow sediment layers, microbial activity), organic geochemistry is intimately associated with petroleum geochemistry."

Philp and Mansuy (1997: 749) point out that petroleum geochemistry is an important discipline for several areas of exploration and production for fossil fuels. Additionally, there have been recent developments alongside other developments in analytical chemistry, including gas chromatography and gas chromatography–mass spectrometry. An important development is the use of new techniques for the detection of usually trace amounts of biomarker compounds, which are compounds in oil with source rock extracts, the analysis of which can provide information on the origin of the oils in the Earth's strata. Philp and Mansuy (1997: 749) argue that at the time of their publication in the mid-90s, major developments in geochemistry included the evolution of methods for reservoir studies (e.g., the use of high-resolution gas chromatography to study reservoir continuity, and high-temperature gas chromatography to study wax deposits), new methods in exploration and biomarker geochemistry (e.g., pyrolysis techniques to study insoluble organic matter in source rocks or asphaltenes in oils), techniques which enable "a far more detailed and comprehensive picture on the origin of fossil fuels than could ever have been imagined a mere two decades ago" (Philp & Mansuy, 1997, 749).

Mukhopadhyay et al. (1995: 86) developed the argument that "the major unresolved issues on hydrocarbon generation, migration, and entrapment, include: (a) the distribution of source rocks in various stratigraphic intervals; (b) variations in maturity for both source rocks and crude oils and condensates; (c) the relation between hydrocarbon generation and over pressuring; and (d) possible oil-oil and oil-condensate source rock correlation." Here the issues concern studies that examined and described some potential source rocks in several stratigraphic intervals, the geochemical properties of some oils and condensates, and some possible oil-oil and oil-source rock correlations (Mukhopadhyay et al., 1995, 86). Related studies examine the possibly ways to study and predict oil bearing sediments and the formation of such varied physical forms. For example, Zou (2017) writes about "unconventional petroleum geology," of which an example is the theory of kerogen thermal degradation. This was initially based on earlier research, such as that of Bray and Van Tuyl (1961) who found that the carbon number distributions of n-alkanes or normal alkanes (alkanes are hydrocarbons in which the carbon atoms are held together by single bonds, with the general formula C_nH_{2n+2}) and further research increased knowledge on "the diagenetic evolution mechanism of organic matter and its relationship with the formation of oil" (Zou, 2017). Phillippi (1965) is cited as one of the founders of such studies, as the research findings indicated that sedimentary organic matter can only be transformed into hydrocarbons if it is at a particular burial depth and temperature. Later studies which examined the links between the generation of petroleum and temperature and time, such as those of Connan (1974) and Tissot and Welte (1978) "established a petroleum generation model that quantitatively determines the hydrocarbon-generating capacity of source rocks" this supporting the theory of kerogen thermal degradation (Zou 2017).

This theory assumes that deposited organic matter changes into kerogen after biochemical and polycondensation processes (Zou 2017). The shrinkage of the kerogen to a greater depth slowly generates oil through thermal degradation. Such organic matter includes bitumen, which is soluble in organic solvents, and kerogen which is insoluble in organic solvents. The character and hydrocarbon potential of kerogen can be measured using these parameters and using theoretical and practical analysis may evaluate the sources and amounts of hydrocarbon development. This study of hydrocarbon generation is linked to that of the concept of petroleum migration, the latter being important for the field of petroleum geology. Petroleum became a topic of primary interest in petroleum geology. Zou (2017) cites some formative literature that explored the relationships of the primary migration of petroleum, including those of Magara (1978) who looked at mudstone compaction and other primary migration issues, Barker (1972) who examined the role of aquathermal pressure, Burst (1969) who investigated the dehydration of clay minerals, Leithauser et al. (1982) who examined hydrocarbon diffusions and Dickey (1975) which looked at the primary migration phase of oil. Zou (2017) argues that these studies "led to the general identification of key scientific issues such as the phase, dynamics, and pathways of the primary migration of petroleum."

Other oil-based geochemical studies include that of Boreham et al. (2021), which examines the monoalkene contents of a case study of Australian oils. A wide range of oils were selected for study, and the authors argue that "thirteen Australian oils and one condensate, covering oil reservoir ages from Mesoproterozoic to Early Cretaceous, show monoalkene contents varying from 0.01 to 22.3 wt% of the whole liquid" (Boreham et al., 2021). The methodology was based on radiolysis of saturated hydrocarbons (radiolysis refers to the molecular damage to a substance by ionizing radiation), which is stated as the "most likely process leading to oils with high alkene contents." Alkanes and alkenes are referred to in this study. Alkanes have single bonds between carbon atoms and are called saturated hydrocarbons. Alkenes have at least one carbon-carbon double bond and are unsaturated hydrocarbons. Boreham et al. (2021) found that the main radiolytic component was an unresolved complex mixture (UCM), with most of the resolved alkene compounds being positional isomers (the basic carbon skeleton remaining the same, but the important groups moved on the skeleton) of n-alkenes), with minor components being methyl branched and cyclohexyl alkenes.

Importantly, the authors note that "the oil with the longest reservoir residence time shows the highest content of internal n-alkenes relative to terminal 1-alkenes as well as the highest trans/cis ratio, suggesting the extended time has resulted in rearrangement to near thermodynamic equilibrium of the congruent monoalkenes" and the "internal n-alkene isomers have a trans configuration dominant over the cis isomer" (Boreham et al., 2021). Here cis isomer is part of Cis–trans isomerism, "where Cis and trans isomers occur when an alkene has two different atoms or groups of atoms attached to each double-bonded carbon atom" (Ouellette & Rawn, 2019). Boreham et al. (2021) found that the "relative proportion of alkene mimics the relative abundance of n-alkanes, suggesting that radiolytic C–C bond cleavage is suppressed when the alkene/alkane ratio is elevated and that the preferred pathway of n-alkane radiolysis favors the production of terminal monoalkenes: and the radiolysis of the alkane UCM with the networking of n-alkane-derived radiolysis products contribute to the higher relative proportion of the alkene UCM, and conclude that the similarity of the carbon and hydrogen isotopic ratios of the n-alkanes and n-alkenes supports a parent–daughter relationship, and radiolytic n-alkenes have a similar distribution to that of the n-alkanes; and additionally an unresolved complex mixture dominates the unresolved complex mixture, there is a similarity of the carbon and hydrogen isotope values of n-alkenes and n-alkanes; and a long term oil residence results in a near-equilibrium n-alkene isomer distribution (Boreham et al., 2021).

Another study by Curiale and Frolov et al. (1998) examined the origin and occurrence of alkenes in petroleum, and primary and secondary sources for the alkene content in oils. The primary sources of alkenes were from the primary migrating phase, expelled from the source rock. The secondary sources of alkenes are migration-contamination, usually contamination from lower maturity organic matter, from the migration pathway or disseminated within the reservoir, and alteration within or near the trapped fluid, and the "latter process can involve either thermal pyrolysis or low temperature radiolytic dehydrogenation of saturated hydrocarbons (Curiale & Frolov, 1998). An igneous intrusion near the hydrocarbons in the reservoir or within an organic matter-rich rock may cause transient high temperature thermal degradation of the oil and condensate, hence producing an alkene-rich migrating fluid (Curiale & Frolov, 1998). The authors conclude that "enrichments of radiogenic elements (U, Th and K) within reservoir rock minerals can result in the radiolytic-generated alkenes making more than 10 wt% of the altered petroleum liquid" (Curiale & Frolov, 1998; see also Frolov et al., 1998).

Apart from oil and petroleum studies, organic geochemistry considers other aspects of organic materials in atmosphere, lithosphere and ocean. An important issue concerns marine dissolved organic matter (DOM), which is a major bioactive reservoir of carbon in the ocean. This reservoir is estimated at a stock of 700 Pg C in the global ocean, which is similar in size to the stock of carbon resident in atmospheric CO_2 (Hansell & Carlson, 2014 xxi). Hansell and Carlson (2014, xxi) note that with the development of studies of the elemental cycles (carbon, nitrogen, and phosphorus), searching research has been conducted on such organic matter accumulation, with certain central questions: "can we accurately, with community wide consistency, measure the concentrations of dissolved organic matter in the ocean; what are the distributions of the dissolved organic C/N/P pools and what processes controls these distributions; what are the rates, biogeographical locations, and controls on elemental cycling through the pools; what are the biological and physicochemical sources and sinks; what is the composition of the pools and what does this tell us about elemental cycling? Finally, do we understand DOM in elemental cycling well enough to accurately represent the processes in numerical models?" Answering such questions has become important studies on DOM.

Catalá et al. (2021, 7225) note that marine-dissolved organic matter (DOM) is a huge and mostly unexplored molecular space, mostly resident in the ocean for millennia. This huge reservoir is very diverse one for the most diverse molecular mixtures yet documented, and comprises millions of individual compounds. Research has been conducted on the role of dissolved organic matter in the processes of biogeochemical cycles, climate dynamics, microorganisms, and the molecular composition of the DOM. Catalá et al. (2021, 7225) nevertheless note that the study of DOM bioactivities is still undeveloped, mostly due to the technical challenges of analyzing such chemically complex material. The current solution is to develop more advanced technologies, for the analysis of DOM bioactivities. The newer screening technologies are slowly enabling more detailed studies, especially the accelerated identification of bioactivities for small molecules from natural products and seek to replace the "laborious chemical fractionation" with the "application of untargeted metabolomics and multiplexed high-throughput molecular-phenotypic screening techniques that are providing first insights on previously undetectable DOM bioactivities" (Catalá et al., 2021, 7225).

In another study, Pötz et al. (2021) note that organic matter held to be an important reductant in some sediment-hosted base metal deposits, but the exact nature of the interactions between OM and hydrothermal fluids is debatable, and the interconnected reactions that develop over the long term, geological timescales are also currently being studied. Their study examined organonitrogen, -sulfur, and -oxygen (NSO) compounds, the justification of this focus being that these compounds are the most affected by organic–inorganic interactions with mineral surfaces, metals, and aqueous fluids (Pötz et al., 2021). The study methodology used ultra-high resolution mass spectrometry (FT–ICR–MS). According to Qin et al. (2021, 347) "FT-ICR MS measures the mass-charge ratio and abundance of all ions at the same time. FT-ICR allows detection of very low ion concentrations due to its ability to amplify signals. This way, the method achieves high sensitivity as well as resolution." In the study by Pötz et al. (2021) the findings indicated a generally homogenous deposition of type II marine organic matter, hydrothermal issues, losses of extractable organic matter and especially long-chain n-alkanes and enrichment in oxygenated compounds and increased molecular oxygen content. Zakaria et al. (2018, 345) refer to n-Alkanes as "saturated hydrocarbons…, hydrogen and carbon with solely single bonds" with the formula C_nH_{2n+2} showing the maximum bonds between carbon and hydrogen atoms. The authors argue that "the main source of global occurrence of n-alkanes is petroleum; however, they also occur naturally. n-Alkanes can be detected in riverine, estuarine, and marine sediments as well as aquatic organisms" Zakaria et al. (2018, 345)." Pötz et al. (2021) conclude that "FT–ICR–MS, combined with determination of stable hydrogen isotope compositions, can provide valuable information about interaction time scales and the origin of mineralizing hydrothermal fluids."

Other geochemical studies include micro-assessments of chemical compound materials in earth or water systems, sometimes with applications to microbiology.

For example, O'Beirne et al. (2022) write about the "characterization of diverse bacteriohopanepolyols in a permanently stratified, hyper-euxinic lake." The authors argue that these bacteriohopanepolyols (BHPs) are a diverse class of bacterial lipids that have potential status as biomarkers of some specific microbes, microbial processes, and environmental conditions (O'Beirne et al., 2022). The subject study of Mahoney Lake in Canada was of a water and sediment environment characterized by hyper-euxinic (usually with no oxygen and higher levels of hydrogen sulfide H_2S) and meromictic (layers of water that do not intermix) conditions (see Wetzel, 2001; Meyer & Kump, 2008). The findings of the study indicated that there were clearly distinct BHP distributions in the oxic mixolimnion (mixolimnion refers to the upper layer of a meromictic lake, above the chemocline, mixed by the wind, with free circulation and low density, oxic refers to oxygen present), the chemocline (this is the zone in a meromictic lake, that transitions between the upper mixolimnion and lower monimolimnion layers, with a change from aerobic to anaerobic conditions, and monimolimnion refers to the lower layer of a meromictic lake, below the chemocline, with dense static water) (O'Beirne et al., 2022).

The findings also indicated that Bacteriohopanetetrol (BHT) and unsaturated BHT are the main BHPs in the oxic mixolimnion and the chemocline, a novel BHP occurs in the euxinic monimolimnion. Additionally, composite BHPs occurred in the euxinic monimolimnion and sediments, showing production by anaerobic bacteria. Bacteriohopanepolyols (BHPs) were produced by anaerobic bacteria and methylated bacteriohopanepolyols are produced from low oxygen and high osmolarity (osmolarity is the number of particles per liter of solution) (O'Beirne et al., 2022). The authors also found that "BHTs are produced within the euxinic sediments in response to low oxygen and high osmotic concentrations, as opposed to being diagnostic biomarkers of cyanobacteria and aerobic metabolisms" (O'Beirne et al., 2022).

Another study by Cooke et al. (2009: 1151), on bacteriohopanepolyols in soils, focuses on a group of BHPs, including adenosylhopane and related compounds, which enable the transport of terrestrial organic matter (terrOM) to the marine environment. The case study was of Russian (estuary surface sediments) and American rivers (river sediments). Using the BHP signatures and high performance liquid chromatography–tandem mass spectrometry (HPLC–MSn), the study found 15 different BHPs from several different bacteria, and a significant mound of OM from terrestrial sources and suggested that increases in terrestrial OM or increased preservation of OM resulted from of shorter periods of permafrost thawing (Cooke et al., 2009: 1151). In addition to variation between the rivers (great differences between the North American rivers in BHP input and higher soil OM contribution, and variations between the Russian rivers in "soil-marker" BHPs and tetrafunctionalized BHPs), "aminobacteriohopanepentol, an indicator of aerobic methane oxidation, was observed in all the sediments, with the source being either the marine environment or methane producing terrestrial environments" (Cooke et al., 2009: 1151).

Organic geochemistry may be closely linked to biogeochemistry, as carbon has strong biological links. Rafferty (2013) describes the relationship between biogeochemistry and organic geochemistry: biogeochemistry is the "study of the behaviour of inorganic chemical elements in biological systems of geologic scope as opposed to organic geochemistry, which is the study of the organic compounds found in geologic materials and meteorites, including those of problematic biological origin." Topics within both biogeochemistry and organic geochemistry are listed as petroleum and life origins, biogeochemical prospecting for mineral deposits, some ore deposit origins, natural water chemistry, soil formation, and coal chemistry (Rafferty, 2013). Organic geochemical studies may apply to biogeochemistry. For example, a study of the organosulfur compound dimethylsulfide (DMS) by Hopkins et al. (2023: 361) titled "The biogeochemistry of marine dimethylsulfide" argued that "advances in molecular genetics and large-scale biogeochemical measurements have revealed the global prevalence of DMS-related processes, including in previously overlooked environments and organisms, such as sediment-dwelling bacteria." The study examined the production and cycling of marine DMS and its entry into the atmosphere and noted that the study of dimethylsulfide in the atmosphere, which may be based on crudely defined biological parameters, including total chlorophyll, require more advanced study of the biogeochemical processes that control DMS production (Hopkins et al., 2023: 361).

Inorganic Geochemistry

Definitions of inorganic geochemistry usually emphasize the role of carbon in organic geochemistry and the lack of carbon in studies in inorganic geochemistry. For example, Geoscience Australia (2022) notes that "inorganic chemistry is concerned with the properties of all the elements in the periodic table and their compounds, with the exception of organic compounds (compounds containing C-H bonds). Inorganic chemistry investigates the characteristics of substances that are not organic, such as nonliving matter and minerals found in the Earth's crust." Inorganic geochemistry is linked to inorganic chemistry: the American Chemical Society (ACS) (2022) defines inorganic chemistry as concerned "with the properties and behavior of inorganic compounds, which include metals, minerals, and organometallic compounds. While organic chemistry is defined as the study of carbon-containing compounds, inorganic chemistry is the study of the remaining (i.e., not carbon-containing) subset of compounds. But there can be overlap between the two fields. For example, organometallic compounds usually contain a metal or metalloid bonded directly to carbon."

Several authors have written concerning the foci of inorganic geochemistry, including the main inorganic elements and minerals, and the techniques for their study. For example, Kumar et al. (2021, 149), write about highly sophisticated analytical methods, including powder X-ray diffraction, scanning electron microscopy, electron probe micro-analysis and other methods, which can be used for the study of inorganic or geological minerals. Elements and minerals studied with these systems include iron (Fe), oxygen (O), silicon (Si), magnesium (mg), quartz (SiO_2), feldspar ($KAlSi_3O_8$ – $NaAlSi_3O_8$ – $CaAl_2Si_2O_8$), among others), amphiboles (SiO_4 tetrahedra, containing ions e.g., iron and/or magnesium, etc.), pyroxene (generally the formula $ABSi_2O_6$), olivine (Mg^{2+}, $(Fe^{2+})_2SiO_4$), kaolinite ($Al_2Si_2O_5(OH)_4$), bentonite ($(Na,Ca)_{0.33}(Al,Mg)_2Si_4O_{10}(OH)_2{\cdot}(H_2O)_n$), also sodium bentonite ($Al_2H_2Na_2O_{13}Si_4$), calcium montmorillonite ($(Na,Ca)0.33(Al,Mg)_2(Si_4O_{10})(OH)_2{\cdot}nH_2O$), vermiculite ($(Mg,Fe^{2+},Fe^{3+})_3$ $[(Al,Si)_4$ $O_{10}](OH)_2{\cdot}4H_2O$), and biotite ($K(Mg,Fe)_3(AlSi_3O_{10})$ $(F,OH)_2$).

Geochemistry (2022) reports that the major elements in Earth's mantle and crust are, in order of abundance: oxygen (44%), magnesium (23%), silicon (21%), iron (8%), calcium (2.5%), and aluminum (2.4%). Iron is one of the main elements described in studies of inorganic geochemistry. For example, Williamson (1998) notes that "iron commonly exists in one of three oxidation states: Fe^0 (elemental iron), Fe^{2+} (ferrous iron), and Fe^{3+} (ferric iron). It is the fourth most abundant element in the Earth's crust, exceeded only by oxygen, silicon and aluminum, but is believed to be the most abundant in the Earth as a whole." It exists at a crustal average of 7–8%, in ultramafic 9.6%, basalt 8.6%, granite 2.2%, limestone 0.33%, and sandstone 0.98% (Taylor & McLennan, 1985; Faure, 1991).

Inorganic geochemistry may also be concerned with oil exploration and analysis. For example, Ivanov et al. (2021) examined the inorganic geochemistry of crude oils of northern Eurasia using mass-spectroscopy with inductively coupled plasma (ICP-MS), which enabled the study of the microelement composition of crude oil and its derivatives "with the limit of detection (LOD) at the ppt level" taking an example from West Siberian and Tatarstan Romashinskoye oilfields. The findings using the ICP-MS method indicated variations in microelements in the oil reservoirs. Fifty rare, rare earth, and other microelements were detected, and the studied crudes were found to have a specific microelement composition. It was argued that the main geochemical feature of the crude oil was the extremely low content of many trace elements. This study of the inorganic geochemistry of West-Siberian and Tatarstan crude oils "using the most modern equipment shows that crude oils (at least from these two provinces, and, apparently, others) have an extremely specific microelement composition with no comparisons" (Ivanov et al., 2021).

The authors note that "according to modern concepts…, the analysis of the distribution of rare and rare earth elements in crude oils makes it possible to identify the sources and characteristics of naphthide-forming fluids (Naphthalene is an organic compound with formula $C_{10}H_8$, and naphthide is a compound of naphthalene or its radical with a metallic element, mercuric naphthide) (Ivanov et al., 2021). It was concluded that there was a specific microelement composition to the studied crudes, and based on the elevated contents of transit elements and platinoids, the study made a conclusion about the "ultrabasic" geochemical–metallogenic specialization of studied petroleum systems and the assumption about its origin was proposed" (Ivanov et al., 2021; see also Marakushev et al., 2004; Ivanov et al., 2008; Gottikh et al., 2009).

In another example of the application of inorganic geochemistry methods to petroleum sources, Ratcliffe et al. (2012, 4) give a case study of oil extraction from shale (the commonest sedimentary rock), arguing that as shale continues to be the most actively explored and developed hydrocarbon plays in North America, their study "demonstrates how inorganic whole-rock geochemical data can be used to help with the development of shale plays." The authors further note that the elemental data which is utilized for chemostratigraphy may be used for modeling mineralogy and total organic carbon, and also to assess paleoredox facies and derive information on the formation brittleness, which are important pieces of information for the exploitation of the shale resource plays (Ratcliffe et al., 2012, 4). Here chemostratigraphy may be defined as "the study of the inorganic/organic chemical variations within the sedimentary sequences, either based on the elemental or isotopic composition of the rock. It provides a useful tool for unconventional resource exploration and development" (Slatt et al., 2021).

Ratcliffe et al. (2012: 4) also argue that recently in the United States, shale has become an important oil source, but "the fine grained, macro-scale homogeneity of many shale plays currently being exploited has negated some of the more traditional approaches to reservoir characterization and stratigraphic correlation, resulting in the search for new methodologies that enable better understanding of shale reservoirs." In this study, the technique used was the application of inorganic whole-rock geochemical data to shale resource plays. Features measured by this chemostratigraphy technique, which help to understand shale reservoirs, include changes in element concentrations through time and using those to model changes with respect to geological events, such as paleoclimate, and relative rock brittleness (see also Pearce et al., 2005; Ratcliffe et al., 2006, 2007, 2010; Tribovillard et al., 2006, 2008; Turgeon & Brumsack, 2006; Negri et al., 2009; Jenkyns, 2010; Hildred et al., 2010; Wright et al., 2010). Ratcliffe et al. (2012: 9) conclude that although until relatively recently, stratigraphic studies were the main objective for deriving whole rock inorganic geochemical data for the petroleum industry, it is now more obvious that these same datasets may be used to assess bulk mineralogy and total organic carbon (TOC) (both semiquantitatively), temporal and lateral variation in paleoredox conditions, changes in rock brittleness, and changes in mineralogical composition for industrial purposes.

Other studies of inorganic geochemistry focus on elements or compounds in the earth's crust, either in solid materials or water. For example, Smith et al. (2009, 2348) examine the occurrence of ammonium (NH_4^+) and nitrogen processing in water from coal-bed natural gas (CBNG) production wells, in a case study of the Powder River Basin, Wyoming, USA. The findings of this study indicated that dissolved ammonium concentrations ranged from 95 to 527 μM, ammonium concentrations decreased with the transport distance (with later increases in nitrite and nitrate concentrations) and the removal efficiency, or uptake, of total dissolved inorganic nitrogen also differed between the channel types. The study also found that the uptake of dissolved inorganic nitrogen was greater in channels in gentle-sloped, vegetated topography, than in the steeper, less vegetated channels. There were also high correlations between dissolved inorganic nitrogen and diel (diel refers to 24 hours; daily diel alternation between night and day) patterns of incident light and dissolved oxygen concentration. Other findings were that in a larger main channel with several discharge inputs ($n = 13$), dissolved inorganic nitrogen (DIN) concentrations were > 300 μM, with pH > 8.5, after 5 km of transport. Ammonium comprised 25–30% of the large-channel DIN, and ammonium concentrations were relatively constant over with time, and there was only a weak diel pattern (Smith et al., 2009, 2348). The authors conclude from the findings that CBNG discharge may sometimes be an important DIN source in western watersheds, and that the net oxidation and/or removal depends upon the amount of contact with sediment and biomass, the type of drainage channel, and the time of day (Smith et al., 2009: 2348).

In another study, Sowder et al. (2010, 1604) studied trace metal occurrence in Coalbed natural gas (CBNG)-produced water, which can accumulate in produced water retention ponds, taking a case study of the Powder River Basin (PRB) of Wyoming. The objective of the study was to ascertain the groundwater pollution potential and to explain the factor for the high Arsenic (As) concentrations (Sowder et al., 2010: 1604). The parameters were the infiltration characteristics, subsurface hydrology, the fall and pond water quality, the isotope signatures, and the trace metal balances. The findings indicated that there was little or no infiltration of pond water, and there was no measurable contamination of the shallow groundwater. The high pond As concentrations were caused by semicontinuous inputs of CBNG-produced water with low As concentrations (0.20–0.48 microg L(−1)), which were affected by the low pond water volumes during low rainfall. Sowder et al. (2010: 1604) concluded that "because of reduced infiltration and high evaporation rates, As became concentrated over time" and "reduced infiltration was most likely caused by the high sodium concentration and high sodium adsorption ratio of the CBNG-produced water, which disrupt soil structure."

In another study, Soldatova et al. (2021: 441) examined nitrogen compound behavior in a shallow groundwater-soil system, which is an important aspect of the nitrogen cycle within agricultural landscapes, and a factor for groundwater chemical composition, with a case study from an agricultural landscape in the Poyang Lake area (Jiangxi Province, China). The methodology involved soil and groundwater samples taken before and after harvesting (spring and fall)., examining the data on the chemical and microbiological composition of the soils and groundwater, and isotopic data on dissolved nitrate, to determine the entry of nitrogen into the studied system. The findings indicated that nitrogen entered through organic nitrogen compounds (mostly through early fertilization), which change into nitrate due to soil nitrifiers before entering the aquifer, where there may be denitrification. The authors conclude that in the case study, the reducing conditions, and the formation of clay minerals from the aqueous solution, may create a geochemical barrier for the accumulation of nitrogen compounds, which may stop the transformation of ammonium to nitrate and providing its sorption (sorption refers to the removal of a compound from solution by solid phase constituents). They also note that "bacterial diversity in the shallow groundwater has a strong relation with the amount of nitrate in the system, whereas in the soil, it is connected with sampling depth" (Soldatova et al., 2021: 441).

Links between biogeochemistry and inorganic geochemistry largely concern the presence of inorganic elements and compounds such as those of iron, oxygen, silicon, magnesium, and mercury in the environment and in organisms, especially in fresh and saltwater (Campbell, 2020). For example, Morel and Price (2003) detail planktonic uptake of metals, which contributes to low concentrations of these metals in surface seawater. The study found that microorganisms release complexing agents and catalyze redox reactions that may change the bioavailability of trace metals and increase cycling in the upper water column, this contributing to the taking up of the micronutrients. It is also argued that the low availability of some metals contributes to photosynthesis rates and the transformation and uptake of nutrients such as nitrogen. The authors conclude that "the extremely low concentrations of several essential metals are both the cause and the result of ultraefficient uptake systems in the plankton and of widespread replacement of metals by one another for various biochemical functions" (Morel & Price, 2003: 944).

Another study examines the role of the sulfur biogeochemical cycle in mercury methylation in marine sediments (Wang et al., 2022). The basis of the study is that "estuaries are sinks for mercury, in which the most toxic mercury form, neurotoxic methylmercury (MeHg), is produced by mercury methylators and accumulates in estuarine sediments" (Wang et al., 2022). In association, sulfate-reducing bacteria may trigger the microbial sulfur cycle and may act as the mercury methylator. The study examined sulfur and mercury speciation in sediments from a case study of 70 estuaries distributed globally. The findings indicated abundant mercury and sulfur species, with mercury levels risky to aquatic ecological systems, and significant correlations between sulfur and MeHg concentrations, with the porewater sulfate concentration positively correlated to MeHg production. The sulfur cycle contributed to MeHg formation by activating mercury methylator activities and reducing the bioavailability of mercury, which may increase or inhibit MeHg formation at variable sulfur speciation concentrations. The authors concluded that the findings suggest "sulfur biogeochemical cycle plays an important role in mercury methylation in estuarine sediments, and the effect of the sulfur cycle on mercury methylation deserves to be further explored in future research."

In another study, Nóbrega et al. (2013: 7393) also examined the role of iron and sulfur, which are "key elements in the biogeochemistry of estuarine soils, in which Fe and sulfate reduction (SR) pathways are important for organic matter decomposition." Taking a case study of the Brazilian coastal mangrove forest with strong weather variations and shrimp farms, the objective was to assess the impacts of shrimp farm effluents, mostly wastewater on iron and sulfur geochemistry in mangrove soils. A comparison was of two mangrove forest soils from a mangrove forest receiving wastewater from shrimp ponds and a control site). The findings indicated a decrease in pyritic iron and degree of pyritization in the soils of the first case study, possibly linked to the anaerobic oxidation of pyrite and nitrate reduction, or the dominance of denitrification over sulfate reduction. The lower TOC contents in the first site suggested that underground decomposition increased in reaction to eutrophication. Seasonal variations impacted the semiarid mangrove soils, as during the drier period, soils had oxidizing conditions with losses of reduced and oxidized iron, which is environmentally significant as iron is biolimiting for marine primary production. The authors conclude that the findings indicate "that both factors (seasonal weather variations and shrimp effluents) play important roles in the geochemical processes that occur in these soils and, thus, may affect their functioning and maintenance" (Nóbrega et al., 2013: 7393).

Isotope Geochemistry

Isotopes are defined as atoms of the same element that have different numbers of neutrons, but the same numbers of electrons and protons, this translating into different physical characteristics. Isotope geology is the study and application of isotopes to geological processes and the time scales of these processes (Kendall et al., 1995; Kendall & Caldwell, 1998). Isotope geochemistry is defined by Allaby (2008) as the "study of the abundance ratios of isotopes (both stable and radioactive) of major and trace elements in rocks (e.g., Rb/Sr, Pb/U, etc.), to elucidate a number of geologic problems and processes. These include the age relationships of rocks, and the age of the Earth itself (see geochronology; isotopic dating; and radiometric dating); paleotemperatures and geothermometry (see oxygen-isotope analysis; and geothermometer); and the provenance of natural waters, ore-forming fluids, and magmas (see isotope fractionation; SMOW-standard mean ocean water; d/h ratio; oxygen–isotope ratio; and stable-isotope studies)."

Isotope geochemistry may be divided into stable isotope geochemistry and radiogenic or radioactive isotope geochemistry. Li et al. (2019) describe isotopes as divided into stable isotopes and radioactive isotopes, where stable isotopes refers to the isotopes whose nuclear remain stable through time, as opposed to radioactive isotopes that decay into daughter isotopes of different elements." Mamedov (2021: 6) compares the stable and unstable types, notes that "in contrast to unstable (radioactive) isotopes or isotopes produced from the decay of another element (radiogenic), stable-isotope geochemistry uses isotopes whose abundances do not change with time." Yadav et al. (2021, 93) point out that "stable isotopes have shown very versatile applications in various fields. Vallero (2014, 139) notes that "stable isotopes do not undergo natural radioactive decay, whereas radioactive isotopes involve spontaneous radioactive decay, as their nuclei disintegrate."

Additionally, the stable isotopes of light elements, including hydrogen, carbon, nitrogen, oxygen, and sulfur have been used in earth science research from the 1950s, and have consequently been described as traditional stable isotopes. The stable isotopes of other elements (note: most of them are metals such as Mg, Cu, Fe, Zn, Hg) have been described as nontraditional stable isotopes, because they have mostly received attention since the beginning of the 2000s (Li et al., 2019). Yadav et al. (2021: 93) have also written that in the case of geochemistry, two types of stable isotopes have been used, namely the traditional stable isotopes including C, N, O, S, and H, and nontraditional stable isotopes such as Zn, Cu, Fe, Pd, Cd, and other metals. Yadev et al. (2021: 93) also contend that traditional isotopes have been used in studies as geothermometers, tracers in hydrological, biological systems, ore deposits, hydrothermal systems, among others, while nontraditional isotopes have increasingly been used in several fields, including the tracing of toxic environmental pollutants like Cr and Hg, application in mining (Yadav et al. (2021: 93).

Traditional stable isotopes are referred to as "those of carbon, hydrogen, nitrogen, oxygen, and sulfur, elements with relatively low atomic masses that play a significant role in geological and biological processes" and the basis of traditional stable isotope geochemistry, which is "based on using small atomic mass differences between isotopes of the same element to probe earth system processes. Isotopes fractionate themselves between different phases or molecules in ways that are related to temperature and reaction mechanism" (Stern & Wieman, 2021,100). Nontraditional isotopes are described by Hu and Teng (2021, 114) as "isotope systems other than those of H, C, N, O, and S"; these authors also point out that "the field of stable isotope geochemistry has advanced rapidly with the developing capability to make high-precision measurements using multi-collector inductively-coupled-plasma mass spectrometry (MC-ICPMS) at high mass-resolution."

ChemEurope (2022, see also Dickin, 2005) describes radioactive isotopes as the parents to the radiogenic daughter isotopes, because a radiogenic nuclide is one that is produced by a process of radioactive decay. Vallero (2014, 139) maintains that the radioactive decay of the radioactive isotopes contributes to the formation of new isotopes or elements, and that the stable product of an element's radioactive decay is termed a radiogenic isotope. The example is cited of lead (Pb), which has four naturally occurring isotopes, with varying masses (204Pb, 206Pb, 207Pb, 208Pb), of which only the first listed is stable. The isotopes 206Pb and 207Pb are classified as "daughter" (or progeny) products from the radioactive decay of uranium (U), while 208Pb is a product from thorium (Th) decay (Vallero, 2014, 139). ChemEurope (2022, see also Dickin, 2005) also cites lead as "perhaps the best example of a radiogenic substance, as it is produced from the radioactive decay of uranium and thorium. Specifically, Pb-206 is formed from U-238, Pb-207 from U-235, and Pb-208 from Th-232." Other adiogenic substances include argon-40, which is formed from radioactive potassium, and nitrogen-14, formed by the decay of carbon-14, radon and helium (ChemEurope (2022).

Radiogenic isotope geochemistry is important for current scientific research on the chronology of rock-forming events, as isotope geochemistry is related to the relative and absolute concentration of elements and their isotopes in Earth (Haldar, 2018). ChemEurope (2022) also states that radiogenic nuclides (more commonly referred to as "radiogenic isotopes") are among the most important tools in the Earth sciences, with two main applications: (1) "in comparison with the quantity of the radioactive 'parent isotope' in a system, the quantity of the radiogenic 'daughter product' is used as a radiometric dating tool (e.g., uranium-lead geochronology)" and (2) "in comparison with the quantity of a nonradiogenic isotope of the same element, the quantity of the radiogenic isotope is used as an isotopic tracer (e.g., 206Pb/204Pb)." Vallero (2014: 139) also adds that "radiogenic isotopes are useful in determining the relative age of materials. The length of time necessary for the original number of atoms of a radioactive element in a rock to be reduced by half (radioactive half-life) can range from a few seconds to billions (109) of years."

Radiogenic isotopes may also be used for studying certain geological formations. For example, Cooke et al. (2014: 364) argue that radiogenic isotopes (mentioning U–Pb, Sm–Nd, Rb–Sr, and Lu–Hf isotopes), "are used extensively to investigate the igneous and hydrothermal evolution of porphyry deposits and the potential sources of contained metals." The authors note that the U–Pb (uranium-lead) isotopic system is one of the most used tools in the study of porphyry deposits (these are deposits of minerals of copper, molybdenum, gold, or other metals, within small veinlets within hydrothermally altered igneous formations), as a geochronologic tool and a isotopic tracer for the evaluation of magma and metal sources in reservoirs, as lead is a common element in some rock-forming silicate minerals and sulfide minerals in porphyry deposits (Cooke et al., 2014 364). It is further argued that lead isotopes are useful for the tracking of metal sources and determination of magmatic and other ore-forming processes. Isotopic evolution histories can be determined by Th/U (Thorium-Uranium, where both uranium and thorium have unstable isotopes) and U/Pb ratios, which reveal geological history (see also, e.g., Chiaradia & Fontbote, 2002; Kamenov et al., 2002; MacFarlane et al., 1990; Wooden et al., 1988).

In the Arizonan case study of copper deposits, Cooke et al. (2014: 364) write that certain distinct Paleoproterozoic basement terranes impart distinguishable Pb isotopic characteristics to the Mesozoic and Cenozoic igneous rocks which intruded into the region and to related hydrothermal systems. The lead isotopic characteristics show the igneous processes, where low Pb concentration magmas which are derived from the mantle assimilate crustal material as they move upward into the shallower crust. As the Paleoproterozoic crystalline crust and the contemporaneous underlying mantle have isotopic compositions which are different from the younger mantle-derived magma, (because of the growth of radiogenic daughter isotopes), the assimilation of a small amount of the "ancient crystalline crust can change the Pb isotopic composition of any magma and derived porphyry Cu hydrothermal system to values reflecting the values of the crustal column" (Cooke et al., 2014: 364; see also Bouse et al., 1999; Titley, 2001; Wooden et al., 1988).

Radiogenic isotopes may also be used for marine geochemistry. For example, Roy-Barman and Jeandel (2016: 129) describe radioactive and radiogenic isotopes as chronometers and source tracers in the ocean, because a radioactive isotope of an element has a constant probability per unit time (the time constant) to decay into a radiogenic isotope of another element. The authors argue that long-lived isotopes continually decay though the life of the Earth; therefore, the abundance of the radiogenic isotopes produced may vary according to the geological history of the rock that contains them and provides the source signatures when they are released into the ocean (Roy-Barman & Jeandel, 2016: 129). The long-lived isotopes are exemplified by ^{87}Rb, ^{147}Sm, ^{238}U, and ^{235}U (rubidium is Rb, samarium is Sm, U is uranium), and produced radiogenic isotopes include ^{87}Sr, ^{143}Nd, ^{206}Pb, and ^{207}Pb (Strontium is Sr, Neodymium is Nd, lead is Pb). The authors also hold that the short-lived isotopes are produced by spallation reactions (note that spallation refers to a reaction when a target is bombarded by very high-energy particles) in the atmosphere (^{10}Be, ^{14}C), during the U and Th radioactive decay chains (^{230}Th, ^{222}Rn, ^{226}Ra) and by anthropogenic nuclear activity (^{14}C, ^{226}Pu). These isotopes are chronometers used to determine rates (sedimentation rate with ^{230}Th, circulation rate with ^{14}C and ^{3}H)." Here, Be refers to beryllium, C refers to carbon, Rn refers to Radon, Ra refers to radium, and Pu refers to plutonium, all these elements having radioactive isotopes.

In practical geological studies, applications of isotope geochemistry include temperature determination and dating. Britannica (2014) writes about "the enrichment or impoverishment of certain isotopic species that results from the influence of differences in mass of molecules containing different isotopes. Measurements of the proportions of various isotopic species can be used as a form of geologic thermometer." Crucially, one method concerns the ratio of oxygen-16 to oxygen-18 in calcium carbonate secreted by marine organisms from solutions of calcium carbonate in the oceans. This is related to the ocean temperature; therefore, measurement of the proportions of oxygen-16 with respect to oxygen-18 in calcareous shells of marine fossils can enable estimations of the ocean temperature during habitation by these fossil species. One application is the estimation of the ocean temperatures during the ice ages, using the isotopic composition of fossils in sea floor sediment (Britannica, 2014).

Another important application of isotopic geochemistry is radiometric age dating. These radiometric dating methods are based on the timed rate of radioactive isotope decay which may produce a daughter isotope. Such radiometric measures are "based on the ratio of the proportion of parent to daughter isotopes, others on the proportion of parent remaining, and still others on the proportion of daughter isotopes with respect to each other" (Britannica, 2014). Cited examples include uranium-238, which decays to a natural lead isotope, lead-206. Time measurements may be based on minerals containing uranium-238, measuring the proportions of lead-206 and uranium-238; as "the older the specimen, the greater the proportion of lead-206 with respect to uranium-238" (Britannica, 2014). Another dating method is based on the decay of potassium-40 to form argon-40, and the isotope of uranium (uranium-235), which decays to form lead-207, and thorium-232, which eventually decays to lead-208. These decay periods are extremely long: the half-life uranium-238 is about 4,510,000,000 years and that of uranium-235 is about 713,000,000 years (Britannica, 2014). For shorter periods, carbon-14, which is a radioactive isotope of carbon (unlike carbon-12 and carbon-13, which are stable isotopes) has a half-life of 5570 years. As carbon-14 is part of all living material, linked to atmospheric carbon dioxide, with its short half-life it can be used for dating recent biological materials and events such as the recent ice ages (Britannica, 2014).

Devriendt (2016) cites the case of the oxygen isotope ratio ($^{18}O/^{16}O$) of inorganic and biogenic carbonate minerals, which is used to measure past temperatures, and sea level, ice volume and hydrologic changes. Devriendt (2016) argues that temperature and temperature-independent factors may be important for the carbonate-water oxygen isotope fractionation ($\alpha_{c/w}$), but "uncertainties surrounding the reasons for $\alpha_{c/w}$ variations complicate paleoenvironmental reconstructions based on carbonate $1^{8}O/^{16}O$" (Devriendt, 2016). Devriendt (2016) argues that an incisive approach is a "new general model of oxygen isotope fractionations in the $CaCO_3$-DIC-H_2O system that quantifies DIC-H_2O and $CaCO_3$-DIC fractionation as a function of temperature, pH, salinity, calcite or aragonite saturation state (Ω), DIC residence time in solution and the activity of the enzyme carbonic anhydrase." This model in the study is used with supporting data "to explore the cause of $\alpha_{c/w}$ variations for inorganic calcite, ostracod calcite and coral aragonite" (Devriendt 2016). "Isotopic fractionation is defined as the relative partitioning of the heavier and lighter isotopes between two coexisting phases in a natural system" (Tiwari et al., 2015). Tiwari et al. (2015) note two types of isotopic fractionation processes, namely, "equilibrium" and "kinetic" where *Equilibrium fractionation* is where "an exchange of isotopes takes place, which affects the equilibrium constant…, isotopes can move to and fro and equilibrium is attained when the isotopic ratios do not change any more with time" and *Kinetic fractionation* is "associated with incomplete and unidirectional processes..," affecting the rate constant of the reaction., with no isotopic equilibrium attained (Tiwari et al., 2015).

Isotope geochemistry is linked to isotope biogeochemistry, the latter referring to the methods described above in biologically related studies (Kendall et al., 2014). For example, Paytan et al. (2020) note that "sulfur isotope biogeochemistry has broad applications to geological, biological, and environmental studies." Stricker (2023) also points out that "biogeochemical cycling is a cornerstone of ecosystem function and structure. Much has been learned about element cycles in a variety of systems using standard geochemical techniques." In a study by Sanders Jr et al. (2011), linked aquatic food webs and groundwater nutrients, in northern Lake Huron. Samples were collected from groundwater vents and analyzed using ^{13}C, ^{15}N, and ^{34}S isotopes, and the findings indicated that dissolved inorganic carbon (DIC) in the groundwater was depleted in ^{13}C. The aqueous sulfate was enriched in ^{34}S, with the mean variation between the groundwater and reference sites were −3.9‰ and +12.0‰, respectively. Benthic primary producers, macroinvertebrates, and benthivorous fish were found to have lower $\delta^{13}C$ values in groundwater environments. Benthivorous fish were depleted (−2.5‰) in $\delta^{34}S$ at the groundwater sites, when compared to the reference sites. The $\delta^{15}N$ values were similar between groundwater and reference sites, and the pelagic ecosystem plankton and planktivorous and piscivorous fish were similar in both $\delta^{13}C$ and $\delta^{15}N$. The authors concluded that the findings indicated benthic metazoan communities near groundwater vents related to benthic primary production from groundwater, but planktivorous and piscivorous communities which are not directly linked with the benthos are not reliant on groundwater nutrients.

Aqueous Geochemistry

Aqueous geochemistry is concerned with the chemical interactions of water (H_2O) and rocks and may therefore be an offshoot of aqueous chemistry, groundwater chemistry, hydrochemistry, or water chemistry (Sen, 2015). As biogeochemistry is strongly concerned with chemical cycling between the hydrosphere, lithosphere, biosphere, and atmosphere, this is an extremely important part of the relationship between geochemistry and biogeochemistry. Nordstrom (1997) defines this subject as "the application of chemistry to reactions between rock and natural water" and further argues that this involves the perspectives of physical, analytical, inorganic and organic chemistry, with examples of studied processes including "the dissolution and precipitation of minerals, adsorption and desorption of ions, oxidation-reduction or redox reactions, gas uptake or production, transformations involving organic matter, complexation and chelation, evaporation, ion exchange, and anthropogenic changes" which may be applied to various aspects of the hydrosphere (e.g., rain, riverine, and marine environments) (Nordstrom, 1997).

White (2010: 231) takes the example of springs, which in this context differ from other water sources in that "springs result from a concentration of groundwater flow paths, so that water issues from a single location instead of diffuse flow into surface streams." The chemistry of spring waters consequently results from the interaction of the groundwater with the aquifer host rock and any possible chemical constituents introduced from surface sources, with little difference between spring water chemistry and groundwater chemistry. This author defines the approach to the study of aqueous chemistry as examining the composition, thermodynamic properties, mineral solubilities, and interactions of the host rock's minerals, from which the water chemistry can be predicted. Common ions in spring water include the cations (positively charged ions) Ca^{2+}, Mg^{2+}, Na^{+}, and K^{+}, and the anions (negatively charged ions) HCO_{3^-}, SO_4^{2-}, and Cl^-. White (2010: 231) points out that these ions are the main indicators for the classification of spring water. There are three main groups in such classifications, these being bicarbonate waters, where Ca and Mg may be the dominant cations; sulfate waters, where Mg levels may be greater than those of Ca, and there are also alkali ions; and chloride waters, where alkali ions may be the dominant ions. In such classifications, the dominant chemical composition of the spring generally depends on the chemistry of the host aquifer rock from which it emerges (White, 2010: 231).

Important concepts of aqueous geochemistry include chemical equilibrium and mineral saturation. Minerals differ in their reaction to water, with some minerals dissolving in the water, and breaking up into their component ions, while other minerals may react chemically, with the formation of soluble ions and sometimes insoluble residues (White, 2010: 231). Examples are cited of halite (NaCl), gypsum ($CaSO_4\,{}_2H_2O$), and calcite ($CaCO_3$), for which the author notes there is a "certain equilibrium concentration, defined thermodynamically by the concentration at which the free energies of the species on both sides of the dissolution reaction balance exactly or kinetically where the rate of dissolution is equal to the rate of precipitation." This ideal equilibrium concentration may not be fully present in the real world, arguably because "the concentrations of dissolved species are often not at equilibrium with the minerals of the host rock. This is particularly true in carbonate rocks, where the nonequilibrium is a useful interpretive tool" (White, 2010: 231). Calculations of mineral solubilities nevertheless may represent a reference for the evaluation of spring water chemistry, as the chemical levels predicted from equilibrium calculations may be compared with water chemical samples, using chemical analysis (White, 2010: 231). Literature relevant to these issues are listed by the author (Hem, 1985; Morel & Hering, 1993; Stumm & Morgan, 1996; Drever, 1997; Langmuir, 1997; Parkhurst & Appelo, 2008).

Aqueous biogeochemistry adds biological themes to aqueous geochemistry, when the elements, compounds and processes that exist in the geochemistry of water intrude into or relate to organisms. One important process is biomineralization, which concerns the precipitation of minerals by living organisms (Hendry et al., 2018). Nature Portfolio (2023) defines biomineralization as the "process by which living organisms produce minerals" which "often lead to the hardening or stiffening of the mineralized materials" and the "formation of silicates in algae and diatoms, carbonates in invertebrates, and calcium phosphates and carbonates in the hard tissues of vertebrates." Examples of biomineralization include the formation of iron oxide by brown algae, which is affected by environmental conditions, and other processes which may be dependent on the metabolism of the organisms under indirect genetic control and the cellular processes that enable biomineral formation. There is also "biologically controlled biomineralization" which "requires direct genetic control, generates characteristically patterned structures, and involves selective uptake and pre-concentration of mineral ions" (Hendry et al., 2018). Biomineralization may also concern iron oxide, strontium sulfate, calcareous compounds, and silica ($SiO_2 \cdot nH_2O$), the last being described as the "taxonomically widespread biomineral" and "present in all eukaryotic supergroups" (Hendry et al., 2018). Hendry et al. (2018) also note that "the most extreme degree of silicification is evident in the diatoms, where almost all species have an obligate requirement for silicon to complete cell wall formation and cell division" and "biogeochemically and ecologically, diatoms are believed to be the most important silicifiers in modern marine ecosystems."

Cosmochemistry

Cosmochemistry has been variously defined, but the consensus is that it studies the chemical composition of the physical constituents of space, including gases, solids and even liquids. Numerous books and research articles have examined topics related to cosmochemistry, such as those of Wasserburg and Hayden (1955), Schmus and Wood (1967), Black and Pepin (1969), Lee et al. (1977), Hoyle and Wickramasinghe (1981), Burke (1986), McKay et al. (1996), Lodders and Fegley Jr. (1998), Railsback (2003), Davis (2004), Lauretta and Killgore (2005) and Lauretta and McSween Jr. (2006). These studies present increasingly more sophisticated analyses of space materials, from the perspectives of chemistry, physics, and astronomy, and document the development of related research methods.

Gounelle (2011: 380) defines cosmochemistry as the study of the formation and evolution of the Solar System and its individual components through the analysis of extraterrestrial samples in the laboratory." MacPherson and Thiemens (2011: 19130) add that "cosmochemistry is the chemical analysis of extraterrestrial materials. This term generally is taken to mean laboratory analysis, which is the cosmochemistry gold standard because of the ability for repeated analysis under highly controlled conditions using the most advanced instrumentation unhindered by limitations in power, space, or environment." Further, "recent advances in the state of the art of cosmochemistry…, range from instrumental analysis of meteorites to theoretical–computational and astronomical observations" (MacPherson & Thiemens, 2011: 19130).

McSween and Huss, G. (2010: 1) define cosmochemistry as "the study of the chemical composition of the universe and the processes that produced those compositions" primarily on the immediate solar system, including the compositions of all the bodies in the Solar System, namely the Sun, planets and their satellites, the asteroids and comets, meteorites, and interplanetary dust particles (IDPs). Cosmochemists study the composition, chemistry and formation chronology of these bodies, using laboratory measurements of samples and/or remote-sensing techniques (McSween Jr & Huss, 2010). The science of cosmochemistry has important relationships with other sciences, especially with geochemistry, and astronomy, astrophysics, and geology (McSween Jr & Huss, 2010: 1). This view is supported by Makishima (2017), who notes that there are important links between cosmochemistry and other Astro sciences, including astronomy, astrobiology, astrogeology, and astrophysics.

McSween Jr and Huss (2010) point out that "traditionally, cosmochemistry has been treated as a branch of geochemistry – usually defined as the study of the chemical composition of the Earth," with geochemistry principally focused on the Earth (hence the prefix "geo"), with generally little coverage of cosmochemistry in geochemistry textbooks. McSween Jr and Huss (2010) further mention that "the line between geochemistry and cosmochemistry has always been somewhat fuzzy. The most prominent technical journal in this discipline, Geochimica et Cosmochimica Acta, has carried both names since its inception in 1950." Also mentioned is the rising field of planetary geochemistry, which is still has the "geo" prefix, despite the focus being the non-Earth solar bodies. McSween Jr and Huss (2010) argue that for the inclusion of solar bodies in the study of geochemistry "a broader and more appropriate definition of geochemistry might be the study of element and isotope behavior during geologic processes, such as occur on and within the Earth and other planets, moons, and planetesimals. Using this definition, we will include planetary geochemistry as an essential part of our treatment of cosmochemistry."

Another important issue with cosmochemistry concerns the source of research materials, which may be derived from direct sourcing in space or from falling meteors. Initially, cosmochemistry studies were dependent on the chance discovery of meteorites (from observed falls or chance discovery on the ground). However, in 1969, Japanese explorers in Antarctica led by Masaru Yoshida found meteorites on bare ice, which influenced the formation of meteorite-collecting expeditions to Antarctica, and meteorite-collecting in the southwestern United States, Western Australia, and North Africa. The meteorites so discovered varied, including carbonaceous chondrites (containing carbon-bearing compounds, with primarily organic molecules), but disputes arose as to whether the carbon organic molecules were from space or Earth surface contamination (McSween Jr & Huss, 2010). Currently, smaller meteorites may be recovered when they fall on bare rock, sand, or ice, while larger meteorites may leave craters or other damage evidence on the Earth's surface.

Research methods in cosmochemistry have also developed. Davis (2014) argues that "cosmochemistry has made tremendous progress since the 1960s, largely because of improvements in analytical technology" with an important boost experienced by the Apollo space program, which acquired physical samples from the moon, which influenced increased investments in analytical laboratories during the 1960s and 1970s. Developments included "advanced analytical instrumentation with isotopic precision (thermal ionization and gas source mass spectrometry), precise microbeam chemical analysis (electron probe microanalysis), and sensitive trace-element analysis (neutron activation analysis)" and the fall of and recovery of the Allende (Chihuahua, Mexico, 1969) and Murchison (Murchison, Australia in 1969) meteorites, which influenced more advanced research on cosmochemistry (Davis, 2014). Koeberl (2007: 1) writes about the geochemistry and cosmochemistry of impact craters and processes, which is argued to be "a rapidly developing research area that encompasses such wide-ranging topics as the simple chemical characterization of the various rock types involved (target rocks, impact breccias, melt rocks, etc.), the identification of extraterrestrial components in impact ejecta, the determination of the impactor (projectile) composition, and the determination of the causes of environmental changes from chemo-lithostratigraphic analyses."

Due to the lack of evidence of life in space, there are few connections between cosmochemistry and biogeochemistry. However, Callahan et al. (2011:13995) describe possibilities of chemical substances recorded on meteorites, which may indicate extraterrestrial sources for compounds detected on the Earth's surface. The basis of this argument is that organisms depend on nucleic acids, namely ribonucleic acid (RNA) and deoxyribonucleic acid (DNA), which encode genetic information using pyrimidine and purine nucleobases. It is speculated that carbon rich meteorites may have been a source of organic compounds during the development of life on Earth. The authors note that "so far, the few nucleobases reported in meteorites are biologically common and lacked the structural diversity typical of other indigenous meteoritic organics" (Callahan et al., 2011:13995). Carbonaceous chondrites described rare meteorites, which may contain organic compounds, including compounds important for biology (see also Sephton, 2002; Schmitt-Kopplin et al., 2010). The study by Callahan et al. (2011:13995) investigated nucleobases and nucleobase analogs in formic acid extracts of meteorites, using liquid chromatography–mass spectrometry, and concluded that the purines in meteorites were consistent with some products of ammonium cyanide chemistry, which might be a mechanism for their synthesis in the asteroids. The authors conclude that as "meteorites provide a record of the chemical processes that occurred in the solar system before life began on Earth" the "discovery of new nucleobase analogs in meteorites also expands the prebiotic molecular inventory available for constructing the first genetic molecules" (Callahan et al., 2011:13995).

Trace-Element Geochemistry

Trace elements are an important focus of geochemical research, and are also examined within the context of biogeochemistry, when trace elements occur in living organisms. A trace element is defined as one with a concentration of less than 1000 ppm or 0.1% of a rock's composition. White (2013: 259) notes that "though trace elements, by definition, constitute only a small fraction of a system of interest, they provide geochemical and geological information out of proportion to their abundance" and "trace element geochemistry has been of enormous use in understanding the evolution of the Earth." Kennedy (1998: 221) define trace elements (TE) as those elements that occur in a mineral in very small amounts, to the extent that they are excluded from the mineral's chemical formula, and they may also be defined as "elements that occur at less than 0.1 wt% in the Earth's crust; however, this is not true of all trace elements" (see Brownlow, 1979). Kennedy (1998: 221) argues that "early interest in trace element geochemistry focused on factors contributing to the occurrence of trace elements in the crystal lattice of various minerals" citing Goldschmidt (1937). Rubidium (Rb) is cited as an example of a common trace element in orthoclase, a feldspar mineral ($KAlSi_3O_8$). Other common trace elements (including rare earth elements (REEs), which may be used in research) are cited, as barium (Ba), Strontium (Sr), zirconium (Zr), chromium (Cr), vanadium (V), rubidium (Rb), zinc (Zn), nickel (Ni), copper (Cu), cobalt (Co), yttrium (Y), niobium (Nb), lithium (Li), scandium (Sc), gallium (Ga), lead (Pb), thorium (Th), hafnium (Hf), uranium (U), lanthanum (La), cerium (Ce), neodymium (Nd), praseodymium (Pr), gadolinium (Gd), dysprosium (Dy), erbium (Er), ytterbium (Yb), europium (Eu), holmium (Ho), terbium (Tb), thulium (Tm), and lutetium (Lu) (Kennedy, 1998; see also Wedepohl, 1995).

Recent research on trace elements has underscored their importance for geochemical analyses, but definitions analyses may be problematic. For example, White (2013: 260) writes about the difficulty of definition, and notes that for igneous, sedimentary, and metamorphic rocks, a possible definition is that trace elements are those elements that "are not stoichiometric constituents of phases in the system of interest" which is a "a bit fuzzy" because "a trace element in one system is not one in another." The example is cited of potassium, which may be a trace element in mid-ocean ridge basalts but is not a trace element in granites. White (2013: 260) further notes that for most silicate rocks, oxygen (O), silicon (Si), aluminum (Al), sodium (Na), magnesium (Mg), calcium (Ca), and iron (Fe) are "major elements," while hydrogen (H), carbon (C), sulfur (S), potassium (K), phosphorus (P), titanium (Ti), chromium (Cr), and manganese (Mn) "are sometimes 'major elements' in the sense that they can be stoichiometric constituents of phases. These are often referred to as 'minor elements'" and "all the remaining elements are always trace elements" (White, 2013: 260).

For identification and assessment of trace elements in geological formations, increasingly sophisticated methods may be employed. For example, Torró et al., 2022 presents a study of the trace-element occurrence in sphalerite (a sulfide mineral with the formula (Zn,Fe)S) and chalcopyrite (a copper iron sulfide mineral, which is the commonest copper ore mineral, with the formula $CuFeS_2$) from four volcanogenic massive sulfide (VMS) deposits. VMS deposits are metal sulfide ore deposits, mostly copper-zinc formations linked to volcanic hydrothermal occurrences in submarine environments. Torró et al. (2022) defines VMS deposits as features which "are exhalative and/or replacive accumulations of sulfide minerals formed at or near the seafloor where uprising hydrothermal fluids driven and in part generated by synchronous magmatism are entrained with cold seawater or porewaters." The research method was laser ablation and inductively coupled plasma mass spectrometry.

The general findings of this study by Torró et al. (2022) were that for the studied sulfide deposits, from the surface downward, for sphalerite there was a progressive increase in the trace elements In, Cu, Mn, and Se, and a decline in Ge (Germanium). It is argued that this distribution pattern correlates with increasing crystallization temperatures and/or volatile magmatic influx near the lower portion of the mineralization (Torró et al., 2022). For chalcopyrite, the distribution of the trace elements was "rather uneven except for a sustained enrichment in Se toward the basal portion of the sulfide body" (Torró et al., 2022). The author argues that the preservation of such trends despite extensive recrystallization during thermal metamorphism in parts of the study area massive sulfide mineralization suggests that it is a closed metamorphic system and that element interdiffusion was largely local (Torró et al., 2022).

The findings also indicated that in sphalerite, Fe, Mn, Cd, Hg, Ag, Sb, Se, In, Ge, and Ga are lattice-bound, but Sn, Tl, Bi, and Pb are at least partly mineral micro-inclusions. Here lattice-bound refers to the regular arrangement of atoms/molecules within a crystal. Frye (1981) defines a lattice as "a three-dimensional array of points, each of which has an identical surrounding of neighboring points... Lattices extending infinitely in three dimensions are used in crystallography to describe the structural patterns found in all crystalline materials." Stoll et al. (2021) define micro-inclusions as small particles, for example micrometer-sized (μm, one millionth of a meter), that occur in a substance. Torró et al. (2022) also found variations in the contents trace elements are observed in sphalerite grains from deposits. There were also substitutions between the elements, with a strong negative correlation between iron and zinc.

For chalcopyrite (a copper–iron sulfide mineral with chemical formula $CuFeS_2$), Zn, Ag, Sn, Cd, Se, In, Ga, and Ge were found to be lattice-bound, and Mo, Au, Tl, Sb, Pb, and Bi may be micro-inclusions (Torró et al., 2022). Compared with sphalerite, chalcopyrite differed in having lower amounts of In and Ga, and higher concentrations of Ge. Zinc had the highest concentrations of the other trace elements in the examined chalcopyrite, with lower of amounts of silver, selenium, tin, and cadmium. The authors noted that "general positive correlation trends between Zn, Cd, In, Ge, and Ga in chalcopyrite suggest varied coupled substitution mechanisms of Fe and Cu with fluctuating valences due to covalent bonding" (Torró et al., 2022).

Another study using trace elements for geological assessments is that of Dudás et al. (2021), which examined trace and major and trace-element data for the

Permian-Triassic boundary (PTB) at Meishan in China, using strata samples covering the last 75,000 years of the Permian (a 47 million year geologic period and stratigraphic system from 298.9 million BP to 251.9 million years BP) and the first 335,000 years of the Triassic Period (another 50.6 million year geological period, lasting from the end of the Permian period, 251.902 million years BP, to 201.36 million years BP) (Dudás et al., 2021).This covers the Permian–Triassic Extinction Event (about 251.9 million years BP), when the majority of the Earth's living species were rendered extinct, possibly due to the environmental impacts of rising temperatures. The study ranks layer beds as from before the extinction event or interval just before the PTB (beds 22 to 24e), the extinction event beginning (bed 24f) to after the extinction event and PTB (beds 28 to 34) (Dudás et al., 2021).

The main rock change documented was from carbonate dominated strata in the Permian to mudstone and marl in the Triassic, which the authors argue indicated "an increase of siliciclastic input and MgO in and above the extinction interval… and silica diagenesis in carbonates below the extinction horizon" with evidence that "siliciclastic input dominates trace element distributions in the Triassic" (Dudás et al., 2021). During the extinction interval, there were increases in As, Mo, U and transition metals, and enrichment of V, Cr, Co, Ni, Cu, Zn, Pb, and Ba. Below the strata representing the extinction interval there was enrichment of V, Cr, Co, Ni, Cu, Zn, Pb, and Ba within a layer diagenetic silicification. The authors conclude that "trace elements thus reflect siliciclastic input, diagenetic redistribution, and responses to redox conditions" and such trace element patterns "suggest either a change in provenance of the detrital component, or a change in the proportion of mechanical to chemical weathering that is coincident with the beginning of the extinction" period (Dudás et al., 2021).

Considering the behavior of the trace elements in the layer samples, Dudás et al. (2021) report anomalous behavior from Ba, Zr, and Zn, with little variation from Ba, despite variations in biological activity and redox conditions. Zinc concentrations decline in the extinction horizon, possibly due to changes in phytoplankton productivity. Zr mobility is suggested, as the enrichment factor for Zr is found to be variable in the carbonates below the extinction layers. Dudás et al. (2021) conclude that a "transition from carbonate-dominated sedimentation in beds 24e and below" (below the extinction levels), "to dolomite or ankerite in beds 24f and above" (the extinction levels and above), strong correlations between trace elements with Al_2O_3 (Sc, Ga, Rb, Ba, Th) within the siliciclastic fraction of the Meishan rocks, and "trace element data suggest that there was a change in either the provenance of the detrital fraction, or in the proportion of mechanical to chemical weathering coincidentally with the extinction beginning at the top bed 24e," which may not be linked to volcanism of the period (Dudás et al., 2021). Note that detrital refers to rock particles derived from the physical breakdown of pre-existing rocks, due to weathering and erosion (see Marshak, 2009).

Trace-element geochemistry has also been applied to archaeological studies, contributing to the field of geoarchaeology, a research field using the methods of the sciences of geomorphology, hydrology, sedimentology and pedology, the research methods of remote sensing, geophysics, geochemistry, and chemical, radiometric, magnetic, biological methods, including geochronology, and exploration geophysics (Elias, 2021). Geoarchaeology's research points include site date and formation, ancient environmental characteristics, and human decisions about site location. Although older geoarchaeological research was mostly on "climate reconstruction, paleodiet analysis, or artifact conservation," recent studies have acknowledged the importance of archeological site sediments, and more advanced dating and remote sensing techniques for site identification and surveying archeological sites (Elias, 2021).

Elias (2021) documents the work of Jenkins (1989) on trace-element geochemical applications to archaeology, which is described as "a foundational paper on this topic," which highlighted the "great potential use of trace element geochemistry in archeological research" and also "highlighted the problem of being able to differentiate between anthropogenic and naturally occurring trace element anomalies, given the inherent geochemical diversity of sediments as well as the superimposed chemical alterations resulting from subsequent pedogenic processes in archeological sites" (Elias, 2021). Trace-element analysis has been useful in the chemical fingerprinting of flints/cherts and obsidian, and the origins of artifacts, including glazed ceramic ware, and charcoal analysis. The study by Jenkins (1989) illustrated the dating of charcoal using trace-element analyses. Elias (2021) notes that the results of this study points to the fact that charcoal accumulates certain trace elements significantly and selectively, and this concentration provides for greater sensitivity than bulk analysis including the nearby soil.

Jenkins (1989: 57) noted that people "can leave a geochemical imprint on an archaeological site in several ways" based on "a selective enrichment of elements in his body tissues which, upon death and burial, may lead to detectable anomalies," with phosphorus being the most obvious, and other possibilities being metals such as tin (Sn), silver (Ag), gold (Au), copper (Cu), lead (Pb) and zinc (Zn) which may have been used by ancient cultures, and may be measured in sites. Jenkins (1989, 57) also states that charcoal is among the commonest finds during excavations, and it can adsorb and concentrate metals from percolating solutions progressively from its burial time. Careful analysis may therefore derive an important historical record. Therefore, "providing care is taken to interpret results in their particular geochemical and pedochemical context, trace element analysis may thus offer a useful insight into the history of man's activities in an archaeological site" (Jenkins, 1989, 57).

Elias (2021) also describes the development of trace-element analyses of charcoal, including the recently designed Fourier transform infrared spectroscopy (FTIR), which can analyze samples more than 100 times smaller than those of the techniques described by Jenkins (1989). Elias (2021) also argues that most trace-element studies have focused on copper (Cu), iron (Fe), manganese (Mn), lead (Pb), and zinc (Zn), as some of these metals can remain stable in soils for long periods, because their ions are adsorbed and precipitated onto clay surfaces, and can form stable compounds, including insoluble oxides, sulfates, or carbonates. Examples include the study of metallic elements (e.g., hematite Fe_2O_3 and cinnabar HgS) in pre-Columbian Mayan sites for pigments and paints, which may indicate locations for pigment processing, and craft workshops. Study methods include trace metal extraction and inductively coupled plasma mass spectrometry (ICP/MS) or atomic emission spectroscopy (ICP-AES) analyses of soil and floor samples at the Guatemalan archeological sites in Cancuén, Piedras Negras, and Aguateca (Elias, 2021; see also Terry, 2017).

Trace-element geochemistry may also be linked to cosmochemistry. For example, a study by Ollila et al. (2013: 255) examined the findings of the ChemCam instrument package on the Mars Rover, Curiosity, which "provides new capabilities to probe the abundances of certain trace elements in the rocks and soils on Mars using the laser-induced breakdown spectroscopy technique." The study focused on detecting and quantifying lithium (Li), barium (Ba), rubidium (Rb) and strontium (Sr). The findings indicated that Li was detected for the first time on Mars, with low concentrations (<15 ppm). There were local areas with higher concentrations of Li, up to ~60 ppm. Bathurst_Inlet (a rock on the surface of Aeolis Palus, between Peace Vallis and Aeolis Mons in Gale crater) is documented as a fine-grained bedrock with some evidence of a decrease in Li and other alkalis with depth, possibly indicating past low-level aqueous alteration drawing alkalis to the surface. Ba (with a concentration of ~1000 ppm) was found in the Akaitcho sand ripple and may correlate with Si, Al, Na, and K, which the authors argue may indicate a "feldspathic composition" Ollila et al. (2013: 255). Rb and Sr were found in concentrations of >100 ppm and >1000 ppm, respectively. The authors note that these areas also tended to have high concentrations of Si and alkali, these combinations being consistent with a feldspar composition. Ollila et al. (2013: 255) conclude that "together, these trace-element observations provide possible evidence of magma differentiation and aqueous alteration."

Trace-element biogeochemistry is derived from the issues, methods, and findings of trace-element geochemistry, which may examine trace elements in inanimate matter, and also in organisms and living tissue (Winkel & Sunderland, 2022). The International Conference on the Biogeochemistry of Trace Elements (ICOBTE) (2023) writes that "trace elements (TEs) play an important role in the biogeochemical cycle of the whole environment. In various ecosystems TEs are enriched, while the sources can be either natural (geogenic) or anthropogenic (human-induced). Natural processes releasing TEs into the biosphere include mineral weathering, volcano eruptions, and natural erosion." In addition to the natural issues of the solid earth mentioned earlier, trace-element biogeochemistry may include chemical artificially introduced and/or increased by human activities. The relevant activities and inputs may include the externalities of industrial production, including mining, electroplating, smelting, wastewater, aerosol deposition, sewage, and fertilizers (ICOBTE, 2023).

Winkel and Sunderland (2022) elaborate on the themed issue of the journal *Environmental Science: Processes & Impacts* by the Royal Society of Chemistry, which aims to document advances in trace-element biogeochemistry research, where trace elements are defined as including metals, metalloids, and nonmetals. They note that this examines "a wide range of biogeochemical processes and environmental impacts of toxic and essential TEs," with a strong focus on research across phase- and compartment interfaces and coupled biogeochemical cycles between TEs and/or major elements." Key issues currently emerging for trace-element biogeochemistry include changes in materials science, electronics and computing, and "global mining and use of TEs has increased massively over the last decades and there has been strong global demand for the production and use of rare earth elements and platinum-group elements" (Winkel & Sunderland, 2022). The authors therefore argue that new issues include elements important for battery technology, electric vehicles, renewable energy, a low-carbon economy, and the need for less environmental contamination in biota, air, water, and human environments. The authors also mention the three main topics of the themed issue as, firstly, the "biotic processes involved in cycling of toxic and essential trace elements in soils and marine and brackish waters," secondly the "sediments and soils and their interfaces to aquatic systems (e.g., groundwater, river water, and the hyporheic zone and lake systems), and thirdly, the "TE distributions at large geographical scales in soils, sediments and porewater, lakes, oceans and in the atmosphere" (Winkel & Sunderland, 2022).

Igneous and Metamorphic Rock Geochemistry

The geochemistry of both igneous and metamorphic rocks largely concerns the elements and compounds that occur in the processes, solid materials and sinks within which these chemicals exist. As igneous rocks are created by magmatism, and metamorphic rocks are usually pre-existing igneous, sedimentary, and metamorphic rocks that have undergone more heating or pressure, there may be differences in the geochemistry of igneous and metamorphic rocks, and the possible links with biogeochemistry (Campbell, 2020). This section looks at the geochemistry of igneous and metamorphic rocks and takes silicon as an example of an element that cycles though the igneous and metamorphic geochemical cycles, and is vital to global biogeochemical cycles, hence illustrating the links between geochemistry and biogeochemistry, the latter when organisms are involved in the cycling.

Igneous rock geochemistry concerns the chemistry of igneous rocks, which are rocks that result from volcanic activity (Campbell, 2020). Kudo and Jahns (2022) define igneous rocks as "any of various crystalline or glassy rocks formed by the cooling and solidification of molten earth material… formed from the solidification of magma, which is a hot (600 to 1,300 °C, or 1,100 to 2,400 °F) molten or partially molten rock material." Hence, the immense heating of volcanic activity comprises an essential aspect of the development of igneous rock. Mather (2013) states that in "igneous geochemistry… processes generally occur at much higher temperatures and pressures" than in other fields such as low temperature, earth surface geochemistry. Kudo and Jahns (2022) argue that most igneous are composed of silicate minerals (compounds including silicon (Si) and oxygen (O), and in some cases carbonates). The example is cited of a sodium carbonate lava (Na_2CO_3), which included 0.05 wt% of silica (silicon dioxide, SiO_2) from the Ol Doinyo Lengai volcano in Tanzania. Generally, the low levels of silica and in soda (Na_2O) and potash

(K_2O) correlate with increased occurrences of magnesium oxide (MgO) and iron oxides (FeO, Fe_2O_3, and Fe_3O_4). High levels of silica also correlate with high levels of calcium oxide (CaO) and alumina (Al_2O_3).

Kudo and Jahns (2022) also write that as silica relates to several other chemical compounds in rocks, the silica content is the feature that is often used to classify igneous rocks. These classifications include "silicic or felsic, rocks having more than 66 percent silica; intermediate, rocks with 55 to 66 percent silica; and subsilicic, rocks containing less than 55 percent silica. The latter may be further divided into two groups: mafic, rocks with 45 to 55 percent silica and ultramafic, those containing less than 45 percent" (Kudo & Jahns, 2022). As the subsilicic rocks are enriched in iron (Fe) and magnesium (Mg), they may be called femic, a term derived from ferrous iron and magnesium. The silicic rocks are called sialic (a combination of silica and aluminum) or salic (also from silica and aluminum). Kudo and Jahns (2022) write that the terms mafic (from magnesium and ferrous iron) and felsic (feldspar and silica) may be used interchangeably with femic and sialic.

Silicate minerals may form from the inclusion of silica from the cooling and crystallization of magma, and the combination of silica with other cationic oxides. Examples include the combination of SiO_2 and MgO to form $MgSiO_3$ (magnesium-rich pyroxene, enstatite), the combination of two moles of SiO_2 with one mole each of CaO and Al_2O_3 to create $CaAl_2Si_2O_8$ (calcium-rich plagioclase, anorthite). Note that pyroxenes are the commonest group of rock-forming ferromagnesian silicates. Kudo and Jahns (2022) point out that there are cases where there are insufficient silica levels in the magma to create pyroxene; hence, compounds requiring less silica may be created (in subsilicic rocks), in addition to pyroxene, such as Mg_2SiO_4, magnesium-olivine (forsterite), and Mg_2SiO_4, in addition to pyroxene. Such subsilicic rocks contain silicate minerals such as magnesium-olivine, sodium-nepheline ($NaAlSiO_4$), leucite ($KAlSi_2O_6$), which may partially substitute for enstatite, albite ($NaAlSi_3O_8$), and orthoclase feldspar ($KAlSi_3O_8$) which require more silica, and there may be no quartz. The first three minerals are termed undersaturated *in silica* and are found in undersaturated subsilicic rocks. There are also cases where there is excess silica in the magma; hence, the excess silica after the formation silicate minerals (which were formed though combination of silica and oxides) may form into quartz (SiO_2) or related minerals, and magnesium-pyroxenes (see Simmons, 2014 for a discussion of pyroxenes). Where rocks have excess silica, the rocks are termed silicic and supersaturated rocks, containing quartz and magnesium-pyroxene, considered saturated minerals (Kudo & Jahns, 2022).

Due to this variation into excess or lower silica content, igneous rocks can generally be divided into two groups. These are the felsic (derived from feldspar and silica) and mafic (derived from magnesium and ferrous iron). Feldspars usually contain sodium, calcium, potassium, orbarium, with the formulas $KAlSi_3O_8$ – $NaAlSi_3O_8$ – $CaAl_2Si_2O_8$ (based on these formulae, Orthoclase or Orthoclase feldspar, Plagioclase or Plagioclase feldspar, Anorthite, respectively). Ferrous iron refers to Iron (II) (Fe^{2+}) (contrasted with iron (III) salts, or Fe^{3+}) (Weiss et al., 2003; Anderson & Anderson, 2010). Felsic minerals include quartz, tridymite, cristobalite, feldspars (plagioclase and alkali feldspar), feldspathoids (nepheline and leucite), muscovite, and corundum. They lack iron and magnesium, hence are usually light colored or leucocratic. Mafic minerals include olivine, pyroxenes, amphiboles, and biotites, which are dark colored or melanocratic. Due to this contrast in color, felsic rocks are usually classified as having a color index of less than 50, and mafic rocks with a color index over 50, and ultramafic rocks with a color index above 90. These numbers refer only to the rock mineralogical content but may not represent chemical terms. Kudo and Jahns (2022) cite the examples of a mineral plagioclase rock, which would be felsic, but chemically would be a subsilicic mafic rock; and a pyroxene igneous rock with a silica content of around 50%, which would be a mineralogically ultramafic rock, but a chemically igneous rock. During the magmatic formation of these rocks, supersaturated minerals (such as quartz) are separated from undersaturated minerals (such as feldspathoids like leucite and nepheline) or magnesium-rich olivine. Other minerals termed accessory minerals are also formed and exist in igneous rocks in minor amounts include monazite, allanite, apatite, garnets, ilmenite, magnetite, titanite, spinel, and zircon (Kudo & Jahns, 2022).

The geochemistry of igneous rocks is relevant to environmental change. For example, Bataille et al. (2017) argue that the chemical weathering of silicates transfers elements from the continental crust to ocean waters and hence affects some biogeochemical cycles. Chemical weathering of silicates transfers calcium (Ca) and magnesium (Mg) to ocean waters and regulates atmospheric carbon dioxide levels and temperature at the surface by impacting on the rate of marine carbonate precipitation. Bataille et al. (2017) further remark that there is insufficient knowledge of the impact of the composition of igneous rocks on the content of chemical weathering products, and the consequent impacts on ocean waters and atmospheric change. The authors use the strontium isotope ratio in seawater [($^{87}Sr/^{86}Sr$) seawater] as a proxy for chemical weathering and test the sensitivity of $^{(87}Sr/^{86}Sr)$ seawater changes to the strontium isotopic composition ($^{87}Sr/^{86}Sr$) in igneous rocks which were generated over time.

The use of the Strontium ($^{87}Sr/^{86}Sr$) isotope is important as a versatile application. For example, Salifu et al. (2018: 42) state that "the $^{87}Sr/^{86}Sr$ ratio is not fractionated by these processes, which refer to the fact that "interpretation of geochemical data based primarily on elemental concentrations often leads to ambiguous results due to multiple potential sources including mineral weathering, atmospheric input, biological cycling, mineral precipitation and exchange processes." Salifu et al. (2018: 43) also argue that "the Strontium ($^{87}Sr/^{86}Sr$) isotope ratio is an important and powerful investigative tool which has been extensively and successfully utilized in petrogenesis, mineral weathering and acidity in soils, discriminating between atmospherically-derived Sr and those of mineral weathering inputs as well as differentiating between carbonate and silicate weathering sources" (the authors also cite Graustein & Armstrong, 1983; Åberg et al., 1989; Jacks et al., 1989; Miller et al., 1993; Nakai et al., 1993; Bullen et al., 1996; Stewart et al., 2001; Jacobson et al., 2002; Chamberlain et al., 2005; Tipper et al., 2006; Shand et al., 2007; Jin et al., 2011; Subías et al., 2015; Kozlik et al., 2016).

Bataille et al. (2017) found that the $^{87}Sr/^{86}Sr$ ratio in igneous rocks was correlated to the epsilon hafnium (εHf) of their hosted zircon grains and used the detrital zircon record to reconstruct the evolution of the $^{87}Sr/^{86}Sr$ ratio in the zircon-bearing igneous rocks. The findings indicated that the reconstructed $^{87}Sr/^{86}Sr$ variations in igneous rocks were correlated with the ($^{87}Sr/^{86}Sr$) seawater variations over the last 1000 million years, which suggested a direct controlling link of the isotopic composition of silicic magmatism on ($^{87}Sr/^{86}Sr$) seawater variations. The study therefore showed that the correlation between $^{87}Sr/^{86}Sr$ ratio in igneous rocks and seawater decreases over long periods, which may indicate changes in the chemical weathering rates which are associated with paleogeographic, climatic, or tectonic events, and there is the strong argument that "for most of the last 1000 million years, the ($^{87}Sr/^{86}Sr$)seawater variations are responding to changes in the isotopic composition of silicic magmatism rather than to changes in the global chemical weathering rate" and that "the ($^{87}Sr/^{86}Sr$)seawater variations are of limited utility to reconstruct changes in the global chemical weathering rate in deep times" (Bataille et al., 2017). The suggestion is that when the global isotopic composition of silicic igneous rocks increases, the fast cycling and later rock weathering contributes to an increase in the $(^{87}Sr/^{86}Sr)_{seawater}$ ratio within a relatively short time period (<20 My)" (Bataille et al., 2017).

Bataille et al. (2017) conclude that their analysis represents "a new method to reconstruct the evolution of the strontium isotopic composition of silicic igneous rocks through time" as the $(^{87}Sr/^{86}Sr)_{i\text{-}zig}$ variations exerted dominant control of ($^{87}Sr/^{86}Sr$)seawater evolution over the last 1000 My, and $(^{87}Sr/^{86}Sr)_{seawater}$ variations were interpreted as tracking the relative proportion of evolved versus less evolved magmas in the continental crust, rather than changes in the chemical weathering rate of radiogenic continental surfaces (Bataille et al., 2017). The changes in the global isotopic composition of silicic magmas may be related to continental mountain-building events, and subduction phases. It is concluded that other $(^{87}Sr/^{86}Sr)_{seawater}$ variations could result from variations in the relative Sr flux from isotopically distinct sources which may be linked to "specific paleogeographic configurations, mountain-building events, emplacement of LIPs, or climate variations" and "the $(^{87}Sr/^{86}Sr)_{seawater}$ variations are of limited utility to reconstruct the long-term chemical weathering rate" (Bataille et al., 2017).

Metamorphic (from the root word metamorphosis, meaning change) rock geochemistry concerns the chemical composition and change, mostly of the metamorphic process where pre-existing igneous, sedimentary, or metamorphic rocks undergo extreme heating (without reaching the melting stage, which would convert the rock to igneous rock), pressure or other change and create new, metamorphic rocks (United States Geological Survey, 2022). The United States Geological Survey (2022) states that "the process of metamorphism does not melt the rocks, but instead transforms them into denser, more compact rocks. New minerals are created either by rearrangement of mineral components or by reactions with fluids that enter the rocks. Pressure or temperature can even change previously metamorphosed rocks into new types… metamorphic rocks do not get hot enough to melt, or they would become igneous rocks." Examples of metamorphic rocks include gneiss (from the igneous rock granite), slate (from the sedimentary rock shale), marble (from the sedimentary rock limestone), greenschist, amphibolite, or eclogite (from the igneous rock basalt), and schist (from mudstone/shale, or some igneous rocks, the result of greater heating and pressure, and with coarser, larger crystals than slate – i.e., a greater degree of schistosity) (University of Auckland, 2005).

The United States Geological Survey (2022) distinguishes two types of metamorphic rocks. Foliated metamorphic rocks, such as granite gneiss and biotite schist, which are banded or foliated, where foliated refers to the parallel arrangement of some mineral grains, giving the rock a striped appearance, as the pressure forced the elongate minerals in the rock into aligned, platy or sheet-like structure in the direction of the pressure that reflects the direction that pressure was applied. Nonfoliated metamorphic rocks have nonplaty or nonsheet-like structure and may result from the metamorphism of rocks such as limestone, composed of minerals that will never be elongated regardless of pressure, or contact metamorphism, which results from heating (e.g., a hot, igneous intrusion into pre-existing rock, without pressure (United States Geological Survey (2022).

Research on metamorphic rock geochemistry contributes to knowledge on a variety of topics on geodynamics. For example, Tang et al. (2007: 48) present of geochemical studies of metamorphism for the understanding of the geodynamics of continental movement. Using the case study of the Jiaobei terrane northeast of the Dabie-Sulu orogenic belt in China, the authors argue that the "tectonic affinity of tectono-lithological units close to ultrahigh-pressure metamorphic belt is a key issue for understanding the geodynamics of continental collision" (Tang et al., 2007: 48). The methodology was based on the use of data from LA-ICPMS zircon U–Pb dating, whole-rock elements and Nd–Sr isotopes, and mineral O isotopes for the study area. The two Triassic collided continents were the North and the South China Blocks. Protolith (the original, unmetamorphosed rock from which the metamorphic rock is formed) ages for TTG gneiss, amphibolite and mafic granulite were ~2.7, ~2.5 and ~2.4 Ga (Ga (for giga annum or billions of years), respectively, and regional metamorphism occurred at ~1.76 Ga (Tang et al., 2007: 48). Here TTG refers to tonalite–trondhjemite–granodiorite (TTG) rocks, which are intrusive rocks with granitic composition (quartz and feldspar), with only a little potassium feldspar (see Winter, 2013).

The findings indicated that the protolith of the TTG gneiss was generated by the partial melting of mantle-derived rocks at the bottom of the crust, the protolith of the amphibolite could be a product of arc-like magmatism, and the protolith of the mafic granulite was derived from a depleted mantle source. The findings indicated that both protoliths were contaminated by supracrustal materials, the protoliths of paragneiss and schist were largely derived from supracrustal sources, but protolith of amphibolite was of mantle-derived signature (Tang et al., 2007: 48). Further findings were that the Jiaobei metamorphic rocks preserved their original mantle-like O isotope compositions, unlike the ultra-high-pressure metamorphism (UHP) metaigneous rocks in the Dabie-Sulu orogenic belt that did not and showed ^{18}O-depletion (^{18}O isotopes). Additionally, the characteristics of the geological events recorded in the metamorphic rocks from the Jiaobei terrane are more like those from the North China Block than the South China Block. The authors therefore argue that "the Jiaobei terrane is concluded to have tectonic affinity to the former but behave like a micro-continent during the Triassic continental collision" and that the "~1.76 Ga regional metamorphism in the Jiaobei terrane is likely related to reworking of the arc-continent collisional orogen in the periphery of the North China Block rather than the ~1.85 Ga collision event between the eastern and western North China Blocks" (Tang et al., 2007: 48). It is also argued that the findings support the established assumption that the suture boundary between the North and South China Blocks in the Sulu orogen is along the Wulian-Yantai fault and that tectonic mingling along the Wulian-Yantai fault maybe linked to subduction erosion during the continental collision (Tang et al., 2007: 48).

Links between biogeochemistry, and igneous and metamorphic geochemistry may be based on specific chemical elements that are relevant to metamorphic rocks and organisms. For example, Aston (1983) describes silicon as occupying a dominant place in biogeochemistry, as it is an essential nutrient and found in animal skeletons. For example, Tréguer et al. (2021: 1269) point out that "the element silicon (Si) is required for the growth of silicified organisms in marine environments, such as diatoms. These organisms consume vast amounts of Si together with N, P, and C, connecting the biogeochemical cycles of these elements. Thus, understanding the Si cycle in the ocean is critical for understanding wider issues such as carbon sequestration by the ocean's biological pump." Silicon "often combines with oxygen or other elements to form silicates which are regarded as the largest class of rock-forming materials" (Geetha Thanuja et al., 2022). Silica is added to soil and taken up by plants after weathering from metamorphic rocks, and "weathering is a complex function of rainfall, runoff, lithology, temperature, topography, and vegetation" (Conley, 2002: 68).

Another study is that of Sutton et al. (2018) which acknowledges the crucial role of silicon in the regulation of primary productivity and carbon cycling in the oceans and continents. The study acknowledges the beginnings of the global Si cycle with the chemical weathering of silicate minerals, which may be transformed and redistributed into waterbodies, soils, and organisms, and into mineral form again. Through these "transformations, Si interacts with numerous other major (e.g., C, N) and minor (e.g., Al, Ge, Zn) elements and, in turn, influences their biogeochemical cycles" (Sutton et al., 2018). The study focuses on the analytical methods that are used to study the sources, sinks, and fluxes of the silicon cycle though the Earth's spheres, these including the elemental and stable isotope ratio data for silicon, zinc, germanium, and other elements (Sutton et al., 2018). The authors note an emphasis on the geochemistry (e.g., Al/Si, Ge/Si, Zn/Si, δ13C, δ15N, δ18O, δ30Si) of dissolved and biogenic Si, present case studies, such as the Silicic Acid Leakage Hypothesis, and investigate the issues associated with the development of these environmental proxies for the global silicon cycle (Sutton et al., 2018). Silicon is a particularly relevant element for research methods that transcend geochemistry and biogeochemistry, as its sources, sinks, and processes are widely distributed in the terrestrial, freshwater, and marine systems, and "exists in major pools in dissolved and solid forms in all reservoirs: extra-terrestrial, continental (e.g., soil, vegetation, hydrothermal), freshwater (e.g., rivers, lakes, groundwater, organisms, sediment), atmospheric (e.g., aerosols), and oceanic (e.g., water column, organisms, sediment and pore-waters, oceanic crust, hydrothermal)" (Sutton et al., 2018).

Sutton et al. (2018) argue that "the chemical weathering of silicate minerals and the eventual cycling of weathered products (clays, dissolved Si) provide the starting point of Si biogeochemistry and its interaction with other elemental cycles such as carbon. Silicate weathering represents an important sink of atmospheric CO_2 over geological time scales." The main transformation processes are the weathering of silicon minerals from rocks, the formation of secondary minerals and dissolved silicon, and the uptake of dissolved Si by organisms, which results in the biomineralization of biogenic silicon, and the remineralization of silicon." The organisms that use dissolved silicon include diatoms, silicoflagellates, radiolarians and sponges, and in some cases plants, which may create biogenic silica ($BSiO_2$) (Sutton et al., 2018). Moriceau et al. (2009: 1381) describe biogenic silica is a key factor in biogeochemistry, with a study based on the role of biogenic silica in the degradation of diatom organic carbon, and the role of organic compounds in biogenic silica dissolution. Sutton et al. (2018) conclude that "the bio-geochemical analyses used to study the global biogeochemical cycling of Si (e.g., stable isotopes and associated trace elements), are emerging as useful tools to examine the influence that silicifying organisms have on the different reservoirs of Si (e.g., atmospheric, terrestrial, freshwater, and marine)." These analyses have been conducted by several tools, including low temperature bio-geochemical tools, which have garnered information on silicon cycling and environmental interactions, including the role of silicifying organisms in other (e.g., nitrogen and carbon) biogeochemical cycles (Sutton et al., 2018).

Photogeochemistry

Photogeochemistry refers to the impact of sunlight on the chemistry of the Earth. Doane (2017) defines photogeochemistry as the "photochemistry of Earth-abundant minerals in shaping biogeochemistry, and this can be extended to the entire interface between photochemistry and geochemistry to include any chemical reaction induced by sunlight among naturally occurring substances" (see also Schrauzer et al., 1983; Kim et al., 2011; Falkowski, 2015). Doane (2017) also notes that "photogeochemistry describes photochemical reactions on Earth that are not facilitated by living organisms. The reactions that comprise photosynthesis in plants and other organisms, for example, are not included, since the physiochemical context for these reactions is installed by the organism and must be maintained in order for the reactions to continue (the photoreactions cease if the organism dies)." A qualification is that "if a certain substance is produced by an organism, and the organism dies but the substance remains (e.g., plant residue or biogenic mineral precipitates), photoreactions involving this substance still contribute to photogeochemistry" (Doane, 2017). Photogeochemistry therefore represents a combined relationship between photochemistry and surface geochemistry. Doane (2017) argues along this line, in that "by overlaying photochemistry and surface geochemistry, complementary approaches can be adopted to identify natural photochemical reactions and discern their significance in the environment... principles of photochemistry can be readily merged with geochemistry" (Doane, 2017).

Regarding the position of photogeochemistry in Earth scientific research, Kim et al. (2011: 10073) argue that "photochemical reactions of minerals are underappreciated processes that can make or break chemical bonds" and "that the photochemistry of Earth-abundant minerals with wide band gaps would have potentially played a critical role in shaping the biogeochemical evolution of early Earth." Their case study of the photooxidation of siderite ($FeCO_3$) by UV radiation to produce hydrogen gas and iron oxides via a two-photon reaction, suggests that the photooxidation of siderite could be a source of molecular hydrogen for the earlier half of Earth's history, which could have resulted in the formation of iron oxides during anoxic conditions, from the Archean to possibly the early phases of the Great Oxidation Event, which could have provided a process for the oxidation of the atmosphere through hydrogen losses to space, while providing a reductant for microbial metabolism (Kim et al., 2011: 10073).

Photogeochemistry is important for biogeochemistry, as some research indicates close linkages (Kim et al., 2011; Doane, 2017). Doane (2017) argued that photogeochemistry may parallel biogeochemistry, with some previous findings based on biological research, in efforts to find "analogous photochemical reactions"; some studies finding analogous, photochemical counterparts to biochemical reactions included in the photochemical disproportionation of acetic acid, which is similar to acetoclastic methanogenesis. The latter is a form of methane production, largely through the action of two methanogenic microbial genera (*Methanosaeta* and *Methanosarcina*) (Kraeutler & Bard, 1978; Miyoshi et al., 1991; Kurade et al., 2019). There are also studies of light-induced depletion of O_2 through catalytic cycles including organic matter and iron, which is similar to microorganism consumption of O_2 (Miles & Brezonik, 1981; Doane, 2017).

Environmental or Low-Temperature Geochemistry

Environmental or low-temperature geochemistry is concerned with the chemical processes occurring in the earth's surficial environments at natural surficial temperatures, and as such may be closely connected to biogeochemistry (Forbes, 2018). White (2017) even defines biogeochemistry as "a subfield of low-temperature geochemistry that focuses on the interaction between life and its environment." Mather (2013) notes that "environmental geochemistry refers to the chemical makeup of rocks, soil, water, air, and organisms near the surface of the earth and the chemical reactions that take place among them. It deals not only with natural processes, but also with interactions between humans and their environment." The Scripps Institution of Oceanography (2022) states that this field observes the chemical processes affecting minerals, water, and air at or just below the earth's surface, typically below about 200 °C. Forbes (2018) describes included topics as "acid-base equilibrium, reduction-oxidation reactions, and solubility," within a broad field, and major areas of study including "mineral precipitation, chemical weathering, soil chemistry, sedimentary processes and diagenesis, biogeochemical cycles, and contaminant transport." Mather (2013) notes that in low-temperature geochemistry, the environment is generally cooler than in fields such as igneous geochemistry, as the latter may concern volcanism and tectonic pressure and friction. Important topics include environmental geochemistry, diagenesis of sediments and geochemical exploration for mineral deposits (Mather, 2013). University of Michigan (2022) also points out that low-temperature geochemistry research includes "geochemistry of sediments, minerals, waters, fossils, soils, and organic matter occurring at temperatures commonly found at or near the Earth's surface" and topics may focus on "noble gas content in groundwater, organic carbon content of modern Arctic permafrost and runoff, trace-element behavior in modern and fossil soils and marine carbonates, and many more."

Recent research results in low-temperature geochemistry are both pure science-based and more related to technological applications in resource extraction. For example, Tranter (2015) describes the development of a wide range of geochemical topics over the recent decades, including the rock cycle, rock-water interactions, and chemical processes (see also Garrels & Christ, 1965; Garrels & Mackenzie, 1971); multiple geochemical cycles on the Earth's surface (see also Holland, 1978, 1984); the modeling of early diagenesis (physical, chemical and/or biological change of sediments into sedimentary rock at relatively low temperatures and pressures, changing the original mineralogy and texture) (see also Berner, 1980); links between atmospheric CO_2 concentrations and geological timescale (millions of years) carbonate and silicate weathering cycles (see also Berner et al., 1983); the oceans as CO_2 reservoirs regulating atmospheric CO_2 levels over glacial-interglacial timescales (alternating cold and warm periods over thousands of years) (see also Broecker & Peng, 1982; Broecker, 2002); fresh and saline water compositional variations (see also Drever, 1997); and the geochemical bases of rock, water and gas interactions (see also Stumm & Morgan, 1996; Langmuir, 1997).

Tranter (2015) notes that these low-temperature geochemistry studies are examples of early developments in the field, and more important contributions are being promoted, citing the example of recent advances in biogeochemical processes in the cryosphere, where "cold environments are now thought of as a biome, rather than as abiotic systems" (see Priscu & Christner, 2004; Anesio & Laybourn-Parry, 2012), and more importantly within geochemistry as "potential fertilizers of the polar oceans" (Tranter, 2015, see also Raiswell et al., 2006; Wadham et al., 2013). Glaciation as a field of study has also advanced, acknowledging the glacial agency in physical erosion, mostly abrasion, and fracture/traction which create transported, suspended, and deposited sediments (see Hallet et al., 1996; Knight, 1999). Tranter (2015) points out the role of glacial erosion in producing eroded bedrock, which liberates trace reactive components of these rocks, including sulfides and carbonates, and mineral phosphates such as apatite which enter water, creating the high rates of P weathering in glaciated areas (see also Föllmi et al., 2009; Tranter &

Wadham, 2013). Tranter (2015) writes that crucially, current understandings acknowledge "most geochemical reactions, apart from hydrolysis, are microbially mediated... This was a step change for the discipline less than 20 years ago since glacier beds were formerly thought to be abiotic" (Sharp et al., 1999; Tranter et al., 2002; Montross et al., 2013).

Other examples of such low-temperature geochemistry include studies of microbial life, and linked sulfide oxidation and organic matter oxidation, which are processes that reduce the Oxidation/Reduction Potential (Eh) of waters distant from inputs of oxygenated surface melt, and REDOX reactions which may restrict the reaction types, linked to the oxidation of sulfides and organic matter, with oxidizing agents including NO_3^-, Fe(III), Mn(IV), and SO_4^{2-} (Tranter, 2015). Tranter (2015) argues that "mass balance constraints are also imposed – reaction extents are ultimately decided by the sulfide and organic matter content of the basal debris and available oxidants." Tranter (2015) also cites links between geochemical activity (which would be dependent on sulfides, O_2, and organic matter) under thick ice sheets overlying crystalline bedrock, as therefore, high geochemical activity is likely to be linked to areas near "legacy organic matter found in overridden soils and sedimentary rocks" rather than the rare or absent recent, rapidly changing (labile) organic matter and O_2 inputs (see Skidmore et al., 2000; Wadham et al., 2010). As the source of organic matter from legacy sources would be limited, the linked microbial systems would be rare and declining, with the erosion of the sedimentary rock, but Tranter (2015) argues that "the production of H_2 by glacial crushing removes this gloomy fate" as it creates the possibility of microbial communities on crystalline bedrocks supported by the energy generated by physical comminution (reduction into smaller particles) of bedrock, this increasing the likelihood subglacial lakes under thick Antarctic ice have microbial life. In research, silicate rocks like those of subglacial environments in Arctic and Antarctica were crushed and wetted in an inert, laboratory atmosphere, with hydrogen (H_2) production resulting at 0 °C, like methanogenesis rates in Greenland and Canadian glacier studies (Tranter, 2015; see also Siegert et al., 2001; Boyd et al., 2010; Telling et al., 2015).

Geochemical Techniques

Geochemical research techniques are complex, due to the wide range of environmental contexts and materials explored, as can be seen from the research foci and methods of the branches of geochemistry described earlier. These may also link to biogeochemical research (Campbell, 2020). These mostly focus on the study of chemical elements and compounds, and their movements within inanimate environments, using methods that may also be used in biogeochemical studies incorporating biological environments. An example is the research at the interface between Environmental Geochemistry and Biogeochemistry, as pursued at the Department of Geology and Environmental Science at the University of Pittsburgh. This research looks at geochemical processes between biological and geological at several scales, using geochemical and isotopic tracers for chemical constituents, and the modeling of processes through soil, plant, and hydrologic systems. The Department notes that "the application of a wide variety of isotopic methods is a fundamental strength of our department's research… our understanding of present-day biogeochemical cycling is extended to the near and distant past through analysis of unique geochemical signals in lacustrine and marine depositional systems, providing constraints on ancient environments and processes" (Department of Geology and Environmental Science, 2023).

Another practical application of such geochemical methods is on mineral (including oil) exploration, which may also be used for biogeochemical studies. Rose (1998) notes that "geochemical exploration is any method of mineral or petroleum exploration that utilizes systematic measurements of one or more chemical properties of a naturally occurring material. The materials analyzed most commonly are rock, soil, stream and lake sediment, natural waters, vegetation, and soil air. Ores of metals such as Cu, Ni and Au are the most common targets of geochemical exploration, but nonmetallic commodities (diamonds, talc, and fluorite) and deposits of oil and gas are also sought by geochemical exploration." Also, as noted earlier, geochemical research may focus on any of the naturally occurring elements in the Earth, and on the inorganic constituents of rocks and rock forming minerals, as well as organic chemistry, and soils, waters, and biological tissues (Trueman et al., 2019).

Geochemical prospecting for minerals, examining chemical constituents is a major area where the techniques of the geochemical sciences have been applied. For example, Hawkes (1957: 226) notes that "geochemical prospecting for minerals, as denned by common usage, includes any method of mineral exploration based on systematic measurement of one or more chemical properties of a naturally occurring material. The chemical property measured is most commonly the trace content of some element or group of elements; the naturally occurring material may be rock, soil, gossan (refers to highly oxidized, weathered or decomposed rock), glacial debris, vegetation, stream sediment, or water." This is based on the definition of geochemistry, which paraphrases as "the determination of the relative and absolute abundance of the elements…in the earth" and "the study of the distribution and migration of the individual elements in the various parts of the earth…with the object of discovering the principles governing this distribution and migration" (Hawkes, 1957).

Sarala (2015: 711) refers to geochemical exploration methods as "mainly based on observations of anomalous concentrations of major or trace elements that are derived from a core part of a mineral deposit itself or a wider halo surrounding the ore body"…in this process the method is the "use of chemical properties of naturally occurring substances (including rocks, glacial debris, soils, stream sediments, waters, vegetation, and air) to find economic deposits of metals, minerals, and hydrocarbons." Further, a geochemical survey can be scaled at continental, regional, targeting, and local scales and have several purposes and consequent designs, including the evaluation of the areal elemental and mineralogical concentrations, the geochemical or mineralogical anomalies within a potential mineralized area, or mineral exploration in a known mineralized area (see also McMartin & McClenaghan, 2001; McClenaghan et al., 2013).

Winterburn et al. (2017: 495) argue that although "geochemistry continues to be a major component of mineral exploration and evaluation programs on a global basis" recent developments have factored a decline in the cutting-edge research on geochemical methods, and "geochemical exploration has required less innovation as traditional techniques have continued to be used with success in shallowly covered or outcropping terrains." In terms of the application of geochemical techniques to mineral exploration, Winterburn et al. (2017, 495) point to the fact that developments in geochemical methods lie within four main areas, namely the study of metal mobility and mechanisms, fast geochemical analyses, effective data use, and innovation in the laboratory-based techniques (see also Winterburn et al., 2020).

SGS (2022) lists some tasks that may be included in their top class geochemical services, including sample preparation. These focus on the classic geochemical foci of chemical ions, elements and compounds, and trace-element analysis. Mentioned techniques include Mobile Metal Ion (MMITM) analysis, Mobile Metal Ion (MMI) orientation surveys, ultratrace (an ultratrace refers to a chemical element that may be less than one microgram per gram of a given organism but is important for its metabolism) and trace-element analysis by Inductively Coupled Plasma Atomic Emission Spectroscopy (ICP-AES) and ICP-MS. As noted by Neikov and Yefimov (2019), "the ICP-AES is an analytical technique based on the principles of atomic spectroscopy for the determination of more than 70 elements with detection limits in the parts per billion to parts per million range." This would enable the detection of elements in biogeochemical analyses. Inductively coupled plasma mass spectrometry (ICP-MS) is a mass spectrometry method, and the sample is ionized by an inductively coupled plasma. The atomization of the sample creates atomic and small polyatomic ions, which may be detected, this allowing the detection of low concentrations of metals and some nonmetals in liquid samples, and different isotopes of the same element (Lee, 2018). Wilschefski and Baxter (2019) describe inductively coupled plasma mass spectrometry (ICP-MS) as an analytical technique that can measure elements at trace levels in biological fluids, which is superseding older techniques including atomic absorption and atomic emission.

Other methods listed by SGS (2022) include X-ray fluorescence, metal speciation studies, carbon and sulfur analysis and reference material services. X-ray fluorescence (XRF) refers to the emission of fluorescent X-rays from materials made unstable by contact with X-rays, with applications to the chemical analysis of metals and other substances and for research in geochemistry (Van Grieken & Markowicz, 2002; Beckhoff et al., 2006; Pessanha et al., 2019). Intertek (2022) notes that "metals speciation testing provides… precise analysis of metal species such as mercury, arsenic, sulfur, and lead in crude oil, chemicals, and many other products, as well as aqueous samples such as seawater."

Applied to biogeochemical research, such methods may be used to the interfaces of biological and inanimate environments, and the presence and movements of ions, elements, and compounds. One biogeochemically-based study is that of Reid and Hill (2010:105), which uses vegetation-based studies for mineral exploration, in this case gold. The authors describe this technique as testing the "ability of plants to show signatures of mineralisation as well as the optimum scale of sampling for first-pass mineral exploration surveys." The study is based in the semiarid Tanami Gold Province in northern Australia, which includes many plant species. The Snappy gum (*Eucalyptus brevifolia* F.Muell.) had a distribution and sulfur and zinc signature, which indicated an underlying geological stratum that contained gold mineralization. Soft spinifex (*Triodia pungens* R.Br. 1810), a species of grass native to northwestern Australia, had gold, arsenic, zinc, sulfur, and cerium, which may indicate underground gold mineralization. The authors point out that in the study area, mineralization occurs at the contact between granite and dolerite, and biogeochemical signatures from snappy gum and dogwood (*Acacia coriacea*) showed higher levels of gold, arsenic, zinc, sulfur, and cerium. Therefore, the biogeochemical methods can be used to locate mineralization, substrate differences, supported by more information on the local soil, hydrology, and geological formations. The effectiveness of the biogeochemical method depended on the coverage of an area (Reid & Hill, 2010:105).

Another important geochemical method is geochemical mapping (Demetriades, 2021: 267). This is defined as a method "to document the spatial distribution of chemical elements and compounds, and physicochemical properties across the Earth's surface" (Demetriades, 2021, 267). The functions of geochemical mapping include mineral exploration, forest assessment, medical research, the location of contaminated areas, and even the mapping of the geochemical environment in urban areas. Demetriades (2021: 267) notes that geochemical mapping initially developed in the Union of Soviet Socialist Republics (USSR) in the 1930s for mineral exploration, a function that eventually developed internationally. Demetriades (2021, 267) argues that "this expansion in the usage of geochemical data has been facilitated by the production of national multielement and multipurpose geochemical atlases compiled at different sample densities and various map scales depending on project objectives. The end-product is always an interpretation of the spatial distribution of chemical elements and compounds, and the processes that control or influence this spatial variation." Systematic geochemical mapping is therefore the most effective method for recording the spatial variation of chemical elements, which can be applied to surface and shallow strata (Demetriades, 2021: 267).

Applied to biogeochemistry, geochemical mapping is termed biogeochemical mapping, with a greater focus on vegetation and soil constituents related to vegetation (Higueras et al., 2019). Higueras et al. (2019) mention soil health measurement as an important application of biogeochemical mapping, with soil health defined as "the continued capacity of the soil to function as a vital living ecosystem that sustains plants, animals and humans" and "an approach to the consideration of soil as a living being." Rinot et al. (2019: 1484) note that "soil health index should reflect soil ability to provide ecosystem services (ES)" and "the relationship between soil attributes and ES should be quantified" and "The soil health (SH) concept has been introduced due to an evolving understanding that soil is not just a growing medium for crops but that it provides a foundation for other essential ecosystem services (ES)." Laishram et al. (2012: 20) present a definition of soil health and/or quality, which states that "degradation or deterioration in soil health or quality implies loss of the vital functions of soil" these functions being those that provide "physical support, water and essential nutrients required for growth of terrestrial plants," the regulation of environmental water flow, and the "elimination of the harmful effects of contaminants by means of physical, chemical and biological processes."

Geochemical applications are also complex and varied. For example, petroleum geochemistry, as an aspect of organic geochemistry has emerged as an important application in current applied technology and industry. Petroleum geochemistry is defined as "the science and application of chemical concepts to understand the origin of petroleum – natural gas, condensate, and crude oil – and its occurrence and fate on the earth's surface and within its crust" (Curiale, 2017). Kvenvolden (2006: 1) argues that "organic geochemistry had its origin in the early part of the twentieth century when organic chemists and geologists realized that detailed information on the organic materials in sediments and rocks was scientifically interesting and of practical importance." It is also noted that "organic geochemistry is now a widely recognized geoscience in which organic chemistry has contributed significantly not only to geology (i.e., petroleum geochemistry, molecular stratigraphy) and biology (i.e., biogeochemistry), but also to other disciplines, such as chemical oceanography, environmental science, hydrology, biochemical ecology, archaeology, and cosmochemistry."

Dembicki Jr. (2017) lists some of the developments of petroleum geochemistry over the past century. While acknowledging that "petroleum geochemistry is a relatively young science, tracing its roots back to the 1934 discovery of chlorophyll-like structures in crude oil by Albert Treibs (1934, 1936), major developments have occurred, including the discovery of the organic origins of crude oil, the establishment of professional societies and research conferences on organic geochemistry, the development of analytical tools, including gas chromatography and mass spectrometry, and of the concept of biological maker compounds, and studies of the links between chemical fossils and oil sources (see also Breger, 1963; Eglinton & Calvin, 1967; Durand, 2003). Other developments in the late twentieth century included the study of the hydrocarbon generation process and the oil window concept, the thermal maturity of sediments and the kerogen composition (a solid, insoluble organic matter in sedimentary rocks, see Vandenbroucke & Largeau, 2007), and the Rock-Eval instrument which used as a standardized pyrolysis method for source-rock characterization and evaluation in petroleum geochemistry, and new treatises on petroleum geology (Dembicki Jr., 2017; see also Tissot & Welte, 1978; Espitalié et al., 1977). Studies included foci on hydrocarbon fingerprinting, forensic geochemistry, and gas chromatography (Hunt, 1979). Later developments included advances and understandings in pyrolysis techniques, biomarker applications, petroleum migration, basin modeling, and the introduction of personal computers for use in geochemical studies (Dembicki Jr., 2017; see also Lopatin, 1971; Waples, 1980; Sluijk & Parker, 1984).

Instructive scientific publications during the late twentieth century included those of Sluijk and Parker (1984) on the value of petroleum geochemistry in exploration, Kaufman et al. (1990) on reservoir continuity, and production problem analysis, Magoon and Dow (1994) on the concept of petroleum systems and the associated role of petroleum geochemistry Passey et al. (2010) on understanding source rocks and reservoirs, and Dembicki Jr. (2014) on the fluid properties and phase behavior for successful oil exploitation. Dembicki Jr. (2017) concludes the study of the historical development of geochemical methods by stating that "whatever problems the future holds in petroleum exploration and production, innovative applications of the concept and methods of petroleum geochemistry will continue to contribute to solutions" (Kvenvolden, 2002, 2006; Hunt et al., 2002; Durand, 2003; Dow, 2014).

The methods of petroleum geochemistry may be applied to biogeochemistry. For example, Head et al. (2014) describe a study of "microbiological, geochemical, and biogeochemical data" that contributes to knowledge of processes regulating deep life in petroleum reservoir ecosystems and of biotic and abiotic factors that determine the biodegradation of petroleum in situ, as this phenomenon may oil exploration and production." This study proposes that reservoir formation water salinity affects the occurrence of biodegraded heavy oil reservoirs. The authors note that in the study of microbial activity in the ground, the important factors are "temperature, pH, salinity, water activity, radiation and availability of resources such as carbon and energy sources, electron acceptors and inorganic nutrients" and for in-reservoir crude oil biodegradation the important factors are "temperature, salinity, and inorganic nutrient availability" (Head et al., 2014). It is also noted that nutrient availability is major control for the degradation of crude oil in aerobic environments, as such degradation depends on hydrocarbon-degrading organisms, with hydrocarbon conversion to biomass for growth requiring nitrogen and phosphorus. Microbial biomass has also been found to correlate with organic carbon content (Head et al., 2014).

Conclusions

This chapter has examined the relationship between the highly multidisciplinary and related sciences of geochemistry and biogeochemistry, taking the former as the background of the latter. Geochemistry, as the science that examines the Earth and sometimes other solar bodies from the approach of the chemical sciences, is plainly one of the roots for the more recently developed discipline of biogeochemistry, which adds more foci on biological roles in chemical cycling. As the evolution of life is based on a chemical foundation, mostly the carbon link with water and living tissue, the relationship between these two sciences sits on the most interesting science issue: the origin and continuing development of life, on a chemical foundation. This chapter has skimmed over the main topics of the extremely broad and deep science of geochemistry, looking at subdisciplines including organic geochemistry, inorganic geochemistry, isotope geochemistry, aqueous geochemistry, cosmochemistry, trace-element geochemistry, igneous rock geochemistry, metamorphic rock geochemistry, photo-geochemistry and low-temperature or environmental geochemistry. More mention will be made in later chapters of the main points of this chapter, to explore further the links between biogeochemistry and the supporting Earth sciences.

References

Åberg, G., Jacks, G., & Hamilton, P. J. (1989). Weathering rates and 87 Sr/86 Sr ratios: An isotopic approach. *Journal of Hydrology, 109*(1), 65–78.

Allaby, M. (2008). *A dictionary of earth sciences* (3rd ed.). Oxford University Press. https://doi.org/10.1093/acref/9780199211944.001.0001

American Association of Petroleum Geologists. (2022). Journals. Retrieved from https://www.aapg.org/publications/journals/bulletin

American Chemical Society. (2022). Inorganic Chemistry. Retrieved from https://www.acs.org/content/acs/en/careers/chemical-sciences/areas/inorganic-chemistry.html

American Geophysical Union. (2022). Journal of Geophysical Research. Retrieved from https://agupubs.onlinelibrary.wiley.com/journal/21562202

American Mineralogist. (2022). American Mineralogist. Mineralogical Society of America. Retrieved from http://www.minsocam.org/MSA/AmMin/Scope_and_Mission_2final.htm

Anderson, R. S., & Anderson, S. P. (2010). *Geomorphology: The mechanics and chemistry of landscapes*. Cambridge University Press.

Anesio, A. M., & Laybourn-Parry, J. (2012). Glaciers and ice sheets as a biome. *Trends in Ecology & Evolution, 27*, 219–225. https://doi.org/10.1016/j.tree.2011.09.012

Aston, S. R. (1983). *Silicon geochemistry and biogeochemistry*. Academic.

Atkins, P., & de Paula, J. (2009). *Elements of physical chemistry* (5th ed.). Oxford University Press.

Barker, C. (1972). Aquathermal pressuring—Role of temperature in development of abnormal-pressure zones. *AAPG Bulletin, 56*, 2068–2071.

Bataille, C. P., Willis, A., Yang, X., & Liu, X. M. (2017). Continental igneous rock composition: A major control of past global chemical weathering. *Science Advances, 3*(3), e1602183. https://doi.org/10.1126/sciadv.1602183

Beckhoff, B., Kanngießer, B., Langhoff, N., Wedell, R., & Wolff, H. (2006). *Handbook of practical X-ray fluorescence analysis*. Springer.

Berner, R. A. (1980). *Early diagenesis. A theoretical approach*. Princeton University Press.

Berner, R. A., Lasaga, A. C., & Garrels, R. M. (1983). The carbonate-silicate geochemical cycle and its effect on atmospheric carbon dioxide over the past 100 million years. *American Journal of Science, 283*, 641–683. https://doi.org/10.2475/ajs.283.7.641

Bianchi, T. S. (2021). The evolution of biogeochemistry: Revisited. *Biogeochemistry, 154*, 141–181. https://doi.org/10.1007/s10533-020-00708-0

Bianchi, T. S., & Canuel, E. A. (2011). *Chemical biomarkers in aquatic ecosystems*. Princeton University Press.

Black, D. C., & Pepin, R. O. (1969). Trapped neon in meteorites II. *Earth and Planetary Science Letters, 6*, 395–405.

Boreham, C., Jinadasa, N., Sohn, J., Hong, Z., & Blake, C. (2021). Characterisation of radiogenic monoalkenes in Australian oils and condensate. *Organic Geochemistry, 163*, 104332. https://doi.org/10.1016/j.orggeochem.2021.104332

Bouse, R. M., Ruiz, J., Titley, S. R., Tosdal, R. M., & Wooden, J. L. (1999). Lead isotope compositions of Late Cretaceous and Early Tertiary igneous rocks and sulfide minerals in Arizona: Implications for the sources of plutons and metals in porphyry copper deposits. *Economic Geology, 94*(2), 211–244. https://doi.org/10.2113/gsecongeo.94.2.211

Boyd, E. S., Skidmore, M., Mitchell, A. C., Bakermans, C., & Peters, J. W. (2010). Methanogenesis in subglacial sediments. *Environmental Microbiology Reports, 2*, 685–692. https://doi.org/10.1111/j.1758-2229.2010.00162.x

Bray, L. A., & Van Tuyl, H. H. (1961). *Laboratory development of a carrier-precipitation process for the recovery of strontium from purex wastes*. General Electric Co. Hanford Atomic Products Operation.

Breger, I. A. (1963). *Organic geochemistry*. Pergamon, Macmillan.

Briggs, D. E. G., & Summons, R. E. (2014). Ancient biomolecules: Their origins, fossilization, and role in revealing the history of life. *BioEssays, 36*, 482–490.

Britannica. (2014). Geochemistry. *Encyclopedia Britannica*. https://www.britannica.com/science/geochemistry

Broecker, W. S. (2002). *The glacial world according to wally*. Eldigio Press.

Broecker, W. S., & Peng, T.-H. (1982). *Tracers in the sea*. Eldigio Press.

Brownlow, A. H. (1979). *Geochemistry*. Prentice-Hall, Inc.

Bullen, T., White, A., Blum, A., Harden, J., & Schulz, M. (1996). Chemical weathering of a soil chronosequence on granitoid alluvium: II. Mineralogic and isotopic constraints on the behaviour of strontium. *Geochimica et Cosmochimica Acta, 61*(2), 291–306.

Burke, J. G. (1986). *Cosmic debris: Meteorites in history*. University of California Press.

Burrows, A., Holman, J., Parsons, A., Pilling, G., & Price, G. (2009). *Chemistry3*. Oxford University Press.

Burst, J. F. (1969). Diagenesis of Gulf Coast clayey sediments and its possible relation to petroleum migration. *Bulletin of the American Association of Petroleum Geologists, 53*(1), 73–93.

Callahan, M. P., Smith, K. E., Cleaves, H. J., 2nd, Ruzicka, J., Stern, J. C., Glavin, D. P., House, C. H., & Dworkin, J. P. (2011). Carbonaceous meteorites contain a wide range of extraterrestrial nucleobases. *Proceedings of the National Academy of Sciences of the United States of America, 108*(34), 13995–13998. https://doi.org/10.1073/pnas.1106493108

Campbell, M. (2020). *The physical geography of El Salvador: A geosphysical and ecological approach*. Nova Science Publishers.

Catalá, T. S., Shorte, S., & Dittmar, T. (2021). Marine dissolved organic matter: A vast and unexplored molecular space. *Applied Microbiology and Biotechnology, 105*, 7225–7239. https://doi.org/10.1007/s00253-021-11489-3

Chamberlain, C. P., Waldbauer, J. R., & Jacobson, A. D. (2005). Strontium, hydrothermal systems, and steady-state chemical weathering in active mountain belts. *Earth and Planetary Science Letters, 238*(3), 351–366.

ChemEurope. (2022). *Radiogenic*. Retrieved from https://www.chemeurope.com/en/encyclopedia/Radiogenic.html

Chiaradia, M., & Fontbote, L. (2002). Lead isotope systematics of Late Cretaceous – Tertiary Andean arc magmas and associated ores between 8°N and 40°S: Evidence for latitudinal mantle heterogeneity beneath the Andes. *Tera Nova, 14*(5), 337–342. https://doi.org/10.1046/j.1365-3121.2002.00426.x

Clarke, F. W., & Washington, H. S. (1924). The composition of the Earth's crust. *United States Geological Survey Professional Paper, 127*, 117.

Commonwealth Scientific and Industrial Research Organisation (CSIRO). (2022). *Environmental Chemistry*. Retrieved from https://www.publish.csiro.au/en/AbouttheJournal

Conley, D. J. (2002). Terrestrial ecosystems and the global biogeochemical silica cycle. *Global Biogeochemical Cycles, 16*(4), 68-1–68-7. https://doi.org/10.1029/2002GB001894

Connan, J. (1974). Time-temperature relation in oil genesis: Geologic notes. *AAPG Bulletin, 58*, 2516–2521. https://doi.org/10.1306/83D91BEB-16C7-11D7-8645000102C1865D

Cooke, M., Dongen, B., Talbot, H., Semiletov, I., Shakhova, N., Guo, L., & Gustafsson, Ö. (2009). Bacteriohopanepolyol biomarker composition of organic matter exported to the Arctic Ocean by seven of the major Arctic rivers. *Organic Geochemistry, 40*, 1151–1159. https://doi.org/10.1016/j.orggeochem.2009.07.014

Cooke, D. R., Hollings, P., Wilkinson, J. J., & Tosdal, R. M. (2014). Geochemistry of porphyry deposits. In H. D. Holland & K. K. Turekian (Eds.), *Treatise on geochemistry* (Vol. 13, 2nd ed., pp. 357–381). https://doi.org/10.1016/B978-0-08-095975-7.01116-5

Curiale, J. A. (2017). Petroleum geochemistry. In R. Sorkhabi (Ed.), *Encyclopedia of petroleum geoscience* (Encyclopedia of earth sciences series). Springer. https://doi.org/10.1007/978-3-319-02330-4_2-1

Curiale, J. A., & Frolov, E. B. (1998). Occurrence and origin of olefins in crude oils. A critical review. *Organic Geochemistry, 29*, 397–408.

Czech Geological Society. (2022). *Journal of Geosciences*. Retrieved from http://www.jgeosci.org/scope

Davis, A. M. (2004). *Treatise on geochemistry, volume 1: Meteorites, comets, and planets*. Elsevier Pergamon.

Davis, A. M. (2014). Meteorites and cosmochemical processes. In H. D. Holland & K. K. Turekian (Eds.), *Treatise on geochemistry* (2nd ed., pp. 1–52). NL Elsevier.

Dembicki, H., Jr. (2014). Challenges to black oil production from shales: AAPG Search and Discovery Article #80355.

Dembicki, H., Jr. (2017). *Practical petroleum geochemistry for exploration and production*. Elsevier Science. https://doi.org/10.1016/C2014-0-03244-3

Demetriades, A. (2021). Geochemical mapping. In D. Alderton & S. A. Elias (Eds.), *Encyclopedia of geology* (2nd ed., pp. 267–280). https://doi.org/10.1016/B978-0-08-102908-4.00059-X

Department of Geology and Environmental Science. (2023). *Environmental geochemistry and biogeochemistry*. Retrieved from https://www.geology.pitt.edu/research-area/environmental-geochemistry-and-biogeochemistry

Devriendt, L. S. J. (2016). Controls on the oxygen isotope ratio of inorganic and biogenic calcium carbonates. Unpublished PhD Thesis, School of Earth and Environmental Sciences, University of Wollongong. https://ro.uow.edu.au/theses/4901

Dickey, P. A. (1975). Possible migration of oil from source rock in oil phase. *American Association of Petroleum Geologists Bulletin, 59*, 337–345.

Dickin, A. P. (2005). *Radiogenic isotope geology*. Cambridge University Press.

Doane, T. A. (2017). A survey of photogeochemistry. *Geochemical Transactions, 18*(1). https://doi.org/10.1186/s12932-017-0039-y

Dow, W. (2014). *Musings on the history of petroleum geochemistry - from my perch*. Retrieved from https://www.searchanddiscovery.com/documents/2014/80375dow/ ndx_dow.pdf

Drever, J. I. (1997). *The geochemistry of natural water surface and groundwater environments* (3rd ed.). Prentice Hall.

Dudás, F. Ö., Zhang, H., Shen, S., & Bowring, A. (2021). Major and trace element geochemistry of the Permian-Triassic boundary section at Meishan, South China. *Frontiers in Earth Science*. https://doi.org/10.3389/feart.2021.637102

Durand, B. (2003). A history of organic geochemistry. *Oil & Gas Science and Technology – - Revue d'IFP Energies nouvelles, 58*(2), 203–231.

Eglinton, G., & Calvin, M. (1967). Chemical fossils. *Scientific American, 16*(1), 32–43. http://www.jstor.org/stable/24931372

Elias, S. A. (2021). Geoarchaeology. In D. Alderton & S. A. Elias (Eds.), *Encyclopedia of geology* (2nd ed., pp. 538–553). Elsevier. https://doi.org/10.1016/B978-0-12-409548-9.12467-4

Espitalié, J., Laporte, J. L., Madec, M., Marquis, F., Leplat, P., Paulet, J., & Boutefeu, A. (1977). Rapid method for characterizing source rocks their oil potential and their degree of development. *Journal of the French Petroleum Institute, 32*, 23–42.

European Geosciences Union. (2022a). *Nonlinear processes in geophysics*. Retrieved from https://www.nonlinear-processes-in-geophysics.net/about/aims_and_scope.html

European Geosciences Union. (2022b). *Solid earth*. Retrieved from https://www.solid-earth.net/about/aims_and_scope.html

Falkowski, P. G. (2015). From light to life. *Origins of Life and Evolution of Biospheres, 45*, 347–350.

Faure, G. (1991). *Principles and application of inorganic geochemistry. A comprehensive textbook for geology students*. Macmillan.

Föllmi, K. B., Hosein, R., Arn, K., & Steinmann, P. (2009). Weathering and the mobility of phosphorus in the catchments and forefields of the Rhône and Oberaar glaciers, Central Switzerland: Implications for the global phosphorus cycle on glacial-interglacial timescales. *Geochimica et Cosmochimica Acta, 73*, 2252–2282. https://doi.org/10.1016/j.gca.2009.01.017

Forbes, T. Z. (2018). Low-temperature geochemistry. In W. M. White (Ed.), *Encyclopedia of geochemistry* (Encyclopedia of earth sciences series). Springer. https://doi.org/10.1007/978-3-319-39312-4_23
Frolov, E. B., Smirnov, M. B., Melikhov, V. A., & Vanyukova, N. A. (1998). Olefins of radiogenic origin in crude oils. *Organic Geochemistry, 29*, 409–420.
Frye, K. (1981). Lattice. In *Mineralogy* (Encyclopedia of earth science). Springer. https://doi.org/10.1007/0-387-30720-6_65
Gaillardet, J., & Galy, A. (2008). Himalaya: Carbon sink or source? *Science, 320*(5884), 1727–1728.
Gaines, S. M., Eglinton, G., & Rullkotter, K. (2009). *Echoes of life. What fossils reveal about earth history*. Oxford University Press.
Garrels, R. M., & Christ, C. L. (1965). *Solution, minerals and equilibria*. Freeman Cooper.
Garrels, R. M., & Mackenzie, F. T. (1971). *Evolution of sedimentary rocks*. Norton.
Geetha Thanuja, K., Reddy Kiran Kalyan, V. S., Karthikeyan, S., & Anthoniraj, S. (2022). Microbial transformation of silicon in soil. In C. J. Hurst (Ed.), *Microbial metabolism of metals and metalloids* (Advances in environmental microbiology) (Vol. 10). Springer. https://doi.org/10.1007/978-3-030-97185-4_15
Geochemical Journal. (2022). *Geochemical Journal*. Retrieved from https://geochemical-journal.jp
Geochemical Society. (2022). *Organic geochemistry division*. Retrieved from https://www.geochemsoc.org/programs/organicgeochemistrydivisio
Geochemistry. (2022). Retrieved from https://www.encyclopedia.com/environment/energy-government-and-defense-magazines/geochemistry
Geological Society of London. (2012). *Journal of the Geological Society*. Retrieved from https://www.geolsoc.org.uk/jgs
Geoscience Australia. (2022). *Geochemistry*. Retrieved from https://www.ga.gov.au/scientific-topics/disciplines/geochemistry#:~:text=Inorganic%20Geochemistry,-Inorganic%20chemistry%20is&text=Inorganic%20chemistry%20investigates%20the%20characteristics,found%20in%20the%20Earth's%20crust
GeoScienceWorld. (2022). *Lithosphere*. Retrieved from https://pubs.geoscienceworld.org/lithosphere
Goldschmidt, V. M. (1937). The principles of distribution of chemical elements in minerals and rocks. *Journal of the Chemical Society*, 655–672.
Goldschmidt, V. M. (1954). *Geochemistry*. Clarendon Press.
Gottikh, R. P., Vinokurov, S. F., & Pisotskii, B. I. (2009). Rare-earth elements as geochemical criteria of endogenous sources of microelements contained in oil. *Doklady Earth Sciences, 425*, 325–329.
Gounelle, M. (2011). Cosmochemistry. In M. Gargaud, R. Amils, J. C. Quintanilla, H. J. Cleaves, W. M. Irvine, D. L. Pinti, & M. Viso (Eds.), *Encyclopedia of astrobiology* (pp. 380–387). Springer. https://doi.org/10.1007/978-3-642-11274-4_1747
Graustein, W. C., & Armstrong, R. L. (1983). The use of strontium-87/strontium-86 ratios to measure atmospheric transport into forested watersheds. *Science, 219*(4582), 289–292. https://doi.org/10.1126/science.219.4582.289
Greenwalt, D. E., Goreva, Y. S., Siljestrom, S. M., Rose, T., & Harbachet, R. E. (2013). Hemoglobin-derived porphyrins preserved in a Middle Eocene blood-engorged mosquito. *Proceedings of the National Academy of Sciences USA, 110*, 18496–18500.
Haldar, S. K. (2018). Exploration geochemistry. In S. K. Haldar (Ed.), *Mineral exploration: Principles and applications* (2nd ed., pp. 85–101). https://doi.org/10.1016/B978-0-12-814022-2.00005-8
Hallberg, R. O. (2009). Biogeochemistry. In V. Cilek (Ed.), *Earth system: History and natural variability - Volume IV* (pp. 435–446). Developed under the Auspices of the UNESCO, Eolss Publishers.
Hallet, B., Hunter, L., & Bogen, J. (1996). Rates of erosion and sediment evacuation by glaciers: A review of field data and their implications. *Global and Planetary Change, 12*(1–4), 213–235. https://doi.org/10.1016/0921-8181(95)00021-6

Hansell, D. A., & Carlson, C. A. (Eds.). (2014). *Biogeochemistry of marine dissolved organic matter*. Academic. https://doi.org/10.1016/C2012-0-02714-7

Hawkes, H. E. (1957). Principles of geochemical prospecting. *USGS Numbered Series, 1000*, 225–355. https://doi.org/10.3133/b1000F

Head, I. M., Gray, N. D., & Larter, S. R. (2014). Life in the slow lane; biogeochemistry of biodegraded petroleum containing reservoirs and implications for energy recovery and carbon management. *Frontiers in Microbiology, 5*, 566. https://doi.org/10.3389/fmicb.2014.00566

Hellweger, F. L. (2008). Biogeochemical models. In *Encyclopedia of ecology* (pp. 386–396). Elsevier B.V.

Hem, J. D. (1985). *Study and interpretation of the chemical characteristics of natural water* (3rd ed.). US Geological Survey Water-Supply Paper 2254, University of Virginia.

Hendry, K. R., Marron, A. O., Vincent, F., Conley, D. J., Gehlen, M., Ibarbalz, F. M., Quéguiner, B., & Bowler, C. (2018). Competition between silicifiers and non-silicifiers in the past and present ocean and its evolutionary impacts. *Frontiers in Marine Science, 5*, 22. https://doi.org/10.3389/fmars.2018.00022

Higueras, P., Campos, J. A., Esbrí, J. M., García-Noguero, E. M., & Elmayel, I. (2019). Biogeochemical mapping: A new tool to assess the soil quality and health. In D. Doronzo, E. Schingaro, J. Armstrong-Altrin, & B. Zoheir (Eds.), *Petrogenesis and exploration of the Earth's interior. CAJG 2018* (Advances in science, technology & innovation). Springer. https://doi.org/10.1007/978-3-030-01575-6_1

Hildred, G. V., Ratcliffe, K. T., Wright, A. M., Zaitlin, B. A., & Wray, D. S. (2010). Chemostratigraphic applications to low-accommodation fluvial incised-valley settings; an example from the Lower Mannville Formation of Alberta, Canada. *Journal of Sedimentary Research, 80*, 1032–1045.

Holland, H. D. (1978). *The chemistry of the atmosphere and oceans*. Wiley.

Holland, H. D. (1984). *The chemical evolution of the atmosphere and oceans*. Princeton University Press.

Hopkins, F. E., Archer, S. D., Bell, T. G., Suntharalingam, P., & Todd, J. D. (2023). The biogeochemistry of marine dimethylsulfide. *Nature Reviews Earth & Environment, 4*, 361–376. https://doi.org/10.1038/s43017-023-00428-7

Housecroft, C. E., & Sharpe, A. G. (2008). *Inorganic chemistry* (3rd ed.). Pearson Education.

Hoyle, F., & Wickramasinghe, C. (1981). Where microbes boldly went. *New Scientist, 91*, 412–415.

Hu, Y., & Teng, F. Z. (2021). Non-traditional stable isotope geochemistry. In D. In Alderton & S. A. Elias (Eds.), *Encyclopedia of geology* (2nd ed., pp. 114–124). Academic.

Hunt, J. M. (1979). *Petroleum geochemistry and geology*. W.H. Freeman and Company.

Hunt, J. M., Philp, R. P., & Kvenvolden, K. A. (2002). Early developments in petroleum geochemistry. *Organic Geochemistry, 33*(9), 1025–1052. https://doi.org/10.1016/S0146-6380(02)00056-6

Imperial College. (2022). *Organic Geochemistry Research (ICOG)*. Retrieved from https://www.imperial.ac.uk/earth-science/research/research-groups/icog/research/

Inamuddin, Ahamed, I. M., & Lichtfouse, E. (2021). Water pollution and remediation: Heavy metals environmental chemistry for a sustainable world. https://doi.org/10.1007/978-3-030-52421-0

Indiana University. (2022). *Department of Earth and Atmospheric Sciences: Geochemistry and biogeochemistry*. Retrieved from https://earth.indiana.edu/research/research-areas/geochemistry-and-biogeochemistry/index.html#:~:text=Geochemistry%20and%20Biogeochemistry%20are%20inherently,that%20can%20constrain%20computational%20models

Intertek. (2022). *Metals speciation analysis and testing: Metal speciation analysis*. Retrieved from https://www.intertek.com/chemicals/metals-speciation/

Ivanov, K. S., Kucherov, V. G., & Fedorov, Y. N. (2008). On the question of the deep origin of oil STATE, trends and problems of development of oil and gas potential of Western Siberia. In *Materials of the international academic conference* (pp. 160–173). ZapSibNIIGG.

Ivanov, K. S., Erokhin, Y. V., & Kudryavtsev, D. A. (2021). Inorganic geochemistry of crude oils of northern Eurasia after ICP-MS data as clear evidence for their deep origin. *Energies, 15*(1), 48. https://doi.org/10.3390/en15010048

Jacks, G., Åberg, G., & Hamilton, P. J. (1989). Calcium budgets for catchments as interpreted by strontium isotopes. *Hydrology Research, 20*(2), 85–96.

Jacobson, A. D., Blum, J. D., Chamberlain, C. P., Poage, M. A., & Sloan, V. F. (2002). Ca/Sr and Sr isotope systematics of a Himalayan glacial chronosequence: Carbonate versus silicate weathering rates as a function of landscape surface age. *Geochimica et Cosmochimica Acta, 66*(1), 13–27.

Jenkins, D. A. (1989). Trace element geochemistry in archaeological sites. *Environmental Geochemistry and Health, 11*, 57–62. https://doi.org/10.1007/BF0178299

Jenkyns, H. C. (2010). Geochemistry of oceanic anoxic events. *Geochemistry Geophysics Geosystems, 11*, 1–30.

Jin, Z., You, C., Yu, J., Wu, L., Zhang, F., & Liu, H. (2011). Seasonal contributions of catchment weathering and eolian dust to river water chemistry, northeastern Tibetan plateau: Chemical and Sr isotopic constraints. *Journal of Geophysical Research: Earth Surface, 116*(F4).

John Wiley & Sons Ltd. (2022). *Geological Journal*. Retrieved from https://onlinelibrary.wiley.com/page/journal/10991034/homepage/productinformation.html

Kamenov, G., MacFarlane, A. W., & Riciputi, L. (2002). Sources of lead in the San Cristobal, Pulacayo, and Potosí mining districts, Bolivia, and a reevaluation of regional ore lead isotope provinces. *Economic Geology, 97*(3), 573–592. https://doi.org/10.2113/gsecongeo.97.3.573

Kaufman, Y. J., Tucker, C. J., & Fung, I. (1990). Remote sensing of biomass burning in the tropics. *Journal of Geophysical Research, 95*, 9927–9939. https://doi.org/10.1029/JD095iD07p09927

Kendall, C., & Caldwell, E. A. (1998). Fundamentals of isotope geochemistry. In C. Kendall & J. J. McDonnell (Eds.), *Isotope tracers in catchment hydrology* (pp. 51–86). Elsevier Science.

Kendall, C., Sklash, M. G., & Bullen, T. D. (1995). Isotope tracers of water and solute sources in catchments. In S. T. Trudgill (Ed.), *Solute modelling in catchment systems* (pp. 261–303). Wiley.

Kendall, C., Doctor, D., & Young, M. (2014). Environmental isotope applications in hydrologic studies. In *Treatise on geochemistry* (Vol. 7, 2nd ed., pp. 273–327). https://doi.org/10.1016/B978-0-08-095975-7.00510-6

Kennedy, M. E. (1998). Elements: Trace. In *Geochemistry. Encyclopedia of earth science* (pp. 221–223). Springer. https://doi.org/10.1007/1-4020-4496-8_1_11

Killops, S. D., & Killops, V. J. (2013). *Introduction to organic geochemistry* (2nd ed.). Wiley-Blackwell.

Kim, J. D., Yee, N., Nanda, V., & Falkowski, P. G. (2011). Anoxic photochemical oxidation of siderite generates molecular hydrogen and iron oxides. *Proceedings of the National Academy of Sciences of the United States of America, 110*(25), 10073–10077. https://doi.org/10.1073/pnas.1308958110

Knight, P. G. (1999). *Glaciers*. Stanley Thornes (Publishers) Ltd.

Koeberl, C. (2007). 1.28 - The geochemistry and cosmochemistry of impacts. In H. D. Holland & K. K. Turekian (Eds.), *Treatise on geochemistry* (Vol. 1, pp. 1–52). https://doi.org/10.1016/B978-008043751-4/00228-55

Kozlik, M., Gerdes, A., & Raith, J. G. (2016). Strontium isotope systematics of sheelite and apatite from the felbertal tungsten deposit, Austria–results of in-situ LA-MC-ICP-MS analysis. *Mineralogy and Petrology, 110*(1), 11–27.

Kraeutler, B., & Bard, A. J. (1978). Heterogeneous photocatalytic synthesis of methane from acetic acid-new Kolbe reaction pathway. *Journal of the American Chemical Society, 100*(7), 2239–2240. https://doi.org/10.1021/ja00475a049

Kudo, A. M., & Jahns, R. H. (2022). Igneous rock. *Encyclopedia Britannica*. https://www.britannica.com/science/igneous-rock

Kumar, S. P., Mandlimath, T. R., & Ramesh, M. (2021). Inorganic geochemistry. In Inamuddin, M. I. Ahamed, R. Boddula, & T. Altalhi (Eds.), *Geochemistry: Concepts and applications* (pp. 149–160). Wiley; Scrivener Publishing. https://doi.org/10.1002/9781119710134.ch9

Kurade, M. B., Saha, S., Salama, E. S., Patil, S. M., Govindwar, S. P., & Jeon, B. H. (2019). Acetoclastic methanogenesis led by Methanosarcina in anaerobic co-digestion of fats, oil and grease for enhanced production of methane. *Bioresource Technology, 272*, 351–359. https://doi.org/10.1016/j.biortech.2018.10.047

Kvenvolden, K. A. (2002). Methane hydrate in the global organic carbon cycle. *Terra Nova, 14*(5), 302–306. https://doi.org/10.1046/j.1365-3121.2002.00414.x

Kvenvolden, K. A. (2006). Organic geochemistry – A retrospective of its first 70 years. *Organic Geochemistry, 37*(1), 1–11. https://doi.org/10.1016/j.orggeochem.2005.09.001

Laishram, J., Saxena, K. G., Maikhuri, R. K., & Rao, K. S. (2012). Soil quality and soil health: A review. *International Journal of Ecology and Environmental Sciences, 38*(1), 19–37.

Langmuir, D. (1997). *Aqueous environmental geochemistry*. Prentice Hall.

Lastovicka, J. (2009). *Geophysics and geochemistry*. Retrieved from https://www.eolss.net/ebooklib/bookinfo/geophysics-geochemistry.aspx#

Lauretta, D. S., & Killgore, M. (2005). *A color atlas of meteorites in thin section*. Golden Retriever Press.

Lauretta, D. S., & McSween, H. Y., Jr. (Eds.). (2006). *Meteorites and the early solar system II*. University of Arizona Press.

Lee, H.-C. (2018). Review of inductively coupled plasmas: Nano-applications and bistable hysteresis physics. *Applied Physics Reviews, 5*(1), 011108.

Lee, T., Papanastassiou, D. A., & Wasserburg, G. J. (1977). Aluminum-26 in the early solar system: Fossil or fuel? *Astrophysical Journal Letters, 211*, L107–L110.

Leithauser, P., Schaefer, R. G., & Yukler, A. (1982). Role of diffusion in primary migration of hydrocarbons. *American Association of Petroleum Geologists Bulletin, 66*, 408–429.

Lenton, T. M., & Daines, S. J. (2017). Biogeochemical transformations in the history of the ocean. *Annual Review of Marine Science, 9*, 31–58.

Li, W., Gou, W., Li, W., Zhang, T., Yu, B., Liu, Q., & Shi, J. (2019). Environmental applications of metal stable isotopes: Silver, mercury and zinc. *Environmental Pollution, 252*(Part B), 1344–1356. https://doi.org/10.1016/j.envpol.2019.06.037

Lodders, K., & Fegley, B., Jr. (1998). *The planetary scientist's companion*. Oxford University Press.

Lopatin, N. V. (1971). Temperature and time as factors of coalification. *Izvestiya Akademii Nauk SSSR, Seriya gelogicheskaya (in Russian), 3*, 95–106.

Lu, X., Ma, C., & Wang, S. (2019). Classifying the geology journals by editorial board interlocks. *Procedia Computer Science, 162*, 682–687. https://doi.org/10.1016/j.procs.2019.12.038

MacFarlane, A. W., Marcet, P., LeHuray, A. P., & Petersen, U. (1990). Lead isotope provinces of the Central Andes inferred from ores and crustal rocks. *Economic Geology, 85*(8), 1857–1880. https://doi.org/10.2113/gsecongeo.85.8.1857

MacPherson, G. J., & Thiemens, M. H. (2011). Cosmochemistry: Understanding the solar system through analysis of extraterrestrial materials. *PNAS, 108*(48), 19130–19134. https://doi.org/10.1073/pnas.1111493108

Magara, K. (1978). *Compaction and fluid migration: Practical petroleum geology fundamentals of numerical reservoir simulation* (Developments in petroleum science 9). Elsevier Scientific Pub. Co.

Magoon, L. B., & Dow, W. G. (1994). The petroleum system. In L. B. Magoon & W. G. Dow (Eds.), *The petroleum system—From source to trap* (Vol. 60, pp. 3–24). AAPG Memoire.

Makishima, A. (2017). *Origins of the earth, moon, and life. An interdisciplinary approach*. Elsevier.

Mamedov, L. (2021). Stable isotope geochemistry. *Journal of Science and Geosciences, 9*(2), 6–7.

Marakushev, A. A., Pisotskii, B. I., Paneyakh, N. A., & Gottikh, R. P. (2004). Geochemical features of oil and the origin of oil fields. *Doklady Earth Sciences, 399*, 1120–1124.

Marshak, S. (2009). *Essentials of geology* (3rd ed.). W. W. Norton & Company.

Mather, T. A. (2013). Reference module in earth systems and environmental sciences || Geochemistry. https://doi.org/10.1016/B978-0-12-409548-9.05929-7

McClenaghan, M. B., Plouffe, A., McMartin, I., Campbell, J., Spirito, W., Paulen, R., Garrett, R., & Hall, G. (2013). Till sampling and geochemical analytical protocols used by the Geological

Survey of Canada. *Geochemistry Exploration Environment Analysis, 13*(4), 285–301. https://doi.org/10.1144/geochem2011-083

McKay, D. S., Gibson, E. K., Jr., Thomas-Keprta, K. L., Vali, H., Romanek, C. S., Clemett, S. J., Chillier, X. D., Maechling, C. R., & Zare, R. N. (1996). Search for past life on Mars: Possible relic biogenic activity in Martian meteorite ALH84001. *Science, 273*(5277), 924–930. https://doi.org/10.1126/science.273.5277.924

McMartin, I., & McClenaghan, M. B. (2001). Till geochemistry and sampling techniques in glaciated shield terrain: A review. Geological Society London Special Publications 185(1):19–43. https://doi.org/10.1144/GSL.SP.2001.185.01.02. In M. B. McClenaghan, P. T. Bobrowsky, G. E. M. Hall, & S. J. Cook (Eds.), *Drift exploration in glaciated terrain* (pp. 19–43). Geological Society of London. https://doi.org/10.1144/GSL.SP.2001.185

McSween, H., Jr., & Huss, G. (2010). Introduction to cosmochemistry. In H. McSween Jr. & G. Huss (Eds.), *Cosmochemistry* (pp. 1–28). Cambridge University Press. https://doi.org/10.1017/CBO9780511804502.002

Meyer, K. M., & Kump, L. R. (2008). Oceanic euxinia in earth history: Causes and consequences. *Annual Review of Earth and Planetary Sciences, 36*(1), 251–288.

Miles, C. J., & Brezonik, P. L. (1981). Oxygen consumption in humic-colored waters by a photochemical ferrous-ferric catalytic cycle. *Environmental Science & Technology, 15*(9), 1089–1095. https://doi.org/10.1021/es00091a010

Miller, E. K., Blum, J. D., & Friedland, A. J. (1993). Determination of soil exchangeable-cation loss and weathering rates using Sr isotopes. *Nature, 362*(6419), 438–441.

Miyoshi, H., Mori, H., & Yoneyama, H. (1991). Light-induced decomposition of saturated carboxylic acids on iron oxide incorporated clay suspended in aqueous solutions. *Langmuir, 7*(3), 503–507. https://doi.org/10.1021/la00051a015

Montross, S. N., Skidmore, M., Tranter, M., Kivimäki, A.-L., & Parkes, R. J. (2013). A microbial driver of chemical weathering in glaciated systems. *Geology, 41*, 215–218. https://doi.org/10.1130/G33572.1

Morel, F. M. M., & Hering, J. G. (1993). *Principles and applications of aquatic chemistry*. Wiley.

Morel, F. M. M., & Price, N. M. (2003). The biogeochemical cycles of trace metals in the oceans. *Science, 300*(5621), 944–947. https://doi.org/10.1126/science.1083545

Moriceau, B., Goutx, M., Guigue, C., Lee, C., Armstrong, R., Duflos, M., Tamburini, C., Charrière, B., & Ragueneau, O. (2009). Si–C interactions during degradation of the diatom Skeletonema marinoi. *Deep Sea Research Part II: Topical Studies in Oceanography, 56*(18), 1381–1395. https://doi.org/10.1016/j.dsr2.2008.11.026

Mukhopadhyay, P. K., Wade, J. A., & Kruge, M. A. (1995). Organic facies and maturation of Jurassic/cretaceous rocks, and possible oil-source rock correlation based on pyrolysis of asphaltenes, Scotian Basin, Canada. *Organic Geochemistry, 22*, 85–104. https://doi.org/10.1016/0146-6380(95)90010-1

Nakai, S., Halliday, A. N., & Rea, D. K. (1993). *Provenance of dust in the Pacific Ocean*. https://doi.org/10.1016/0012-821X(93)90012-X.

Nature Portfolio. (2023). *Biomineralization articles from across Nature Portfolio*. Retrieved from https://www.nature.com/subjects/biomineralization

Negri, A., Ferretti, A., Wagner, T., & Meyers, P. A. (2009). Organic-carbon-rich sediments through the Phanerozoic: Processes, progress, and perspectives. *Palaeogeography, Palaeoclimatology, Palaeoecology, 273*, 302–328.

Neikov, O. D., & Yefimov, N. A. (2019). Chapter 1 - Powder characterization and testing. In O. D. Neikov, S. S. Naboychenko, & N. A. Yefimov (Eds.), *Handbook of non-ferrous metal powders* (Technologies and applications) (2nd ed., pp. 3–62). Elsevier. https://doi.org/10.1016/C2014-0-03938-X

Nóbrega, G. N., Ferreira, T. O., Romero, R. E., Marques, A. G., & Otero, X. L. (2013). Iron and sulfur geochemistry in semi-arid mangrove soils (Ceará, Brazil) in relation to seasonal changes and shrimp farming effluents. *Environmental Monitoring and Assessment, 185*(9), 7393–7407. https://doi.org/10.1007/s10661-013-3108-

Nordstrom, D. K. (1997). Some fundamentals of aqueous geochemistry. In G. S. Plumlee, M. J. Logsdon, & L. F. Filipek (Eds.), *The environmental geochemistry of mineral deposits: Part A: Processes, techniques, and health issues part B: Case studies and research topics* (Vol. 6). Society of Economic Geologists, Inc. https://doi.org/10.5382/Rev.06

O'Beirne, M., Sparkes, R., Hamilton, T., Dongen, B., Gilhooly, W., III, & Werne, J. (2022). Characterization of diverse bacteriohopanepolyols in a permanently stratified, hyper-euxinic lake. *Organic Geochemistry, 168*, 104431. https://doi.org/10.1016/j.orggeochem.2022.104431

Ollila, A., Newsom, H., Clark, B., Wiens, R., Cousin, A., Blank, J., Mangold, N., Sautter, V., Maurice, S., Clegg, S., Gasnault, O., Forni, O., Tokar, R., Lewin, E., Dyar, M., Lasue, J., Anderson, R., McLennan, S., Bridges, J., & Team, M.S.L. (2013). Trace element geochemistry (Li, Ba, Sr, and Rb) using curiosity's ChemCam: Early results for gale crater from Bradbury landing site to rocknest. *Journal of Geophysical Research: Planets, 119*(1), 255–285. https://doi.org/10.1002/2013JE004517

Organic Geochemistry. (2022). *Organic geochemistry*. Retrieved from https://www.journals.elsevier.com/organic-geochemistry

Ouellette, R. J., & Rawn, J. D. (2019). *Organic chemistry: Structure, mechanism, synthesis* (2nd ed.). Elsevier. https://doi.org/10.1016/C2016-0-04004-4

Parkhurst, D. L., & Appelo, C. A. J. (2008). *PHREEQC interactive 2.15.0*. United States Geological Survey.

Passey, Q. R., Bohacs, K. M., Esch, W. L., Klimentidis, R., & Sinha, S. (2010). From oil-prone source rock to gas-producing shale reservoir – Geologic and petrophysical characterization of unconventional shale-gas reservoirs. In *International oil and gas conference and exhibition in China, Beijing, China*. https://doi.org/10.2118/131350-MS

Paytan, A., Yao, W., & Faul, K., & Gray, E. T. (2020). Sulfur isotope. *Stratigraphy, 1*, 259–278. https://doi.org/10.1016/B978-0-12-824360-2.00009-7

Pearce, T. J., Wray, D. S., Ratcliffe, K. T., Wright, D. K., & Moscariello, A. (2005). Chemostratigraphy of the upper carboniferous schooner formation, Southern North Sea. In J. D. Collinson, D. J. Evans, D. W. Holliday, & N. S. Jones (Eds.), *Carboniferous hydrocarbon geology: The southern North Sea and surrounding onshore areas* (Occasional publications series 7) (pp. 147–164). Yorkshire Geological Society.

Pessanha, S., Queralt, I., Carvalho, M. L., & Sampaio, J. M. (2019). Determination of gold leaf thickness using X-ray fluorescence spectrometry: Accuracy comparison using analytical methodology and Monte Carlo simulations. *Applied Radiation and Isotopes, 152*, 6–10.

Peters, K. E., Walters, C. C., & Moldowan, J. M. (2005). *The biomarker guide* (2nd ed.). Cambridge University Press.

Phillippi, G. T. (1965). On the depth, time and mechanism of petroleum generation. *Geochimica et Cosmochimica Acta, 29*, 1021–1104.

Philp, R. P., & Mansuy, L. (1997). Petroleum geochemistry: Concepts, applications, and results. *Energy & Fuels, 11*(4), 749–760. https://doi.org/10.1021/ef960174v

Pötz, S., Liu, Y., Vieth-Hillebrand, A., Yang, S., Magnall, J. M., Gleeson, S. A., & Schulz, H.-M. (2021). The Kupferschiefer in Spremberg, East Germany: Signals of organic matter alteration from hydrogen isotopes and Nso compounds - conference proceedings. In *30th international meeting on organic geochemistry (IMOG 2021) (Online 2021)*. https://doi.org/10.3997/2214-4609.202134210

Priscu, J. C., & Christner, B. C. (2004). Earth's icy biosphere. In A. T. Bull (Ed.), *Microbial diversity and prospecting* (pp. 130–145). ASM Press.

Qin, Z., Liu, H. M., Ma, Y. X., & Wang, X. D. (2021). Chapter 9: Developments in extraction, purification, and structural elucidation of proanthocyanidins (2000–2019). *Studies in Natural Products Chemistry, 68*, 347–391. https://doi.org/10.1016/B978-0-12-819485-0.00008-6

Rafferty, J. P. (2013). *Biogeochemistry*. Retrieved from https://www.britannica.com/science/biogeochemistry

Railsback, L. B. (2003). An earth scientist's periodic table of the elements and their ions. *Geology, 31*, 737–740.

Raiswell, R., Tranter, M., Benning, L. G., Siegert, M., De'Ath, R., Huybrechts, P., & Payne, T. (2006). Contributions from glacially derived sediment to the global iron (oxyhydr)oxide cycle: Implications for iron delivery to the oceans. *Geochimica et Cosmochimica Acta, 70*(11), 2765–2780. https://doi.org/10.1016/j.gca.2005.12.027

Rass, I. T., Aranovich, L. Y., Korpechkov, D. I., & Kozlovskii, V. M. (2014). Geochemistry of metamorphic processes in mafic rocks of the Krasnaya Guba area, Belomorian Mobile Belt. *Geochemistry International, 52*(8), 670–686. https://doi.org/10.1134/S0016702914060068

Ratcliffe, K. T., Hughes, A. D., Lawton, D. E., Wray, D. S., Bessa, F., Pearce, T. J., & Martin, J. (2006). A regional chemostratigraphically defined correlation framework for the late Triassic TAG-I in Blocks 402 and 405a, Algeria. *Petroleum Geoscience, 12*, 3–12.

Ratcliffe, K. T., Morton, A., Ritcey, D., & Evenchick, C. E. (2007). Whole rock geochemistry and heavy mineral analysis as exploration tools in the Bowser and Sustut Basins, British Colombia, Canada. *Journal of Canadian Petroleum Geology, 55*, 320–237.

Ratcliffe, K. T., Wright, A. M., Montgomery, P., Palfrey, A., Vonk, A., Vermeulen, J., & Barrett, M. (2010). Application of chemostratigraphy to the Mungaroo formation, the gorgon field, offshore Northwest Australia. *Australian Petroleum Production and Exploration Association Journal, 50*, 371–385.

Ratcliffe, K. T., Wright, A. M., & Schmidt, K. (2012). Application of inorganic whole-rock geochemistry to shale resource plays: An example from the Eagle Ford Shale Formation, Texas. *Sedimentary Record, 10*(2), 4–9. https://doi.org/10.2110/sedred.2012.2.4

Reid, N., & Hill, S. (2010). Biogeochemical sampling for mineral exploration in arid terrains: Tanami Gold Province, Australia. *Journal of Geochemical Exploration, 104*(3), 105–117. https://doi.org/10.1016/j.gexplo.2010.01.004

Rinot, O., Levy, G. J., Steinberger, Y., Svoray, T., & Eshel, G. (2019). Soil health assessment: A critical review of current methodologies and a proposed new approach. *The Science of the Total Environment, 648*, 1484–1491. https://doi.org/10.1016/j.scitotenv.2018.08.259

Rose, A. W. (1998). Geochemical exploration. In *Geochemistry. Encyclopedia of earth science*. Springer. https://doi.org/10.1007/1-4020-4496-8_137

Roy-Barman, M., & Jeandel, C. (2016). Radioactive and radiogenic isotopes. In M. Roy-Barman & C. Jeandel (Eds.), *Marine geochemistry: Ocean circulation, carbon cycle and climate change* (pp. 129–161). Oxford University Press. https://doi.org/10.1093/acprof:oso/9780198787495.003.0004

Sahai, N., Kaddour, H., & Dalai, P. (2016). The transition from geochemistry to biogeochemistry. *Elements, 12*(6), 389–394. https://doi.org/10.2113/gselements.12.6.389

Salifu, M., Aiglsperger, T., Hällström, L., Martinsson, O., Billström, K., Ingri, J., Dold, B., & Alakangas, L. (2018). Strontium ($^{87}Sr/^{86}Sr$) isotopes: A tracer for geochemical processes in mineralogically-complex mine wastes. *Applied Geochemistry, 99*, 42–54. https://doi.org/10.1016/j.apgeochem.2018.10.022

Sanders, T. G., Jr., Biddanda, B. A., Stricker, C. A., & Nold, S. C. (2011). Benthic macroinvertebrate and fish communities in Lake Huron are linked to submerged groundwater vents. *Aquatic Biology, 12*, 1–11. https://doi.org/10.3354/ab00318

Sarala, P. (2015). Chapter 10.1: Surficial geochemical exploration methods. In W. D. Maier, R. Lahtinen, & H. O'Brien (Eds.), *Mineral deposits of Finland* (pp. 711–731). https://doi.org/10.1016/B978-0-12-410438-9.00027-3

Schlesinger, W. H., & Bernhardt, E. S. (2013). *Biogeochemistry: An analysis of global change* (Third ed.). Academic.

Schmitt-Kopplin, P., Gabelica, Z., Gougeon, R. D., Fekete, A., Kanawati, B., Harir, M., Gebefuegi, I., Eckel, G., & Hertkorn, N. (2010). High molecular diversity of extraterrestrial organic matter in Murchison meteorite revealed 40 years after its fall. *Proceedings of the National Academy of Sciences of the United States of America, 107*(7), 2763–2768. https://doi.org/10.1073/pnas.0912157107

Schmus, W. R., & Wood, J. A. (1967). A chemical–petrologic classification for the chondritic meteorites. *Geochimica et Cosmochimica Acta, 31*, 747–765. https://doi.org/10.1016/S0016-7037(67)80030-9

Schrauzer, G. N., Strampach, N., Hui, L. N., Palmer, N. R., & Salehi, J. (1983). Nitrogen photoreduction on desert sands under sterile conditions. *Proceedings of the National Academy of Sciences of the United States of America, 80*, 3873–3876.

Science Direct. (2022a). *Organic geochemistry*. Retrieved from https://www.sciencedirect.com/journal/organic-geochemistry/about/aims-and-scope

Science Direct. (2022b). *Applied clay science*. Retrieved from https://www.sciencedirect.com/journal/applied-clay-science

Science Direct. (2022c). *Applied geochemistry*. Retrieved from https://www.sciencedirect.com/journal/applied-geochemistry/about/aims-and-scope

Science Direct. (2022d). *Applied geochemistry*. Retrieved from https://www.sciencedirect.com/journal/applied-geochemistry

Science Direct. (2022e). *Chemical geology*. Retrieved from https://www.sciencedirect.com/journal/chemical-geology

Science Direct. (2022f). *Continental shelf research*. Retrieved from https://www.science direct.com/journal/continental-shelf-research

Science Direct. (2022g). *Earth and planetary science letters*. Retrieved from https://www.sciencedirect.com/journal/earth-and-planetary-science-letters/about/aims-and-scope

Science Direct. (2022h). *Engineering geology*. Retrieved from https://www.sciencedirect.com/journal/engineering-geology/about/aims-and-scope

Science Direct. (2022i). *Geochemistry*. Retrieved from https://www.sciencedirect.com/journal/geochemistry/about/aims-and-scope

Science Direct. (2022j). *Geochimica et Cosmochimica Acta*. Retrieved from https://www.sciencedirect.com/journal/geochimica-et-cosmochimica-acta/about/aims-and-scope

Science Direct. (2022k). *Geoscience Frontiers*. Retrieved from https://www.sciencedirect.com/journal/geoscience-frontiers/about/aims-and-scope

Science Direct. (2022l). *International Journal of Coal Geology*. Retrieved from https://www.sciencedirect.com/journal/international-journal-of-coal-geology/about/aims-and-scope

Science Direct. (2022m). *Journal of Asian Earth Sciences: X*. Retrieved from https://www.sciencedirect.com/journal/journal-of-asian-earth-sciences-x/about/aims-and-scope

Science Direct. (2022n). *Journal of Geochemical Exploration*. Retrieved from https://www.sciencedirect.com/journal/journal-of-geochemical-exploration/about/aims-and-scope

Science Direct. (2022o). *Journal of Geodynamics*. Retrieved from https://www.science direct.com/journal/journal-of-geodynamics/about/aims-and-scope

Science Direct. (2022p). *Journal of Palaeogeography*. Retrieved from https://www.sciencedirect.com/journal/journal-of-palaeogeography/about/aims-and-scope

Science Direct. (2022q). *Journal of South American Earth Sciences*. Retrieved from https://www.sciencedirect.com/journal/journal-of-south-american-earth-sciences

Science Direct. (2022r). *Journal of Volcanology and Geothermal Research*. Retrieved from https://www.sciencedirect.com/journal/journal-of-volcanology-and-geothermal-research

Science Direct. (2022s). *Lithos*. Retrieved from https://www.sciencedirect.com/journal/lithos/about/aims-and-scope

Science Direct. (2022t). *Marine and petroleum geology*. Retrieved from https://www.sciencedirect.com/journal/marine-and-petroleum-geology/about/aims-and-scope

Science Direct. (2022u). *Physics and Chemistry of the Earth, Parts A/B/C*. Retrieved from https://www.sciencedirect.com/journal/physics-and-chemistry-of-the-earth-parts-a-b-c/about/aims-and-scope

Science Direct. (2022v). *Precambrian Research*. Retrieved from https://www.science direct.com/journal/precambrian-research/about/aims-and-scope

Science Direct. (2022w). *Results in Geochemistry*. Retrieved from https://www.sciencedirect.com/journal/results-in-geochemistry/about/aims-and-scope

Scott, S. D. (2014). Volume Editor's introduction. In K. Turekian & H. Holland (Eds.), *Treatise on geochemistry* (Vol. 13, 2nd ed., pp. xxiii–xxv).

Scripps Institution of Oceanography. (2022). *Earth low temperature geochemistry*. Retrieved https://scripps.ucsd.edu/research/topics/low-temperature-geochemistry

Sen, Z. (2015). *Practical and applied hydrogeology*. Elsevier. https://doi.org/10.1016/C2013-0-14020-2

Sephton, M. A. (2002). Organic compounds in carbonaceous meteorites. *Natural Product Reports, 19*(3), 292–311. https://doi.org/10.1039/b103775g

SGS. (2022). *Mining geochemistry*. Retrieved from https://www.sgs.ca/en/mining/analytical-services/geochemistry

Shand, P., Darbyshire, D. F., Gooddy, D., & Haria, A. H. (2007). $^{87}Sr/^{86}Sr$ as an indicator of $^{87}Sr/^{86}Sr$ ratios flow paths and weathering rates in the Plynlimon experimental catchments, Wales, UK. *Chemical Geology, 236*(3), 247–265.

Sharp, M., Parkes, J., Cragg, B., Fairchild, I. J., Lamb, H., & Tranter, M. (1999). Widespread bacterial populations at glacier beds and their relationship to rock weathering and carbon cycling. *Geology, 27*(2), 107–110. https://doi.org/10.1130/0091-7613(1999)027<0107:WBPAGB>2.3.CO;2

Siegert, M. J., Ellis-Evans, J. C., Tranter, M., Mayer, C., Petit, J.-R., Salamatin, A., & Priscu, J. C. (2001). Physical, chemical, and biological processes in Lake Vostok and other Antarctic subglacial lakes. *Nature, 414*, 603–609. https://doi.org/10.1038/414603a

Simmons, W. B. (2014). Pyroxene. *Encyclopedia Britannica*. https://www.britannica.com/science/pyroxene

Sinha, S. (2013). *Biogeochemistry*. Retrieved from https://www.britannica.com/science/biogeochemistry#ref1068660

Skidmore, M. L., Foght, J. M., & Sharp, M. J. (2000). Microbial life beneath a high arctic glacier. *Applied and Environmental Microbiology, 66*(8), 3214–3220. https://doi.org/10.1128/AEM.66.8.3214-3220.2000

Slatt, R. M., Zhang, J., Torres-Parada, E., Brito, R., Duarte, D., & Milad, B. (2021). Unconventional petroleum resources. In D. Alderton & S. A. Elias (Eds.), *Encyclopedia of geology* (pp. 783–807). Academic. https://doi.org/10.1016/B978-0-08-102908-4.00023-0

Sluijk, D., & Parker, J. R. (1984). Comparison of predrilling predictions with post drilling outcomes. *AAPG Bulletin, 68*(4), 528. https://doi.org/10.1306/AD4611E2-16F7-11D7-8645000102C1865D

Smith, R. L., Repert, D. A., & Hart, C. P. (2009). Geochemistry of inorganic nitrogen in waters released from coal-bed natural gas production wells in the Powder River Basin, Wyoming. *Environmental Science & Technology, 43*(7), 2348–2354. https://doi.org/10.1021/es802478p

Soldatova, E., Dong, Y., Li, J., Liu, Y., Zan, J., Boeckx, P., & Sun, Z. (2021). Nitrogen transformation and pathways in the shallow groundwater-soil system within agricultural landscapes. *Environmental Geochemistry and Health, 43*(1), 441–459. https://doi.org/10.1007/s10653-020-00733-w

Sowder, J. T., Kelleners, T. J., & Reddy, K. J. (2010). The origin and fate of arsenic in coalbed natural gas-produced water ponds. *Journal of Environmental Quality, 39*(5), 1604–1615. https://doi.org/10.2134/jeq2009.0428

Springer Nature. (2022a). *Bulletin of Volcanology*. Retrieved from https://www.springer.com/journal/445/aims-and-scope

Springer Nature. (2022b). *Contributions to mineralogy and petrology*. Retrieved from https://www.springer.com/journal/410#:~:text=Contributions%20to%20Mineralogy%20and%20Petrology%20is%20an%20international%20journal%20that,metamorphic%20petrology%2C%20geochemistry%20and%20mineralogy

Springer Nature. (2022c). *Geochemistry International*. Retrieved from https://www.springer.com/journal/11476

Springer Nature. (2022d). *Physics and Chemistry of Minerals*. Retrieved from https://www.springer.com/journal/269

Stern, J. C., & Wieman, S. T. (2021). Traditional stable isotope geochemistry. In D. Alderton & S. A. Elias (Eds.), *Encyclopedia of geology* (2nd ed., pp. 100–113). Academic.

Stewart, B. W., Capo, R. C., & Chadwick, O. A. (2001). Effects of rainfall on weathering rate, base cation provenance, and Sr isotope composition of Hawaiian soils. *Geochimica et Cosmochimica Acta, 65*(7), 1087–1099.

Stoll, N., Eichler, J., Hörhold, M., Erhardt, T., Jensen, C., & Weikusat, I. (2021). Microstructure, micro-inclusions, and mineralogy along the EGRIP ice core – Part 1: Localisation of inclusions and deformation patterns. *The Cryosphere, 15*, 5717–5737. https://doi.org/10.5194/tc-15-5717-2021

Stricker, C. (2023). *Biogeochemistry: Stable isotope applications*. Retrieved from https://www.usgs.gov/centers/fort-collins-science-center/science/biogeochemistry-stable-isotope-applications#overview

Strunz, H. (1941). *Mineralogische Tabellen*. Akademische Verlagsgesellschaft.

Stumm, W., & Morgan, J. J. (1996). *Aquatic chemistry, chemical equilibria and rates in natural waters* (3rd ed.). Wiley.

Subías, I., Fanlo, I., & Billström, K. (2015). Ore-forming timing of polymetallic-fluorite low temperature veins from central Pyrenees: A Pb, Nd and Sr isotope perspective. *Ore Geology Reviews, 70*, 241–251.

Sutton, J. N., André, L., Cardinal, D., Conley, D. J., de Souza, G. F., Dean, J., Dodd, J., Ehlert, C., Ellwood, M. J., Frings, P. J., Grasse, P., Hendry, K., Leng, M. J., Michalopoulos, P., Panizzo, V. N., & Swann, G. E. A. (2018). A review of the stable isotope biogeochemistry of the global silicon cycle and its associated trace elements. *Frontiers of Earth Sciences, 5*, 112. https://doi.org/10.3389/feart.2017.00112

Tang, J., Zheng, Y., Wu, Y. B., Gong, B., & Liu, X. (2007). Geochronology and geochemistry of metamorphic rocks in the Jiaobei terrane: Constraints on its tectonic affinity in the Sulu orogen. *Precambrian Research, 152*, 48–82. https://doi.org/10.1016/j.precamres.2006.09.001

Taylor and Francis Online. (2022a). *Australian Journal of Earth Sciences*. Retrieved from https://www.tandfonline.com/action/journalInformation?show=aimsScope&journalCode=taje20

Taylor and Francis Online. (2022b). *Tellus B: Chemical and physical meteorology*. Reterived from https://www.tandfonline.com/action/journalInformation?show=aimsScope&journalCode=zelb20

Taylor, S. R., & McLennan, S. M. (1985). *The continental crust: Its composition and evolution*. Blackwell Scientific Publishers.

Telling, J., Boyd, E. S., Bone, N., Jones, E. L., Tranter, M., Macfarlane, J. W., Martin, P. G., Wadham, J. L., Lamarche-Gagnon, G., Skidmore, M. L., Hamilton, T. L., Hill, E., Jackson, M., & Hodgson, D. A. (2015). Rock comminution as a source of hydrogen for subglacial ecosystems. *Nature Geoscience, 8*, 851–855. https://doi.org/10.1038/ngeo2533

Terry, R. E. (2017). Field geochemistry. In A. S. Gilbert (Ed.), *Encyclopedia of geoarchaeology* (Encyclopedia of earth sciences series). Springer. https://doi.org/10.1007/978-1-4020-4409-0_165

The International Conference on the Biogeochemistry of Trace Elements (ICOBTE). (2023). Retrieved from https://icobte-ichmet-2023.com/frontend/index.php

Tipper, E. T., Bickle, M. J., Galy, A., West, A. J., Pomiès, C., & Chapman, H. J. (2006). The short term climatic sensitivity of carbonate and silicate weathering fluxes: Insight from seasonal variations in river chemistry. *Geochimica et Cosmochimica Acta, 70*(11), 2737–2754. https://doi.org/10.1016/j.gca.2006.03.005

Tissot, B. P., & Welte, D. H. (1978). *Petroleum formation and occurrence*. Springer. https://doi.org/10.1007/978-3-642-96446-6

Titley, S. R. (2001). Crustal affinities of metallogenesis in the American Southwest. *Economic Geology, 96*(6), 1323–1342. https://doi.org/10.2113/gsecongeo.96.6.1323

Tiwari, M., Singh, A. K., & Sinha, D. K. (2015). Stable isotopes: Tools for understanding past climatic conditions and their applications in chemostratigraphy. In M. Ramkumar (Ed.), *Chemostratigraphy: Concepts, techniques, and applications* (pp. 65–92). https://doi.org/10.1016/C2013-0-09872-6

Torró, L., Benites, D., Vallance, J., Laurent, O., Ortiz Benavente, B. A., Chelle-Michou, C., Proenza, J., & Fontboté, L. (2022). Trace element geochemistry of sphalerite and chalcopyrite in arc-hosted VMS deposits. *Journal of Geochemical Exploration, 232*, 106882. https://doi.org/10.1016/j.gexplo.2021.106882

Tranter, M. (2015). Grand challenge for low temperature and pressure geochemistry—Sparks in the dark, on Earth, Mars, and throughout the Galaxy. *Frontiers in Earth Science, 3*, 69. https://doi.org/10.3389/feart.2015.00069

Tranter, M., & Wadham, J. L. (2013). Geochemical weathering in glacial and proglacial environments. In J. I. Drever (Ed.), *Treatise on geochemistry* (Vol. 5, Surface and ground water, weathering, and soils, 2nd ed., pp. 153–174). Elsevier.

Tranter, M., Sharp, M. J., Lamb, H. R., Brown, G. H., Hubbard, B. P., & Willis, I. C. (2002). Geochemical weathering at the bed of Haut Glacier d'Arolla, Switzerland - A new model. *Hydrological Processes, 16*, 959–993. https://doi.org/10.1002/hyp.309

Trask, P., & Patnode, D. (1942). *Source beds of petroleum*. American Association of Petroleum Geologists.

Tréguer, P. J., Sutton, J. N., Brzezinski, M., Charette, M. A., Devries, T., Dutkiewicz, S., Ehlert, C., Hawkings, J., Leynaert, A., Liu, S. M., Llopis Monferrer, N., López-Acosta, M., Maldonado, M., Rahman, S., Ran, L., & Rouxel, O. (2021). Reviews and syntheses: The biogeochemical cycle of silicon in the modern ocean. *Biogeosciences, 18*, 1269–1289. https://doi.org/10.5194/bg-18-1269-2021

Treibs, A. E. (1934). Chlorophyll- und Häminderivate in bituminösen Gesteinen, Erdölen, Erdwachsen und Asphalten. *Ein Beitrag zur Entstehung des Erdöls, 510*(1), 42–62.

Treibs, A. E. (1936). Chlorophyll- und Ha¨minderivate in organischen Mineralstoffen. *Angewandte Chemie, 49*, 682–686.

Tribovillard, N., Algeo, T. J., Lyons, T., & Riboulleau, A. (2006). Trace metals as paleoredox and paleoproductivity proxies: An update. *Chemical Geology, 232*, 12–32.

Tribovillard, N., Bout-Roumazeilles, V., Algeo, T., Lyons, T. W., Sionneau, T., Montero-Serrano, J. C., Riboulleau, A., & Baudin, F. (2008). Paleodepositional conditions in the Orca Basin as inferred from organic matter and trace metal contents. *Marine Geology, 254*, 62–72.

Trueman, C., Rodgers, K., McLellan, I., & Hursthouse, A. (2019). Geochemistry: Inorganic. In P. Worsfold, C. Poole, A. Townshend, & M. Miró (Eds.), *Encyclopedia of analytical science: Reference module in chemistry, molecular sciences and chemical engineering* (3rd ed., pp. 271–282). Elsevier Limited. https://doi.org/10.1016/B978-0-12-409547-2.14362-8

Turgeon, S., & Brumsack, H.-J. (2006). Anoxic vs. dysoxic events reflected in sediment geochemistry during the Cenomanian–Turonian Boundary Event (Cretaceous) in the Umbria–Marche Basin of central Italy. *Chemical Geology, 234*, 321–339.

United States Geological Survey. (2022). *What are metamorphic rocks?* Retrieved from https://www.usgs.gov/faqs/what-are-metamorphic-rocks

University of Auckland. (2005). *Geology: Rocks and minerals*. https://rocksminerals.flexiblelearning.auckland.ac.nz/rocks/schist.html

University of Michigan. (2022). *Low temperature geochemistry*. Retrieved from https://lsa.umich.edu/earth/research/geochemistry.html

Utrecht University. (2022). *Organic geochemistry*. Retrieved from https://www.uu.nl/en/research/department-of-earth-sciences/organic-geochemistry

Vallero, D. (2014). Inherent properties of air pollutants. In D. Vallero (Ed.), *Fundamentals of air pollution* (pp. 139–195). Academic. https://doi.org/10.1016/B978-0-12-401733-7.00006-2

Van Grieken, R. E., & Markowicz, A. A. (2002). *Handbook of X-ray spectrometry* (2nd ed.). Marcel Dekker.

Vandenbroucke, M., & Largeau, C. (2007). Kerogen origin, evolution and structure. *Organic Geochemistry, 38*(5), 719–833. https://doi.org/10.1016/j.orggeochem.2007.01.001

Volkman, J. K. (1998). Organic geochemistry. In *Geochemistry* (Encyclopedia of earth science). Springer. https://doi.org/10.1007/1-4020-4496-8_226

Wadham, J. L., Tranter, M., Skidmore, M., Hodson, A. J., Priscu, J., Lyons, W. B., Sharp, M., Wynn, P., & Jackson, M. (2010). Biogeochemical weathering under ice: Size matters. *Global Biogeochemical Cycles, 24*, GB3025. https://doi.org/10.1029/2009GB003688

Wadham, J. L., De'ath, R., Monteiro, F., Tranter, M., Ridgwell, A., Raiswell, R., & Tulaczyk, S. (2013). The potential role of the Antarctic Ice Sheet in global biogeochemical cycles. *Earth and Environmental Science Transactions of the Royal Society of Edinburgh, 104*(1), 55–67. https://doi.org/10.1017/S1755691013000108

Wang, J., Dai, J., Chen, G., & Jiang, F. (2022). Role of sulfur biogeochemical cycle in mercury methylation in estuarine sediments: A review. *Journal of Hazardous Materials, 423*(Pt A), 126964. https://doi.org/10.1016/j.jhazmat.2021.126964

Waples, D. W. (1980). Time and temperature in petroleum generation-application of Lopatin's technique to petroleum exploration. *AAPG Bulletin, 64*, 916–926.

Wasserburg, G. J., & Hayden, R. J. (1955). Age of meteorites by the 40Ar–40K method. *Physics Review, 97*, 86–87. https://doi.org/10.1103/PhysRev.97.86

Wedepohl, K. H. (1995). The composition of the continental crust. *Geochimica et Cosmochimica Acta, 59*, 1217–1232. https://doi.org/10.1016/0016-7037(95)00038-2

Weiss, J. V., Emerson, D., Backer, S. M., & Megonigal, P. (2003). Enumeration of Fe(II)-oxidizing and Fe(III)-reducing bacteria in the root zone of wetland plants: Implications for a rhizosphere iron cycle. *Biogeochemistry, 64*, 77–96. https://doi.org/10.1023/A:1024953027726

Wetzel, R. G. (2001). *Limnology: Lake and river ecosystems* (Third ed.). Academic.

White, W. B. (2010). Chapter 6: Springwater geochemistry. In N. Kresic & Z. Stevanovic (Eds.), *Groundwater hydrology of springs: Engineering, theory, management, and sustainability* (pp. 231–268). Butterworth-Heinemann. https://doi.org/10.1016/C2009-0-19145-6

White, W. M. (2013). *Geochemistry*. Wiley-Blackwell.

White, W. M. (2017). Geochemistry. In W. White (Ed.), *Encyclopedia of geochemistry*(Encyclopedia of earth sciences series). Springer. https://doi.org/10.1007/978-3-319-39193-9_294-1

White, W. M. (2020). *Geochemistry*. Wiley.

Whiteside, J. H., & Grice, K. (2016). Biomarker records associated with mass extinction events. *Annual Review of Earth and Planetary Sciences, 44*, 581–516.

Williamson, M. A. (1998). Iron. In *Geochemistry* (Encyclopedia of earth science). Springer. https://doi.org/10.1007/1-4020-4496-8_172

Wilschefski, S. C., & Baxter, M. R. (2019). Inductively coupled plasma mass spectrometry: Introduction to analytical aspects. *The Clinical Biochemist. Reviews, 40*(3), 115–133. https://doi.org/10.33176/AACB-19-00024

Winkel, L. H. E., & Sunderland, E. M. (2022). Introduction to the biogeochemistry of the trace elements themed issue. *Environmental Science: Processes & Impacts, 24*, 1277–1278. https://doi.org/10.1039/D2EM90031A

Winter, J. D. (2013). *Principles of igneous and metamorphic petrology*. Pearson Education.

Winterburn, P. A., Noble, R. R. P., & Lawie, D. (2017). Advances in exploration geochemistry, 2007 to 2017 and beyond. *Geochemistry. Paper, 34*, 495–505.

Winterburn, P. A., Noble, R. R. P., & Lawie, D. (2020). Advances in exploration geochemistry, 2007 to 2017 and beyond. *Geochemistry: Exploration, Environment, Analysis, 20*(2), 157–166. https://doi.org/10.1144/geochem2019-030

Wooden, J. L., Stacey, J. S., Howard, K. A., Doe, B. R., & Miller, D. M. (1988). Pb isotopic evidence for the formation of Proterozoic crust in the southwestern United States. In *Metamorphism and crustal evolution of the western United States* (pp. 69–86). Prentice-Hall.

Wright, A. M., Ratcliffe, K. T., Zaitlin, B. A., & Wray, D. S. (2010). The application of chemostratigraphic techniques to distinguish compound incised valleys in low-accommodation incised-valley systems in a foreland-basin setting: An example from the Lower Cretaceous Mannville Group and Basal Colorado Sandstone (Colorado Group), Western Canadian Sedimentary Basin. In K. T. Ratcliffe & B. A. Zaitlin (Eds.), *Application of modern stratigraphic techniques: Theory and case histories* (Vol. 94, pp. 93–109). SEPM Special Publication.

Yadav, P. K., Mauraya, A. K., Kochar, C., Taneja, L., & Swarupa Tripathy, S. (2021). Isotope geochemistry. In Inamuddin, M. I. Ahamed, R. Boddula, & T. Altalhi (Eds.), *Geochemistry: Concepts and applications* (pp. 93–109). Wiley; Scrivener Publishing.

Yale University. (2022). *The Department of Earth & Planetary Sciences*. Retrieved from https://earth.yale.edu/geochemistry

Zakaria, M., Bong, C. W., & Vaezzadeh, V. (2018). Fingerprinting of petroleum hydrocarbons in Malaysia using environmental forensic techniques. In S. A. Stout & Z. Wang (Eds.), *Oil spill environmental forensics case studies* (pp. 345–372). Butterworth-Heinemann.

Zou, C. (2017). *Unconventional petroleum geology*. Retrieved from Unconventional Petroleum Geology. Retrieved from https://www.sciencedirect.com/book/9780128122341/unconventional-petroleum-geology

Chapter 3
Earth Systems Science (ESS) and Systems Ecology

Michael O'Neal Campbell

Abstract Earth systems science (ESS) is a science that is strongly linked to biogeochemistry, through its study of the earth systems (lithosphere, hydrosphere, biosphere, and atmosphere) through which the chemical elements and compounds studied by biogeochemistry flow. The two disciplines developed largely separately, but this chapter explores common ground, in terms of origins, developments, and future possibilities. A basic definition of ESS is that it is the application of systems science to the earth's surface. Systems science and its subsets, systems ecology and earth systems science, provide methodologies that can document, describe, analyze, and understand spatial and networked ecological relations, within the larger, complex disciplines of the environmental sciences. ESS is a recent development, within physical geography and the earth and environmental sciences, with the objective of studying the integrated relations, physical, energy, and chemical, to link the contributions of the increasingly polarized and segmented earth and environmental sciences. This science is concerned with, but not limited to, the relations between the global to local contexts of the biosphere, hydrosphere, lithosphere, and atmosphere. The developmental trend of ESS is particularly relevant to the field of biogeochemistry, which itself is also a multidisciplinary field seeking to override the generally artificial disciplinary boundaries between biological, chemical, and physical sciences to derive answers for complex environmental questions. This chapter examines the basics of ESS and then looks at the approach and methodologies of systems ecology and the links to biogeochemistry, using recent literature sources on the definition, application, and status of ESS and related sciences. It is argued that ESS must battle on two fronts: the question for broader knowledge to solve the increasingly complex, multidisciplinary environmental issues, and the requirement for deep specialization to understand the issues in the first place, some at microscopic level. Institutional barriers also creep in, as the topics of ESS may be scattered across departments, sometimes different from those of biogeochemistry. The linking between ESS and biogeochemistry must also be measured against the changes in the environmental

M. O. Campbell (✉)
Simon Fraser University, Burnaby, BC, Canada
e-mail: evaz@torontomu.ca

M. O. Campbell, *Biogeochemistry and the Environment*,
https://doi.org/10.1007/978-3-031-47017-2_3

focus of the basic sciences (chemistry, physics, mathematics, biology) and the fortunes of the environmentally applied progeny of these disciplines (geochemistry, environmental chemistry, biochemical and chemical engineering, biochemistry, geophysics, environmental and atmospheric physics, civil, geological, and environmental engineering, oceanography, statistics, etc.). The understanding of these complex issues contributes to the development of biogeochemical studies.

Introduction

Earth systems science (ESS) relates to biogeochemistry, as the former examines all the Earth systems that include water, earth materials, air, and living organisms, while biogeochemistry studies the chemical element and compound flows within these systems, adding the study of organisms to the study of the inanimate systems (the field of geochemistry). ESS recently emerged from an amalgamation of systems science, physical geography, and related environmental and earth sciences (Campbell, 2017, 2018). Cornell (2013: 85) contends that "the field of Earth Systems Science is just a little over 50 years old. This area of research analyzes the interlinked dynamics of the physical and living 'components' of Earth, bridging local to global-scale processes." Steffen et al. (2020: 60) also assert that "ESS emerged in the early to mid-20th century from conceptualizations of the Earth that emphasized its systemic nature, such as Vernadsky's observation that life has a strong influence on the chemical and physical properties of Earth and the Gaia hypothesis of Lovelock and Margulis that Earth functions as a single organism, with self-regulating processes and feedbacks that maintain homeostasis." It is further advanced that "ESS then developed rapidly, from the 'new science of the Earth's movement in the 1980s to the global research efforts of international programs such as the IGBP" (International Geosphere-Biosphere Program) (Steffen et al., 2020, 54–63; see also Vernadsky, 1926; Lovelock, 1979).

But what precisely is ESS? ESS has been variously defined, but the main points refer to its broad yet detailed approach, links to systems science, and the physical, biological, and sometimes aspects of the social sciences, and its recent development through the current scientific dynamics (Jacobson et al., 2000; Stanley, 2005). Human systems may be included in the study of Earth systems: as noted by Jacobson et al. (2000), ESS "is an integrated discipline that has been rapidly developing over the last two decades" and "is the basis for understanding all aspects of anthropogenic global change, such as chemically forced global climate change." Environmental dynamics in ESS include material and energy fluxes and connections, feedback, holistic, spatial, and temporal views, and in terms of subjects, geology, meteorology, oceanography, ecology, physical geography, climatology, and space science, among others. The Earth's environmental systems are also named, namely the atmosphere, hydrosphere, biosphere, lithosphere, and geosphere, and sometimes even the cryosphere (ice) and pedosphere (soils) (Merritts et al., 1998; Ohtani et al., 2000; Hergarten, 2002; Neugebauer & Simmer, 2003; Butz, 2004; Kump et al., 2004; Harris & Murton, 2005; Ehlers and Krafft, 2006; Wilkinson, 2006; Tsonis & Elsner, 2007; Martin, 2011; Cornell et al., 2012; Marsh & Kaufman,

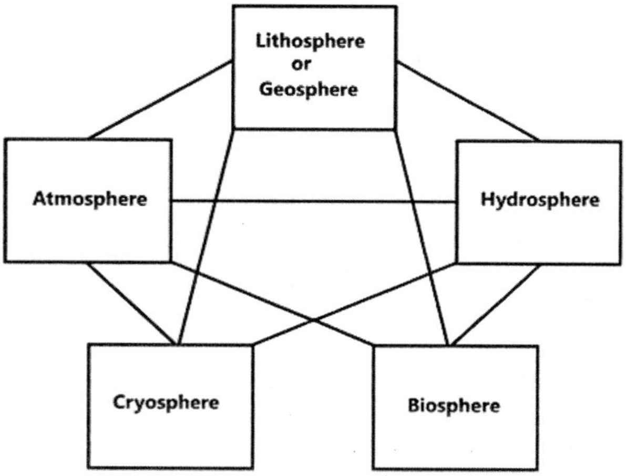

Fig. 3.1 The earth systems. Some classifications use the lithosphere or geosphere, but not both. Some others include the cryosphere (solid ice and frozen ground) with the hydrosphere. This diagram shows the links between the systems, as each is related to the others, sometimes through yet others

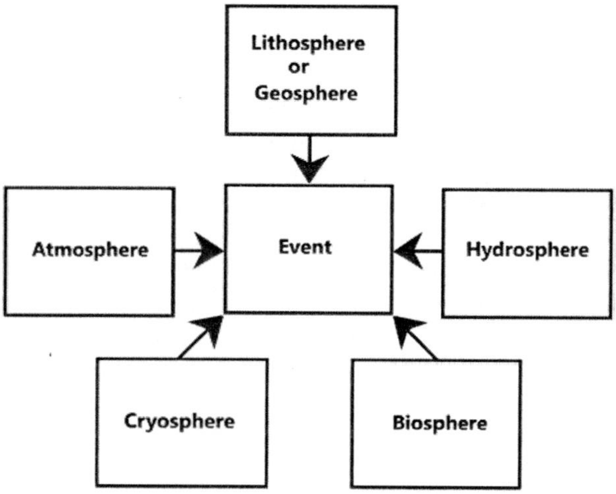

Fig. 3.2 Interactions of earth systems. Some events are the result of interactions between two or more of the systems. Examples include volcanic eruptions, the ice age, mass biological extinctions and climatic changes, which engage with various systems

2013; Christiansen & Hamblin, 2014). Figures 3.1 and 3.2 show the topics and format of earth systems science.

Definitions of ESS may emphasize subject areas, considering functions, subject matter, applications and/or interdisciplinary relations. For example, Lenton (2016: 1) describes ESS as "the research field that seeks to understand how our planet functions as a whole system. Its scope is broad; it spans 4.5 billion years of Earth history, how the system functions now, projections of its future state, and its ultimate fate. It also considers how humans as a species are reshaping the planet. ESS is a deeply interdisciplinary field, which synthesizes elements of geology, biology, chemistry, physics, and mathematics." The interactions within the system may also be mentioned: "due to an almost closed system, all the elements needed by life must be efficiently recycled within the Earth system, which then need energy to transform materials chemically and to move them physically around the planet. 'Recycling' introduces the life-sustaining global biogeochemical cycles of matter between the biosphere, atmosphere, ocean, land, and crust" (Lenton, 2016, 18). Additionally, "regulation introduces the ways in which the biogeochemical cycles of the Earth system are self-regulated, how they are coupled to the Earth's climate, and how scientists study this regulation" (Lenton, 2016: 38).

McGill (2022) defines ESS as a discipline that "concentrates on the links between the Earth's biological, chemical, physical, and human subsystems. It examines the cycling of energy and matter through the biosphere, the atmosphere, the cryosphere, the hydrosphere, and the solid Earth." This science examines "the six 'Grand Challenges' fundamental to our understanding of how the Earth operates," which are listed as:" Global biogeochemical cycles, climate variability and change, land use and land cover change, energy and resources, earth hazards, such as volcanoes, earthquakes and hurricanes, earth-atmosphere observation, analysis, and prediction" (McGill, 2022). A similar definition is given by National Geographic (2022a), which claims that the subject matter of earth systems science covers "the five systems of Earth (geosphere, biosphere, cryosphere, hydrosphere, and atmosphere)" that interact to produce the familiar environmental systems, describing the geosphere as the "interior and surface of Earth, both of which are made up of rocks," the biosphere as the "limited part of the planet that can support living things," the hydrosphere as the "areas of Earth that are covered with enormous amounts of water," the atmosphere as the "envelope of gas that keeps the planet warm and provides oxygen for breathing and carbon dioxide for photosynthesis," and the cryosphere "which contains huge quantities of ice at the poles and elsewhere, constituting the cryosphere." This point is supported by Berkeley (2022), who adds human issues and argues that ESS studies the interconnected components of the Earth's environment (the atmosphere, hydrosphere, lithosphere, cryosphere, and biosphere) and the integrated whole that emerges from their interconnections. This science "utilizes the fundamental disciplines of mathematics, physics, chemistry, and biology and applies them in the context of human activities and landscapes to understand the Earth, at scales ranging from single watersheds to the entire globe" (Berkeley, 2022).

Ruzek (2018) also mentions the human dimension and describes ESS as embracing the basic sciences beyond disciplinary boundaries to treat the Earth as an integrated system and seeks a deeper understanding of the phenomena that determine the history, current and future dynamics of the Earth's environment. Furthermore, "Earth Systems Science builds upon the fundamental disciplines, which form the scientific integrating basis for any discussion of the system as a whole. This representation shows the critical relationship of ESS concepts as a bridge between the disciplines and the increasingly interpretive and integrative endeavors of human social processes. The future of our planet and the destiny of humankind are dependent upon this interdisciplinary pyramid of earth system processes and phenomena"

(Ruzek, 2018). Ruzek (2018) further argues that "knowledge from the traditional earth science disciplines… is being gleaned and integrated to form a physical basis for Earth System Science" and "Earth Systems Science fosters the synthesis of disciplinary knowledge into a holistic model of Earth with broader interdisciplinary relevance… The intersection of disciplinary specialties often provides the most fertile and interesting fields for study but is easily sidetracked by traditional disciplinary interests."

The National Aeronautics and Space Administration (NASA) (2003: 8) maintains that in ESS, research scientists may employ a contextual approach to research, and may examine the cause and effect dynamics that may be behind extreme earth events, such as weather, volcanic or forest change, which may have strong effects on people and may also be influenced by human action. The example is cited of ocean temperatures and circulation to create relatively moderate climate of northern Europe, despite its location in the northern latitudes. Another example is that of the links between the annual variations of ozone concentration over Antarctica and the production of industrial refrigerants in the Northern Hemisphere. The physics and chemistry of the atmosphere may also be linked to air quality and freshwater availability (NASA, 2003).

ESS may also be defined within the links and areas of related, holistic disciplines (Campbell, 2017, 2020). For example, Kalipeni (2001) avers that "physical geographers increasingly refer to their subject matter as earth systems science or global change instead of physical geography" and "the emerging idea of earth systems science as an integrative science has made the old-fashioned term 'physical geography' obsolete so that people who used to call themselves physical geographers are now able to cross traditional disciplinary boundaries with ease." ESS is also related to systems science, which is a broader field, encompassing systems and networks applicable to technical, natural, social, and applied sciences (Campbell, 2020). Hieronymi (2013) described systems science as "the ordered arrangement of knowledge acquired from the study of systems in the observable world, together with the application of this knowledge to the design of man-made systems." In these studies, and applications, the key foci are interdisciplinarity, networking, modeling, applicability, and artificial representation. Applications at the more detailed level include systems design and engineering, social systems theory, systems biology, systems psychology, system dynamics, cybernetics, dynamical systems theory, information theory, linguistics or systems theory, control theory and operations research (Ison, 2017; Mobus & Kalton, 2015).

Some definitions of ESS also examine the recent development of this discipline. For example, Steffen et al. (2020: 54) declare that, developmentally, "Earth System Science (ESS) is a rapidly emerging transdisciplinary endeavour aimed at understanding the structure and functioning of the Earth as a complex, adaptive system" arguing that "inspired by early work on biosphere–geosphere interactions and by novel perspectives such as the Gaia hypothesis, ESS emerged in the 1980s following demands for a new 'science of the Earth'." More recent developments include the International Geosphere-Biosphere Program, with associated international and interdisciplinary developments, new concepts such as the Anthropocene, part of the discourse of global-change, the "grand challenge for ESS is to achieve a deep integration of biophysical processes and human dynamics to build a truly unified understanding of the Earth System" (Steffen et al. (2020: 54).

The history of ESS is summarized based on geological developments and increasingly sophisticated military developments during the Cold War era, the consequent development of geophysics research, and the International Geophysical Year (IGY) 1957–1958 (mostly focused on the geosphere, especially oceanography, glaciology, and meteorology) (Steffen et al., 2020). Steffen et al. (2020: 54) add that "one of the key impacts of the IGY was a lasting transformation in the practices used to understand how the Earth works. The interpretative and qualitative geological and climatological research based on field observations – as classically studied by geographers – was replaced by field instrumentation, continuous and quantitative monitoring of multiple variables and numerical models" with a resulting focus on modern climatology and plate tectonics, with parallel developments in ecosystem ecology and environmental sciences, the International Biological Program (IBP), and the Gaia hypothesis in 1972. This "led to a new understanding of the Earth, challenging a purely geophysical conception of the planet, and transforming our view of the environment and nature. The stage was now set for the introduction of a new science – a more formal and well-organized" (Steffen et al., 2020: 56; see also Lovelock, 1979).

The development of ESS during the latter half of the twentieth century was based on the acknowledgment of the integrated nature of the earth's systems, the participation of powerful stakeholders such as NASA (the organization that originated the term Earth Systems Science), theoretical developments from these, and common agreement among stakeholders on the internationality and interdisciplinarity of ESS, the subject matter of which disregarded national borders and disciplinary boundaries. Steffen et al. (2020: 56) noted that the global and interdisciplinary approach (physical, chemical, and biological) "created a significant challenge in bringing different disciplines together to study the Earth System as a whole." Reactive developments included the International Council for Science (ICSU), the International Geosphere-Biosphere Program (IGBP), the World Climate Research Program (WCRP, 1980), and the World Commission on Environment and Development (1987), which tackled biogeochemical issues (such as terrestrial ecosystems, ocean carbon cycles, atmospheric chemistry, the hydrological cycle, etc.) and the change from an interdisciplinary approach with multiple disciplines, a transdisciplinary approach where the disciplinary boundaries may disappear. Steffen et al. (2020: 56) further notes that "by the end of the 1980s, ESS had emerged as a powerful new scientific endeavor, triggered by the growing recognition of global change and built on the rapid development of interdisciplinary research methods."

During the 1990s, the international research program DIVERSITAS was created (1991) to complement the work of the IGBP, the International Human Dimensions Program (IHDP) on Global Environmental Change (1996) was also founded, with a stronger focus on social issues, and at the beginning of the twenty-first century, sustainability science was developed, as did ideas concerning the possibilities of nonlinear multidirectional human-participated environmental change, including the concept of the Anthropocene. An important meeting was the 1999 IGBP Congress, and another was the 2001 conference, "Challenges of a Changing Earth" which included IGBP, WCRP, IHDP and DIVERSITAS), which influenced the formation of the ESS Partnership (ESSP) and the formulation the concept of the "Earth System" which included human, as well as biological, physical, and chemical processes. By 2015, the IGBP, IHDP and DIVERSITAS were integrated into Future Earth, which emphasized greater connections with policy makers, and private organizations.

Recent projects included the International Global Atmospheric Chemistry (IGAC) project, PAGES ((Past Global Changes, a project is for management of past global change research) and the ESS Partnership (ESSP) Global Carbon Project. Relevant research centers included the US National Center for Atmospheric Research (NCAR), the Potsdam Institute for Climate Impact Research (PIK), the Stockholm Resilience Centre (SRC), and the International Institute for Applied Systems Analysis (IIASA). At the theoretical level, mathematical models have merged as vital tools for analyses and understandings, these including general circulation models (GCMs), and more human centered integrated assessment models (IAMs, which may integrate economic and climate models), both relevant to topics such as greenhouse gas emissions, and the Integrated Global System Model (IGSM), which further integrate human and climate issues. Earth System Models of Intermediate Complexity (EMICs) are also touted as "arguably the most powerful tools for exploring the complex dynamics of the Earth System, particularly at long timescales" which "include the same main processes as GCMs but have a lower spatial resolution and a greater number of parameterized processes, allowing them to run longer timescale simulations that include nonlinear forcings and feedbacks between components of the Earth System" (Steffen et al., 2020, 59). Steffen et al. (2020, 59) further notes that EMICs, "can be run at timescales of up to hundreds of thousands of years, allowing the models to be tested against palaeo observations and to explore possible climates of the far future. Taken together, GCMs, IAMs and EMICs create powerful ways to explore Earth-System dynamics at numerous space scales and timescales."

Earth Systems Science (ESS) and Systems Ecology

Systems ecology is closely related to ESS, through its theoretical and methodological approach, although it may be more biological and hence be a narrower disciplinary cluster (Campbell, 2017, 2018; Odum, 1964). Systems ecology is also vital to biogeochemistry, as systems that describe global to local ecologies represent the pathways and reservoirs of geochemical and biogeochemical cycling. Odum (1964) originated the term "systems ecology" in his publication, "The New Ecology" (Odum, 1964). The underlying position of systems ecology is the reliance on broad interacting systems and knowledge, while still emphasizing detail. For example, Van Dyne (1969, 31, 35) notes that a systems ecologist must be "conversant with specialists" but also retain "a holistic or systems viewpoint." Wiegert (1988) reviewed these points from the perspective of Reductionism Versus Holism, an enduring debate within ecology and related sciences. Van Dyne (1966) defines systems ecology, according to the background envisaged by Odum (1964) as related to the fact that ecology is inherently a system-based subject: "We have been taught that ecology is the study of the relationships between organisms and their environment, and that ecology may be subdivided into autecology (of individuals or species), population ecology, and community ecology. Systems ecology in a sense approximates communities' ecology. The terms system and ecology both imply a holistic viewpoint."

Van Dyne (1966) also elaborates on the role of mathematics in systems ecology, holding that many researchers may define systems ecology as the field of application to mathematics to ecological systems. This, however, may be a narrow definition, as "systems ecology can be broadly defined as the study of the development, dynamics, and disruption of ecosystems" and has "two main phases – a theoretical and analytical phase and an experimental phase" and the "study of problems in systems ecology requires three groups of tools and processes: conceptual, mechanical, and mathematical" (Van Dyne, 1966). The complexity of systems ecology requires different analytical tools than the simpler science of conventional ecology, this complexity including multifactor events, the use of computer technology, logic, and complex scientific and statistical methods, including hypothesis testing, and modeling (mathematical abstractions of environmental reality). Van Dyne (1966) also posits two ways of experimenting with ecosystems. The first involves the process of hypothesis formulation, and experimental design and practice, and the analysis and interpretation of the findings. The second abstracts the system parameters and components into a model, applies mathematical arguments, and interprets the mathematically derived conclusions. Mathematical modeling is argued to be "somewhat new to many conventional ecologists and, in part, is just as much an art as a science" (Van Dyne, 1966). These issues are still important in the current scene (Campbell, 2017).

Montague (2014) defines systems ecology as "an approach to ecosystem study based on formal procedures of systems thinking, synthesis, and modeling. Its goals are those of ecosystem ecology in general: develop and test theory of ecosystem organization; detect and manage emergent properties; and predict responses to disturbance." This discipline includes applications from Mathematical Ecology and Simulation Modeling, and ecosystem management aspects of Human Ecology and Applied Ecology (Montague, 2014). Systems ecology evolved through computer development, and the advancement of technology in engineering projects, both in civilian and military systems. These included cybernetics (Marinescu, 2017 defines cybernetics as "concerned with concepts at the core of understanding complex systems such as learning, cognition, adaptation, emergence, communication, and efficiency") systems and the ecosystem.

Systems ecology emerged with methodologies based on computerized solutions in mathematical modeling, applied to the complex issues of ecosystems. Computer technology enabled more sophisticated, accurate tools, for the testing of hypotheses and the measurement of problematic issues, and gradually systems ecology was integrated into general ecosystem ecology. Montague (2014) notes that "common traits of successful systems ecologists include familiarity with systems methodology; ability to synthesize a system from a breadth of information; and fascination with self-organization, response to disturbance, and emergent properties of ecosystems." Systems ecology eventually comprised two "visions," namely "soft systems and hard science" (Montague, 2014). Montague (2014) describes the soft systems vision as acknowledging ecosystem complexity, continual evolution, and the possibilities scientific testing (observations, hypotheses, testing, analyses), based on tools such as cybernetics, general systems theory, epistemology, anthropology, industrial management, and evolutionary theory (Montague, 2014). The hard science vision is based on laws applicable to all ecosystems, reliably predictable for equilibria even without empirical support, based on thermodynamics, evolutionary genetics, and mathematical analyses of population and trophic interactions (Montague, 2014). Montague (2014) concludes that both of these visions need the use of field observations, experiments, and computer models. The integration of these two visions contributes to advances such as adaptive ecosystem management, ecological engineering and economics, the theory of complex ecosystems and resilience theory, landscape ecology, and field studies of ecological and biodiversity assessment. Literature cited and relevant to the early developments of this topic include Odum (1964) Van Dyne (1969), Dale (1970), Shugart and O'Neill (1979), and Wiegert (1988).

There are also attempts to link systems ecology more closely to human or human integrated systems (Campbell, 2017, 2018). For example, Evans et al., 2013 emphasize both greater predictability for ecological models and links with human parameters. There is a close dependency of human societies on the state of the surrounding and linked ecosystems, which may be undergoing rapid change, usually in reaction to anthropogenically derived modifications. In consequence, research must develop the paradigm of predictive systems ecology, to uncover, analyze, and predict the properties and behavior of ecological systems. Prediction is necessary for the enhanced utility of ecology in biological and social studies, but prediction is difficult, due to the inherent complexity of ecological systems, and their nonlinear, and variable outcomes. Additionally, "future conditions often lie outside the envelope of parameters used to develop models" (Evans et al., 2013). An example is the UK National Ecosystem Assessment (NEA), which despite being extremely rigorous, with in-depth information for the UK. It was acknowledged by its authors to have few relevant assessments (Evans et al. (2013). Evans et al. (2013) nevertheless state that "the societal imperative to predict the impacts of environmental change on ecosystems (usually, but not exclusively, the result of anthropogenic pressures) should add impetus to this endeavour."

Accurate projections generated by effective models are necessary for ecological predictions, with process-based models required for predicting biological responses to environmental change (Evans, 2012; Grafen & Hails, 2002; Kitano, 2002; Moss et al., 2010; Nicholson et al., 2009; Purves et al., 2008; Purves & Pacala, 2008; Schmidt, 2007; Strigul et al., 2008; Watt, 1966). Evans et al. (2013) argue that point out that the process-based models of ecosystems would require adequate information from "lower levels of organization (populations, individuals, genes and the abiotic environment) to allow accurate and realistic ecosystem behaviour to emerge" and the emergent properties of these models would be "predictions of ecosystem behaviour (the way in which the properties in an ecosystem change over time or in response to perturbation) and/or ecosystem services (which are products of ecosystems and often are valuable to society)" (see also Nicholson et al., 2009; Evans et al., 2012).

Points to be considered for systems approaches include uncertainty (due to stochastic effects, and greater instability than biological or climate based models) (Wynne, 1992; Forster & Sober, 1994; Beven & Freer, 2001; Regan et al., 2002; Van Asselt & Rotmans, 2002; Brentnall et al., 2003; Gneitling & Raftery, 2005; Parry & Bithell, 2011; Polasky et al., 2011; Spiegelhalter & Riesch, 2011; Kujala et al., 2013), complexity (due the enormous amount of data involved in modeling, and the complicated requirements of ecosystem representation, requiring representation of components, and effective mathematics) (Eisinger & Thulke, 2008; Weisberg, 2007; Wimsatt, 2006) and constraining models with data (requiring constraining models, hypothesis testing multiple generation of multilevel projections) for predictive systems ecology (Brown et al., 2005; Grimm & Railsback, 2012; Rounsevell et al., 2012; Evans et al., 2013).

Currently, there are several types of models in systems ecology type studies (Evans et al., 2013). These include terrestrial vegetation models (Kurz et al., 2008; Medvigy et al., 2009; Medvigy & Moorcroft, 2012; Moorcroft, 2006; Moorcroft et al., 2001; Pacala et al., 1996; Purves et al., 2008; Purves & Pacala, 2008; Seidl et al., 2005; Strigul et al., 2008), ocean ecosystem models, including models of biogeochemical cycles (Fach et al., 2002, 2006; Murphy et al., 2004; Buitenhuis et al., 2006; Follows et al., 2007; Christensen et al., 2009; Allen & Fulton, 2010; Lehodey et al., 2010; Christensen, 2011, 2013; Fulton et al., 2011; Moloney et al., 2011; Dueri et al., 2012; Mork et al., 2012; Murphy & Hofmann, 2012; Murphy et al., 2012; Maury & Poggiale, 2013; Yool et al., 2013), global ecological models (Purves et al., 2013), and humans and ecosystems and human–ecological systems models (Bithell & Brasington, 2009; Brown et al., 2012; Carpenter et al., 2009; Dunlop et al., 2009; Enberg et al., 2009; Fulton, 2010; Fulton et al., 2011; Griffith et al., 2012; Jørgensen et al., 2007; Milner-Gulland, 2012; Ostrom, 2009; Pahl-Wostl et al., 2013; Rounsevell et al., 2012).

Evans et al. (2013) also advance important challenges for systems ecology, including scale, as biological processes have different scales and variably interact with variable physical and chemical processes, requiring scaling rules from local to global, and in biological systems from genes to ecosystems (Murphy et al.1988; Murphy et al., 2007; Nicholson et al., 2009; Murphy & Hofmann, 2012; Murphy et al., 2012), hence "developing models that resolve the appropriate physical, chemical, biological and social processes at different scales presents a major challenge, but scaling from individual behaviours to changes in population sizes at a regional scale is being attempted" (Evans et al. (2013, see also De Young et al., 2004; Robson

& Webster, 2006; Travers et al., 2007; Dutkiewicz et al., 2009; Allen & Fulton, 2010; Bocedi et al., 2012; Purves et al., 2013).

Another challenge is evolution, as evolution is fundamental to biological systems, affecting the development and relevance of models (Pacala et al., 1996; Coulson et al. 2006a, b; Kokko & Lopez-Sepulcre, 2007; Purves et al., 2008; Matthews et al., 2011). Ecological and evolutionary change are argued to be intertwined, as population dynamics relate to life histories of "individuals within the population, while the strength of selection is modified by properties of the population" (Benton, 2012; Cameron et al., 2013; Kokko & Lopez-Sepulcre, 2007; Matthews et al., 2011), this requiring more attention to evolution systems modeling (Evans et al., 2013, see also Grafen, 1999; Beckerman et al., 2006; Whitham et al., 2006; Grafen, 2007; Schoener, 2011; Cameron et al., 2013; Diaz et al., 2013; Moustakas & Evans, 2013).

Smith et al. (2012: 311) point out the "great potential" for systems approaches in applications to predictive ecology, these systems approaches range from "Bayesian calibration techniques at plot scale, through data assimilation methods at regional to continental scales, to multi-disciplinary numerical model applications at country to global scales." Ecology, as a science is defined as "almost by definition" a systems science, and "ecologists, and other scientists involved in modeling ecological or ecosystem interactions, have long considered themselves to be systems scientists." Systems ecology has been contrasted with biological sciences that emphasize reductionist specialization, and it is argued that both approaches are needed for the full understanding of ecological and earth dynamics (Smith et al., 2012).

The Lithosphere or Geosphere

The lithosphere is the source of most of the chemical elements and compounds involved in biogeochemical cycling, including those also found in the other spheres (Campbell, 2017). The lithosphere is generally considered as the same as the geosphere (but the geosphere is sometimes referred to as including the lithosphere, hydrosphere, and the atmosphere), that is the solid, outer surface of the Earth including the crust and the upper mantle, above the asthenosphere (also part of the upper mantle) (United States Geological Survey, 2022). The lithosphere includes the continental lithosphere and the denser oceanic lithosphere (Artemieva, 2011). The United States Geological Survey (2022) describes the lithosphere as about 100 km thick, but "its thickness is age dependent (older lithosphere is thicker). The lithosphere below the crust is brittle enough at some locations to produce earthquakes by faulting, such as within a subducted oceanic plate." Encyclopaedia Britannica (2022) writes that the lithosphere is broken into "about a dozen separate, rigid blocks, or plates. Slow convection currents deep within the mantle, generated by radioactive heating of the interior, are believed to cause the lateral movements of the plates (and the continents that rest on top of them) at a rate of several inches per year." The lithosphere contrasts with the asthenosphere in that most of the composing materials (apart from magma) are brittle and nonviscous, while the latter is mostly composed of viscous materials; hence, the lithosphere–asthenosphere boundary (LAB) marks the difference in ductility between these two layers (here, ductility refers to the ability of material to deform under force (Encyclopaedia Britannica, 2022).

Plate tectonics, the movement of the huge plates (including the North American, Caribbean, South American, Scotia, Antarctic, Eurasian, Arabian, African, Indian, Philippine, Australian, Pacific, Juan de Fuca, Cocos, and Nazca) that comprise the crust, is a major characteristic of the lithosphere (Artemieva, 2011). The major dynamics of the lithosphere include tectonic activity at the boundaries of the plates, including earthquakes, volcanoes, and mountain-building. Rychert et al. (2020) describe plate tectonic theory as the framework that Earth processes from the formation of the continents to natural smaller scale disasters including volcanoes, earthquakes, and tsunamis. They argue that accuracy and relatedness characterize studies of Earth systems; for example, climate change estimates over geologic timescales must depend on the accurate plate tectonic reconstructions analyze the paleo-oceans. Rychert et al. (2020) also note that plate tectonic theory, developed during the late twentieth century, is basic to the study of the Earth's evolution, and "explains observations of magnetic lineations on the seafloor, linear volcanic island chains, large transform fault systems, and deep earthquakes near deep sea trenches" (Rychert et al., 2020). The moving tectonic plates are composed of "chemically distinct crust and some amount of rigid mantle, which move over a weaker mantle beneath," and are "the conductively cooling part of the mantle convection system" (Rychert et al., 2020). Rychert et al. (2020) contend that the lithosphere is gradually thickening with age, as evidenced by the evidence of changes in heat flow, seafloor bathymetry, seismic imaging, and magnetotelluric (MT), suggesting "that temperature is an important factor in determining lithospheric thickness" despite other complex issues.

However, Rychert et al. (2020) also argue that a purely thermal model cannot explain certain observations, such as discontinuities from teleseismic scattered waves, active source reflections, anomalies from surface and body wave tomography and MT imaging. The authors suggest the presence of melt defines the lithosphere–asthenosphere boundary (LAB), and that "asthenospheric melt interpretations include a variety of forms: in small or large melt triangles beneath spreading ridges, in channels, in layers, along a permeability boundary leading to the ridge, at a depth of neutral buoyancy, punctuated, or pervasively over broad areas and either sharply or gradually falling off with depth." The authors make the main points that temperature defines the LAB depth to first order, but the LAB is laterally variable in depth and possibly compositional form, with clear discontinuities overlying strong seismic and magnetotelluric differences, which suggest there is in some locations a melt-defined LAB, and this is determined by the mantle character changes, including those of melting and migration, this having implications for the evolution of the Earth (Rychert et al., 2020).

Mukherjee et al. (2022), writing about the lithosphere–asthenosphere boundary (LAB), argue that this "is a fundamental element of the plate tectonic hypothesis that accommodates the differential motion of rigid lithosphere over the weaker asthenosphere" and take the example of the lithospheric structure beneath the Eastern Dharwar Craton (EDC) of the Indian Shield, with their findings indicating or suggesting a lithospheric thickness of 98–118 and 94–118 km, respectively, and with a sharp transition across the LAB. There is also an indication of a thick thermal lithosphere of approximately 200 km for the normal mantle solidus with cold geotherm. Mukherjee et al. (2022) conclude that "to reconcile our observations, we invoke partial melts or melts enriched in volatiles, which significantly lowers the viscosity of mantle rocks inducing a zone of weakness between the rigid lithosphere (~125 km) and the convective asthenosphere. Further, we favour the view that the thick lithosphere beneath the Indian plate has been thinned by a plume during the Gondwanaland breakup at ~ 130 Ma."

The Hydrosphere

The hydrosphere is a huge and complex geochemical entity, and is vital for biogeochemical cycling as the waters of the oceans and fresh water bodies carry many chemicals that also enter the atmosphere and move between the solid earth and water (Campbell, 2020). Speight (2020) notes that "the hydrosphere (often referred to as the aquasphere) is generally defined by geochemists as the vapor, liquid, and solid water present at and near the land surface, and its dissolved constituents." An inclusive definition is that the hydrosphere comprises the "discontinuous layer of water at or near Earth's surface" including all the free liquids, frozen surface waters, groundwater in rocks and soils, and atmospheric water vapour," thus including about 1.4 billion cubic km (326 million cubic miles) of water in the oceans, lakes, streams, glaciers, and groundwaters. Britannica (2021a, b) notes that a central issue to the definition of the hydrosphere is the concept of the water cycle (or hydrologic cycle), which us defined as the complex, networked group of reservoirs containing water, the processes by which water is transferred from one reservoir to another, or transformed from one form to another), and the rates of transfer linked with such processes. These water transfer networks move through the whole hydrosphere, extending to the altitude of about 15 km in the atmosphere and to the depth of 5 km in the crust." Tables 3.1 and 3.2 show the percentages of water in some of the more important reservoirs on the Earth, and Tables 3.3, 3.4, 3.5, 3.6, 3.7, 3.8 and 3.9 show the water bodies of the Earth.

Origin of Water on Earth

The origin of water on Earth has been debated, with the acknowledgment that water is essential to life and the biogeochemical cycles (Campbell, 2020). Sarafian et al. (2014: 623) note that "determining the origin of water and the timing of its accretion

Table 3.1 Water on the Earth's surface

Reservoir	Volume (cubic kilometers)	Percent of total
Oceans	1,338,000,000	96.5
Ice caps, glaciers, permanent snow	24,064,000	1.74
Groundwater (fresh and saline)	23,400,000	1.69
Ground ice, permafrost	300,000	0.22
Lakes (fresh and saline)	176,400	0.013
Soil moisture	16,500	0.001
Atmosphere (water vapor)	12,900	0.001
Rivers and swamps	13,590	0.0010
Biota	1120	0.0001
Total	1,385,984,510	100.0

Source: Adapted from Shiklomanov (1993). The differences between the totals for the table and that below may be due to the inclusion or exclusion of some ice

Table 3.2 Volume of the oceans

Ocean	Volume (cubic kilometers)	Percent of total
Pacific	669,880,000	50.1
Atlantic	310,410,900	23.3
Indian	264,000,000	19.8
Southern	71,800,000	5.4
Arctic	18,750,000	1.4
Total	1,335,000,900	100

Source: Adapted from Eakins and Sharman (2010)

within the inner solar system is important for understanding the dynamics of planet formation. The timing of water accretion to the inner solar system also has implications for how and when life emerged on Earth." Stierwalt (2019) points out that "understanding how water arrived on Earth is a key part of understanding how and when life evolved here as well. But we don't even know how it where it came from. Scientists are still actively researching how our planet got to be so wet in the first place." Alexander, 2021 argues that "the origin of Earth's water is an open question and quite controversial. It really is part of a larger question about how Earth got its very volatile elements that include not just the hydrogen in water, but also carbon, nitrogen, and the noble gases (helium, neon, argon, krypton, and xenon) amongst others."

Alexander (2021) further argues that currently majority opinion on the source of the Earth's waters is of an origin based on water-rich objects (planetesimals) that composed part of its building blocks, and that such water-rich planetesimals were either comets or asteroids. The author cautions that "from what we know about comets, they are not a good match to either the volatile element abundances or isotopic compositions of the Earth. It may also have been difficult to deliver enough comets to the Earth" (Alexander (2021). Therefore, at present, "asteroids up to a few hundred kilometers across seem the most likely sources of most of Earth's

Table 3.3 Discharge of the Earth's largest rivers (above 5000 cubic meters per second – m^3/s)

River	Continent	Discharge (m^3/s)	Length (km)	Drainage area (km^2)
Amazon	South America	230,000	6500	6,112,000
Congo	Africa	41,200	4370	4,014,500
Ganges	Asia	38,129	2704	1,731,334
Orinoco	South America	37,740	2250	989,000
Yangtze	Asia	30,146	6300	1,808,500
Río de la Plata	South America	27,225	290	3,182,064
Yenisey	Asia	20,000	3487	2,554,482
Mississippi	North America	18,434	3766	3,248,000
Lena	Asia	17,067	4294	2,490,000
Saint Lawrence	North America	16,800	3058	1,344,200
Mekong	Asia	16,000	4350	795,000
Irrawaddy	Asia	15,112	2210	411,000
Ob	Asia	12,692	3697	2,990,000
Amur	Asia	11,330	2824	1,855,000
Mackenzie	North America	10,338	1738	1,790,000
Pearl (Xi)	Asia	9500	2400	437,000
Volga	Europe	8256	3531	1,380,000
Columbia	North America	7504	2044	668,217
Fly	Oceania	7500	1060	76,000
Magdalena	South America	7200	1540	257,438
Sepik	Oceania	7000	1146	80,321
Salween	Asia	6600	2815	320,000
Danube	Europe	6546	2860	801,463
Kapuas	Asia	6500	1143	98,749
Yukon	North America	6428	3187	854,696
Essequibo	South America	5650	1014	150,769
Niger	Africa	5589	4167	2,090,000
Indus	Asia	5533	3180	1,081,718
Mamberamo	Oceania	5500	1175	78,992
Barito	Asia	5497	1090	81,675
Detroit	North America	5300	44	1800
Mahakam	Asia	5000	980	77,095

Sources: Bossche and Bernacsek (1990), Kammerer (1990), Povisen (1993), International Union for the Conservation of Nature (1995), Guerrero et al. (1997), Morales (1999), Brabets et al. (2000), Loy et al. (2001), Penn (2001), Firmansyah (n.d.), Water Resources Management Branch of the Office of Environment (2002), Mackenzie River Basin Board (2003), Revenga et al. (2003), Kumar et al. (2005), Kuusisto (2005), Mekong River Commission (2005), Nilsson et al. (2005), Kimbrough et al. (2006), Gupta (2007), Joesron (2008), Lehner et al. (2008), Likens (2010), Natural Resources Canada (2010), Fei (2011), Ra (2011), Gâştescu and Ţuchiu (2012), Dahl et al. (2013), Britannica (2013), Panagiotopoulos et al. (2014), Córdova and González (2015), Yang et al. (2015), Muranov et al. (2016), Huang et al. (2017), Radhika et al. (2017), Dongue et al. (2018), Holmes et al. (2018), Kalugin (2018), Micklin and Kuzin (2019), Towner et al. (2019), Baronas et al. (2020), Brunnschweiler et al. (2020), Osadchiev et al. (2020), Rivers Network (2020), Wang et al. (2020), Dobler et al. (2021), Prabhu (2021), Stadnyk et al. (2021), Denevan et al. (2022), Muranov and Greer (2022), Penčev and Pinka (2022) and tl-myubi-tv (2022)

Table 3.4 Discharge of the Earth's largest rivers (above 2500 cubic meters per second – m^3/s)

River	Continent	Discharge (m^3/s)	Length (km)	Drainage area (km^2)
Ogooué	Africa	4935	1200	223,856
Pechora	Europe	4533	1809	289,532
Red	Asia	4300	1149	143,600
Zambezi	Africa	4134	2574	1,331,000
Kolyma	Asia	4130	2129	644,000
Digul	Oceania	4000	853	45,600
Godavari	Asia	3505	1465	312,812
Fraser	North America	3475	1375	220,000
Kaladan	Asia	3468	450	40,000
Northern Dvina	Europe	3332	744	357,052
Khatanga	Asia	3320	1636	364,000
Kikori	Oceania	3274	405	23,309
São Francisco	South America	2943	2830	641,000
Rhine	Rhine	2900	1233	198,735
Nile	Africa	2830	6653	3,349,000
Koksoak	North America	2800	137	133,400
Yellow	Asia	2571	5464	752,546

Sources: Energy, Mines and Resources Canada (1985), Beilfuss and Dos Santos (2001), Coleman and Huh (2004), British Columbia Ministry of the Environment (2008), Price (2010), Rivers Network (2020), R-Arctic Net (2001), Pilarczyk (2003), Nilsson et al. (2005), World Bank (2010), Britannica (2013), Fekete et al. (2014), United Nations Environment Program (2016), Dongue et al. (2018), Shibru (2019), Firmansyah (n.d.), Osadchiev et al. (2020), Kumar (2020), Canadian Heritage Rivers System (2022), Popov and Greer (2022), Zhu et al. (2022)

Table 3.5 Largest lakes of the world (ranked by volume, above 1000 km^3)

Name	Continent	Surface area (km^2)	Water volume (km^3)
Caspian Sea	Europe/Asia	371,000	78,200
Baikal	Asia	31,722	23,600
Tanganyika	Africa	32,900	18,900
Superior	North America	82,100	12,100
Malawi	Africa	29,600	8400
Michigan	North America	57,800	4920
Huron	North America	59,600	3540
Victoria	Africa	69,484	2750
Great Bear	Canada	31,153	2236
Issyk-Kul	Asia	6236	1730
Ontario	North America	18,960	1640
Great Slave	North America	27,200	1580

Sources: Kodayev (1973), Johnson (1975), Herdendorf (1982), Serruya and Pollinger (1983), Savvaitova and Petr (1992), Dumont (1998), Hofer et al. (2002), INTAS Project 99-1669 (2002), Hebert (2007), Yohannes (2008), Britannica (2020, 2021a, b), Leontiev et al. (2021), Environmental Protection Agency (2022), World Lake Database (2022)

Table 3.6 Largest lakes of the world (ranked by volume, below 1000 km³)

Name	Continent	Surface area (km^2)	Water volume (km^3)
Ladoga	Europe	17,680	908
Titicaca	South America	8372	893
General Carrera	South America	1764	705
Van	Asia	3755	607
Kivu	Africa	2700	569
Khövsgöl or Hovsgol	Asia	2770	381
Onega	Europe	9890	280
Winnipeg	North America	23,750	284
Nipigon	North America	4510	248
Toba	Asia	1100	240
Nueltin	North America	2279	228
Argentino	South America	1466	220
Athabasca	North America	7900	204
Turkana	Africa	6750	204
Vänern	Europe	5648	153
Tahoe	North America	496	151
Mistassini	North America	2115	150
Dead Sea	Asia	810	147
Albert	Africa	5300	132
Viedma	South America	1193	119
Erie	North America	25,700	116
Iliamna	North America	2590	115
Nettilling	North America	5542	114
Baker	North America	1887	113
Nicaragua	North America	8150	108
Balkhash	Asia	18200	106
Qinghai	Asia	4583	85

Sources: Degens et al. (1984), Van der Leeden et al. (1990), Murdie et al. (1999), Mathisen et al. (2002), Natural Resource Canada (2002), Coskun and Musaoğlu (2004), Evans (2021), The Nature Conservancy/World Wildlife Fund (2019), Environmental Protection Agency (2022), International Lake Environment Committee (2022), World Atlas (2022), World Lake Database (2022)

water, specifically the types of asteroids that dominate the outer asteroid belt between Mars and Jupiter" (Alexander, 2021).

The next stage of water evolution on Earth concerns the development of the oceans, rainfall and fresh surface water bodies including lakes, rivers, and surface runoff, which may not have existed millions of years ago (Campbell, 2020). Cavosie (2016) contends that "for many it would be difficult to envision an Earth without its blue blanket of oceans. However, this is precisely what the early stages of our planet were like. An ocean-free Earth existed, perhaps for several hundred million years as a consequence of extremely high surface temperatures following planetary accretion." The National Ocean Service (2021) adds that a "primitive ocean" formed over "vast periods of time" with water existing as a gas until the Earth's temperature

Table 3.7 Types of lakes

Lake type	Formation	Examples
Tectonic	Lateral, vertical movements of crust, e.g., faulting, folding	Caspian, Baikal, Dead Sea
Volcanic	Volcanism: craters, lava flow restriction	Toba
Glacial	Glaciated movements, including erosion of crust	Superior, Michigan, Huron
Fluvial	Flow of water from rivers	Great Slave
Solution	In basin formed by surface dissolution of rock.	Ohrid
Landslide	River blockage of a river valley slippage materials	Quake Lake
Aeolian	Water trapped by wind-blown sand deposits/ dunes.	Moses
Shoreline	Blockage of estuaries or beach ridge development.	Shoreline
Organic/Peat	Formed in peatland areas, for example, in bogs	Peat
Meteorite	Impacts of extraterrestrial objects on crust	Pingualuit crater
Anthropogenic	Dams and excavations by people	Volta, Ethiopian Blue Nile

Sources: Hutchinson (1957), Håkanson and Jansson (1983, 2002), Cohen (2003), Håkanson (2012), Mitra et al. (2022)

Table 3.8 Types of lakes, by temperature (according to Hutchinson, 1957)

Temperature	Description
Amictic	Perennially ice-covered.
Cold monomictic	Water temperatures below 4 °C, and some circulation in summer at or below 4 °C.
Dimictic	Biannual circulation in the spring and fall, stratified in summer, inverse stratification in winter.
Warm monomictic	Circulate freely once a year in the winter at or above 4 °C and are stably stratified for the remainder of the year; not ice-covered.
Oligomictic	Thermally stratified much of the year but cooling sufficiently for rare circulation periods at irregular intervals; not ice-covered.
Polymictic	Common annual mixing per year, with no ice cover. Cold polymictic is circulation around 4 °C, in hot regions, high wind, low humidity, minimal seasonal air temperature, with diurnal heating in cold, highlands and nocturnal losses. Warm polymictic, usually tropical lakes include temperatures above 4 °C, with little variation, including circulation periods, short heating periods, rapid cooling periods, and disruption of stratification from convectional circulation and wind.

Sources: Ruttner (1963) and Wetzel (2001)

cooled to below 100 °C, about 3.8 billion years BP, when water condensed into rain creating the oceans. It is arguable that "most scientists agree that the atmosphere and the ocean accumulated gradually over millions and millions of years with the continual 'degassing' of the Earth's interior…According to this theory, the ocean formed from the escape of water vapor and other gases from the molten rocks of the

Table 3.9 Types of lakes, by ability to support life (according to Fuller & Taricska, 2012)

Classification	Total phosphorus (μg/L)[1]	Chlorophyll-a μg/L)[1]	Secchi transparency (meters)
Oligotrophic	<10	<2.2	>4.6
Mesotrophic	10–20	2.2–6	2.3–4.6
Eutrophic	>20	>6	< 2.3

Sources: Fuller and Taricska (2012)

Earth to the atmosphere surrounding the cooling planet... The forces of gravity prevented the water from leaving the planet" (National Ocean Service, 2021).

Cavosie (2016) argues that "the timing of this transition – which has implications for when surface conditions necessary for the development were established – is poorly known. Part of the uncertainty of the timing of this transition is due to the fragmentary nature of the rock record for the first ~500 million years of Earth history." There is a rock record and a mineral record. For the rock record, there is the so-called Isua BIF, which refers to "the oldest direct evidence for the presence of surface waters" which are sedimentary rocks about ~3800 million years old called banded iron formation (BIF), exposed in southwest Greenland at a location called Isua (Cavosie, 2016). These formations are not the oldest recorded, preceded by 4-billion-year-old metamorphosed and deformed granitoids from northwestern Canada termed the Acasta gneiss, which may have crystallized 4030 million years BP, but do not record evidence of Earth surface conditions during the rock formation. Cavosie (2016) adds that "the very existence of the Isua BIF requires the presence of stable surface water, at least locally for the chemical deposition of the sedimentary components at ca. 3800 MA." For the mineral record, there are detrital zircons, as the "oldest known Earth materials are actually not rocks. Sand grains comprised of the mineral zircon (ZrSiO4) have been discovered that are almost 400 million years older than the oldest rocks in the rock record" (Cavosie, 2016). It is pointed out that "Zircon is a very useful mineral that is mechanically resistant to erosion, chemically resistant to fluids, and can be 'dated' with the U-Pb method owing to the ubiquitous presence of trace amounts of radioactive U and Th that are incorporated in most zircons at the time of crystallization" (Cavosie, 2016).

Although these ancient zircons give evidence of igneous rock at about 4400 million years BP, the evidence of ocean development is yet more complex (National Ocean Service, 2021). For example, Deming (2002, 137) explains that "the key idea necessary for the development of a unified theory that explains the origin of the ocean and continents is the recognition that the timing of the growth of continental crust is an earmark for the presence of abundant water on the surface of the Earth and its entry into the mantle by subduction." Cavosie (2016) uses the evidence of oxygen isotopes in geologic materials to assess ocean formation. The main points are that temperatures created by rock and mineral changes affect oxygen isotopes, the oxygen isotope ratio (the ratio of 18O-to-16O or δ18O) of the mantle has little variation due to the high temperatures (e.g., about ~5.5%), and the oxygen isotope ratio of crustal surface materials by contrast has higher variation due to the variable, lower temperatures of oceans, etc. (e.g., from 5–6 to 30%). This indicates lower δ18O values for mantle materials, and higher δ18O values for crustal surface materials (Cavosie, 2016).

Cavosie (2016) argues that due to these relationships, δ18O(zircon) reliably indicates whether the parental δ18O(magma) was "mantle-equilibrated" or whether the parental δ18O(magma) was "crustal" (a component of its oxygen budget was derived from crustal materials, the latter resulting in higher δ18O values in the rock and minerals). Therefore, "to address our question of when oceans first formed on Earth we can investigate the δ18O values of the ancient zircons to see if they record "mantle-equilibrated" values, meaning no evidence of a crustal component is detectable in the oxygen or if they record "crustal" δ18O values meaning that the early magmas assimilated crustal materials that were affected by low-temperature interaction with water prior to melting" (Cavosie, 2016). With the dating of the igneous zircon grains, a sequence emerges: (1) the oldest zircons recorded lower, mantle-equilibrated values for the oxygen isotope ratios (δ18O) (about ~5.3 to 5.4‰) dated from 4400 to ~4325 million years BP; (2) some more recent zircons, dated from 4325 to ~4200 million years BP, have δ18O values slightly higher (up to 6.3‰) which are in the upper ranges higher than values produced in a mantle melt, with some overlaps; and (3) still more recent igneous zircons, dated after 4200 million years BP, have higher isotope ratios up to 7.3‰, probably excluding a mantle source. These inferences or findings may imply that the igneous protolith of the zircon assimilating or remelting crustal materials changed lower temperature processes on the Earth's crust. This may indicate that surface waters existed around 4200 million years BP (Cavosie, 2016).

The supporting evidence indicates that relevant information is derived from two forms: "preserved sediments up to ~3800 million years old" and "oxygen isotope studies of detrital zircons up to 4400 million years old" (Cavosie, 2016). Two hypotheses are posed: (1) the formation of the oceans occurred around 3800 million years BP, with evidence from the Isua banded iron formation (BIF, the Isua Greenstone Belt is an Archean greenstone belt in Greenland, dated to 3.7 and 3.8 billion year BP), although there is no final evidence for the Isua BIF being deposited in the Earth's first ocean; (2) there was an earlier ocean formation, at least 4200 million years BP, with the example being the Jack Hills detrital zircons, which show a period prior to low-temperature weathering recorded in magmas before around 4200 million years BP, and change in magmatic oxygen with higher 18O (zircon), after 4200 million years BP; therefore, the "detrital zircons actually record a boundary condition that marks when surface weathering, and hence the presence of oceans, occurred" (Cavosie, 2016; see also Cavosie et al., 2004, 2005; Nutman, 2006; Valley et al., 2005; Valley, 2006).

The oceans further developed due to crustal tectonic movements. As noted by the National Ocean Service (2022), 250 million years ago "there was only one continent, called Pangaea, and one ocean, called Panthalassa. As Earth's mantle heated and cooled over many millennia, the outer crust broke up and commenced the plate motion that continues today. The huge continent eventually broke apart, creating new and ever-changing land masses and oceans." The oceans also became the fundamental building block of the hydrological cycle. Webb (2019) postulates that "as the early Earth cooled, the water vapor in the atmosphere condensed and fell as rain. By about 4 billion years ago, the first permanent accumulations of water were present on Earth, forming the oceans and other bodies of water. Water moves between these different reservoirs through the hydrological cycle. Water is evaporated from the oceans, lakes, streams, the surface of the land, and plants (transpiration) by solar energy."

The Cryosphere

The cryosphere, composed of frozen water, may also be included in the hydrosphere (National Ocean Service, 2022), and also act as agents of biogeochemical cycling, much as other water forms, though perhaps on a slower scale as ice moves more slowly than water. The World Meteorological Organization (WMO) (2022) defines the cryosphere as "the part of the Earth's climate system that includes solid precipitation, snow, sea ice, lake and river ice, icebergs, glaciers and ice caps, ice sheets, ice shelves, permafrost, and seasonally frozen ground. Kumar (2022: 47) notes that the "cryosphere comprises all regions on and beneath the surface of the Earth and ocean where water is in solid form, including sea ice, lake ice, river ice, snow cover, glaciers and ice sheets, and frozen ground, including permafrost." Henshaw et al. (2000: 114) point out that the "cryosphere – the portion of the Earth's water frozen in ice caps, glaciers, and sea ice – contains the largest reserves of freshwater on Earth." It is argued that most of the ice-covered areas are remote; hence, the estimates of water storage in such areas are at best imprecise. Estimates (based on modern remote sensing technology) measure the ice storage at 3 × 107 km^3, with the Antarctic ice sheet comprising about 90%, with small inflows and outflows, contributing to the cryosphere as a static component of the hydrosphere (Henshaw et al., 2000: 114).

The cryosphere interacts strongly other aspects of the hydrosphere and may be differentiated from these other systems. Kumar (2022) argues that "the role of the cryosphere in the climate system varies from that of the liquid hydrosphere owing to the fact that ice is comparatively immobile and because it reflects almost all of the solar energy that falls on it." Henshaw et al. (2000: 114) note that "ice-covered portions of the planet do contribute significantly to albedo and can affect atmospheric circulation, making them important links in climatic feedbacks." Kumar (2022) adds that "the cryosphere draws its importance to the climate system from its high reflectivity, also known as albedo, for solar radiation, its low thermal conductivity, its large thermal inertia and, especially, its critical role in driving deep ocean water circulation; and for the reason that the ice sheets store a large amount of water, variations in their volume are a potential source of sea level variations." The National Ocean Service (2022) argues that "the components of the cryosphere play an important role in the Earth's climate. Snow and ice reflect heat from the sun, helping to regulate our planet's temperature. Because polar regions are some of the most sensitive to climate shifts, the cryosphere may be one of the first places where scientists are able to identify global changes in climate." Kumar (2022) adds that "the role of the cryosphere in the climate system varies from that of the liquid hydrosphere owing to the fact that ice is comparatively immobile and because it reflects almost all of the solar energy that falls on it."

The World Meteorological Organization (2022) examines the main components of the cryosphere in turn. Snow (formed from ice crystals, and comprising soft precipitates to large, hard hail), which falls as precipitation, "has the largest geographic extent of the cryosphere components. It covers nearly 50 million km^2 of the Northern Hemisphere in winter," with the high albedo reducing solar energy absorption and promoting lower surface temperatures, while also insulating the land from energy losses in winter and influencing the land surface energy budget, and atmospheric circulation. It is also noted that "the high sensitivity of snow to changes in temperature and precipitation makes it a primary indicator of climate change and implicates it in climate change hypotheses concerning the redistribution and acceleration of the water cycle" (World Meteorological Organization, 2022).

Sea ice is another component, which is formed on the sea surface at the freezing point of seawater (lower than that of fresh water, ~1.8 °C vs. 0 °C), and does not include icebergs, glaciers, ice sheets and ice shelves, which are formed on land from fresh water. Sea ice varies in thickness, which may be positively correlated with age, from new ice (<5 cm thick) to thick first-year ice (120~180 cm thick). World Meteorological Organization (2022) notes that "sea ice limits exchanges of heat and moisture between the ocean and the atmosphere, acting as a "thermal blanket." Because of the large difference in reflectivity between ice (bright) and ocean (dark), a reduction in the extent of sea ice will result in more heat being absorbed by the ocean instead of reflected back into the atmosphere. This is likely to amplify the effect of warming in high latitudes, making the extent of sea ice a potential early and sensitive indicator of climate change." In terms of ecology, "the seasonal sea-ice zone is highly productive biologically, which makes sea ice a key component in the carbon cycle" and is an "important habitat for wildlife, and the shear zone between fast and drifting ice frequently results in open water in these regions, to the benefit of the biota" (World Meteorological Organization, 2022).

There is also lake and river ice, which is important as lakes cover 15 to 40% of the surface of temperate regions, making such lacustrine and riverine ice an important aspect of the cryosphere, and a factor for temporal flooding. In the Northern Hemisphere ice develops on 29% of river length and seasonal ice occurs on 58% (World Meteorological Organization, 2022). The World Meteorological Organization (2022) writes that "seasonal ice cover grows and decays in response to heat transfer through the ice surface layer that is affected by net radiation, surface albedo, on-ice snow depth and density, air temperature, wind speed and water heat flux. Although freshwater ice formation and decay processes are influenced by numerous physical and climatological factors, it has been determined that the timing of break-up and freeze-up correlates best with air temperature during the preceding weeks to months of the event (World Meteorological Organization, 2022).

Permafrost, subsurface ice, and other earth materials that remain at a temperature below 0 °C for over two years, exists in polar and subpolar regions, covering about 23 million km^2, with between 52 and 74% of the land area underlain by permafrost. The permafrost upper layer is defined as the active layer, where there is seasonal thawing. Factors such as climate, soil type, plants, snow, ground gradient, and water bodies, affect permafrost conditions. Pingos (mound-shaped ice under the land surface), and ice wedge polygons (smaller ice formations that wedge into soil and grow with freezing and thawing) which may grow into pingos. Permafrost also serves as store of carbon dioxide, methane, and some other greenhouse gasses. The World Meteorological Organization (2022) points out that "the current estimate on the amount of carbon stored in permafrost is almost twice what is currently in the Earth's atmosphere. Yedoma, a type of permafrost that has existed since the Pleistocene period and is found mostly around Russia and Siberia, contains huge amounts of carbon that could be released into the atmosphere if it were to thaw."

Glaciers are another component of the cryosphere, formed from snow compressed into ice, in areas of perennial snow cover and below-freezing temperatures, such as higher latitudes and altitudes (e.g., for the latter, the Himalayas and Andes). Meier (2022) defines a glacier as "any large mass of perennial ice that originates on land by the recrystallization of snow or other forms of solid precipitation and that shows evidence of past or present flow." The National Snow and Ice Data Center (2022) defines a glacier as "a large accumulation of ice and snow that slowly moves over land. At higher elevations, more snow typically falls than melts, adding to its mass. Eventually, the surplus of built-up ice begins to flow downhill. At lower elevations, there is usually a higher rate of melt or icebergs break off that removes ice mass." Importantly, "when mass gain (snow accumulation) in the glacier's accumulation area outpaces mass loss (ablation) in the glacier's ablation area, the glacier is 'surging' forward; vice versa, and the glacier is 'retreating'" (World Meteorological Organization, 2022). The World Meteorological Organization (2022) further notes that although glaciers cover about 10% of the Earth's surface, it vary greatly in size from hundreds of meters to hundreds of kilometers in length, and "average glacier thickness can be estimated at about half of the surface width of the glacier."

Glaciers may be variably classified. For example, Meier (2022) classifies glaciers into three main groups: (1) "glaciers that extend in continuous sheets, moving outward in all directions, are called ice sheets if they are the size of Antarctica or Greenland and ice caps if they are smaller"; (2) "glaciers confined within a path that directs the ice movement are called mountain glaciers"; and (3) "glaciers that spread out on level ground or on the ocean at the foot of glaciated regions are called piedmont glaciers or ice shelves, respectively." Meier (2022) argues that "glaciers in the third group are not independent and are treated here in terms of their sources: ice shelves with ice sheets, piedmont glaciers with mountain glaciers. A complex of mountain glaciers burying much of a mountain range is called an ice field." Another classification is given by the National Snow and Ice Data Center (2022), in which there are two categories: alpine glaciers and ice sheets (and ice caps and ice fields, comparable to small ice sheets). The alpine glaciers are ice rivers that flow slowly, under the force of gravity from mountains through valleys. Ice sheets are large, such as those on Antarctica and Greenland and Antarctica, while ice caps are like that Iceland, and move over land unrestricted.

The World Meteorological Organization (2022) distinguishes between the "two most well-recognized types of glaciers," which piedmont glaciers formed in steep valleys and into a flat plain, and create "a bulb-like lobe," and tidewater glaciers that flow into the sea. There are also glaciers that are "more difficult to differentiate," including mountain glaciers and hanging glaciers (World Meteorological Organization, 2022). Glaciers may also be distinguished by their speed of flow, as "some glaciers may hardly move over the course of a decade, while others, such as the Kutiah Glacier in Pakistan, have been recorded to surge as much as 110 meters (360 feet) per day" (World Meteorological Organization, 2022).

The ice sheets are the largest forms of ice flow, covering part of continents. The size of the ice sheets is determined by surface snow additions, the margin meltwater runoff, the rate of ice flow, and the calving of icebergs, and their long-term existence allows the study of past climates through the collection of ice cores (Bradley, 2015). Examples of ice sheets include the prehistoric Cordilleran Ice Sheet and the Laurentide Ice Sheet (Waitt & Thorson, 1983; Fulton & Prest, 1987). In modern times, the main ice sheets are the Antarctica and Greenland ice sheets, which contain nearly 70% of the Earth's and almost all the freshwater ice (World Meteorological Organization, 2022). These ice sheets rest on bedrock, which in Antarctica has been depressed by the weight of the ice, up to 2.4 kilometers. Records indicate that the ice sheets at the divides (the boundary on an ice sheet, which separates opposing flow directions of ice, the thickest parts of the ice sheet) is at least four kilometers thick in East Antarctica (averaging 1.9 kilometers thick across the continent, with a volume of (27 million km^3), and three kilometers in parts of Greenland. At the edges of the ice sheets, the ice may be only 200 meters thick, and at the coast may form ice shelves (Fretwell et al., 2013; Aschwanden et al., 2019; Morlighem et al., 2020; World Meteorological Organization, 2022).

The Atmosphere

The atmosphere contains numerous chemical elements and compounds, as gases, aerosols, or vapors, and is vital for global biogeochemistry. Definitions of the atmosphere emphasize its position on Earth, its structure or composition, and may include comparisons between the Earth and other planets (Thomas, 2016). For example, Marty (2011) defines the Earth's atmosphere as "a thin gaseous layer made of oxygen, nitrogen, and trace gases. It permits a mild temperature and provides a protection against cosmic rays, solar wind, and meteorite falls." National Geographic (2022b) defines the atmosphere in terms of its position on Earth, and a planetary phenomenon, and its composition as "the layers of gases surrounding a planet or other celestial body" and "one of the main components of Earth's interdependent physical systems." It is further noted that atmosphere is composed of about 78% nitrogen, 21% oxygen, and other gases, which occur mixed in layers, termed from the lowest in altitude, the troposphere, stratosphere, mesosphere, thermosphere, and exosphere, differentiated by altitude, temperature, and pressure.

Another similar definition is provided by the American Meteorological Society (2017), which writes that the atmosphere is a "gaseous envelope gravitationally bound to a celestial body (e.g., a planet, its satellite, or a star)." Interesting comparisons are made between the Earth's atmosphere and those of other planets, the latter having sometimes more extreme and opposite characteristics: "Different atmospheres have very different properties. For instance, the atmosphere of Venus is very thick and cloudy and is responsible for producing the very high surface temperatures on that planet by virtue of its greenhouse effect. On the other hand, the Martian atmosphere is very sparse. Earth's atmosphere is intermediate between these two extremes" (American Meteorological Society, 2017).

The Earth's atmosphere is however not merely in a continuum between extremes; it is also unique, in that it has a "very active hydrologic cycle," which results from heating, the thermodynamics of water vapor, water condensation, the effects of planetary rotation, and consequent atmospheric circulation (American Meteorological Society, 2017). There are also atmospheric layers, according to thermal and ionization structures. In the troposphere, just above the Earth's surface, temperature decreases due to the upward heat flux. In the layer above it, the stratosphere, the temperature increases upward due to ozone absorption of solar radiation. In the next layer, the mesosphere, the temperature decreases, and above this layer there is the thermosphere, where radiation increases temperature toward the exosphere, at the outer edge of the Earth's atmosphere, the exosphere. Solar radiation ionizes gases in the mesosphere and thermosphere, creating the ionosphere (American Meteorological Society, 2017).

Another comparative and compositional definition of the is given by Pielke (2022), who notes that the atmosphere is the "gas and aerosol envelope that extends from the ocean, land, and ice-covered surface of a planet outward into space." Pielke (2022) adds that "the density of the atmosphere decreases outward, because the gravitational attraction of the planet, which pulls the gases and aerosols (microscopic, suspended particles of dust, soot, smoke, or chemicals) inward, is greatest close to the surface." Comparison is made between the Earth's atmosphere (which has water, in solid, liquid and gas phases), the almost nonexistent atmospheres of some planets (e.g., Mercury) and stronger atmospheres (still mostly lacking water like the Earth's atmospheric hydrodynamics) of some other planets (Venus, Mars, etc.)

Part of the definition of the Earth's atmosphere concerns its evolution (Ozima, 1975; Pepin, 1991; Kasting, 1993; Shaw, 2008). Hayes (2020) argues that "the process by which the current atmosphere arose from earlier conditions is complex; however, evidence related to the evolution of Earth's atmosphere, though indirect, is abundant. Ancient sediments and rocks record past changes in atmospheric composition due to chemical reactions with Earth's crust and, in particular, to biochemical processes associated with life." Pielke (2022) argues that "the evolution of Earth's current atmosphere is not completely understood. It is thought that the current atmosphere resulted from a gradual release of gases both from the planet's interior and from the metabolic activities of life-forms – as opposed to the primordial atmosphere, which developed by outgassing (venting) during the original formation of the planet." It is argued that at the early stages of the evolution of the Earth's atmosphere, there was water in the oceans, and "carbon dioxide and perhaps ammonia (NH_3) must have been present in order to retard the loss of infrared radiation into space" (Pielke, 2022). Currently, and possibly in some of the evolutionary stages, there are gaseous, mostly emissions into the atmosphere, including water vapor, carbon dioxide, carbon monoxide, sulfur dioxide, hydrogen sulfide, chlorine, fluorine, and diatomic nitrogen (Pielke, 2022). Regarding life forms, the earliest forms may have been anaerobic (adapted to the lack of oxygen), and resistant to ultraviolet radiation, but later photosynthesis capable organisms produced oxygen, which enabled the evolution of the ozone layer (Pielke, 2022).

Marty (2011) concurs with these assessments, stating that "the Earth's atmosphere originated probably from the addition of bodies from the asteroid belt between Mars and Jupiter toward the end of terrestrial formation," and "originally, the atmosphere consisted of water, carbon dioxide, nitrogen, sulfur, and noble gases, and it evolved by condensation of water to form the oceans, trapping of carbon in carbonates and organic matter, and biogenic production of oxygen." Zahnle et al. (2010) adds that "the origin of Earth's atmosphere is a profound question of comparative planetology. The basic alternatives are a primary atmosphere captured from the solar nebula or a secondary atmosphere degassed from condensed materials accreted by Earth. Although the debate has long appeared settled in favor of secondary atmospheres, primary atmospheres probably did exist and may have had a minor role to play."

Kasting (1993: 920) looks at the evolution of knowledge of the early Earth's atmosphere, noting that despite an increase in knowledge, "many uncertainties still remain." Current knowledge and agreed positions are that the initial atmosphere included minimal or no free oxygen, which oxygen increasing around 2 billion years BP, (the latter "precise timing of and reasons for its rise remain unexplained), the levels of atmospheric carbon dioxide and other greenhouse gases are largely unknown in relation to the atmospheric greenhouse effect, which "must have been higher in the past to offset reduced solar luminosity," and "a better understanding of past atmospheric evolution is important to understanding the evolution of life and to predicting whether Earth-like planets might exist elsewhere in the galaxy" (Kasting, 1993: 920).

Shaw (2008: 235) also points to an interesting history, as the geological record for this period 2–2.5 billion years BP is "sparse," and "presents several difficulties for attempts to model the atmosphere and its changes through time." Knowledge sources for the last half-century have pointed to atmospheric period as dominated by "a moderately oxidized atmosphere of CO_2 and N_2," supported recent geochemical techniques, but such knowledge has also "left several nagging questions unanswered" (Shaw, 2008: 235). Shaw (2008: 235) notes several unanswered questions: "how was a sufficient reservoir (and concentration) of prebiotic molecules produced? What were the major reservoirs for carbon, and how did they develop their isotopic signatures? Is there a solution to the problem of the "faint young sun?" Why was surface oxidation delayed following the advent of oxygenic photosynthesis?" A recent idea was "the importance of more reducing capacity at the early Earth's surface, but without abandoning the idea of a mainly CO_2–N_2 atmosphere" (Shaw, 2008: 235). In such a case reducing capacity refers to chemical reduction, which is the gain of electrons or a decrease in the oxidation state of a chemical or the atoms within it (Haustein, 2019).

Schlesinger and Bernhardt (2020) offer a definition of the atmosphere based on its definition of nitrogen, oxygen and argon, and trace gases, particles, and aerosols, with ozone screening ultraviolet in the stratosphere. Pielke (2022), supports this composition of the atmosphere, including the inert gases such as neon (Ne), helium (He), and krypton (Kr). This composition may be compared with past compositions, as Hayes (2020), notes that "Earth's original atmosphere was rich in methane, ammonia, water vapour, and the noble gas neon, but it lacked free oxygen. It is likely that hundreds of millions of years separated the first biological production of oxygen by unicellular organisms and its eventual accumulation in the atmosphere." Hayes (2020) argues that "the atmosphere that developed after primordial gases had been lost or had failed to accumulate is termed secondary. Although the chemical composition of the atmosphere has changed significantly in the billions of years since its origin, the inventory of volatile elements on which it is based has not."

Hayes (2020) further argues that "the most critical parameter pertaining to the chemical composition of an atmosphere is its level of oxidation or reduction. At one end of the scale, an atmosphere rich in molecular oxygen (O_2) – like Earth's present atmosphere – is termed highly oxidizing, while one containing molecular hydrogen (H_2) is termed reducing." The example is cited of current volcanic gases, which are near the oxidized end of the scale, with no O_2, and contain hydrogen, carbon, and sulfur in oxidized forms (water vapor (H_2O), carbon dioxide (CO_2), sulfur dioxide (SO_2)) and nitrogen as molecular nitrogen (N_2), rather than ammonia (NH3). Hayes (2020) notes that "a relationship prevails between the oxidation or reduction of outgassing volatiles and the inorganic material with which they come in contact: any hydrogen, carbon, or sulfur brought into contact with modern crustal rocks at volcanic temperatures will be oxidized by that contact." Regarding the origins of the Earth's atmosphere, Hayes (2020) also notes that "the abundance of hydrogen in the solar nebula, the common occurrence of metallic iron in meteorites (representative of primitive solids), and other lines of geochemical evidence all suggest that Earth's early crust was much less oxidized than its modern counterpart. Although all iron in the modern crust is at least partly oxidized (to Fe2+ or Fe3+), metallic iron may have been present in the crust as outgassing began." Table 3.10 is an adaptation of the information presented by Hayes (2020) on the abundances of elements in the Earth and the solar system.

Table 3.10 Abundances of elements

Element	Solar system	Earth
Hydrogen	27,000,000,000	9500
Helium-4	2,200,000,000	0.00005
Carbon	12,000,000	360
Nitrogen	2,500,000	79
Oxygen	20,000,000	3,400,000
Neon-20	3,300,000	0.000093
Magnesium	1,100,000	1,100,000
Sulfur	520,000	98,000
Argon-36	88,000	0.00018
Argon-40	0.55	0.053
Iron	900,000	1,200,000
Krypton-84	26	0.0000036

Source: Hayes (2020). The abundances show the number of atoms of each element (or, in the case of the noble gases, isotopes) accompanying one million silicon (Si) atoms. An example is of the abundance of nitrogen (N) in the solar system, which is 2.5 times than that of Si, but its "abundance on Earth is less than that of Si by a factor of 0.000079. The table includes the eight most abundant volatile elements, together with others" (Hayes, 2020)

The Biosphere

The biosphere includes all living systems on the Earth surface (Campbell, 2017, 2018, 2020), and as an agent for chemical cycling, is the key actor in the global to local biogeochemical cycling. Gates et al. (2022) defines the biosphere as "the thin life-supporting stratum of Earth's surface, extending from a few kilometers into the atmosphere to the deep-sea vents of the ocean. It is composed of living organisms and nonliving factors from which the organisms derive energy and nutrients," with the role of supporting "all life on Earth, 3 to 30 million species of plants, animals, fungi, single-celled prokaryotes such as bacteria, and single-celled eukaryotes such as protozoans" (Gates et al., 2022). The living part of the biosphere comprises the biotic components (plant producers, animal consumers, and animal decomposers). The nonliving part of the biosphere is composed of the abiotic components, which include water, gases, energy flows, chemical nutrients, and organic and inorganic substances. Levchenko et al. (2012: 3) defined the biosphere as "a specific envelope of the Earth, comprising totality of all living organisms and that part of planet matter which is in constant material exchange with these organisms. Biosphere includes geospheres, i.e., lower part of the atmosphere, hydrosphere, and upper levels of lithosphere. Virtually, it is a spherical layer 6–12 km thick."

Gates et al. (2022) further note that "the biosphere is a global ecosystem that can be broken down into regional or local ecosystems, or biomes," "organisms in the biosphere are classified into trophic levels (see food chain) and communities" and "the major cycles that occur in the biosphere are the carbon cycle, nitrogen cycle, sulfur cycle, phosphorus cycle, and hydrological (water) cycle." The biosphere is contrasted with the connected geosphere, which refers to the inanimate lithosphere (rocks, soils), hydrosphere (water in gaseous, solid, and liquid states), and the atmosphere (air and gases). Gates et al. (2022), referring to the origins of the biosphere, notes that "before the coming of life, Earth was a bleak place, a rocky globe with shallow seas and a thin band of gases – largely carbon dioxide, carbon monoxide, molecular nitrogen, hydrogen sulfide, and water vapour."

Gutiérrez-Hernández and García (2021: 239) provide a fundamental definition of the biosphere and its relationship with the rest of the Earth systems, arguing that "the biosphere can be defined as the entire living organisms on planet Earth and their interrelations; this is a life-supporting stratum of Earth's surface among the geosphere, pedosphere, atmosphere, and the hydrosphere It is also called ecosphere because it is the worldwide sum of all ecosystems, and this is characterized by both biotic components (biota) and abiotic factors (chemical and physical parts of the environment) working together as a unique ecosystem" (see also Huggett, 1999; Lutgens et al., 2017; Strahler, 2013). Gutiérrez-Hernández and García (2021: 239) further argue that biosphere oriented studies are traditional foci of both geography and biology, noting that the biosphere as the relatively thin life-supporting stratum of Earth's surface and is historically studied by biogeographers (see also Campbell, 2017, 2018, 2020; Jones, 1980).

A natural subject for biosphere studies is biogeography, which examines "the distribution of life (such as plants, animals) across different spatiotemporal scales," the "biodiversity patterns and the causal factors of variations in the distribution of life on Earth" and the "the role of historical factors in the construction of biomes or the existence of contrasting patterns of range evolution in animals and plants" (Gutiérrez-Hernández & García, 2021: 239; see also Sanmartín, 2012; Lomolino et al., 2016). Sanmartín (2012: 555–568) defines biogeography as "the discipline of biology that studies the present and past distribution patterns of biological diversity and their underlying environmental and historical causes." The origins of biogeography and the sister science of ecology, and the integrative and multiscale format of biogeographical information (syntheses of ecology, evolutionary biology, taxonomy, general geology, physical geography, paleontology, and climatology) are also noted (Cox et al., 2016; Fisher, 2002; Kent, 2005, 2007; Ricklefs & Jenkins, 2011).

Huggett (1999: 425) in attempting a definition of terms, mentions the terms biosphere, ecosphere, and Gaia, as frequently used names for the global ecosystem, but notes that the usage of the three terms may indicate different connotations: "biosphere can mean the totality of living things residing on the Earth, the space occupied by living things, or life and life-support systems (atmosphere, hydrosphere, lithosphere, and pedosphere)"; ecosphere is used as a synonym of biosphere and as a term for zones in the universe where life as we know it should be sustainable" and Gaia is similar to biosphere (in the sense of life and life-support systems) and ecosphere (in the sense of biosphere as life and life-support systems), but, in its most extreme form, refers to the entire planet as a living entity." Huggett (1999: 425) argues that "a case is made for avoiding the term Gaia (at least as a name for the planetary ecosystem), restricting biosphere to the totality of living things, and adopting the ecosphere as the most apt name for the global ecosystem."

Other definitions of biosphere refer it is vital role in the global system (Campbell, 2017, 2018, 2020). For example, Ludwig and Steffen (2018: 45) define the biosphere as "the biosphere is the second high-level component, along with the geosphere, that together comprise the Earth System. The biosphere has probably been changed even more than the geosphere by human activities and so plays a key role in defining the Anthropocene." The central theme of the biosphere is biodiversity, which may be measured by existing species and species extinctions, and Ludwig and Steffen (2018: 45) note that current biodiversity losses are extreme, but "although the present rate of extinctions has not yet reached the level of a global mass extinction, it is following the pattern of past mass extinctions and if continued over the next few centuries, the Earth would experience its sixth great mass extinction event" (see also Barnosky et al., 2011).

Therefore, an important aspect of the biosphere concerns the extent to which the biosphere has been affected by human action, "perhaps even representing the third fundamental stage in the evolution of the biosphere, following the microbial and metazoan stages" (Ludwig & Steffen, 2018: 45; see also Williams et al., 2015). Important developments include the extent to which human action has change species distributions, including invasive species (neobiota) and consequent ecosystem change (these are neobiota, which are organisms that occur in an area only through human influence). Ludwig and Steffen (2018: 45) argue that "this rapid mixing of a large number of species has occurred especially since the mid-20th century and represents a sharp contrast to the slow evolution of biogeographically distinct species that has typified biosphere change in the past."

Human action has also taken over much of the biosphere productivity, up to 40%, by altering landcover for agricultural, urban, and industrial landuse, altering oceanic biodiversity (Ludwig & Steffen, 2018; see also Williams et al., 2015). Domestication has also affected biodiversity, where there is "a concentration of animal biomass into just a few domesticated species that are directly useful for humans – e.g., cattle, sheep, dogs, etc. as well as the focus of industrialized fishing in the oceans on just a few species" (Ludwig & Steffen (2018: 45). Ludwig and Steffen (2018: 45) conclude "the magnitude and rate of changes in the structure and functioning of the Earth System processes, both in the geosphere (climate) and the biosphere, clearly show that the Earth has left the relatively stable regime of the Holocene. Human forcing is now overriding the negative feedback mechanisms that would otherwise keep the Earth System in the Holocene state" (Barnosky et al., 2011; Williams et al., 2015).

Von Bloh (2019: 20) explains that the biosphere played an important role in the global carbon cycle during Earth's evolution from the Archean to the present and future. Importantly, inert organic carbon is developed from the decay of organic matter, contributing to up to a fifth of the present surface reservoirs of carbon, which is complemented by the increase of oxygen in the atmosphere. An important event in biosphere development was the Cambrian explosion, that is, the development of 0.54 billion years BP, which resulted from interactions between components of the geosphere and biosphere. Von Bloh (2019: 20) notes that that "after the Cambrian explosion the environment itself has been actively changed by the biosphere maintaining the temperature conditions for its existence. The ultimate life span of the biosphere is defined by the extinction of prokaryotes in about 1.6 Gyr (1.6 billion years BP) because of CO_2 starvation."

The development of the biosphere is based on its changing and continually interacting systems, which also interact with the geosphere (Campbell, 2017, 2018). For example, Gates et al. (2022) notes that "the biosphere is a system characterized by the continuous cycling of matter and an accompanying flow of solar energy in which certain large molecules and cells are self-reproducing. Water is a major predisposing factor, for all life depends on it. The elements carbon, hydrogen, nitrogen, oxygen, phosphorus, and sulfur, when combined as proteins, lipids, carbohydrates, and nucleic acids, provide the building blocks, the fuel, and the direction for the creation of life. Energy flow is required to maintain the structure of organisms by the formation and splitting of phosphate bonds."

The development of the biosphere is through the processes of biological evolution. Gates et al. (2022) mention the processes of evolution as natural selection (variable successes of the reproduction of hereditary variations, based on environmental impacts), gene flow (genetic movement in species populations), and random genetic drift (chance genetic change in small populations). Natural selection is very important as it influences ecological change by favoring members of the population more suited to ecological change, thereby encouraging ecological adaptation. Genetic adaptation may facilitate the evolution of new mutations within populations, which can also lead to the "the recombination of genes during sexual reproduction, and the migration of and interbreeding with individuals from other populations. In very small populations, however, some of that variation is lost by chance alone through random genetic drift," but the total effect may be species creation (speciation) and interspecies adaptation and coevolution (Gates et al., 2022).

Energy flows are important for the biosphere to function. For example, solar energy and associated photosynthesis are converted to chemical energy, and living tissue. Gates et al. (2022) notes that "in the photosynthetic process, light energy is absorbed by the chlorophyll molecules of plants to convert carbon dioxide and water into carbohydrates and oxygen gas. Proteins, fats, nucleic acids, and other compounds also are synthesized during the process, as long as elements such as nitrogen, sulfur, and phosphorus are available." There is a pyramid of energy, with the highest concentrations in the photosynthetic plants at the lowest level of the food chain, and progressively lower energy levels with the flow the high trophic levels of herbivores and carnivores. Energy flows also diverge as they flow up the food chain; "some of the remaining energy does not pass directly into the plant–herbivore–carnivore food chain but instead is diverted into the detritus food chain. Bacteria, fungi, scavengers, and carrion eaters that consume detritus (detritivores) are all eventually consumed by other organisms" (Gates et al., 2022).

The transfer of energy from through trophic levels is termed ecological efficiency (which may be as low as 10%). Primary productivity is the rate of energy conversion to organic substances by photosynthetic producers (photoautotrophs), energy sources including sunlight, and chemosynthetic producers (chemoautotrophs), which obtain chemical energy through oxidation (Britannica, 2022a, b). Secondary productivity refers to the rate at which consumers convert chemical energy from food into their biomass. Assimilation efficiency refers to the efficiency of the conversion of animal food into energy for development and reproduction. Gates et al. (2022) points out that "herbivores assimilate between 15 and 80 percent of the plant material they ingest, depending on their physiology and the part of the plant that they eat" (usually 15 to 80%) while "carnivores generally have higher assimilation efficiencies than herbivores, often between 60 and 90 percent, because their food is more easily digested." Therefore, "the overall productivity of the biosphere is therefore limited by the rate at which plants convert solar energy (about 1 percent) into chemical energy and the subsequent efficiencies at which other organisms at higher trophic levels convert that stored energy into their own biomass (approximately 10 percent)" (Gates et al., 2022).

The origin of the biosphere is debated, disputed, and constitutes a vital paradigm-based argument with the biological sciences (Levchenko & Starobogatov, 2010). For example, Levchenko et al. (2012: 3) note that "traditionally, the problem of the origin of life on Earth is the study of how biological life arose from inorganic matter and primary living organisms spread around the planet" and while some researchers and philosophers have argued for life origins in the cosmos (a naïve view, considering modern scientific knowledge), a better view is that "the process of the origin and evolution of living organisms is indivisible from the process of the origin and evolution of biosphere, the global planetary system." The authors therefore argue that the "origin of earthly life and the origin of the biosphere are aspects of a whole indivisible process" and "the appearance of such conditions on the early planet, which guaranteed the origin and survivorship of organic life. We suppose also that primary organisms should be incorporated in natural geological processes, accelerating them, and transforming surroundings in directions favorable for the creation of higher forms of life" (Levchenko et al. (2012: 3).

The development of the biosphere is charted through long periods of life evolution (Campbell, 2015). Stratigraphy is the main science used for the documentation of biosphere change in the geologic time scale (GTS). Stratigraphy is defined as the study of sedimentary and volcanic rock layers and layering (stratification). There are two main branches: lithologic stratigraphy or lithostratigraphy, which examines physical and chemical differences in rock layers; and biostratigraphy or paleontological stratigraphy, which examines the fossil record deposited in the rock layers. For studies of the biosphere, biostratigraphy is particularly important as it provides information about the animals that lived during the various geologic time periods. Soil patterns may also be investigated using lithostratigraphy, which may be relevant to plant associations and their links to ecological change. Tables 3.11, 3.12, 3.13 and 3.14 show the stages of the evolution of the biosphere.

Table 3.11 Classification of geochronologic units (time)

Geochronologic unit (time)	Time span
Eon	Several hundred million years
Era	Tens to hundreds of millions of years
Period	Millions to tens of millions of years
Epoch	Hundreds of thousands to tens of millions of years
Subepoch	Thousands to millions of years
Age	Thousands to millions of years

Source: Campbell (2015, 2022)

Table 3.12 Evolution of life on earth: Paleozoic Era

Time name	1 Million years BP	Evolution of life and main events
Precambrian Period	4600	Life forms evolved from unicellular to multicellular forms.
Phanerozoic Eon	540 to Present	
Paleozoic Era	541 to 252.2	
Cambrian Period	541.0 ± 1.0 to 485.4 ± 1.9	Animal life, the "Cambrian Explosion." Change from unicellular forms to aquatic, marine multicellular forms of molluscs and arthropods, land perhaps with unicellular animals and plants.
Ordovician Period	485.4 ± 1.9 to 443.4 ± 1.5	More sophisticated, marine mollusks and arthropods. Few new fish ancestors
Silurian Period	443.4 ± 1.5	Marine bony fish appeared, and terrestrial arthropods and small moss-like land plants.
Devonian Period	419.2 ± 3.2	More vascular plants, spores replaced with seed bearing plants. More sophisticated fish, tetrapod colonization of land, the fins changing into legs
Carboniferous Period	358.9	Amphibians dominate terrestrial life, reptiles evolve later, with larger arthropods such as the Meganeura (giant dragonfly-like insects)
Permian Period	298.9 ± 0.2 to 252.2 ± 0.5	Permian–Triassic (P–Tr) extinction event (Great Dying), 70% of terrestrial, 96% of marine species extinct. Archosaurs, ancestral birds, crocodilians, true mammals.

Sources: Kurten and Anderson (1980), Alvarez et al. (1980),; Olson and Parris (1987), Kurten and Werdelin (1990), Van Valkenburgh and Hertel (1993), Feduccia (1995), Wink (1995), Martin et al. (1996), Turner (1987), Turner and Antón (1997), Jefferson (2001), Grayson and Meltzer (2002), Van Valkenburgh and Sacco (2002), Coltrain et al. (2004), International Commission on Stratigraphy (2004), Christiansen and Harris (2005), Fennec (2005), Shaw and Cox (2006), Christiansen (2007), De Castro and Langer (2008), Rincón et al. (2011)

The Cenozoic Era is the primary concern of the present biosphere, as it spans the period recent zoological evolution (Campbell, 2015). One of the most important developments in recent biological history concerns the presence of avian and mammalian megafauna during the Age of mammals and birds, which created much of the

Table 3.13 Evolution of life on earth: Mesozoic Era

Time name	1 Million years	Evolution of life and main events
Precambrian Period	4600	Life forms evolved from unicellular to multicellular forms.
Phanerozoic Eon	540 to Present	
Mesozoic Era	252 to 66 million years BP	Age of reptiles
Triassic Period	251.902 to 201.36 million years BP	Three categories of organisms: extinction event survivors, temporary new groups, later permanent groups. Archosaurs, and other reptiles the main terrestrial vertebrates
Jurassic Period	201.3 to 145 million years BP	First real birds or Avialans, evolved from Triassic forms. Large dinosaurs, herbivorous sauropods and carnivorous Theropods dominant
Cretaceous Period	45 to 66 million years BP	Bird and mammal development, large reptile extinction. Cretaceous–Paleogene extinction event began (66 million years BP), possibly due to a massive comet/asteroid impact

Sources: Kurten and Anderson (1980), Alvarez et al. (1980), Olson and Parris (1987), Kurten and Werdelin (1990); Vanvalkenburgh and Hertel (1993), Feduccia (1995), Wink (1995), Martin et al. (1996), Turner (1987), Turner and Antón (1997), Jefferson (2001), Grayson and Meltzer (2002), Van Valkenburgh and Sacco (2002), Coltrain et al. (2004), International Commission on Stratigraphy (2004), Christiansen and Harris (2005), Fennec (2005), Shaw and Cox (2006), Christiansen (2007), De Castro and Langer (2008), Rincón et al. (2011)

Table 3.14 Evolution of life on earth: Cenozoic Era

Time name	1 Million years	Evolution of life and main events
Precambrian Period	4600	Life forms evolved from unicellular to multicellular forms.
Phanerozoic Eon	540 to Present	
Cenozoic Era	66 million years BP to present	Age of Mammals (possibly due to the extinction event that ended the Age of Reptiles of the Mesozoic)
Paleogene	66 to 23.03 million years BP	Paleocene, Eocene and Oligocene Epochs: More birds and mammals, increasing in size possibly because of the large reptile extinction.
Neogene	23 to 2.59 million years BP, including the	Miocene and Pliocene Epochs: Bird and mammal species almost similar to modern species. Increased size compared with previous epoch.
Quaternary	2.59 million years BP to present	Pleistocene and Holocene Epoch of the Quaternary Period (2,588,000 to 11,700 years BP). Some mammals, birds larger than modern, human/hominids like Homo sapiens.

Sources: Kurtén and Anderson (1980), Alvarez et al. (1980), Olson and Parris (1987), Kurtén and Werdelin (1990), Vanvalkenburgh and Hertel (1993), Feduccia (1995), Wink (1995), Martin et al. (1996), Turner (1987), Turner and Antón (1997), Jefferson (2001), Grayson and Meltzer (2002), Van Valkenburgh and Sacco (2002), Coltrain et al. (2004), International Commission on Stratigraphy (2004), Christiansen and Harris (2005), Fennec (2005), Shaw and Cox (2006), Christiansen (2007), De Castro and Langer (2008), Rincón et al. (2011)

current ecosystems and food webs vital for biogeochemical cycling (Campbell, 2017). Such prehistoric megafauna includes the large species of mammals that mostly became extinct during Quaternary extinction events, which may have resulted from increased human hunting, climatic change, disease, and/or an asteroid or comet contact with the Earth, which occurred mostly in the Americas, northern Eurasia, and Australia, although sub-Saharan Africa and southeast Asia retained some larger mammals, such as elephants and rhinoceroses (Campbell, 2017). The details of such extinctions are debated, and the extinct species are studied in detail (Feduccia, 1995; Wink, 1995; Grayson & Meltzer, 2002; Grayson, 2006; Ostende et al., 2006; Campbell, 2015).

For example, birds first appeared in the late Jurassic/early Cretaceous Periods (150 million years BP), and there appeared to have been massive bird extinctions during the late Mesozoic a few million years later. The diversification of bird species occurred at the beginning of the Cretaceous/Tertiary (about 65 million years BP) (Feduccia, 1995; Wink, 1995). There is fossil evidence from the Eocene and Oligocene (35 million years BP) of the modern avian Orders (except Passerines). The appearance of most modern orders, termed "an extraordinary explosive evolution" may have occurred over 5 to 10 million years (Feduccia (1995). In North America, 35 genera of primarily large mammals were extinct by the end of the Pleistocene (Grayson, 2006). Although it is commonly assumed that most of these extinctions took place during the period between 10,000 and 12,000 BP, some researchers have argued that only 16 of these extinct genera are conclusively dated to this period and, hence, the extinctions may have occurred over a longer period (Grayson & Meltzer, 2002).

Herbivorous mammal species that became extinct in the last few million years include large members of the Order *Cingulata* (SuperOrder *Xenarthra*), including the Glyptodonts, which were large heavily armored armadillos, which arose in South America about 48 million years BP and spread to southern North America after the continental connection. As well as the huge Glyptodonts (*Glyptodontidae*, Burmeister 1879 or *Chlamyphoridae*, Bonaparte 1850), the extinct mammals included members of the Families *Pampatheriidae* (Paula Couto 1954, plantigrade armored armadillos), *Megalonychidae* (Gervais 1855, ground sloths) *Mylodontidae* (Ameghino 1889, ground sloths), and *Megatheriidae* (J.E. Gray 1821, larger ground sloths, weighing up to 2 tonnes). Herbivores of the Order Artiodactyla included members of the Families *Tayassuidae* (Palmer, 1897, e.g., larger versions of modern peccaries), *Camelidae* (Gray, 1821, usually larger versions of modern camels), *Cervidae* (Goldfuss, 1820, e.g., modern deer); *Antilocapridae* (J. E. Gray, 1866, e.g., modern pronghorns *Antilocapra americana* Ord, 1815), *Bovidae* (Gray, 1821, e.g., the modern musk ox (*Ovibos moschatus,* Zimmermann 1780) and mountain goat (*Oreamnos americanus*, Blainville 1816)), and *Mammutidae* (Hay, 1922, including the American mastodon (*Mammut americanum*, Kerr, 1792) and the Woolly mammoth (*Mammuthus primigenius*, Blumenbach, 1799)).

Carnivorous mammal species that faced extinction in the case study of Pleistocene North America included members of the families *Canidiae* (Fischer de Waldheim, 1817, e.g., the dire wolf, Aenocyon dirus Leidy, 1858), *Ursidae* (G. Fischer de

Waldheim, 1817, including the huge short faced bear (*Arctodus simus*, Cope 1897, and the Florida spectacled bear (*Tremarctos floridanus*, Gidley 1928)), *Mephitidae* (Bonaparte 1845, the extinct skunk, *Brachyprotoma obtusata* Cope, 1899), and *Felidae* (*Fischer von Waldheim* 1817, including the huge cats of the extinct subfamily *Machairodontinae*, Gill, 1872, which existed from about 16 million to 11,000 years BP). The Machairodontinae line of big cats and similar species began in the early or middle Miocene of Africa (Ostende et al., 2006). The three groups of machairdontines were the massive Smilodontini (with huge, upper canines, and the smaller Machairodontini or Homotherini, and the Metailurini, with smaller canines (Barnett et al., 2005; Martin, 1989; Meloro & Slater, 2012; Salesa et al., 2012; Slater & Van Valkenburgh, 2008; Van Valkenburgh, 2007). The last machairodontine genera were Smilodon and Homotherium, which became extinct during the late Pleistocene in the Americas, about 10,000 years BP (Berta, 1985; Kurtén & Anderson, 1980; Kurten & Werdelin, 1990; Rawn-Schatzinger, 1992; Turner & Antón, 1997; Antón et al., 2004; Ascanio & Rincón, 2006; Andersson et al., 2011).

Smilodon was a massive apex predator, preying on mastodons, bison, camels, ground sloths, and pig-like *Platygonus* (LeConte 1848) and the llama-like *Hemiauchenia* (Gervais & Ameghino, 1880), perhaps competing with the American Cave lion (*Panthera leo atrox*, Leidy, 1853) and the Dire Wolf (Vanvalkenburgh & Hertel 1993; Coltrain et al., 2004; Fennec, 2005; Christiansen and Harris (2005). The Smilodons comprised *S. populator* (Lund 1842), *S. fatalis* (Leidy 1869), and *S. gracilis* (Cope 1880). *Smilodon gracilis*, the oldest and smallest, lived from 2.5 million to 500,000 BP and weighed 55 to 100 kg (Christiansen & Harris, 2005). *Smilodon fatalis* lived from 1.6 million–10,000 years ago, weighed 160 to 280 kg, and replaced *S. gracilis* in North America and ranged over western South America during the Great American Interchange (Kurten & Werdelin, 1990; Turner, 1987; Jefferson, 2001; Van Valkenburgh & Sacco, 2002; Christiansen & Harris, 2005; Shaw & Cox, 2006; Christiansen, 2007; Rincón et al., 2011). *Smilodon populator* was the largest (220 to 400 kg) and lived from 1 million to 10,000 years BP, in eastern South America (Kurten & Werdelin, 1990; Christiansen & Harris, 2005; De Castro & Langer, 2008; Sorkin, 2008). The Smilodons species became extinct in Africa about 1.5 million years BP, Eurasia about 30,000 years BP, and North America about 10,000 years BP (Turner, 1987; Turner & Antón, 1997; Campbell, 2015). Another species group, *Megantereon* (Genus *Megantereon*, Croizet & Jobert, 1828), weighing 60 to 150 kg, possibly ancestral to the Smilodon, lived in Africa, North America, and Eurasia, with fossil evidence dated to 4.5 million years from the Pliocene of North America (Turner & Antón, 1997), about 3–3.5 million BP in Africa 2.5 to 2 Million years BP in Asia and about 2–2.5 million years BP in Europe (Martin, 1989; Turner & Antón; Barnett et al., 2005; Slater & Van Valkenburgh, 2008). Megantereon evolved into the larger Smilodon at the end of the Pliocene in North America, and lived Africa and Eurasia until the Middle Pleistocene (Hemmer, 2002; Turner & Antón, 1997).

In the current era, most of the animal species have become extinct, with arguably a few related species still in existence (Campbell, 2015, 2017, 2018, 2022). Table 3.15 lists the number of species of birds, mammals reptiles, etc., that live on

Table 3.15 Current animal species numbers

Animal category	Number of species
Mammals	6578
Birds	11,162
Reptiles	11,820
Amphibians	8395
Fish	36,058
Crustaceans	80,122
Molluscs	83,706
Arachnids	110,615
Insects	1.05 million

Source: Richie and Roser (2021), National Wildlife Federation (2022), Reptile Database (2022), Richie and Roser (2021) note these are the number of identified and named species, as of 2021. As some species have not yet been described, this may be an underestimate of the world's total number of species

the Earth today. Most of these Orders, Families, and Genera are related to past, in some cases larger forms that are currently extinct. Research is ongoing as to the number of extinct species, the factors for their extinction and the trajectory of current populations. Although past extinctions are largely blamed on climate change, evolutionary dead ends, interspecies competition and habitat change, current extinctions at local and global levels are mostly blamed on human action. To understand both prehistoric and current threats to animal life, an ecosystem approach is required, based on all the components of ESS. This will possibly uncover the beginnings of relevant ecological changes.

Conclusions

This chapter has examined the beginning of the discipline and focus of Earth systems science (ESS), which has increasingly taken on the research and disciplinary areas of physical geography and mostly related multidisciplinary subjects. ESS is examined in this chapter as a discipline strongly related to biogeochemistry, with contributions that strengthen the latter science. Earth systems science is a relatively new disciplinary development, but the studied systems (the hydrosphere, lithosphere, biosphere, etc.) have been surveyed and researched for decades and issues concern not only the findings, but also how to integrate and synthesize knowledge about the different but related spheres of the Earth's system. As argued in the introduction and evidenced by the information presented, the study of these Earth spheres involves the analysis of their form and composition, dynamics and interrelations, evidence of origins, and the possibilities of the integrated analyses to answer searching questions concerning this background for biogeochemical studies. The methodology of Earth Systems Science is vital for biogeochemical studies, as it avoids the artificial disciplinary boundaries of the physical and life sciences. The evidence indicates that a major problem concerns the lack of consensus on the origin of these spheres and the need for more detail in studies to derive more definitive conclusions. The origin of the biosphere, which has bedeviled the biological sciences for centuries, and the relation of the biosphere with the hydrosphere, lithosphere, and atmosphere are intrinsic aspects of the development of biogeochemical knowledge. The following chapters will examine further the development of knowledge on aspects of biogeochemistry, including that shared with cognate disciplines.

References

Alexander, C. (2021). *Where did Earth's water come from?* Retrieved from https://epl.carnegiescience.edu/news/where-did-earths-water-come

Allen, J. I., & Fulton, E. A. (2010). Top-down, bottom-up or middle-out? Avoiding extraneous detail and over-generality in marine ecosystem models. *Progress in Oceanography, 84*(1–2), 129–133. https://doi.org/10.1016/j.pocean.2009.09.016

Alvarez, L. W., Alvarez, W., Asaro, F., & Michel, H. V. (1980). Extraterrestrial cause for the cretaceous-tertiary extinction. *Science (New York, N.Y.), 208*(4448), 1095–1108. https://doi.org/10.1126/science.208.4448.1095

Andersson, K., Norman, D., & Werdelin, L. (2011). Sabretoothed Carnivores and the killing of Large Prey. *PLoS ONE, 6*(10), e24971. https://doi.org/10.1371/journal.pone.0024971

Antón, M., Galobart, A., & Turner, A. (2004). Co-existence of scimitar-toothed cats, lions and hominins in the European Pleistocene. Implications of the post-cranial anatomy of Homotherium latidens (Owen) for comparative palaeoecology. *Quaternary Science Reviews, 24*, 1287–1301.

Artemieva, I. (2011). *The lithosphere: An interdisciplinary approach*. Cambridge University Press. https://doi.org/10.1017/CBO9780511975417

Ascanio, D., & Rincón, R. (2006). A first record of the Pleistocene saber-toothed cat Smilodon populator Lund, 1842 (Carnivora: Felidae: Machairodontinae) from Venezuela. *Asociación Paleontologica Argentina, 43*(2).

Aschwanden, A., Fahnestock, M. A., Truffer, M., Brinkerhoff, D. J., Hock, R., Khroulev, C., Mottram, R., & Khan, S. A. (2019). Contribution of the Greenland Ice Sheet to sea level over the next millennium. *Science Advances, 5*(6), eaav9396.

American Meteorological Society. (2017). *Atmosphere*. Retrieved from https://glossary.ametsoc.org/wiki/Atmosphere

Barnett, R., Barnes, I., Phillips, M. J., Martin, L. D., Harington, C. R., Leonard, J. A., & Cooper, A. (2005). Evolution of the extinct Sabretooths and the American cheetah-like cat. *Current Biology: CB, 15*(15), R589–R590. https://doi.org/10.1016/j.cub.2005.07.052

Barnosky, A. D., Matzke, N., Tomiya, S., Wogan, G. O., Swartz, B., Quental, T. B., Marshall, C., McGuire, J. L., Lindsey, E. L., Maguire, K. C., Mersey, B., & Ferrer, E. A. (2011). Has the Earth's sixth mass extinction already arrived? *Nature, 471*(7336), 51–57. https://doi.org/10.1038/nature09678

Baronas, J. J., Stevenson, E. I., Hackney, C. R., Darby, S. E., Bickle, M. J., Hilton, R. G., Larkin, C. S., Parsons, D. R., Myo Khaing, A., & Tipper, E. T. (2020). Integrating suspended sediment flux in large Alluvial River channels: Application of a synoptic rouse-based model to the Irrawaddy and Salween Rivers. *Journal of Geophysical Research: Earth Surface, 125*(9). https://doi.org/10.1029/2020JF005554

Beckerman, A. P., Petchey, O. L., & Warren, P. H. (2006). Foraging biology predicts food web complexity. *Proceedings of the National Academy of Sciences of the United States of America, 103*(37), 13745–13749. https://doi.org/10.1073/pnas.0603039103

Beilfuss, R., & Dos Santos, D. (2001). *Patterns of hydrological change in the Zambezi Delta, Mozambique*. Working Papers of the Program for the Sustainable Management of Cahora Bassa Dam and the Lower Zambezi Valley. Retrieved from https://web.archive.org/web/20081217180850/ http://files.gorongosa.net/filestore/348-patterns_hydrological_change_zambezi_delta.pdf

Benton, T. G. (2012). Individual variation and population dynamics: Lessons from a simple system. *Philosophical transactions of the Royal Society of London. Series B, Biological Sciences, 367*(1586), 200–210. https://doi.org/10.1098/rstb.2011.0168

Berkeley. (2022). *Berkeley geography*. Retrieved from https://geography.berkeley.edu/geography-studies/earth-system-science

Berta, A. (1985). The status of Smilodon in North and South America. *Contributions in Science, Natural History Museum of Los Angeles County, 370*, 1–15.

Beven, K., & Freer, J. (2001). Equifinality, data assimilation, and uncertainty estimation in mechanistic modelling of complex environmental systems using the GLUE methodology. *Journal of Hydrology, 249*(1–4), 11–29. https://doi.org/10.1016/S0022-1694(01)00421-8

Bithell, M., & Brasington, J. (2009). Coupling agent-based models of subsistence farming with individual-based forest models and dynamic models of water distribution. *Environmental Modelling & Software, 24*(2), 173–190. https://doi.org/10.1016/j.envsoft.2008.06.016

Bocedi, G., Pe'er, G., Heikkinen, R. K., Matsinos, Y., & Travis, J. M. J. (2012). Projecting species' range expansion dynamics: Sources of systematic biases when scaling up patterns and processes. *Methods in Ecology and Evolution, 3*(6), 1008–1018. https://doi.org/10.1111/j.2041-210X.2012.00235.x

Bossche, J. P. V., & Bernacsek, G. M. (1990). *Source Book for the Inland Fishery Resources of Africa, Volume 1*. Food and Agriculture Organization of the United Nations.

Brabets, T. P., Wang, B., & Meade, R. H. (2000). *Environmental and hydrologic overview of the Yukon River Basin, Alaska, and Canada*. Retrieved from https://pubs.usgs.gov/wri/wri994204/pdf/wri994204.pdf

Bradley, R. S. (2015). *Paleoclimatology: Reconstructing Climates of the Quaternary*. Academic Press.

Brentnall, S. J., Richards, K. J., Brindley, J., & Murphy, E. (2003). Plankton patchiness and its effect on larger-scale productivity. *Journal of Plankton Research, 25*(2), 121–140. https://doi.org/10.1093/plankt/25.2.121

Britannica, T. Editors of Encyclopaedia. (2021a). Hydrosphere. *Encyclopedia Britannica*. https://www.britannica.com/science/hydrosphere

Britannica, T. Editors of Encyclopaedia. (2021b). Lake Victoria. *Encyclopedia Britannica*. https://www.britannica.com/place/Lake-Victoria

Britannica, T. Editors of Encyclopaedia. (2022a). Lithosphere. *Encyclopedia Britannica*. https://www.britannica.com/science/lithosphere

Britannica, T. Editors of Encyclopaedia. (2020). Lake Nyasa. *Encyclopedia Britannica*. https://www.britannica.com/place/Lake-Nyasa

Britannica, T. Editors of Encyclopaedia. (2013). Red River. *Encyclopedia Britannica*. https://www.britannica.com/place/Red-River-Asia

Britannica, T. Editors of Encyclopaedia. (2022b). Primary productivity. *Encyclopedia Britannica*. https://www.britannica.com/science/primary-productivity

British Columbia Ministry of the Environment. (2008). *Comprehensive Review of Fraser River at Hope: Flood Hydrology and Flows - Scoping Study Final Report*. Retrieved from https://www.env.gov.bc.ca/wsd/public_safety/flood/pdfs_word/review_fraser_flood_flows_hope.pdf

Brown, C. J., Fulton, E. A., Possingham, H. P., & Richardson, A. J. (2012). How long can fisheries management delay action in response to ecosystem and climate change? *Ecological Applications, 22*(1), 298–310. https://doi.org/10.1890/11-0419.1

Brown, D. G., Page, S., Riolo, R., Zellner, M., & Rand, W. (2005). Path dependence and the validation of agent-based spatial models of land use. *International Journal of Geographical Information Science, 19*(2), 153–174. https://doi.org/10.1080/13658810410001713399

Brunnschweiler, D., Denevan, W. M., & Gómez, M. F. (2020). Orinoco River. *Encyclopedia Britannica*. https://www.britannica.com/place/Orinoco-River

Buitenhuis, E., Le Quere, C., Aumont, O., Beaugrand, G., Bunker, A., Hirst, A., Ikeda, T., O'Brien, T., Piontkovski, S., & Straile, D. (2006). Biogeochemical fluxes through mesozooplankton. *Global Biogeochemical Cycles, 20*(2). https://doi.org/10.1029/2005GB002511

Butz, S. D. (2004). *Science of Earth Systems*. Thomson/Delmar Learning.

Cameron, T. C., O'Sullivan, D., Reynolds, A., Piertney, S. B., & Benton, T. G. (2013). Eco-evolutionary dynamics in response to selection on life histories. *Ecology Letters, 16*(6), 754–763. https://doi.org/10.1111/ele.12107

Campbell, M. (2015). *Vultures: Their Evolution, Ecology and Conservation*. Taylor and Francis (CRC Press).

Campbell, M. (2017). *Biological Conservation in the 21st Century: A Conservation Biology of Large Wildlife*. Nova Science Publishers.

Campbell, M. (2018). *Geomatics and Conservation Biology*. Nova Science Publishers.

Campbell, M. (2020). *The Physical Geography of El Salvador: A Geophysical and Ecological Approach*. Nova Science Publishers.

Campbell, M. (2022). *The Great Eagles: Their Evolution, Ecology and Conservation*. Taylor and Francis (CRC Press).

Canadian Heritage Rivers System. (2022). *Fraser River*. Retrieved from https://chrs.ca/en/rivers/fraser-river

Carpenter, S. R., Mooney, H. A., Agard, J., Capistrano, D., Defries, R. S., Díaz, S., Dietz, T., Duraiappah, A. K., Oteng-Yeboah, A., Pereira, H. M., Perrings, C., Reid, W. V., Sarukhan, J., Scholes, R. J., & Whyte, A. (2009). Science for managing ecosystem services: Beyond the Millennium Ecosystem Assessment. *Proceedings of the National Academy of Sciences of the United States of America, 106*(5), 1305–1312. https://doi.org/10.1073/pnas.0808772106

Cavosie, A. J., Wilde, S. A., Liu, D., Weiblen, P. W., & Valley, J. W. (2004). Internal structure and U-Th-Pb chemistry of Jack Hills detrital zircons: A mineral record of Early Archean to Mesoproterozoic 4348-1576 Ma magmatism. *Precambrian Research, 135*, 251–279.

Cavosie, A. J., Valley, J. W., Wilde, S. A., & E.I.M.F. (2005). Magmatic $\delta^{18}O$ in 4400-3900 Ma detrital zircons: A record of the alteration and recycling of crust in the early Archean. *Earth and Planetary Science Letters, 235*(3–4), 663–681. https://doi.org/10.1016/j.epsl.2005.04.028

Cavosie, A. (2016). *When did oceans form on Earth? What evidence is preserved in the rock record?* https://serc.carleton.edu/NAGTWorkshops/earlyearth/questions/formation_oceans.html

Christensen, V. (2013). Ecological networks in fisheries: Predicting the future? *Fisheries, 38*, 76–81. https://doi.org/10.1080/03632415.2013.757987

Christensen, V. (2011). In V. Christensen & J. Maclean (Eds.), *Introduction: Toward ecosystem-based management of fisheries. In Ecosystem approaches to fisheries: A global perspective* (pp. 1–8). Cambridge University Press. '.

Christensen, V., Walters, C. J., Ahrens, R. N., Alder, J., Buszowski, J., Christensen, L. B., Cheung, W. W., Dunne, J. P., Froese, R., Karpouzi, V. S., Kaschner, K., Kearney, K. A., Lai, S. X., Lam, V. W., Palomares, M. L., Peters-Mason, A., Piroddi, C., Sarmiento, J. L., Steenbeek, J., Sumaila, U. R., Watson, R. A., Zeller, D., & Pauly, D. (2009). Database-driven models of the world's Large Marine Ecosystems. *Ecological Modelling, 220*, 1984–1996. https://doi.org/10.1016/j.ecolmodel.2009.04.041

Christiansen, E. H., & Hamblin, W. K. (2014). *Dynamic Earth*. Jones & Bartlett Learning.

Christiansen, P. (2007). Comparative bite forces and canine bending strength in feline and sabre-tooth felids: Implications for predatory ecology. *Zoological Journal of the Linnean Society, 151*(2), 423–437. https://doi.org/10.1111/j.1096-3642.2007.00321.x

Christiansen, P., & Harris, J. M. (2005). Body size of Smilodon (Mammalia: Felidae). *Journal of Morphology, 266*(3), 369–384. https://doi.org/10.1002/jmor.10384

Cohen, A. S. (2003). *Paleolimnology: The History and Evolution of Lake Systems*. Oxford University Press.
Coleman, J. M., & Huh, O. K. (2004). *Major deltas of the world: A perspective from space*. Coastal Studies Institute, Louisiana State University.
Coltrain, J. B., Harris, J. M., Cerling, T. E., Ehleringer, J. R., Dearing, M.-D., Ward, J., & Allen, J. (2004). Rancho La Brea stable isotope biogeochemistry and its implications for the palaeoecology of late Pleistocene, coastal southern California. *Palaeogeography, Palaeoclimatology, Palaeoecology, 205*(3–4), 199–219. https://doi.org/10.1016/j.palaeo.2003.12.008
Córdova, J., & González, M. (2015). La geografía del agua. In A. Gabaldón, A. Rosales, E. Buroz, J. R. Córdova, G. Uzcátegui, & L. Iskandar (Eds.), *Agua en Venezuela: Una riqueza escasa* (pp. 103–140). Fundación Empresas Polar.
Cornell, S. E. (2013). Chapter 8 - Earth System Services—A Global Science Perspective on Ecosystem Services. In *Ecosystem Services: Global Issues, Local Practices* (pp. 85–89). https://doi.org/10.1016/B978-0-12-419964-4.00008-1
Cornell, S. E., Prentice, I. C., House, J. I., & Downy, C. J. (2012). *Understanding the Earth System: Global Change Science for Application*. Cambridge University Press.
Coskun, M., & Musaoğlu, N. (2004). Investigation of Rainfall-Runoff Modelling of the Van Lake Catchment by Using Remote Sensing and GIS Integration. In *Proceedings of the 20th Congress of the International Society for Photogrammetry and Remote Sensing*.
Coulson, T., Benton, T. G., Lundberg, P., Dall, S. R., Kendall, B. E., & Gaillard, J. M. (2006a). Estimating individual contributions to population growth: Evolutionary fitness in ecological time. *Proceedings. Biological Sciences, 273*(1586), 547–555. https://doi.org/10.1098/rspb.2005.3357
Coulson, T., Benton, T. G., Lundberg, P., Dall, S. R. X., & Kendall, B. E. (2006b). Putting evolutionary biology back in the ecological theatre: A demographic framework mapping genes to communities. *Evolutionary Ecology Research, 8*(7), 1155–1171.
Cox, C., Moore, P., & Ladle, R. (2016). *Biogeography: An Ecological and Evolutionary Approach* (9th ed.). John Wiley & Sons.
Dahl, C., Richards, S., & Novotny, V. (2013). The Sepik River (Papua New Guinea) is not a dispersal barrier for lowland rain-forest frogs. *Journal of Tropical Ecology, 29*(6), 477–483. https://doi.org/10.1017/S0266467413000527
Dale, M. B. (1970). Systems analysis and ecology. *Ecology, 51*, 1–16. https://doi.org/10.2307/1933596
De Castro, M. C., & Langer, M. C. (2008). New postcranial remains of Smilodon populator Lund, 1842 from South-Central Brazil. *Revista Brasileira de Paleontologia, 11*(3), 199–206. https://doi.org/10.4072/rbp.2008.3.06
Degens, E. T., Wong, H. K., Kempe, S., & Kurtman, F. (1984). A geological study of Lake Van, eastern Turkey. *International Journal of Earth Sciences, Springer, 73*(2), 701–734. https://doi.org/10.1007/BF01824978
Deming, D. (2002). Origin of the Ocean and Continents: A Unified Theory of the Earth. *International Geology Review, 44*(2), 137–152. https://doi.org/10.2747/0020-6814.44.2.137
Denevan, W. M., Oliveira, W. F., Oteiza, D., & Stewart, N. R. (2022). Río de la Plata. *Encyclopedia Britannica*. https://www.britannica.com/place/Rio-de-la-Plata
De Young, B., Heath, M., Werner, F., Chai, F., Megrey, B., & Monfray, P. (2004). Challenges of modeling ocean basin ecosystems. *Science (New York, N.Y.), 304*(5676), 1463–1466. https://doi.org/10.1126/science.1094858
Díaz, S. M., Purvis, A., Cornelissen, J. H., Mace, G. M., Donoghue, M. J., Ewers, R. M., Jordano, P., & Pearse, W. D. (2013). Functional traits, the phylogeny of function, and ecosystem service vulnerability. *Ecology and Evolution, 3*(9), 2958–2975. https://doi.org/10.1002/ece3.601
Dobler, D., Martinez, E., Rahmania, R., Gautama, B. D., Riza, F. A., Maes, C., & Dounias, E. (2021). *Floating marine debris along Indonesian coasts: An atlas of strandings based on Lagrangian modelling*. Jakarta.

Dongue, S., Frappart, F., Paris, A., Blarel, F., Niño, F., Seyler, F., Mahe, G., Onguéné, R., Etame, J., Paiz, M. C., & Braun, J. J. (2018). Modeling the Ogooué river discharge based on multi-missions altimetry data. In *Conference: AGU Chapman Conference on Hydrologic Research in the Congo BasinAt.*

Dueri, S., Faugeras, B., & Maury, O. (2012). Modelling the skipjack tuna dynamics in the Indian Ocean with APECOSM-E: Part 1. Model formulation. *Ecological Modelling, 245*, 41–54. https://doi.org/10.1016/j.ecolmodel.2012.02.007

Dumont, H. J. (1998). The Caspian Lake: History, biota, structure, and function. *Limnology and Oceanography, 43*(1), 44–52. https://doi.org/10.4319/lo.1998.43.1.0044

Dunlop, E. S., Baskett, M. L., Heino, M., & Dieckmann, U. (2009). Propensity of marine reserves to reduce the evolutionary effects of fishing in a migratory species. *Evolutionary Applications, 2*(3), 371–393. https://doi.org/10.1111/j.1752-4571.2009.00089.x

Dutkiewicz, S., Follows, M. J., & Bragg, J. G. (2009). Modeling the coupling of ocean ecology and biogeochemistry. *Global Biogeochemical Cycles, 23*(4). https://doi.org/10.1029/2008GB003405

Eakins, B. W., & Sharman, G. F. (2010). *Volumes of the World's Oceans from ETOPO1*. NOAA National Geophysical Data Center.

Ehlers, E., & Krafft, T. (2006). *Earth System Science in the Anthropocene: Emerging Issues and Problems*. Springer.

Eisinger, D., & Thulke, H. H. (2008). Spatial pattern formation facilitates eradication of infectious diseases. *Journal of Applied Ecology, 45*(2), 415–423. https://doi.org/10.1111/j.1365-2664.2007.01439.x

Enberg, K., Jørgensen, C., Dunlop, E. S., Heino, M., & Dieckmann, U. (2009). Implications of fisheries-induced evolution for stock rebuilding and recovery. *Evolutionary Applications, 2*(3), 394–414. https://doi.org/10.1111/j.1752-4571.2009.00077.x

Energy, Mines and Resources Canada. Surveys and Mapping Branch. (1985). Canada - Drainage Basins, The National Atlas of Canada, 5th ed. Map Scale 1:7 500 000. Ottawa.

Evans, M. R. (2012). Modelling ecological systems in a changing world. *Philosophical Transactions of the Royal Society of London. Series B, Biological sciences, 367*(1586), 181–190. https://doi.org/10.1098/rstb.2011.0172

Evans, M. R., Norris, K. J., & Benton, T. G. (2012). Predictive ecology: Systems approaches. *Philosophical transactions of the Royal Society of London. Series B, Biological Sciences, 367*(1586), 163–169. https://doi.org/10.1098/rstb.2011.0191

Evans, M. R., Bithell, M., Cornell, S. J., Dall, S. R., Díaz, S., Emmott, S., Ernande, B., Grimm, V., Hodgson, D. J., Lewis, S. L., Mace, G. M., Morecroft, M., Moustakas, A., Murphy, E., Newbold, T., Norris, K. J., Petchey, O., Smith, M., Travis, J. M., & Benton, T. G. (2013). Predictive systems ecology. *Proceedings. Biological Sciences, 280*(1771), 20131452. https://doi.org/10.1098/rspb.2013.1452

Evans, D. (2021). Nettilling Lake. The Canadian Encyclopedia. https://www.thecanadianencyclopedia.ca/en/article/nettilling-lake

Environmental Protection Agency. (2022). *The Great Lakes*. Retrieved from https://www.epa.gov/greatlakes

Fach, B. A., Hofmann, E. E., & Murphy, E. J. (2002). Modeling studies of antarctic krill Euphausia superba survival during transport across the Scotia Sea. *Marine Ecology Progress Series, 231*, 187–203. https://doi.org/10.3354/meps231187

Fach, B. A., Hofmann, E. E., & Murphy, E. J. (2006). Transport of Antarctic krill (Euphausia superba) across the Scotia Sea. Part ll: Krill growth and survival. *Deep-Sea Research Part I: Oceanographic Research Papers, 53*(6), 1011–1043. https://doi.org/10.1016/j.dsr.2006.03.007

Feduccia, A. (1995). Explosive evolution in tertiary birds and mammals. *Science (New York, N.Y.), 267*(5198), 637–638. https://doi.org/10.1126/science.267.5198.637

Fei, Y. (2011). Chinese scientists verify source of four international rivers, including Yarlung Tsangpo. *Xinhua News Agency (in Chinese)*. Retrieved from https://news.sciencenet.cn/htmlnews/2011/8/251272.shtm

Fennec, R. S. (2005). Growth rate and duration of growth in the adult canine of Smilodon gracilis and inferences on diet through stable isotope analysis. *Feranec Bull FLMNH, 45*(4), 369–377. https://doi.org/10.1016/j.palaeo.2004.01.009

Fekete, B. M., Vörösmarty, C. J., & Grabs, W. (2014). *Dvina Basin: Station: Ust-Pinega.* Retrieved from https://www.compositerunoff.sr.unh.edu/html/Polygons/P6970250.html

Firmansyah, R. (n.d.). *Grafik Neraca Air Pulau Papua.*

Fisher, H. J. (2002). Macroecology: New, or biogeography revisited? *Nature, 417*(6891), 787. https://doi.org/10.1038/417787c

Follows, M. J., Dutkiewicz, S., Grant, S., & Chisholm, S. W. (2007). Emergent biogeography of microbial communities in a model ocean. *Science (New York, N.Y.), 315*(5820), 1843–1846. https://doi.org/10.1126/science.1138544

Forster, M., & Sober, E. (1994). How to tell when simpler, more unified, or less ad hoc theories will provide more accurate predictions. *British Journal for the Philosophy of Science, 45*(1), 1–35. https://doi.org/10.1093/bjps/45.1.1

Fretwell, P., Pritchard, H. D., Vaughan, D. G., Bamber, J. L., Barrand, N. E., Bell, R., Bianchi, C., Bingham, R. G., Blankenship, D. D., Casassa, G., Catania, G., Callens, D., Conway, H., Cook, A. J., Corr, H. F. J., Damaske, D., Damm, V., Ferraccioli, F., Forsberg, R., Fujita, S., Gim, Y., Gogineni, P., Griggs, J. A., Hindmarsh, R. C. A., Holmlund, P., Holt, J. W., Jacobel, R. W., Jenkins, A., Jokat, W., Jordan, T., King, E. C., Kohler, J., Krabill, W., Riger-Kusk, M., Langley, K. A., Leitchenkov, G., Leuschen, C., Luyendyk, B. P., Matsuoka, K., Mouginot, J., Nitsche, F. O., Nogi, Y., Nost, O. A., Popov, S. V., Rignot, E., Rippin, D. M., Rivera, A., Roberts, J., Ross, N., Siegert, M. J., Smith, A. M., Steinhage, D., Studinger, M., Sun, B., Tinto, B. K., Welch, B. C., Wilson, D., Young, D. A., Xiangbin, C., & Zirizzotti, A. (2013). Bedmap2: Improved ice bed, surface and thickness datasets for Antarctica. *The Cryosphere, 7*(1), 375–393. https://doi.org/10.5194/tc-7-375-2013

Fuller, L. M., & Taricska, C. K. (2012). *Water-quality characteristics of Michigan's inland lakes, 2001–10: U.S. Geological Survey Scientific Investigations Report 2011-5233.* http://pubs.usgs.gov/sir/2011/5233/

Fulton, R. J., & Prest, V. K. (1987). Introduction: The Laurentide Ice Sheet and its Significance. *Géographie physique et Quaternaire, 41*(2), 181–186.

Fulton, E. A. (2010). Approaches to end-to-end ecosystem models. *Journal of Marine Systems, 81*(1–2), 171–183. https://doi.org/10.1016/j.jmarsys.2009.12.012

Fulton, E. A., Link, J. S., Kaplan, I. C., Savina-Rolland, M., Johnson, P., Ainsworth, C., Horne, P., Gorton, R., Gamble, R. J., Smith, A. D. M., & Smith, D. C. (2011). Lessons in modelling and management of marine ecosystems: The Atlantis experience. *Fish and Fisheries, 12*(2), 171–188. https://doi.org/10.1111/j.1467-2979.2011.00412.x

Gâştescu, P., & Ţuchiu, E. (2012). The Danube River in The Pontic Sector Hydrological Regime. In P. Gâştescu, W. Lewis Jr., & P. Breţcan (Eds.), *Water Resources and Wetlands, Petre Breţcan Conference Proceedings, 14–16 September 2012, Tulcea – Romania.*

Gates, D. M., Thompson, Michael B. & Thompson, J. N. (2022). Biosphere. *Encyclopedia Britannica.* https://www.britannica.com/science/biosphere

Gneiting, T., & Raftery, A. E. (2005). Atmospheric science. Weather forecasting with ensemble methods. *Science (New York, N.Y.), 310*(5746), 248–249. https://doi.org/10.1126/science.1115255

Grafen, A. (1999). Formal Darwinism, the individual–as–maximizing–agent analogy and bet–hedging. *Proceedings of the Royal Society of London. Series B: Biological Sciences*, 266. https://doi.org/10.1098/rspb.1999.0708

Grafen, A., & Hails, R. S. (2002). *Modern Statistics for the Life Sciences.* Oxford University Press.

Grafen, A. (2007). The formal Darwinism project: A mid-term report. *Journal of Evolutionary Biology, 20*(4), 1243–1254. https://doi.org/10.1111/j.1420-9101.2007.01321.x

Grayson, D. (2006). Holocene underkill. *Proceedings of the National Academy of Sciences of the United States of America, 105*, 4077–4078. https://doi.org/10.1073/pnas.0801272105

Grayson, D. K., & Meltzer, D. J. (2002). Clovis Hunting and Large Mammal Extinction: A Critical Review of the Evidence. *Journal of World Prehistory, 16*, 313–359. https://doi.org/10.1023/A:1022912030020

Griffith, G. P., Fulton, E. A., Gorton, R., & Richardson, A. J. (2012). Predicting interactions among fishing, ocean warming, and ocean acidification in a marine system with whole-ecosystem models. *Conservation Biology: The Journal of the Society for Conservation Biology, 26*(6), 1145–1152. https://doi.org/10.1111/j.1523-1739.2012.01937.x

Grimm, V., & Railsback, S. F. (2012). Pattern-oriented modelling: A 'multi-scope' for predictive systems ecology. *Philosophical transactions of the Royal Society of London. Series B, Biological Sciences, 367*(1586), 298–310. https://doi.org/10.1098/rstb.2011.0180

Guerrero, R. A., Acha, E. M., Framin, M. B., & Lasta, C. A. (1997). Physical oceanography of the Río de la Plata Estuary. *Argentina. Continental Shelf Research, 17*(7), 727–742. https://doi.org/10.1016/S0278-4343(96)00061-1

Gupta, A. (2007). *Large Rivers: Geomorphology and Management*. John Wiley and Sons.

Gutiérrez-Hernández, O., & García, L. V. (2021). Relationship between precipitation and species distribution. In J. Rodrigo-Comino (Ed.), *Precipitation: Earth Surface Responses and Processes* (pp. 239–259). https://doi.org/10.1016/C2019-0-04852-2

Håkanson, L., & Jansson, M. (1983). *Principles of Lake Sedimentology* (1st ed.). Springer.

Håkanson, L. & Jansson, M. (2002). *Principles of lake sedimentology*. Caldwell: The Blackburn Press.

Håkanson, L. (2012). Lakes on Earth, Different Types. In L. Bengtsson, R. W. Herschy, & R. W. Fairbridge (Eds.), *Encyclopedia of Lakes and Reservoirs* (Encyclopedia of Earth Sciences Series) (pp. 471–472). Springer. https://doi.org/10.1007/978-1-4020-4410-6_202

Harris, C., & Murton, J. B. (2005). *Cryospheric Systems: Glaciers and Permafrost*. Geological Society of London.

Haustein, C. H. (2019). Oxidation-reduction reaction. In K. L. Lerner & W. Lerner (Eds.), *The Gale Encyclopedia of Science* (5th ed.). Gale.

Hayes, J. (2020). Evolution of the atmosphere. *Encyclopedia Britannica*. https://www.britannica.com/topic/evolution-of-the-atmosphere-1703862

Hebert, P. (2007). *Great Slave Lake, Northwest Territories. Encyclopedia of Earth*. Environmental Information Coalition, National Council for Science and the Environment.

Hemmer, H. (2002). Die Fäden aus dem Epivillafranchium von Untermassfeld. *In Kahlke, R.-D. (Ed.)*. Das Pleistozän von Untermassfeld Bei Meiningen (Thuringen). *Romisch-Germanisches Zentralmuseum Bande, 40*(3), 699–782. https://doi.org/10.11588/propylaeum.560.

Henshaw, P. C., Charlson, R. J., & Burges, S. J. (2000). 6 - Water and the Hydrosphere. In M. C. Jacobson, R. J. Charlson, H. Rodhe, & G. H. Orians (Eds.), *International Geophysics* (Vol. 72, pp. 109–131). https://doi.org/10.1016/S0074-6142(00)80112-6

Herdendorf, C. E. (1982). Large Lakes of the World. *Journal of Great Lakes Research, 8*(3), 379–412.

Hergarten, S. (2002). *Self-Organized Criticality in Earth Systems*. Springer.

Hieronymi, A. (2013). Understanding Systems Science: A Visual and Integrative Approach. *Systems and Behavioral Research, 30*(5), 580–595. https://doi.org/10.1002/sres.2215

Hofer, M., Peeters, F., Aeschbach-Hertig, W., Brennwald, M., Holocher, J., Livingstone, D. M., Romanovski, V., & Kipfer, R. (2002). Rapid deep-water renewal in Lake Issyk-Kul (Kyrgyzstan) indicated by transient tracers. *Limnology and Oceanography, 4*(47), 1210–1216. https://doi.org/10.4319/lo.2002.47.4.1210

Holmes, R. M., Shiklomanov, A. I., Suslova, A., Tretiakov, M., McClelland, J. W., Spencer, R. G. M., & Tank, S. E. (2018). *River discharge. Arctic report card: Update for 2018*. https://www.arctic.noaa.gov/Report-Card/Report-Card-2018/ArtMID/7878/Article ID/786/Riser-Discharge

Huang, T. H., Chen, C. T. A., Tseng, H. C., Lou, J. Y., Wang, S. L., Yang, L., Kandasamy, S., Gao, X., Wang, J. T., Aldrian, E., Jacinto, G. S., Anshari, G. Z., Sompongchaiyakul, P., & Wang,

B. J. (2017). Riverine carbon fluxes to the South China Sea. Journal of Geophysical Research. *Biogeosciences, 122*(5), 1 1239-1259. https://doi.org/10.1002/2016JG003701

Huggett, R. J. (1999). Ecosphere, biosphere, or Gaia? What to call the global ecosystem. *Global Ecology and Biogeography, 8*(6), 425–431. https://doi.org/10.1046/j.1365-2699.1999.00158.x

Hutchinson, G. E. (1957). *A Treatise on Limnology. Vol.1, Geography, Physics, and Chemistry*. Wiley.

INTAS Project 99-1669. (2002). *A new computer-generated bathymetry map of Lake Baikal*. Retrieved from https://users.ugent.be/~mdbatist/intas/intas.htm

International Lake Environment Committee. (2022). Lago Titicaca (Lake Titicaca). World Lake Database. Retrieved from https://wldb.ilec.or.jp/Display/html/3389

International Commission on Stratigraphy. (2004). *International stratigraphic chart*. Retrieved from https://timescalefoundation.org/ICS_chart_archive/Div_GeolTime2004.pdf

International Union for the Conservation of Nature. (1995). *The Fly River Catchment, Papua New Guinea: A Regional Environmental Assessment*. Published in Collaboration with the Department of Environment and Conservation, Boroko, Papua New Guinea.

Ison, R. (2017). *Systems Practice: How to Act. In situations of uncertainty and complexity in a climate-change world* (2nd ed.). Springer & The Open University. https://doi.org/10.1007/978-1-4471-7351-9

Jacobson, M., Charlson, R. J., Rodhe, H. & Orians, G. H. (2000). *Earth system science, from biogeochemical cycles to global changes* (2nd ed.). London, UK: Elsevier Academic Press.

Jefferson, G. (2001). The American Lion. In Harris, J. (Ed.) Rancho La Brea: Death Trap and Treasure Trove, 28. *Terra, 31*(1).

Joesron, L. (2008). *Country report implementation of demonstration project Mamberamo River Basin,* Indonesia. Retrieved from https://www.restec.or.jp/geoss_ap2/pdf/0415/wg2/sr1/06.pdf

Johnson, L. (1975). Physical and chemical characteristics of Great Bear Lake. *Journal of the Fisheries Research Board of Canada, 32*(11), 1971–1987. https://doi.org/10.1139/f75-234

Jones, R. L. (1980). *Biogeography: Structure, process, pattern, and change within the biosphere*. Hulton Educational Publications.

Jørgensen, C., Enberg, K., Dunlop, E. S., Arlinghaus, R., Boukal, D. S., Brander, K., Ernande, B., Gardmark, A., Johnston, F., Matsumura, S., Pardoe, H., Raab, K., Silva, A., Vainikka, A., Dieckmann, U., Heino, M., & Rijnsdorp, A. D. (2007). Ecology: Managing evolving fish stocks. *Science (New York, N.Y.), 318*(5854), 1247–1248. https://doi.org/10.1126/science.1148089

Kalipeni, E. (2001). International Encyclopedia of the Social & Behavioral Sciences. *African Studies: Geography. International Encyclopedia of the Social & Behavioral Sciences*, 240–244 https://doi.org/10.1016/B0-08-043076-7/03241-1

Kalugin, A. S. (2018). Variations of the Present-Day Annual and Seasonal Runoff in the Far East and Siberia with the Use of Regional Hydrological and Global Climate Models. *Water Resources, 45*(Suppl. 1), S102–S111. https://doi.org/10.1134/S0097807818050366

Kammerer, J. C. (1990). *Largest Rivers in the United States*. Retrieved from https://pubs.usgs.gov/of/1987/ofr87-242/

Kasting, J. F. (1993). Earth's early atmosphere. *Science (New York, N.Y.), 259*(5097), 920–926. https://doi.org/10.1126/science.11536547

Kent, M. (2005). Biogeography and macroecology. *Progress in Physical Geography, 29*(2), 256–264. https://doi.org/10.1191/0309133305pp447pr

Kent, M. (2007). Biogeography and landscape ecology. *Progress in Physical Geography, 31*(3), 345–355. https://doi.org/10.1177/0309133307077905

Kimbrough, R. A., Ruppert, G. P., Wiggins, W. D., Smith, R. R., & Kresch, D. L. (2006). Water Data Report WA-05-1: Klickitat and White Salmon River Basins and the Columbia River from Kennewick to Bonneville Dam. In *Water Resources Data-Washington Water Year 2005*. United States Geological Survey.

Kitano, H. (2002). Systems biology: A brief overview. *Science, 295*(5560), 1662–1664. https://doi.org/10.1126/science.1069492

Kodayev, G. V. (1973). Morphometry of Lake Issyk-Kul. *News of the All-Union Geographic Society (Izvestiya VGO).*

Kokko, H., & Lopez-Sepulcre, A. (2007). The ecogenetic link between demography and evolution: Can we bridge the gap between theory and data? *Ecology Letters, 10*(9), 773–782. https://doi.org/10.1111/j.1461-0248.2007.01086.x

Kujala, H., Burgman, M. A., & Mollanen, A. (2013). Treatment of uncertainty in conservation under climate change. *Conservation Letters, 6*(2), 73–85. https://doi.org/10.1111/j.1755-263X.2012.00299.x

Kumar, R., Singh, R. D., & Sharma, K. D. (2005). Water Resources of India. *Current Science, 89*(5), 794–811. https://doi.org/10.5040/9781501302343.ch-002

Kumar, S. (2020). *River Engineering*. Khanna Publishing House.

Kumar, A. (2022). *Ecosystem-Based Adaptation: Approaches to Sustainable Management of Aquatic Resources*. Elsevier B.V.

Kump, L. R., Kasting, J. F., & Crane, R. G. (2004). *The Earth System* (2nd ed.). Prentice Hall.

Kurtén, B., & Werdelin, L. (1990). Relationships between North and South American Smilodon. *Journal of Vertebrate Paleontology, 10*(2), 158–169. https://doi.org/10.1080/02724634.1990.10011804

Kurtén, B., & Anderson, A. (1980). *Pleistocene mammals of North America*. Columbia University Press.

Kurz, W. A., Dymond, C. C., Stinson, G., Rampley, G. J., Neilson, E. T., Carroll, A. L., Ebata, T., & Safranyik, L. (2008). Mountain pine beetle and forest carbon feedback to climate change. *Nature, 452*(7190), 987–990. https://doi.org/10.1038/nature06777

Kuusisto, E. (2005). World water resources and problems. In M. Berglund (Ed.), *International Environmental Law-making and Diplomacy Review* (pp. 153–166). University of Joensuu.

Lehner, B., Verdin, K., & Jarvis, A. (2008). New global hydrography derived from spaceborne elevation data. *Eos, Transactions, AGU, 89*(10), 93–94. https://doi.org/10.1029/2008EO100001

Lehodey, P., Murtugudde, R., & Senina, I. (2010). Bridging the gap from ocean models to population dynamics of large marine predators: A model of mid-trophic functional groups. *Progress in Oceanography, 84*(1–2), 69–84. https://doi.org/10.1016/j.pocean.2009.09.008

Lenton, T. (2016). *Earth System Science: A Very Short Introduction*. Oxford University Press.

Leontiev, O. K., Kosarev, A. N., & Owen, L. (2021). Caspian Sea. *Encyclopedia Britannica*. https://www.britannica.com/place/Caspian-Sea

Levchenko, V. F., & Starobogatov, Y. I. (2010). Origin of biosphere. *Journal of Evolutionary Biochemistry and Physiology, 46*, 213–226. https://doi.org/10.1134/S0022093010020122

Levchenko, V. F., Kazansky, A. B., Sabirov, M. A., & Semenova, E. M. (2012). Early Biosphere: Origin and Evolution. In N. Ishwaran (Ed.), *The Biosphere*. IntechOpen. https://doi.org/10.5772/35301

Likens, G. (2010). *Biogeochemistry of Inland Waters*. Academic Press.

Lomolino, M. V., Riddle, B. R., & Whittaker, R. J. (2016). *Biogeography*. Oxford University Press.

Lovelock, J. (1979). *Gaia: A New Look at Life on Earth*. Oxford University Press.

Loy, W. G., Allan, S., Buckley, A. R., & Meacham, J. E. (2001). *Atlas of Oregon*. University of Oregon Press.

Ludwig, C., & Steffen, W. (2018). The 1950s as the Beginning of the Anthropocene. In D. A. Dellasala & M. I. Goldstein (Eds.), *Encyclopedia of the Anthropocene* (Vol. 1, pp. 45–56). Elsevier B.V. https://doi.org/10.1016/B978-0-12-809665-9.09940-7

Lutgens, F. K., Tarbuck, E. J., & Tasa, D. (2017). *Essentials of Geology* (13th ed.). Pearson.

Mackenzie River Basin Board. (2003). *Mackenzie River Basin. State of the Aquatic Ecosystem Report 2003. 1. Whole Basin Overview*. Retrieved from https://www.mrbb.ca/uploads/media/5c1aaf80799a9/whole-basin-overview.pdf?v1

Marinescu, D. C. (2017). *Complex Systems and Clouds: A Self-Organization and Self-Management Perspective*. Elsevier. https://doi.org/10.1016/C2015-0-00979-0

Marsh, W. M., & Kaufman, M. M. (2013). *Physical Geography: Great Systems and Global Environments*. Cambridge University Press.

Martin, R. (2011). *Earth's Evolving Systems: The History of Planet Earth*. Jones & Bartlett Learning.

Martin, L. D. (1989). Fossil History of the Terrestrial Carnivora. In J. L. Gittleman (Ed.), *Carnivore Behavior, Ecology, and Evolution* (pp. 536–568). Springer. https://doi.org/10.1007/978-1-4757-4716-4_20

Martin, R. E., Wehmiller, J. F., Harris, M. S., & Liddell, W. D. (1996). Comparative taphonomy of bivalves and foraminifera from Holocene tidal flat sediments, Bahia la Choya, Sonora, Mexico (Northern Gulf of California): Taphonomic grades and temporal resolution. *Paleobiology, 22*(1), 80–90. https://doi.org/10.1017/S0094837300016031

Marty, B. (2011). Earth's Atmosphere, Origin and Evolution of. In *Encyclopedia of Astrobiology*. Springer. https://doi.org/10.1007/978-3-642-11274-4_470

Mathisen, O. A., Sands, N. J., & Haubenstock, N. (2002). Trophic ranking of biota in Iliamna Lake, Alaska. *Verhandlungen des Internationalen Verein Limnologie (Stuttgart), 28*, 1060–1065.

Matthews, B., Narwani, A., Hausch, S., Nonaka, E., Peter, H., Yamamichi, M., Sullam, K. E., Bird, K. C., Thomas, M. K., Hanley, T. C., & Turner, C. B. (2011). Toward an integration of evolutionary biology and ecosystem science. *Ecology Letters, 14*(7), 690–701. https://doi.org/10.1111/j.1461-0248.2011.01627.x

Maury, O., & Poggiale, J. C. (2013). From individuals to populations to communities: A dynamic energy budget model of marine ecosystem size-spectrum including life history diversity. *Journal of Theoretical Biology, 324*, 52–71. https://doi.org/10.1016/j.jtbi.2013.01.018

McGill. (2022). *Earth system science*. Retrieved from https://www.mcgill.ca/undergraduate-admissions/program/earth-system-science

Medvigy, D., & Moorcroft, P. R. (2012). Predicting ecosystem dynamics at regional scales: An evaluation of a terrestrial biosphere model for the forests of northeastern North America. *Philosophical transactions of the Royal Society of London. Series B, Biological Sciences, 367*(1586), 222–235. https://doi.org/10.1098/rstb.2011.0253

Medvigy, D., Wofsy, S. C., Munger, J. W., Hollinger, D. Y., & Moorcroft, P. R. (2009). Mechanistic scaling of ecosystem function and dynamics in space and time: Ecosystem demography model version 2. *Journal of Geophysical Research, 114*, G1. https://doi.org/10.1029/2008JG000812

Meier, M. F. (2022). Glacier. *Encyclopedia Britannica*. https://www.britannica.com/science/ glacier

Mekong River Commission. (2005). *Overview of the Hydrology of the Mekong Basin*. MRC.

Meloro, C., & Slater, G. J. (2012). Covariation in the skull modules of cats: The challenge of growing saber-like canines. *Journal of Vertebrate Paleontology, 32*(3), 677–685. http://www.jstor.org/stable/41515287

Merritts, D., De Wet, A., & Menking, K. (1998). *Environmental Geology: An Earth System Science Approach*. W.H. Freeman.

Micklin, P. P., & Kuzin, P. S. (2019). Volga River. *Encyclopedia Britannica*. https://www.britannica.com/place/Volga-River

Milner-Gulland, E. J. (2012). Interactions between human behaviour and ecological systems. *Philosophical transactions of the Royal Society of London. Series B, Biological Sciences, 367*(1586), 270–278. https://doi.org/10.1098/rstb.2011.0175

Mitra, A., Fazli, P., Zaman, S., Pramanick, P., & Mitra, A. (2022). *Lakes and their origin*. Retrieved from https://www.researchgate.net/profile/Abhijit-Mitra-4/publication/265339171_LAKES_AND_THEIR_ORIGIN/links/6086a65d907dcf667bc4a4a9/ LAKES-AND-THEIR-ORIGIN.pdf

Mobus, G. E., & Kalton, M. C. (2015). *Principles of systems science*. Springer.

Moloney, C. L., St John, M. A., Denman, K. L., Karl, D. M., Köster, F. W., Sundby, S., & Wilson, R. P. (2011). Weaving marine food webs from end to end under global change. *Journal of Marine Systems, 84*(3–4), 106–116. https://doi.org/10.1016/j.jmarsys.2010.06.012

Montague, C. L. (2014). *Systems ecology*. Oxford Bibliographies. Retrieved from https://www.oxfordbibliographies.com/view/document/obo-9780199830060/obo-9780199830060-0078.xml

Moorcroft, P. R. (2006). How close are we to a predictive science of the biosphere?*Trends in. Ecology and Evolution, 21*(7), 400–407. https://doi.org/10.1016/j.tree.2006.04.009

Moorcroft, P. R., Hurtt, G. C., & Pacala, S. W. (2001). A method for scaling vegetation dynamics: The ecosystem demography model (ED). *Ecological Monographs, 71*(4), 557–586. https://doi.org/10.1890/0012-9615(2001)071[0557:AMFSVD]2.0.CO;2

Morales, F. (1999). *Geografía física del territorio en reclamación: Guayana Esequiba*. Fondo Editorial Humanidades.

Mork, K. A., Gilbey, J., Hansen, L. P., Jensen, A. J., Jacobsen, J. A., Holm, M., Holst, J. C., Maoiléidigh, Ó. N., Vikebø, F., McGinnity, P., Melle, W., Thomas, K., Verspoor, E., & Wennevik, V. (2012). Modelling the migration of post-smolt Atlantic salmon (Salmo salar) in the Northeast Atlantic. *ICES Journal of Marine Science, 69*, 1616–1624. https://doi.org/10.1093/icesjms/fss108

Morlighem, M., Rignot, E., Binder, T., Blankenship, D., Drews, R., Eagles, G., Eisen, O., Ferraccioli, F., Forsberg, R., Fretwell, P., & Goel, V. (2020). Deep glacial troughs and stabilizing ridges unveiled beneath the margins of the Antarctic ice sheet. *Nature Geoscience, 13*(2), 132–137.

Moss, R. H., Edmonds, J. A., Hibbard, K. A., Manning, M. R., Rose, S. K., van Vuuren, D. P., Carter, T. R., Emori, S., Kainuma, M., Kram, T., Meehl, G. A., Mitchell, J. F., Nakicenovic, N., Riahi, K., Smith, S. J., Stouffer, R. J., Thomson, A. M., Weyant, J. P., & Wilbanks, T. J. (2010). The next generation of scenarios for climate change research and assessment. *Nature, 463*(7282), 747–756. https://doi.org/10.1038/nature08823

Moustakas, A., & Evans, M. R. (2013). Integrating evolution into ecological modelling: Accommodating phenotypic changes in agent based models. *PloS One, 8*(8), e71125. https://doi.org/10.1371/journal.pone.0071125

Mukherjee, S., Ray, L., Maurya, S., Shalivahan, & Kumar, P. (2022). Nature of the lithosphere-asthenosphere boundary beneath the Eastern Dharwar Craton of the Indian Shield. *Journal of Asian Earth Sciences, 227*. https://doi.org/10.1016/j.jseaes.2021.105071

Muranov, A. P. & Greer, C. E. (2022). Yangtze River. *Encyclopedia Britannica*. https://www.britannica.com/place/Yangtze-River

Muranov, A. P., Greer, C. E. & Owen, L. (2016). Amur River. *Encyclopedia Britannica*. https://www.britannica.com/place/Amur-River

Murdie, R. E., Pugh, D. T., Styles, P., & Muñoz, M. (1999). Heatflow, Temperature and Bathymetry of Lago General Carrera and Lago Cochrane, Southern Chile. In *Extended Extracts of the Fourth International Symposium on Andean Geodynamics, Gottingen, Germany* (pp. 539–542). Institut De Recherche Pour Le Développement.

Murphy, E. J., Watkins, J. L., Trathan, P. N., Reid, K., Meredith, M. P., Thorpe, S. E., Johnston, N. M., Clarke, A., Tarling, G. A., Collins, M. A., Forcada, J., Shreeve, R. S., Atkinson, A., Korb, R., Whitehouse, M. J., Ward, P., Rodhouse, P. G., Enderlein, P., Hirst, A. G., Martin, A. R., et al. (2007). Spatial and temporal operation of the Scotia Sea ecosystem: A review of large-scale links in a krill centred food web. *Philosophical transactions of the Royal Society of London. Series B, Biological Sciences, 362*(1477), 113–148. https://doi.org/10.1098/rstb.2006.1957

Murphy, E. J., & Hofmann, E. E. (2012). End-to-end in Southern Ocean ecosystems. *Current Opinion in Environmental Sustainability, 4*(3), 264–271. https://doi.org/10.1016/j.cosust.2012.05.005

Murphy, E. J., Cavanagh, R. D., Hofmann, E. E., Hill, S. L., Constable, A. J., Costa, D. P., Pinkerton, M. H., Johnston, N. M., Trathan, P. N., Klinck, J. M., Wolf-Gladrow, D. A., Daly, K. L., Maury, O., & Doney, S. C. (2012). Developing integrated models of Southern Ocean food webs: Including ecological complexity, accounting for uncertainty and the importance of scale. *Progress in Oceanography, 102*, 74–92. https://doi.org/10.1016/j.pocean.2012.03.006

Murphy, E. J., Thorpe, S. E., Watkins, J. L., & Hewitt, R. P. (2004). Modeling the krill transport pathways in the Scotia Sea: Spatial and environmental connections generating the seasonal distribution of krill. *Deep-sea Research Part Ii-topical Studies in Oceanography, 51*(12–13), 1435–1456. https://doi.org/10.1016/j.dsr2.2004.06.019

NASA. (2003). *Earth science enterprise strategy*. Retrieved from https://web.archive.org/web/20150923061559/ http://science.nasa.gov/media/medialibrary/2010/03/31/ESE_Strategy2003.pdf

National Geographic. (2022a). *Earth systems*. Retrieved from https://education.nationalgeographic.org/resource/earths-systems

National Geographic. (2022b). *Atmosphere*. Retrieved from https://education.nationalgeographic.org/resource/resource-library-atmosphere

National Ocean Service. (2021). *Why do we have an ocean?* Retrieved from https://oceanservice.noaa.gov/facts/why_oceans.html#:~:text=The%20ocean%20formed%20billions%20of,know%20as%20our%20world%20ocean

National Ocean Service. (2022). *What is the cryosphere?* Retrieved from https://oceanservice.noaa.gov/facts/cryosphere.html

National Snow and Ice Data Center. (2022). *Glaciers*. Retrieved from https://nsidc.org/learn/parts-cryosphere/glaciers

National Wildlife Federation. (2022). *Mammals*. Retreived from https://www.nwf.org/Educational-Resources/Wildlife-Guide/Mammals

Natural Resource Canada. (2002). *Lakes*. Retrieved from https://web.archive.org/web/20070410230512/ http://atlas.nrcan.gc.ca/site/english/learningresources/facts/lakes.html

Natural Resources Canada. (2010). *The Atlas of Canada: Rivers*. Retrieved from https://web.archive.org/web/20130122002203/, http://atlas.nrcan.gc.ca/site/English/learningresources/facts/rivers.html

Neugebauer, H. J., & Simmer, C. (2003). *Dynamics of Multiscale Earth Systems*. Springer.

Nicholson, E., Mace, G. M., Armsworth, P. R., Atkinson, G., Buckle, S., Clements, T., Ewers, R. M., Fa, J. E., Gardner, T. A., Gibbons, J., Grenyer, R., Metcalfe, R., Mourato, S., Muûls, M., Osborn, D., Reuman, D. C., Watson, C., & Milner-Gulland, E. J. (2009). Priority Research Areas for Ecosystem Services in a Changing World. *Journal of Applied Ecology, 46*(6), 1139–1144. https://doi.org/10.1111/j.1365-2664.2009.01716.x

Nilsson, C., Reidy, C. A., Dynesius, M., & Revenga, C. (2005). Fragmentation and flow regulation of the world's large river systems. *Science (New York, N.Y.), 308*(5720), 405–408. https://doi.org/10.1126/science.1107887

Nutman, A. P. (2006). Antiquity of the oceans and continents. *Elements, 2*(4), 223–227.

Odum, E. P. (1964). The new ecology. *BioScience, 14*, 14–16. https://doi.org/10.2307/1293228

Ohtani, S., Fujii, R., Hesse, M., & Lysak, R. L. (2000). *Magnetospheric Current Systems*. American Geophysical Union.

Olson, S. L., & Parris, D. C. (1987). *The Cretaceous Birds of New Jersey*. Smithsonian Institution. https://doi.org/10.5479/si.00810266.63.1

Osadchiev, A., Medvedev, I., Shchuka, S., Kulikov, M., Spivak, E., Pisareva, M., & Semiletov, I. (2020). Influence of estuarine tidal mixing on structure and spatial scales of large river plumes. *Ocean Science, 16*(4), 781–798. https://doi.org/10.5194/os-16-781-2020

Ostende, L. W. H., van den Morlo, M., & Nagel, D. (2006). Fossils explained 52: Majestic killers: The sabre-toothed cats. *Geology Today, 22*(4), 150–157. https://doi.org/10.1111/j.1365-2451.2006.00572.x

Ostrom, E. (2009). A general framework for analyzing sustainability of social-ecological systems. *Science (New York, N.Y.), 325*(5939), 419–422. https://doi.org/10.1126/science.1172133

Ozima, M. (1975). Ar isotopes and Earth-atmosphere evolution models. *Geochimica et Cosmochimica Acta, 39*(8), 1127–1134. https://doi.org/10.1016/0016-7037(75)90054-X

Pacala, S. W., Canham, C. D., Saponara, J., Silander, J. A., Kobe, R. K., & Ribbens, E. (1996). Forest Models Defined by Field Measurements: Estimation. *Error Analysis and Dynamics. Ecological Monographs, 66*(1), 1–43. https://doi.org/10.2307/2963479

Pahl-Wostl, C., Giupponi, C., Richards, K., Binder, C., de Sherbinin, A., Sprinz, D., Toonen, T., & van Bers, C. (2013). Transition towards a new global change science: Requirements for methodologies, methods, data and knowledge. *Environmental Science & Policy, 28*, 36–47. https://doi.org/10.1016/j.envsci.2012.11.009

Panagiotopoulos, C., Sempéré, R., Jacq, V., & Charrière, B. (2014). Composition and distribution of dissolved carbohydrates in the Beaufort Sea Mackenzie margin (Arctic Ocean). *Marine Chemistry, 166*, 92–102. https://doi.org/10.1016/j.marchem.2014.09.004

Parry, H., & Bithell, M. (2011). Large scale agent-based modelling: A review and guidelines for model scaling. In A. Heppenstall, A. T. Crooks, L. M. See, & M. Batty (Eds.), *Agent-based Models for Geographical Systems* (pp. 271–308). Springer.

Penčev, P. G., & Pinka, P. G. (2022). Danube River. *Encyclopedia Britannica*. https://www.britannica.com/ place/Danube-River

Penn, J. R. (2001). *Rivers of the World: A Social, Geographical, and Environmental Sourcebook*. ABC-CLIO.

Pepin, R. O. (1991). On the origin and early evolution of terrestrial planetary atmospheres and meteoritic volatiles. *Icarus, 92*(1), 1–79. https://doi.org/10.1016/0019-1035(91)90036-S

Pielke, R. A. (2022). Atmosphere. *Encyclopedia Britannica*. https://www.britannica.com/science/atmosphere

Pilarczyk, K. (2003). *Bank erosion Mekong Delta and Red River*. Mission report – 23 November-6 December 2003. Delft.

Polasky, S., Carpenter, S. R., Folke, C., & Keeler, B. (2011). Decision-making under great uncertainty: Environmental management in an era of global change. *Trends in Ecology & Evolution, 26*(8), 398–404. https://doi.org/10.1016/j.tree.2011.04.007

Popov, I. V., & Greer, C. E. (2022). Yellow River. *Encyclopedia Britannica*. https://www.britannica.com/place/Yellow-River

Povisen, A. F. (1993). *Papua New Guinea -Fisheries survey of the upper Purari River Part 2 - Results and discussion. A report prepared for the Sepik River Fish Stock Enhancement Project, PNG/85/001*. Food and Agriculture Organisation Of The United Nations.

Prabhu, T. L. (2021). *Agricultural Engineering: An Introduction to Agricultural Engineering*. NestFame Creations Pvt. Ltd.

Price, S. (2010). *River discharge database*. Retrieved from https://web.archive.org/web/20100621153903/ http://www.sage.wisc.edu/riverdata/scripts/station_table.php?qual=32&filenum=865

Purves, D., & Pacala, S. (2008). Predictive models of forest dynamics. *Science (New York, N.Y.), 320*(5882), 1452–1453. https://doi.org/10.1126/science.1155359

Purves, D., Scharlemann, J., Harfoot, M., Newbold, T., Tittensor, D. P., Hutton, J., & Emmott, S. (2013). Ecosystems: Time to model all life on Earth. *Nature, 493*(7432), 295–297. https://doi.org/10.1038/493295a

Purves, D. W., Lichstein, J. W., Strigul, N., & Pacala, S. W. (2008). Predicting and understanding forest dynamics using a simple tractable model. *Proceedings of the National Academy of Sciences of the United States of America, 105*(44), 17018–17022. https://doi.org/10.1073/pnas.0807754105

Ra, K. (2011). *Water quality management at River Basin in Myanmar*. Retrieved from http://www.wepa-db.net/pdf/1203forum/10.pdf

Radhika, R., Firmansyah, R., & Hatmoko, W. (2017). Computation of surface water availability in Indonesia based on satellite data. *Jurnal Sumber Daya Air*. https://doi.org/10.32679/jsda.v13i2.206

R-Arctic Net. (2001). *Kolyma At Kolymskoye: Anadyr Kolyma*. Retrieved from https://www.r--arcticnet.sr.unh.edu/v4.0/ViewPoint.pl?Point=5952

Rawn-Schatzinger, V. (1992). The scimitar cat Homotherium serum Cope: Osteology, functional morphology, and predator behavior. *Illinois State Museum Reports of Investigation, 47*, 1–80. Retrieved from https://illinoisstatemuseum.org/content/scimitar-cat-homotherium-serum-cope-osteology-functional-morphology-and-predatory-behavior

Regan, H. M., Colyvan, M., & Burgman, M. A. (2002). A taxonomy and treatment of uncertainty for ecology and conservation biology. *Ecological Applications, 12*(2), 618–628. https://doi.org/10.1890/1051-0761(2002)012[0618:ATATOU]2.0.CO;2

Reptile Database. (2022). *What's new? 2022*. Retrieved from http://www.reptile-database.org/db-info/news.html

Revenga, C., Nackoney, J., Hoshino, E., Kura, Y. & Maidens, J. (2003). *Water resources eAtlas: Watersheds of the World*. 2003.

Ritchie, H. & Roser, M. (2021). *Biodiversity*. Published online at OurWorldInData.org. Retrieved from https://ourworldindata.org/biodiversity.

Ricklefs, R. E., & Jenkins, D. G. (2011). Biogeography and ecology: Towards the integration of two disciplines. *Philosophical transactions of the Royal Society of London. Series B,*

Biological sciences, 366(1576), 2438–2448. https://royalsocietypublishing.org/doi/10.1098/rstb.2011.0066

Rincón, A. D., Prevosti, F. J., & Parra, G. E. (2011). New saber-toothed cat records (Felidae: Machairodontinae) for the Pleistocene of Venezuela, and the Great American Biotic Interchange. *Journal of Vertebrate Paleontology, 31*(2), 468–478. http://www.jstor.org/stable/25835839

Rivers Network. (2020). Retrieved from https://www.riversnetwork.org/

Robson, B., & Webster, I. (2006). Representing the effects of subgrid-scale variations in bathymetry on light and primary production. *Environmental Modelling & Software, 21*(6), 802–811. https://doi.org/10.1016/j.envsoft.2005.02.007

Rounsevell, M. D., Pedroli, B., Erb, K., Gramberger, M. R., Busck, A. G., Haberl, H., Kristensen, S. B., Kuemmerle, T., Lavorel, S., Lindner, M., Lotze-Campen, H., Metzger, M. J., Murray-Rust, D., Popp, A., Pérez-Soba, M., Reenberg, A., Vădineanu, A., Verburg, P. H., & Wolfslehner, B. (2012). Challenges for land system science. *Land Use Policy, 29*(4), 899–910. https://doi.org/10.1016/j.landusepol.2012.01.007

Ruttner, F. (1963). *Fundamentals Of Limnology*. University of Toronto Press.

Ruzek. M. (2018). *Earth system science in a nutshell.* Retrieved from https://serc.carleton.edu/introgeo/earthsystem/nutshell/index.html

Rychert, C. A., Harmon, N., Constable, S., & Wang, S. (2020). The Nature of the Lithosphere-Asthenosphere Boundary. JGR Solid. *Earth, 125*(10), e2018JB016463. https://doi.org/10.1029/2018JB016463

Salesa, M. J., Pesquero, M. D., Siliceo, G., Antón, M., Alcalá, L., & Morales, J. (2012). A rich community of Felidae (Mammalia, Carnivora) from the late Miocene (Turolian, MN 13) site of Las Casiones (Villalba Baja, Teruel, Spain). *Journal of Vertebrate Paleontology, 32*(3), 658–676. https://doi.org/10.1080/02724634.2012.649816

Sanmartín, I. (2012). Historical Biogeography: Evolution in Time and Space. *Evolution: Education and Outreach, 5*, 555–568. https://doi.org/10.1007/s12052-012-0421-2

Sarafian, A. R., Nielsen, S. G., Marschall, H. R., McCubbin, F. M., & Monteleone, B. D. (2014). Early solar system. Early accretion of water in the inner solar system from a carbonaceous chondrite-like source. *Science (New York, N.Y.), 346*(6209), 623–626. https://doi.org/10.1126/science.1256717

Savvaitova, K., & Petr, T. (1992). Lake Issyk-Kul, Kirgizia. *International Journal of Salt Lake Research, 1*(2), 21–46. https://doi.org/10.1007/BF02904361

Schlesinger, W. H., & Bernhardt, E. S. (2020). The Atmosphere. *Biogeochemistry*, 51–97. https://doi.org/10.1016/B978-0-12-814608-8.00003-7

Schmidt, G. A. (2007). The physics of climate modeling. *Physics Today, 60*(1), 72–73. https://doi.org/10.1063/1.2709569

Schoener, T. W. (2011). The newest synthesis: Understanding the interplay of evolutionary and ecological dynamics. *Science (New York, N.Y.), 331*(6016), 426–429. https://doi.org/10.1126/science.1193954

Seidl, R., Lexer, M. J., Jäger, D., & Hönninger, K. (2005). Evaluating the accuracy and generality of a hybrid patch model. *Tree Physiology, 25*(7), 939–951. https://doi.org/10.1093/treephys/25.7.939

Serruya, C., & Pollinger, U. (1983). *Lakes of the Warm Belt*. Cambridge University Press.

Shaw, C., & Cox, S. (2006). The large carnivorans: Wolves, bears, and big cats. In G. Jefferson & L. Lindsay (Eds.), *Fossil Treasures of the Anza-Borrego Desert: The Last Seven Million Years* (pp. 187–188). Sunbelt Publications.

Shaw, G. H. (2008). Earth's atmosphere – Hadean to early Proterozoic. *Geochemistry, 68*(3), 235–264. https://doi.org/10.1016/j.chemer.2008.05.001

Shibru, T. (2019). *Nile River is "taker of life", say Egyptians but not with Grand Ethiopian Renaissance Dam.* Retrieved from https://borkena.com/2020/02/24/nile-river-is-taker-of-life-say-egyptians-but-not-with-grand-ethiopian-renaissance-dam/

Shiklomanov, I. (1993). World freshwater resources. In P. H. Gleick (Ed.), *Water in Crisis: A Guide to the World's Fresh Water Resources*. Oxford University Press.

Shugart, H. H., & O'Neill, R. V. (Eds.). (1979). *Systems Ecology*. Dowden, Hutchinson & Ross.
Slater, G., & Van Valkenburgh, B. (2008). Long in the tooth: Evolution of sabertooth cat cranial shape. *Paleobiology, 34*(3), 403–419. https://doi.org/10.1666/07061.1
Smith, P., Albanito, F., Bell, M., Bellarby, J., Blagodatskiy, S., Datta, A., Dondini, M., Fitton, N., Flynn, H., Hastings, A., Hillier, J., Jones, E. O., Kuhnert, M., Nayak, D. R., Pogson, M., Richards, M., Sozanska-Stanton, G., Wang, S., Yeluripati, J. B., Bottoms, E., Brown, C., Farmer, J., Feliciano, D., Hao, C., Robertson, A., Vetter, S., Wong, H. M., & Smith, J. (2012). Systems approaches in global change and biogeochemistry research. *Philosophical transactions of the Royal Society of London. Series B, Biological Sciences, 367*(1586), 311–321. https://doi.org/10.1098/rstb.2011.0173
Sorkin, B. (2008). A biomechanical constraint on body mass in terrestrial mammalian predators. *Lethaia, 41*(4), 333–347. https://doi.org/10.1111/j.1502-3931.2007.00091.x
Spiegelhalter, D. J., & Riesch, H. (2011). Don't know, can't know: Embracing deeper uncertainties when analysing risks. *Philosophical transactions. Series A, Mathematical, Physical, and Engineering Sciences, 369*(1956), 4730–4750. https://doi.org/10.1098/rsta.2011.0163
Speight, J. G. (2020). 3 - Water chemistry. *Natural Water Remediation: Chemistry and Technology, 91-129*. https://doi.org/10.1016/B978-0-12-803810-9.00003-6
Stadnyk, T. A., Tefs, A., Broesky, M., Dery, S. J., Myers, P. G., Ridenour, N. A., Koenig, K., Vonderbank, L., & Gustafsson, D. (2021). Changing freshwater contributions to the Arctic: A 90-year trend analysis (1981–2070). *Elementa Science of the Anthropocene, 9*(1), 00098. https://doi.org/10.7939/r3-bdg4-mz73
Stanley, S. M. (2005). *Earth System History*. W.H. Freeman.
Steffen, W., Richardson, K., Rockström, J., Schellnhuber, H. J., Dube, O. P., Dutreuil, S., Lenton, T. M., & Lubchenco, J. (2020). The emergence and evolution of Earth System Science. *Nature Reviews Earth & Environment, 1*, 54–63. https://doi.org/10.1038/s43017-019-0005-6
Stierwalt, S. (2019). *How did water get on earth?* Retrieved from https://www.scientific american.com/article/how-did-water-get-on-earth/
Strahler, A. H. (2013). *Introducing Physical Geography* (6th ed.). Wiley.
Strigul, N. S., Pristinski, D., Purves, D., Dushoff, J., & Pacala, S. W. (2008). Scaling from trees to forests: Tractable macroscopic equations for forest dynamics. *Ecological Monographs, 78*(4), 523–545. https://doi.org/10.1890/08-0082.1
Thomas, S. G. (2016). *The Dictionary of Physical Geography* (4th ed.). Wiley. https://doi.org/10.1002/9781118782323
Tl-myubi-tv. (2022). *Rivers*. Retrieved from https://tl-myubi-tv.translate.goog/18967-where-is-the-xi-jiang-river-located-on-a-map?_x_tr_sl=tl&_x_tr_tl=en&_x_tr_hl=en&_x_tr_pto=sc
Towner, J., Cloke, H. L., Zsoter, E., Flamig, Z., Hoch, J. M., Bazo, J., Coughlan de Perez, E., & Stephens, E. M. (2019). Assessing the performance of global hydrological models for capturing peak river flows in the Amazon basin. *Hydrol Earth System Science, 23*, 3057–3080. https://doi.org/10.5194/hess-23-3057-2019
Travers, M., Shin, Y.-J., Jennings, S., & Cury, P. (2007). Towards end-to-end models for investigating the effects of climate and fishing in marine ecosystems. *Progress in Oceanography, 75*(4), 751–770. https://doi.org/10.1016/j.pocean.2007.08.001
Tsonis, A. A., & Elsner, J. B. (2007). *Nonlinear Dynamics in Geosciences*. Springer.
Turner, A. (1987). Megantereon cultridens (Cuvier) (Mammalia, Felidae, Machairodontinae) from Plio-Pleistocene Deposits in Africa and Eurasia, with Comments on Dispersal and the Possibility of a New World Origin. *Journal of Paleontology, 61*(6), 1256–1268. http://www.jstor.org/stable/1305213
Turner, A., & Antón, M. (1997). *The Big Cats and Their Fossil Relatives: An Illustrated Guide to Their Evolution and Natural History*. Columbia University Press.
United Nations Environment Program. (2016). Water System Information Sheets: Pacific Island Countries, Australia & Antarctica. In L. Talaue-McManus (Ed.), *Transboundary Waters: A Global Compendium, Volume 6-Annex K*. United Nations Environment Program (UNEP).

United States Geological Survey. (2022). *Earthquake glossary*. Retrieved from https://earthquake.usgs.gov/learn/glossary/?term=lithosphere

Valley, J. W. (2006). Early Earth. *Elements, 24*, 201–204.

Valley, J. W., Lackey, J. S., Cavosie, A. J., Clechenko, C. C., Spicuzza, M. J., Basei, M. A. S., Bindeman, I. N., Ferreira, V. P., Sial, A. N., King, E. M., Peck, W. H., Sinha, A. K., & Wei, C. S. (2005). 4.4 billion years of crustal maturation: Oxygen isotope ratios of magmatic zircon. *Contributions to Mineralogy and Petrology, 150*, 561–580. https://doi.org/10.1007/s00410-005-0025-8

Van Asselt, M. B. A., & Rotmans, J. (2002). Uncertainty in integrated assessment modelling. *Climatic Change, 54*, 75–105. https://doi.org/10.1023/A:1015783803445

Van der Leeden, F., Troise, F. L., & Todd, D. K. (1990). *The Water Encyclopedia* (2nd ed., pp. 198–200). Lewis Publishers.

Van Dyne, G. M. (1966). *Ecosystems, systems ecology, and systems ecologists*. Retrieved from https://www.osti.gov/servlets/purl/4516278

Van Dyne, G. M. (1969). Ecosystems, systems ecology, and systems ecologists. In G. W. Cox (Ed.), *Readings in conservation ecology* (pp. 21–47). Appleton-Century-Crofts.

Von Bloh, W. (2019). Climate Change 2: Long-Term Dynamics. In B. Fath (Ed.), *Encyclopedia of Ecology (Second Edition)* (pp. 20–28). Amsterdam, Netherlands.

Vanvalkenburgh, B., & Hertel, F. (1993). Tough times at la brea: tooth breakage in large carnivores of the late pleistocene. *Science (New York, N.Y.), 261*(5120), 456–459. https://doi.org/10.1126/science.261.5120.456

Van Valkenburgh, B., & Sacco, T. (2002). Sexual dimorphism, social behavior, and intra-sexual competition in large Pleistocene carnivorans. *Journal of Vertebrate Paleontology, 22*(1), 164–169. https://doi.org/10.1671/0272-4634(2002)022[0164:SDSBAI]2.0.CO;2

Van Valkenburgh, B. (2007). Deja vu: The evolution of feeding morphologies in the Carnivora. *Integrative and Comparative Biology, 47*(1), 147–163. https://doi.org/10.1093/icb/icm016

Vernadsky, V. I. (1926). *The Biosphere*. Scientific Chemico-Technical Publishing).

Waitt, R. B., Jr., & Thorson, R. M. (1983). The Cordilleran Ice Sheet in Washington, Idaho, and Montana. In S. C. Porter & H. E. Wright Jr. (Eds.), *Late-Quaternary Environments of the United States* (Vol. 1, pp. 53–70). University of Minnesota Press.

Wang, K., Niu, J., Li, T., & Zhou, Y. (2020). Facing Water Stress in a Changing Climate: A Case Study of Drought Risk Analysis Under Future Climate Projections in the Xi River Basin, China. *Frontiers in Earth Science, 8*, 86. https://doi.org/10.3389/feart.2020.00086

Water Resources Management Branch of the Office of Environment. (2002). Papua New Guinea: PNG-3: Sepik Wara. In R. Ibbitt, K. Takara, M. Desea, & H. Pawitan (Eds.), *Catalogue of Rivers for the Southeast Asia and the Pacific* (Vol. IV, pp. 243–253). The UNESCO-IHP Regional Steering Committee for Southeast Asia and the Pacific.

Watt, K. E. F. (1966). *Systems Analysis in Ecology*. Academic Press.

Webb, P. (2019). *Introduction to oceanography*. Retrieved from http://rwu.pressbooks.pub/webboceanography

Weisberg, M. (2007). Three kinds of idealization. *Journal of Philosophy, 104*(12), 639–659. https://doi.org/10.5840/jphil20071041240

Wetzel, R. G. (2001). *Limnology*. Academic Press. https://doi.org/10.1016/C2009-0-02112-6

Whitham, T. G., Bailey, J. K., Schweitzer, J. A., Shuster, S. M., Bangert, R. K., LeRoy, C. J., Lonsdorf, E. V., Allan, G. J., DiFazio, S. P., Potts, B. M., Fischer, D. G., Gehring, C. A., Lindroth, R. L., Marks, J. C., Hart, S. C., Wimp, G. M., & Wooley, S. C. (2006). A framework for community and ecosystem genetics: From genes to ecosystems. Nature reviews. *Genetics, 7*(7), 510–523. https://doi.org/10.1038/nrg1877

Wiegert, R. G. (1988). Holism and Reductionism in ecology: Hypotheses, scales, and systems models. *Oikos, 53*, 267–269. https://doi.org/10.2307/3566072

Wilkinson, D. M. (2006). *Fundamental Processes in Ecology: An Earth Systems Approach*. Oxford University Press.

Williams, M., Zalasiewicz, J., Haff, P., Schwägerl, C., Barnosky, A. D., & Ellis, E. C. (2015). The Anthropocene biosphere. *The Anthropocene Review, 2*(3), 196–219. https://doi.org/10.1177/2053019615591020

Wimsatt, W. C. (2006). Reductionism and its heuristics: Making methodological reductionism honest. *Synthese, 151*, 445–475. https://doi.org/10.1007/s11229-006-9017-0

Wink, M. (1995). Phylogeny of Old and New World vultures (Aves: Accipitridae and Cathartidae) inferred from nucleotide sequences of the mitochondrial cytochrome b gene, Zeitschrift fur Naturforschung. *C, Journal of Biosciences, 50*(11–12), 868–882. https://doi.org/10.1515/znc-1995-11-1220

World Atlas. (2022). *The largest Lakes in Africa.* Retrieved from https://www.worldatlas.com/articles/the-largest-lakes-in-africa.html

World Bank. (2010). *The Zambezi River Basin a multi-sector investment opportunities analysis. Volume 4: Modeling, analysis, and input data.* http://web.worldbank.org/archive/website01321/WEB/IMAGES/ZAMBEZ-4.PDF

World Climate Research Program (WCRP) (1980). World Meteorological Organization (WMO), International Council for Science (ICSU).

World Commission on Environment and Development (1987). *Our common future*. Oxford: Oxford University Press

World Lake Database. (2022). *Lake Victoria.* International Lake Environment Committee Foundation. Retrieved from https://wldb.ilec.or.jp/Display/html/3586

World Meteorological Organization (WMO). (2022). *Cryosphere elements*. Retrieved from https://public.wmo.int/en/our-mandate/focus-areas/cryosphere/elements#:~:text=The%20cryosphere%20is%20the%20part,for%20frost%20or%20ice%20cold

Wynne, B. (1992). Uncertainty and environmental learning: Reconceiving science and policy in the preventive paradigm. *Global Environmental Change, 2*(2), 111–127. https://doi.org/10.1016/0959-3780(92)90017-2

Yang, D., Shi, X., & Marsh, P. (2015). Variability and extreme of Mackenzie River daily discharge during 1973–2011. *Quaternary International, 380–381*, 159–168. https://doi.org/10.1016/j.quaint.2014.09.023

Yohannes, O. (2008). *Water resources and inter-riparian relations in the Nile basin.* SUNY Press.

Yool, A., Popova, E., & Anderson, T. R. (2013). MEDUSA-2.0: An intermediate complexity biogeochemical model of the marine carbon cycle for climate change and ocean acidification studies. *Geoscientific Model Development Discussions, 6*, 1259–1365. https://doi.org/10.5194/gmdd-6-1259-2013

Zahnle, K., Schaefer, L., & Fegley, B. (2010). Earth's earliest atmospheres. *Cold Spring Harbor Perspectives in Biology, 2*(10), a004895. https://doi.org/10.1101/cshperspect.a004895

Zhu, G., Liu, Y., Shi, P., Jia, W., Zhou, J., Liu, Y., Ma, X., Pan, H., Zhang, Y., Zhang, Z., Sun, Z., Yong, L., & Zhao, K. (2022). Stable water isotope monitoring network of different water bodies in Shiyang River basin, a typical arid river in China. *Earth System Science Data, 14*, 3773–3789. https://doi.org/10.5194/essd-14-3773-2022

Chapter 4
Biogeochemistry, Biogeography, and Geomatics

Michael O'Neal Campbell

Abstract Biogeochemistry and biogeography are both vital, expanding, interconnected, hybrid, multidisciplinary, interdisciplinary, and transdisciplinary branches of the life, earth, and conservation sciences, with spatially and temporally oriented methodologies, and increasing relevance for current global realities and human survival. Both discipline clusters seek spatial relevance, from local to global scales, interconnectedness, linking multiple ecosystems, and use increasingly sophisticated research methodologies. Both may also use the tools of geomatics including geodesy, radio detection and ranging (radar), and light detection and ranging (LiDAR), to uncover more exact, reliable, and definitive information from researched data. For biogeochemistry, the main use for such techniques is the identification and measurement of the relevant parameters (organisms, habitat variables, earth structures that facilitate and contain chemical flows and reservoirs). For biogeography, spatial distributions assume prominence, including areal extents and differentiation. However, few studies have sought to document the increasingly complex relationship between these research clusters. This chapter uses a literature-based research methodology to uncover these issues, especially biogeography, as biogeochemistry has been defined in the earlier chapters. Selectivity is necessary, as this relationship is as broad as the Earth itself. Case studies are taken from the global scale of biomes and earth systems to the local contexts of ecological change. The role of human action is also explored, in addition to the possibilities of multidirectional change. Research methods within the field of geomatics, as the premier tools for the analysis of spatial ecologies and environmental variables, are also explored. It is argued that these three research clusters, when linked, form the basis of the environmental sciences. This contributes to the development of the environmental sciences and strengthens the status of biogeochemistry as a root science of the study of the Earth.

M. O. Campbell (✉)
Simon Fraser University, Burnaby, BC, Canada
e-mail: evaz@torontomu.ca

M. O. Campbell, *Biogeochemistry and the Environment*,
https://doi.org/10.1007/978-3-031-47017-2_4

Introduction

Biogeochemistry, biogeography, and geomatics are increasingly interlinked as research disciplines (Campbell, 2017, 2018). As mentioned in Chap. 1, one of the most incisive definitions of biogeochemistry is that of Karl and Schlesinger (2014): "biogeochemistry is the term used to describe one of the most important subdisciplines of Biogeosciences, the holistic study of the Earth system. Biogeochemical research includes theoretical, observational, experimental, and modeling components, and now spans from molecular studies to the largest scales in space and time that affect our planet. Therefore, as a scientific discipline, biogeochemistry is both integrative and transdisciplinary, and it is attracting a new and diverse spectrum of investigators each year." Biogeography is similarly broad, as it is a spatial and temporal science, examining the distribution and related issues of biological phenomena, and is closely related to ecology, especially concerning the relations and spatial issues and the factors from which they emerge. Definitions include that of Dansereau (2015): "Biogeography studies all aspects of the adaptations of an organism to its environment, considering systematically the origins, migrations, and associations of living things. Hence, it aims for a synthesis of data from nonbiological and biological disciplines: geology, physical geography, geomorphology, climatology, and meteorology, on the one hand; biology, taxonomy, genetics, and physiology on the other. Biogeographical relationships cannot be understood except from an ecological perspective which tries to explain the exchanges between an organism and its environment."

As biogeography relates to all life and its contexts, several definitions have been stated, reflecting this vast milieu under study (Campbell, 2017). Such definitions include the following examples, from simple to complex: "biogeography is a biological as well as a geographical discipline" (Mueller-Hohenstein, 2001); "biogeography is a scientific approach to understanding the distribution and abundance of living things, the biota, on our planet" (Mueller-Dombois, 2001, 565); biogeography is "the study of spatial and temporal distributions of organisms and their habitats, and the historical and biological factors that produced them" (Monaco et al., 2003); and most broadly, biogeography is relevant to "GIS-based land-use studies, to GPS-based studies of animal tracking, to molecular biology-based studies of plants, animals and insects" and the "distribution of species including our own in space and time and the geological, physical, biological and even cultural forces that explain them" (Smithsonian Tropical Research Institute, 2020).

A vital topic for biogeography as a discipline concerns it relationship with ecology (Campbell, 2017). Some scholars and disciplinary authorities have criticized the foundations of biogeography as duplicating much of the related biological subdiscipline of spatial ecology, while some specialists defend biogeography as vital, unique, and an equal science. Some of the tension between these related disciplines is their institutional distinction: university ecology courses and research are usually based in biology departments, while equivalent biogeography studies are often based in geography departments. Current evidence nonetheless indicates a spatialization and applied trend for ecological topics and priorities, increasingly with applications from spatially oriented technologies like geomatics, and increased sharing of research with the social sciences, in topics within political ecology, environmental economics, and natural resource management, as the human actions studied by the latter disciplines explain much of biogeographical and ecological change (Campbell, 2015, 2017, 2018, 2020, 2022a, b, c).

Bianchi (2021: 156) asks the pertinent question "So, why is biogeography important to biogeochemistry in the twenty-first century?" The reasons are that as biogeography expanded into different subdisciplines, it breadth became more relevant to the breadth of biogeochemistry, with two main branches (ecological biogeography and historical biogeography) (Bianchi, 2021: see also Sanmartín, 2012; Linder, 2005). Relevant topics included global environmental changes and associated organism distribution evolutions, species migration, including gene flows and physical barriers related to Earth developments, recent molecular research methods in understanding adaptions to environmental change, the evolution of plant communities, and new research links between biogeography and ecological developments. Regarding the increased attention to studies of biodiversity, Bianchi (2021: 157) notes that "while contentiously debated at times, it is now generally agreed that organismal diversity can impact biogeochemical cycles (see also Chapin et al., 1997; Tilman et al., 2012). The rest of this chapter will examine the roots, character, and problems of biogeography; the links with biogeochemistry; and the geomatics techniques relevant to this relationship.

The Development of Biogeography

Biogeography as a discipline owes its origin to the events of the nineteenth, twentieth, and twenty-first centuries, although prior relevant work has been cited and the role of historical works is disputed (Campbell, 2020). Riddle (2017) notes that biogeography "emerged as a recognizable science as plant and animal collections grew during the eighteenth-century age of exploration. The foundation of biogeography, that geographically disjunct regions with similar environments have distinct plants and animals, appeared as Buffon's Law in 1761." Nineteenth-century developments included the works of Wallace, Sclater, and Darwin, and the eventual development of biogeography "along the two often distinct and separate trajectories of historical (evolution, systematics, paleontology) and ecological (population, community, ecosystem) biogeography" (Riddle, 2017).

Biogeography further developed during the late twentieth century, based on plate tectonic model of continental drift, equilibrium theory of island biogeography (Riddle, 2017) and disequilibrium theory, the "new" ecology, stronger links with ecology and the social sciences, and geomatics research technologies (Campbell, 2017, 2018, 2022a, b, c). Regarding the increased links with the social sciences, Dansereau (2015) writes that "fortunately, in recent years the ever-swinging oscillation of physical/cultural anthropology and the emergence of human ecology as an increasingly autonomous discipline" have rendered biogeographical topics "more relevant to human studies, environmental planning and landscape management, and "human activity is proving to be a very significant factor in the distribution of plant and animal species."

Currently, biogeography has advanced and, according to Riddle (2017), integrates four "persistent themes (biogeographic regionalization, historical reconstruction, turnover among areas and across gradients, and geographic variation within and among closely related species)" within emerging subdisciplines that include phylogeography, macroecology, and conservation biogeography. Dansereau (2015) points out that currently, biogeography "in its widest compass, biogeography concerns itself with the evolution of species, with changes in their ranges and with their extinctions. The principal factors influencing evolutionary development are climatic and edaphic (i.e., soil-related) constraints, genetic adaptation, and social integration." It is also concerned with spatial scales, the smallest to the biggest being designated as niches, ecotypes, communities, ecosystems, landscapes, and bioclimates. Reflecting on these recent developments, Ebach (2015) writes that "twenty-first century biogeography has multiple origins, most of which are in the twentieth century. Few, if any methodologies, theories, and implementations of twenty-first century biogeography go back to the nineteenth century, let alone the eighteenth. What is more, twenty-first century biogeography has many different practitioners that hail from different backgrounds, very much like the practitioners in the eighteenth and nineteenth centuries." Additionally, eighteenth and nineteenth century plant and animal geography were multidisciplinary professions and in as much conflict as twentieth and twenty-first century biogeography (Ebach, 2015).

Links Between Biogeography and Other Sciences

Biogeography currently includes strands that emphasize the intersection of biology and geography. This may cover the application of spatial analyses to entire field of biology. This is based on the rationale that biology studies all life forms, these forms are located in spatial contexts, spatial distribution is universally relevant, and hence, biogeography is related to any biological science that studies parameters that may influence and be influenced by spatial analyses (Campbell, 2017). This broad focus derives from the links between biogeography and other sciences, manifested through the links between biology and these sciences (see Fig. 4.1). Biology as the science of all life necessarily links with the more basic sciences of mathematics, physics, and chemistry, which also form the backbone of the Earth and environmental sciences (Campbell, 2020).

Examples include interfaces of biology and physics (the field of biophysics, the main techniques including electric currents, temperature, stress and entropy, which are used to study biological phenomena, which may be regionalized); bioclimatology (a vital field using multiple techniques to study of interactions between climate and biological regions, spatially and temporarily); biological oceanography (which

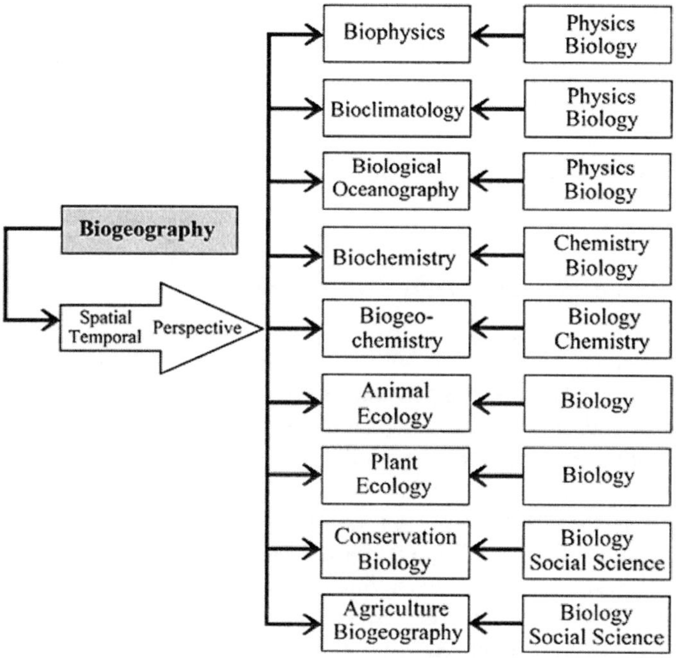

Fig. 4.1 Supporting sciences of biogeography. (See also Campbell (2020)). Biogeography lends a spatiotemporal perspective to the biological sciences listed in the left column. All these more applied disciplines and subdisciplines are also supported by the basic sciences listed in the right column. Each of the sciences in the left column is related to each other, and biogeochemistry in the center is strongly interactive

is a more environmentally focused science than marine biology); biochemistry (the application of the chemistry techniques to the study of biological phenomena and processes) and biogeochemistry (the study of the chemical, physical, geological, and biological processes that link the physical and biological spheres of the earth). These hybrid disciplines (combining the techniques of one discipline with the subject matter of another) may uncover spatial and temporal differentiation and which is the core of biogeographical study (Campbell, 2020).

Biogeography has also expanded toward human geography and the more human centered sciences, as current knowledge increasing sees the role of human actions and consequent landcover change as factor for biogeographical change (Campbell & Jha, 2021). Topics in this vein include deforestation, savanna expansion, mixed landscape development, urbanization, human proximity studies, and the regionalization of conservation issues (Campbell, 2022a). Globalization, manifested by related networks, interconnections, and regionalism is related to ecological change. Research methods including geomatics have derived a space in these studies, abstracting spatial processes and variation, spatial statistics, and supportive analytic computer software as the backbone of cutting-edge biogeographical studies. Biogeography, unlike the other biological sciences, retains connections and supports from human geography as part of the overall discipline of geography and obtains a methodological advantage, linking with new paradigms within the developing sciences of the New Ecology, Environmental Science and Environmental Studies, which justify their relevance on the truism that human action is increasingly overtaking natural forces as factors for biogeographical ecological and biogeochemical change (Campbell, 2015).

Links Between Biogeography and Ecology

Ecology is arguably the science with closest links to biogeography, these links including the biogeochemical flows and reservoirs that are essential for organisms and this micro and macro ecologies (Campbell, 2018). Biogeography and ecology have historically been and still are disciplines, with differences mostly based on the parameters of scale, time, and human action (Campbell, 2017, 2020). Jenkins and Ricklefs (2011: 2331) argue that "both biogeography and ecology seek to understand the processes that determine patterns in nature but do so at different spatial and temporal scales. The two disciplines were not always so different and are recently converging again at regional spatial scales and broad temporal scales." The difference may lie in the emphases in topics, as ecology firstly focuses on adaptive evolution, population processes, and abiotic and species interactions, which explain the coexistence of species in limited areas. Modern ecology also has a strong focus on experimentation and statistical models, and recently advances in computation, with most studies having restricted spatial scales and limited temporal extents (ibid.). Biogeography, by contrast has more focus on spatial relations. Particularly, biogeography examines evolutionary, climatic, and geological processes that may explain the distribution of biodiversity over the land and seas. Although biogeography considered mostly descriptive, but recent advances in technology may incorporate phylogenetic (historical) information systems to attempt more advanced testing of hypotheses about the processes and networks (e.g., cladistic biogeography)" (ibid.) Jenkins and Ricklefs (2011: 2331) conclude that "as such, the two disciplines represent recently overlapping ends of a spatial-temporal gradient in our views of the natural world and how it works (Jenkins & Ricklefs, 2011: 2331). Both sciences relate common themes, as shown in the model diagrams (Fig. 4.1), derived from Campbell (2020).

Ecology may therefore emphasize narrower or smaller spatial contexts than biogeography. Ecological topics include linear, mechanistic change exemplified by "natural" vegetation, mechanistic succession, climatic and vegetation climaxes, succession, ecological equilibrium, and human-nature separation and associated links with animal ecology (Burrows, 1990; Pimm, 1991; Tallis, 1991; Tudge, 1991; Pickett et al., 1992; Beeby, 1993; Simmons, 1993; Campbell, 2017). Definitions of ecology may reflect this: "the scientific study of the processes influencing the distribution and abundance of organisms, the interactions among organisms, and the interactions between organisms and the transformation and flux of energy and matter" (Cary Institute of Ecosystem Studies, 2020); ecosystem issues (Odum, 1971; Stock & Allsopp, 1992); and narrower topics such as microscale issues such as plant reproduction (Le Maitre & Midgley, 1992); plant/animal relationships (Johnson, 1992); organism competition (Campbell, 2022a) and plant structure and function (Campbell, 2020). New developments in ecology led to definitions and topics aware of the recent "nonequilibrium" hypotheses or "new" ecology, that emphasized more unpredictable, "chaotic" relations and changes, rather than linear, overgeneralization (Campbell, 2017, 2018). These include interrelationships within complex and unpredictable contexts, and chaotic fluctuations in environmental contexts (Campbell, 1998).

Biogeography ecological paradigm changes, but usually uses a more spatial approach to biological studies. Biogeographical studies also generally include the impacts of human action on the biosphere, and the broader issues of history, vegetation form and associations, plant diversity and endemism and time regimes (Campbell, 2022a). Such differences may relate to the position of biogeography as both biological and geographical science, the geographical topics being location, spatial distribution and statistics, ecological variation, geographical information systems, and nature–society relations and more human-centered ecologies (Campbell, 2017, 2018). Therefore, an incisive definition of biogeography would be that of Encyclopædia Britannica (2020); "Biogeography, the study of the geographic distribution of plants, animals, and other forms of life. It is concerned not only with habitation patterns but also with the factors responsible for variations in distribution."

Biogeography's relationship with ecology attained greater complexity, partly because both biologists and geographers may study biogeography. Mueller-Hohenstein (2001) argues that "biogeography is a biological as well as a geographical discipline. Biologists prefer a more historical view and underline the importance of the evolution of species and communities. Geographers accentuate more the ecological background of the distribution of plants and animals, including the manifold influences of man, and they underline the importance of scales in time and space. Both try to understand the distribution of patterns of organisms and biocenosis." Geographical research tools, such as geomatics are usually included in biogeographical studies, and more recently ecological studies, which change animal and plant ecology to a more spatial view and transposition into zoogeography (mostly spatial animal ecology) and phytography (mostly spatial plant ecology), from the perspective of both biologists and geographers (Crisci et al., 2003; Campbell, 2017, 2018, 2020; Katinas & Crisci, 2018).

Significant overlaps have developed between the work of biogeographers (both geographers and biologists) and ecologists (mostly biologists) due to these joint interests and duplication of research efforts (Campbell, 2017, 2018). Vitt and Caldwell (2014: 353) argued that "ecology, biogeography, and conservation biology are difficult topics to separate because of the integrated nature of each." Jenkins and Ricklefs (2011, 2331) also noted that "both biogeography and ecology seek to understand the processes that determine patterns in nature but do so at different spatial and temporal scales" and "the two disciplines were not always so different and are recently converging again at regional spatial scales and broad temporal scales" (Jenkins & Ricklefs (2011, 2331). An evaluation of the shared foci of biogeography and ecology may be undertaken using the four common components of niche (an environmental context favored by a species), comparative ecology (a field comparing defined ecologies), macroecology (ecological relationships at large spatial scales), community assembly (processes shaping the identity and abundance of species in ecological communities), and diversity (Campbell, 2018, 2020).

Tables 4.1, 4.2, 4.3, and 4.4 illustrate spatial, temporal, study units, and related biological processes (Jenkins & Ricklefs, 2011, see also Campbell, 2020). These compared parameters are dynamic and subject to change, as both biogeography and ecology are evolving. Jenkins and Ricklefs (2011, 2332–2333) describe changes in biogeography, which include increased focus on evolutionary, climatic and geological processes as factors for biodiversity, and developments in phylogenetic (historical) information for hypothesis testing of phenomena such as processes in cladistic biogeography (the study of historical and evolutionary relationships between species, using their particular distribution patterns), and GIS and remote sensing based techniques to study spatial patterns in biological diversity or macroecology

Table 4.1 Biogeography and ecology: key approaches to spatial scales

Attribute	Biogeography (larger scale)	Ecology (smaller scale)
Spatial scales	*Global,* e.g., range of all the species in an Order, Family, or Genus. *Regional,* e.g., the range of the species through continental or subcontinental scales.	*Regional,* e.g., the range of the species through continental or subcontinental scales. *Local,* e.g., town, patch or field sized studies, such as studies of parks.
	Examples from biogeography journals	Examples from ecology journals
	Guo et al., 2020. Activity niches outperform thermal physiological limits in predicting global ant distributions. Weigelt et al. 2020. GIFT – A Global Inventory of Floras and Traits for macroecology and biogeography. Biber et al., 2019. A comparison of macroecological and stacked species distribution models for ecological prediction	Zhang et al., 2020. Climate change-induced migration patterns and extinction risks of Theaceae species in China. Krupa et al., 2020. Plant-animal interactions between carnivorous plants, sheet-web spiders, and ground-running spiders as guild predators in a wet meadow community. Sharma et al., 2020. Estimating occupancy of Chinese pangolin (Manis pentadactyla) in a protected and nonprotected area of Nepal.

Adapted from Jenkins and Ricklefs (2011), examples from Campbell (2015, 2017, 2018, 2020), Journal of Biogeography; Annual Review of Ecology, Evolution, and Systematics; Proceedings of the Royal Society B Biological Sciences, Nature Ecology and Evolution, Frontiers in Ecology and the Environment, Global Environmental Change and Global Ecology and Biogeography

Table 4.2 Biogeography and ecology: key approaches to temporal scales

Attribute	Biogeography (larger scale)	Ecology (smaller scale)
Temporal scales	*Million years* +, e.g., the extinction the mammal species of North America *Thousand years* +, e.g., the decline in mammal populations in Western Europe, human action, and habitat destruction.	Generation cycles: population cycles with a period of roughly one generation. Population cycles: For example, longer population cycles for larger mammals
	Examples from biogeography journals	Examples from ecology journals
	Leone et al. 2020. Pliocene colonization of the Mediterranean by Great White Shark inferred from fossil records. Marin & Hedges, 2016. Time best explains global variation in species richness of amphibians, birds and mammals.	Spagopoulou, 2020. Transgenerational maternal age effects in nature: Lessons from Asian elephants. Lahdenperä et al., 2019. Capture from the wild has long-term costs on reproductive success in Asian elephants.

Sources: Adapted from Begon et al. (2006), Hartl and Clark (2007), Jenkins and Ricklefs (2011), Engler et al. (2015)

Table 4.3 Biogeography and ecology: key approaches to units of study

Attribute	Biogeography (larger scale)	Ecology (smaller scale)
Units of study	Clade or a monophyletic group of organisms with a common ancestor.	Individuals, one member of a species, with variable or generalizable behavior.
	Species, the largest group where opposite sexes can produce fertile offspring, interspecies differences	Populations, all organisms of one species, in a geographical area, interbreeding possible between any pair
	Range, the geographical area in which a species can be found, which may have requirements for the species existence/health	Communities, an assemblage of populations of different species, with intra- and interspecies competition, predation and mutualism.
	Distribution, how the members of a species is arranged, within the range, usually dependent habitat features	Note: individuals, populations of each species are influenced by individuals, populations of other species.

Sources: Adapted from Begon et al. (2006), Hartl and Clark (2007), Jenkins and Ricklefs (2011), Campbell (2015), and Engler et al. (2015)

(Campbell, 2018). Changes in ecology include increased focus on "adaptive evolution, population processes, and abiotic and species interactions, to explain coexistence of species in local areas" (Jenkins & Ricklefs, 2011: 2332–2333). Both ecology and biogeography include experimentation, statistical models, and computation, but ecological studies have a strong focus on smaller case studies and shorter time periods (Campbell, 1998, 2017, 2018, 2020). Jenkins and Ricklefs (2011: 2333) conclude that "the two disciplines represent recently overlapping ends of a spatiotemporal gradient in our views of the natural world and how it works."

Biogeography and ecology also share integrative subdisciplines, such as ecological biogeography, defined as the study of "the relationship between the ecology of clades and their distribution" (Heads, 2015: 456). Ecological biogeography looks at

Table 4.4 Biogeography and ecology: key approaches to processes of interest

Attribute	Biogeography (larger scale)	Ecology (smaller scale)
Processes of interest	Speciation, the evolutionary process where populations evolve into 2+ species, by geographical separation.	Abiotic and biotic interactions, density, i.e., nonliving variables (temperature, wind, water, soil) and living organisms affecting species or individual density
	Extinction, local (extirpation), wild land (still in zoos or domesticated) or functional (too few for breeding) total global extinction of a species	Abiotic and biotic interactions, distribution, i.e., nonliving variables (temperature, wind, water, soil) and living organisms
	Range expansion or retraction, increase or reduction of species habitats, due to biological, physical or human factors.	Note: density refers to the number of individuals or species per unit area.

Sources: Adapted from Begon et al. (2006), Hartl and Clark (2007), Jenkins and Ricklefs (2011), Campbell (2015), and Engler et al. (2015)

geographical factors such as climate, as instruments for biological clade distributions, a vital area as "in many cases, species distribution models – ecological niche models – based on the current environment of a species (its environmental envelope) fail to predict the actual distribution of the species" and may even "overpredict distributions" (Heads, 2015: 456). Ecological biogeography, by using wider patterns and larger case studies, may complement some ecological studies limited by a focus small, niche defined areas, which may present insufficient material for reliable generalization. Smaller case studies may lack sufficient evidence for mapping biogeographical and evolutionary processes, and links between local patterns and the wider, more generalized systems. Ecological biogeography, using both wider patterns and case studies, and the links between these, may function as an integrating discipline filling this gap of knowledge (Heads, 2015: 456).

Links Between Biogeochemistry and Biogeography

The relationship between biogeochemistry and biogeography is complex, with the main link being the common, shared enterprise of global, spatialized environmental and ecological assessment (Campbell, 2018). For example, the Marin-Spiotta Biogeography and Biogeochemistry (BiogeoLab) Research Group (2022) defines the relationship between biogeography and biogeochemistry by their research interests, which are "how landscape disturbances due to changes in climate and land use alter the accumulation and loss of biomass and biologically active elements within and across ecosystem boundaries…" the "research draws from biogeochemistry, ecosystem ecology, soil science and biogeography… across different spatial scales: from microscopic interactions between organic matter and mineral surfaces in soils, to stand-level forest dynamics, to regional soil carbon inventories."

Krinner et al. (2005) also argue for the synthesis of biogeography and biochemical studies, referring to an emerging dynamic global vegetation model, which is designed as an extension of a pre-existing surface–vegetation–atmosphere transfer scheme that was included in a coupled ocean-atmosphere general circulation model. This global vegetation model is designed to simulate the main biosphere processes that create the global carbon cycle, defined as "photosynthesis, autotrophic and heterotrophic respiration of plants and in soils, fire, etc." and "latent, sensible, and kinetic energy exchanges at the surface of soils and plants" (Krinner et al., 2005). The authors argue that there are two main axes of research: The first major axis relates to the impact of vegetation on land-atmosphere exchanges and consequently on climate, while the second major research axis focuses on the role of terrestrial vegetation in the global carbon cycle (Krinner et al., 2005). Krinner et al. (2005) contend that the two research axes are strongly as in general, the biogeographical models that determine vegetation distributions in response to climate change are also biogeochemical models that represent the continental sized carbon cycle (see also Cramer et al., 2001). The second research axis concerns vegetation carbon sequestration and release for climate changes, and carbon sinks for some emissions of greenhouse gases (Fung et al., 2005; Krinner et al., 2005).

The importance of the relationship between vegetation systems (biogeography) and climate systems (biogeochemistry) is revealed in this study, as the authors argue for the emerging recognition of the vitality of land surface processes in the global climate system. The processes that engage with the terrestrial biosphere are very important because vegetation of the terrestrial biosphere covers most of the Earth's land surface and the terrestrial biosphere has two main impacts on the global climate system. The first is that vegetation cover directly influences land-atmosphere heat, momentum, and moisture fluxes. Second, biospheric processes exert a strong control over atmospheric composition, especially on greenhouse gas concentrations on interannual and longer timescales" (Krinner et al., 2005). Knowledge of the biospheric (biogeographic) processes on global and regional scales is required to effectively analyze the global climate system (Krinner et al., 2005).

Krinner et al. (2005) promote a model that links vegetation and climate, a "new terrestrial biosphere model, called ORCHIDEE (ORganizing Carbon and Hydrology in Dynamic EcosystEms), which is a SVAT coupled to a biogeochemistry and a dynamic biogeography model. As such, it explicitly simulates the phenomena of the terrestrial carbon cycle that are linked to vegetation and soil decomposition processes, but also changes in vegetation distributions in response to climate change as well as short-timescale interactions between the vegetated land surface and the atmosphere" (Krinner et al., 2005; note that SVAT refers to surface–vegetation–atmosphere transfer schemes, used to simulate exchanges of sensible, latent, and kinetic energy at the surface). ORCHIDEE is promoted as part of the new IPSL-CM4 Atmosphere-Ocean-Vegetation coupled general circulation model (see Marti et al., 2005). This model is described as "capable of simulating water, energy, and carbon exchanges and stocks on the continental surface in line with current best estimates, representing vegetation processes on both short (diurnal cycle) and long (multiannual) timescales, and vegetation-atmosphere exchanges on the local scale as well as global mean vegetation characteristics (extent, morphological properties, carbon stocks, and fluxes)...The diurnal and seasonal dynamics of carbon fluxes are correctly represented by ORCHIDEE" (Krinner et al., 2005).

Melillo et al. (1995: 407) also present a comparative study of three biogeography models (BIOME2, Dynamic Global Phytogeography Model (DOLY), and Mapped Atmosphere-Plant Soil System (MAPSS)) and three biogeochemistry models (BIOME-BGC (BioGeochemistry Cycles), CENTURY, and Terrestrial Ecosystem Model (TEM)), with the case study of the contiguous United States (the 48 adjoining U.S. states and the Federal District of the United States of America, excluding Alaska and Hawaii). The simulations assumed contemporary conditions of atmospheric CO_2 and climate, and also assumptions of doubled CO_2 and other climate scenarios. The findings of the study were that "for contemporary conditions, the biogeography models successfully simulate the geographic distribution of major vegetation types and have similar estimates of area for forests (42–46% of the conterminous United States), grasslands (17–27%), savannas (15–25%), and shrublands (14–18%) (Melillo et al., 1995: 407). For the "biogeochemistry models estimate similar continental-scale net primary production (NPP; 3125 to 3772 × 1012 gC year^{-1}) and total carbon storage (108 to 118 × 1015 gC) for contemporary conditions" (Melillo et al., 1995: 407).

For the models of assumed doubled CO_2 and the other climate scenarios, the three biogeography models exhibited gains and losses of total forest area depending on the scenario (ranging between 38% and 53% of the US area) (Melillo et al., 1995: 407). The authors noted that the only consistent expanded forest areas with all the three models (BIOME2, DOLY, and MAPSS) were in the GFDL scenario because of significant increases in precipitation. The forest losses were from the MAPSS model, which lost forest area under UKMO, and DOLY which lost under OSU, and BIOME2 under both UKMO and OSU. The authors argue that the variations in the forest area estimates were due to the different sensitivities to increases in temperature and CO_2, by the hydrologic cycles of the biogeography models. There was little difference between the biogeography models when both climate change and elevated CO_2 concentrations were incorporated. Melillo et al. (1995: 407) record that "for these scenarios, the NPP estimated by the biogeochemistry models increases between 2% (BIOME-BGC with UKMO climate) and 35% (TEM with UKMO climate). Changes in total carbon storage range from losses of 33% (BIOME-BGC with UKMO climate) to gains of 16% (TEM with OSU climate). The CENTURY responses of NPP and carbon storage are positive and intermediate to the responses of BIOME-BGC and TEM."

Additionally, carbon cycle responses varied because the hydrologic and nitrogen cycles of the different biogeochemistry models had different sensitivities to increases in temperature and CO_2. The biogeochemistry models were run with the vegetation distributions of the biogeography models, and the findings suggested "NPP ranges from no response (BIOME-BGC with all three biogeography model vegetations for UKMO climate) to increases of 40% (TEM with MAPSS vegetation for OSU climate)" (Melillo et al. (1995: 407). The total carbon storage response also ranged from a decrease of 39% (BIOME-BGC with MAPSS vegetation for UKMO climate) to an increase of 32% (TEM with MAPSS vegetation for OSU and GFDL climates). It is noted that the dominant cause of the UKMO responses of BIOME-BGC with MAPSS vegetation were decreases in forested areas and temperature-induced water stress. However, the OSU and GFDL responses of TEM with MAPSS vegetations were mostly caused by increased forest cover and nitrogen cycling enhanced by temperature (Melillo et al., 1995: 407).

Global Biomes and Carbon Sinks

The division of the terrestrial surface of the Earth into ecological regions is a major aspect of both biogeography (defining contributing factors for the distribution of life, including zoogeography and phytogeography) and biogeochemistry (defining the factors for the global chemical cycling) (Campbell, 2015). Olson et al. (2001) introduce this topic, describing the Terrestrial Ecoregions of the World (TEOW) classification as a "biogeographic regionalization of the Earth's terrestrial biodiversity," with the biogeographic units as ecoregions, which are defined as "relatively large units of land or water containing a distinct assemblage of natural communities sharing a large majority of species, dynamics, and environmental conditions." The classification included 867 terrestrial ecoregions, classified into 14 different biomes, with variable coverage of vegetation types such as forests, grasslands, or deserts. Such ecoregions "represent the original distribution of distinct assemblages of species and communities" (Olson et al., 2001). The World Wildlife Fund (2006) defines an ecoregion as "a relatively large area of land or water that contains a geographically distinct assemblage of natural communities. These communities share a large majority of their species, dynamics, and environmental conditions." Olson et al. (2001) further note that "ecoregions are nested within two higher-order classifications: biomes (14) and biogeographic realms (8). Together, these nested classification levels provide a framework for comparison among units and the identification of representative habitats and species assemblages." Olson et al. (2001) conclude that ecoregions are increasingly used by stakeholders as models for the analysis of biodiversity patterns, and the assessment of conservation management. The ecological characteristics of these biomes are listed in Table 4.5.

The National Geographic Society (2022) defines "Five Major Types of Biomes," where "a biome is a large area characterized by its vegetation, soil, climate, and wildlife" or a "a large community of vegetation and wildlife adapted to a specific climate." There are five major types of biomes: aquatic, grassland, forest, desert, and tundra, though some of these biomes can be further divided into more specific categories, such as freshwater, marine, savanna, tropical rainforest, temperate rainforest, and taiga." According to Beaumont and Duursma (2012), the 14 biomes are Tropical/Subtropical Moist Broadleaf Forest; Tropical/Subtropical Dry Broadleaf Forest; Tropical/Subtropical Coniferous Forest; Temperate Broadleaf and Mixed Forest; Temperate Coniferous Forest; Boreal Forest/Taiga; Tropical/Subtropical Grassland, Savanna and Shrubland; Temperate Grassland, Savanna and Shrubland;

Table 4.5 Biomes and their attributes

Biome designation	Vegetation	Precipitation (mm/year)	Temp. (°C)
Tropical/Subtropical Moist Broadleaf Forest	Trees, high diversity, closed canopy	2000	32
Tropical/Subtropical Dry Broadleaf Forest	Trees, high diversity, semiclosed canopy	200–2000	>17
Tropical/Subtropical Coniferous Forest	Coniferous-evergreen trees, variable density	720–1500	18–25
Temperate Broadleaf and Mixed Forest	Closed/semi closed canopy	600–1500	3–23
Temperate Coniferous Forest	Coniferous-evergreen trees, variable density	300–900	10–40
Boreal Forest/Taiga	Conifers, deciduous, canopy, open woods	200–750	9–10
Tropical/Subtropical Grass, Savanna, Shrubs	Mixed trees, shrubs, and grasses	900–1500	20–30
Temperate Grassland, Savanna, Shrubland	Grass, mixed grass/shrubs/trees	300–900	3–25
Flooded Grassland and Savanna	Grassland seasonally or permanently flooded	400–450	3–30
Montane Grassland and Shrubland	Grass or shrub dominants higher elevations	500–3000	<5
Tundra	Shrubs, sedges, grasses, mosses, lichens	150–250	10–28
Mediterranean Forest, Woodland, Shrubland	Small trees, shrubs, scrubland	726	16.3
Desert and Xeric Shrubland	Open land, no canopy	<200	>40–<40
Mangrove	Variable, season canopy	1500	20–35

Sources: Murphy and Lugo (1986), Oliver (2008), Quinn (2008), Beaumont and Duursma (2012), Higuera et al. (2011), Sheppard, 2013, Sampson and Hamilton (2014), Campbell (2015, 2017, 2018, 2022a, b, c), Sternberg et al. (2015), Faber-Langendoen et al. (2016), Wilson (2016), Filho (2017), Kucuksezgin et al., 2019, Berlanga-Robles and Ruiz-Luna (2020), Earth Observatory (2022), Petruzzello (2022), and University of Arizona (2022)

Flooded Grassland and Savanna; Montane Grassland and Shrubland; Tundra; Mediterranean Forest, Woodland and Shrubland; Desert and Xeric Shrubland; and Mangrove. There are also seven biogeographic realms, which are Nearctic, Neotropic, Palearctic, Afrotropic, Indo-Malaya, Australasia, and Oceania. The ecological characteristics of these biomes are listed in Table 4.5.

All the above vegetation types trend from dense, wet forest through light dry forest, savanna, grassland semi desert, tundra, and desert. Due to human action, many such vegetation types are no longer pure, but are patch dynamic (Campbell, 1998) or are composed of modified seminatural vegetation, sometimes mixed with agricultural or other human dominated landcover (Campbell, 2017, 2018). Nevertheless, all these biogeographical regions and vegetation types, including transitional ecosystems may be relevant to global biogeochemistry (Campbell, 2020). As observed by Baudena (2015: 1833), "The forest, savanna, and grassland biomes, and the transitions between them, are expected to undergo major changes in the future due to global climate change." Additionally, "tropical grassy biomes (TGBs) are now changing rapidly through high rates of land clearance, increasing land-use intensity, woody encroachment and disruption of the disturbance regimes that maintain ecosystem function" and both forests and tropical grassy biomes are "subject to a complex set of pressures as a result of human actions" (Lehmann & Parr, 2016) as are temperate grasslands (Van Oijen et al., 2020) and temperate forests (Ohte & Tokuchi, 2011).

Forest Vegetation and Biogeochemistry

Forest vegetation here refers to the forest classifications in Table 4.5, namely: Tropical/Subtropical Moist Broadleaf Forest; Tropical/Subtropical Dry Broadleaf Forest; Tropical/Subtropical Coniferous Forest; Temperate Broadleaf and Mixed Forest; Temperate Coniferous Forest; and Boreal Forest/Taiga. Mangrove is also included (Beaumont & Duursma, 2012). The main distinguishing characteristic of forest vegetation is tree stands of varying density and height, from very dense stands with a closed canopy where light may not reach the ground underneath, to less dense tree stands with varyingly sized breaks in the canopy, where shade intolerant vegetation (grasses and/or shrubs) may grow (Campbell, 1998, 2017, 2018). Dense forest stands may be composed of up to four layers: the emergent, canopy, understory, and forest floor layers. The emergent layer is composed of emergents, which are a few very tall trees growing above the canopy; the canopy, which is a mass of connected tree crowns of roughly similar height; the understory composed of smaller, shrub-sized plants between the canopy and the forest floor; and plants (usually shade tolerant) on the forest floor (Campbell, 2020).

In regard to biogeochemistry and biogeography, the tropical forest is one of the most important regions (Campbell, 2018; Campbell & Jha, 2021). For example, Reed et al. (2016) note that "tropical forests represent one of the planet's most active biogeochemical engines. They account for the dominant proportion of Earth's live terrestrial plant biomass, nearly one-third of all soil carbon, and exchange more CO_2 with the atmosphere than any other biome. In the coming decades, the tropics will experience extraordinary changes in temperature, and our understanding of how this warming will affect biogeochemical cycling remains notably poor." The authors also argue that the lack of understanding of the large quantities of carbon that passes through tropical forests is important for the future of the Earth's climate (Reed et al., 2016).

Natural Resources Canada (2022) points out that forests can perform as either carbon sources or carbon sinks. A forest is a carbon source if it releases more carbon than it absorbs. This forest carbon is released from burning or decaying dead (from age, fire, insect attack) trees. A forest is a carbon sink if it absorbs more carbon than it releases, with carbon being absorbed from the atmosphere through photosynthesis, and being deposited in forest biomass and dead organic matter (trunks, branches, roots and leaves, and dead litter and wood). This process of carbon absorption and deposition is termed carbon sequestration (Natural Resources Canada, 2022). Harris and Gibbs (2021) state that the global forest stands emitted an average of 8.1 billion metric tonnes of carbon dioxide into the atmosphere annually, and absorbed 16 billion metric tonnes of CO_2 annually, with tropical rainforests (the South American Amazon, the African Congo River basin, and the Southeast Asian forests) sequestering carbon more carbon than the temperate or boreal forests and being the most important ecosystems affecting climate change (Harris & Gibbs, 2021).

For temperate or boreal forests, although held to sequester less carbon from the atmosphere than tropical forests, carbon cycling is important (Lorenz & Lal, 2010). For example, in temperate forests including broad-leaved deciduous, broad-leaved evergreen, coniferous, and mixed stands, the "total ecosystem C pool in biomes and soils of temperate forest is equivalent to, and sometimes even more, than that of the tropical rainforest ecosystems" (Lal & Lorenz, 2012: 187). Lal and Lorenz (2012: 187) note that the soils of the temperate forests also play a role in this status: The main soils of the temperate forests are Alfisols, Inceptisols, Mollisols, Spodosols, and Ultisols, which are usually fertile soils with a high soil organic C (SOC) pool. Typical temperate forest soils frequently contain over 100 Mg C ha^{-1} in the soil profile, but the authors note that there must be more research achieve understanding of the processes for the sequestering of carbon in soils and vegetation of temperate forests, and the development of methods to measure, monitor, and verify the carbon pool changes over the short term of a few years (Lal & Lorenz, 2012: 187).

The Mediterranean wooded ecosystems (forest, woodland, shrubland), are composed of vegetation of varied density, from forests to woodlands, savannas, shrublands, and grasslands, and stands of mixed vegetation. The forest is mostly oak, pines and Sclerophyll (hard leaved) trees such as olives adapted to dry weather (Gil-Tena et al., 2016). For the forests of this biome (for the shrubland, see the next section on savanna, etc.) there is incisive research detailing the role of the ecosystems in carbon biogeochemistry. For example, Chiesi et al. (2010: 850) applied a simulation model based on use of the BIOME-BGC model to yield estimates of carbon fluxes and pools in Tuscany forests of central Italy. This simulation assessed the impact of realistic climate changes (+2 °C and increased CO_2 levels on forest carbon processes including gross primary production (GPP), net primary production (NPP), and related issues. The findings indicated the temperature change tended to hinder production and allocation processes, which were enhanced by the increased CO_2 concentrations. The combination of the two factors leads to an increase in vegetation carbon, little change in soil carbon, and an increase in GPP and NPP, that is higher for deciduous oaks and chestnut trees (Chiesi et al., 2010: 850).

In another study of Mediterranean forests, Manca et al. (2009) examined the status of the Mediterranean forest ecosystem (*Pinus pinaster*, L.) in San Rossore, Pisa, Italy, as a CO_2 sink. The values considered included Net Ecosystem Exchange NEE, to estimate the Ecosystem Respiration (Reco) and the Gross Ecosystem Productivity (GEP), which represents the photosynthetic uptake of CO_2 without respiratory losses. The ecosystem showed reductions of carbon uptake during the hot, dry summers, when trees close stomata to avoid water losses. Carbon uptake is mostly the result of a reduction of photosynthesis, while ecosystem respiration experiences a lower decrease. The authors cite the wet summer 2003 as "an extreme example of this pattern" (the forest reaction to the wet summer of 2002, when rainfall was 506 mm, 300 mm above the 1999–2007 level), and the high winter temperature of 2007 as another example. For the summer of 2002, GEP and Reco were above average, and GEP had the higher increase. For the winter of 2007 (3 °C above the average for 1999–2007), there was an increase of ecosystem respiration, which was not balanced by the little increase of GEP (Manca et al., 2009).

Mangrove forests are also important for carbon biogeochemistry (Alongi, 2014; Soper et al., 2019). For example, Alongi (2014: 195) points out that "mangroves… are highly productive ecosystems with rates of primary production equal to those of tropical humid evergreen forests and coral reefs. Although mangroves occupy only 0.5% of the global coastal area, they contribute 10 to 15% (24 Tg C y(−1)) to coastal sediment carbon storage and export 10 to 11% of the particulate terrestrial carbon to the ocean." This excess contribution to carbon sequestration is seen as advantageous for conservation and restoration and contributing to possibilities for the amelioration of greenhouse gas emissions. Given their above average contribution to carbon sequestration, it is argued that a key problem is the potential carbon losses to deforestation, which may be greater than the rates of carbon storage (Alongi, 2014: 195).

In addition to the role of mangrove forests in carbon cycling, such vegetation also plays a crucial role in climate change adaptation and protecting coastline elevations and coastal infrastructure from ocean level rise and storm impacts (Soper et al., 2019: 4315). Arguing that "determining the efficacy of mangroves in achieving climate goals can be complicated by the difficulty in quantifying C inputs (i.e., differentiating newer inputs from younger trees from older residual C pools), and mitigation assessments rarely consider potential offsets to CO_2 storage by methane (CH_4) production in mangrove sediments" Soper et al. (2019: 4315) present a case study of nonnative Red mangrove (*Rhizophora mangle* L.) along Hawaiian coastlines over the last century. The study "quantified total ecosystem C storage, sedimentation, accretion, sediment organic C burial and CH_4 emissions from ~70 year old R. mangle stands and adjacent uninvaded mudflats" (Soper et al., 2019: 4315).

The findings indicated that the ecosystem carbon stocks of the mangrove stands exceeded those of mudflats and mangrove stands increased the average coastal accretion (accumulation of sediments). Additionally, the sediment organic carbon burial increased markedly, to twice the global average for old growth mangrove forests. This important finding suggests that carbon accumulation from younger trees may be more rapid than indicated by previous studies, which may be relevant for mangrove restoration. The simulations also indicated that increased CH_4 emissions from sediments only minimally offset ecosystem CO_2 storage (2–4%, which would be 30–60 Mg CO_2 -eq/ha over mangrove lifetime). The authors conclude that the "results highlight the importance of mangroves as novel systems that can rapidly accumulate C, have a net positive atmospheric greenhouse gas removal effect, and support shoreline accretion rates that outpace current sea level rise" and the sequestration potential of novel mangrove forests should be considered mangrove forest removal is contemplated, considering climate issues (Soper et al., 2019: 4315).

Donato et al. (2012: 89) present a similar study, justified by the lack of knowledge of forest carbon stocks, especially for tropical islands and coastal areas where climate change impacts may be severe. This study "presents the first field estimate of island-wide carbon storage in ecosystems of Oceania, with special attention to the regional role of coastal mangroves, which occur on islands and coastal zones throughout the tropics" (Donato et al., 2012: 89). The methodology was based on samples of the above- and below-ground carbon pools (both soil and vegetation) within the three major vegetation structural types: mangroves, upland forests, and open, sometimes forest-derived savannas. The findings indicated that "mangroves contained by far the largest per-hectare C pools… with deep organic-rich soils alone storing more C… than all pools combined in upland systems. Despite covering just 12–13% of land area, mangroves accounted for 24–34% of total island C stocks" (Donato et al., 2012: 89). Additionally, comparing savannas and forests, the authors found that the savannas had lower C stocks than upland forests, "suggesting that reforesting savannas where appropriate has high potential for carbon-based funding to aid restoration objectives" (Donato et al., 2012: 89). These findings indicated the crucial role of mangroves for carbon storage in island and coastal areas and justified the sustainable management of mangrove forests for climate change issues (Donato et al., 2012: 89).

A similar study was presented by Cameron et al. (2019: 419), who maintain that despite mangrove forests being "extremely productive, with rates of growth rivaling some terrestrial tropical rainforests" the current knowledge of the processes "underpinning carbon exchange with the atmosphere and near shore waters, the allocation of carbon in mangroves, and fluxes of non-CO_2 greenhouse gases (GHGs) are limited to a handful of studies. This constrains the scientific basis from which to advocate for greater support for and investment in mangrove restoration and conservation." Their study was based on two high and low productivity systems on the island of Sulawesi, Indonesia, which were compared with established mangroves, to provide a "holistic synthesis of the net potential GHG mitigation benefits resulting from rehabilitating mangroves and established forests" (Cameron et al., 2019: 419). The study presents a carbon budget, allocation, and pathways model, with the inputs being carbon sequestration, and the outputs being GHG emissions of CO_2, N_2O, and CH_4, for estimation of Net Ecosystem Production (NEP) and Net Ecosystem Carbon Balance (NECB). The findings indicated that some of the high production systems represented significant carbon sinks. The authors contended that if Net Ecosystem Production from mangrove-allied primary producer communities (e.g., benthic algae) and the portion of dissolved inorganic carbon exported from mangroves (EXDIC) drives overall removals upward in established forests, and as these values are higher than those of terrestrial forests, there is a strong argument for more mangrove restoration (Cameron et al., 2019: 419).

A study by Charles et al. (2020) looked at changing mangrove cover on coasts, as "coastal wetlands are globally important carbon sinks, yet the effects of shifts in mangrove cover on organic-carbon (OC) storage remains uncertain" and "despite overall global declines, mangroves are expanding into and within many subtropical wetlands, leading to heterogeneous cover of marsh-mangrove coastal vegetation communities near the poleward edge of mangroves' ranges." The case study was of Black mangrove (*Avicennia germinans*, L.) or marsh vegetation in patches along a gradient in mangrove cover from zero cover to 100% cover, with measurements of changes in organic-carbon stocks and fluxes. Within the patches, the above- and below-ground biomass organic-carbon was measured at 61% greater for mangroves than for the recolonized marshes, and the soil organic-carbon under mangrove vegetation was 30% greater than the vegetation under marsh vegetation. The above- and below-ground biomass increased linearly at the plot scale, while the soil organic carbon was much more variable, with no correlation with mangrove cover (Charles et al., 2020).

Charles et al. (2020) found that: after 11 months, the surface organic carbon accretion was negatively related to plot-scale mangrove cover after a high-wrack deposition period (wrack refers to a coastal feature where organic and other debris is deposited at high tide); after 22 months, accretion was 54% higher in mangrove patches, with no link to plot-scale mangrove cover; for marsh plants (saltwort or beachwort, *Batis maritima* L.) leaf and root litter, breakdown rates were 1000% and 35% faster than mangrove leaf and root litter. Other findings were that mangrove cover "enhanced OC storage by increasing biomass, creating more recalcitrant organic matter and reducing k on the soil surface by altering microclimate, despite increasing wood k below-ground and decreasing allochthonous OC subsidies" (Charles et al., 2020). The authors conclude that the "results illustrate the importance of mangroves in maintaining coastal OC storage, but also indicate that the impacts of vegetation change on OC storage may vary based on ecosystem conditions, organic-matter sources, and the relative spatiotemporal scales of mangrove vegetation change" (Charles et al., 2020).

Savanna/Grassland Vegetation and Biogeochemistry

Savanna vegetation here refers to the savanna, grassland, and desert classifications of the global biomes, namely: Tropical/Subtropical Grassland, Savanna and Shrubland; Temperate Grassland, Savanna and Shrubland; Flooded Grassland and Savanna; Montane Grassland and Shrubland; Tundra; Mediterranean Forest, Woodland and Shrubland; Xeric Shrubland; and Desert (Beaumont & Duursma, 2012). Grassland (also termed the prairie or steppe) and shrubland (sometimes defined as similar or equivalent to scrubland) are easily defined, as land where the vegetation is dominated by either grasses or shrubs (Campbell, 1998). Savanna is more difficult to define, as its details are frequently disputed (Campbell, 1998). Savanna vegetation may vary in the relative proportions of trees, shrubs, and grasses, due to human and natural factors elements (Campbell, 2017, 2018). Scholes and Walker (1993: 3) argue that for the term savanna, "the central concept – a tropical mixed tree-grass community – is widely accepted, but the delimitation of the boundaries has always been a problem" (Scholes & Walker, 1993). Savanna may thus define a "continuum of vegetation types which have trees and grasses as their main constituents," although "at the extremes, the distinction between savannas and woodlands is unavoidably arbitrary" (Scholes & Walker, 1993). Their definition is therefore a "tropical vegetation type in which ecological processes, such as primary production, hydrology, and nutrient cycling, are strongly influenced by both woody plants and grasses, and only weakly influenced by plants of other growth forms" (Scholes & Walker, 1993: 4).

Savanna may occur in a broad range of climatic types, with average annual rainfall ranging from 250 to 2000 mm, approximately between semidesert and dry forest (Campbell, 1998, 2017). For example, in Africa, the driest savanna Sahel or Sahelian savanna, which transitions from the Sahara Desert, followed by the wetter Sudan or Sudanian savanna, which in its turn transitions into the wetter Guinea savanna (sometimes defined as derived savanna), which also merges with dry forest (Campbell, 1998). These grades, roughly from the drier north to the wetter south, are characterized by increasing trees southward (Campbell, 1998). The temperate grasslands, savannas and shrublands cover parts of North (e.g., the Prairies) and South America (e.g., the Pampas and Llanos), and Eurasia (e.g., the Steppe) (Campbell, 2017, 2018).

Desert occurs in areas with lower rainfall than the driest savannas and grasslands and may consequently be partially or totally devoid of vegetation. Commonly, xeric grassland, which is composed of small grass clumps, may be a transition zone between desert and dry savanna/grassland (Campbell, 2017). Deserts may be natural, due to lack of rainfall (for example, land in rain shadow areas behind mountains) of human influenced due to continual removal of vegetation through livestock grazing or crop cultivation in a dry zone, which may convert dry savanna/grassland to desert (Campbell, 1998). Definitions of deserts and desertification vary: Geist (2005) argues that a widely accepted definition is that of the Princeton University Dictionary which defines it as "the process of fertile land transforming into desert typically as a result of deforestation, drought or improper/inappropriate agriculture." Geist and Lambin (2004: 817) point to the definition of the desertification by the Convention to Combat Desertification: "land degradation in arid, semiarid and dry subhumid areas resulting from various factors, including climatic variations and human activities' (see United Nations Environment Programme UNEP, 1994).

Research on deserts and desertification owes much work to attempts to define and understand the Sahara Desert (Mortimore, 1989). Desertification as a human constructed process has been described as the final stage of deforestation, the end result of a transition passing through the stages of dry forest, savanna woodland, tree savanna, grassland (also termed steppe in Europe and prairie in the United States) (Campbell, 1998). Geist and Lambin (2004) listed factors for desertification, including agricultural activities (nomadic/extensive grazing, intensive production), crop production (annuals, perennials); infrastructure extension (watering/irrigation, hydrotechnical installations, dams, canals, boreholes, etc.); transport (roads); human settlements; industrial developments (oil, gas, mining, quarrying); wood extraction and related activities (fuelwood or pole wood cutting), digging and collection of plant products) and human population issues (migration, natural increment) (Geist & Lambin, 2004: 819).

Savanna and grassland are important for the global ecosystem due to their areal extent (Campbell, 1998). For example, Scurlock and Hall (1998: 229) describe grasslands as among the most widespread, global vegetation types, covering nearly 20% of the Earth's land surface or 15 million km^2 in the tropics (as much as tropical forests) and a further 9 million km^2 in temperate regions (Scurlock & Hall (1998: 229). In terms of biogeochemistry, the role of savannas in for example carbon cycling is disputed, as interactions between climate change and grassland biogeochemistry by the end of the twentieth century, savanna ecosystems were less studied than forests (Scurlock & Hall, 1998: 229). Scurlock and Hall (1998: 229–233) also cite recent studies suggesting that tropical grasslands and savannas contribute more to the "missing sink" than was previously understood, possibly around 0.5 Pg (5 0.5 Gt) carbon annually: hence, "tropical and temperate natural grasslands play a significant but poorly recognized role in the global carbon cycle," and "their carbon fluxes may have been previously underestimated, especially in the tropics" (Scurlock & Hall, 1998: 229–233). Additionally, "grassland soil carbon stocks amount to at least 10% of the global total but other sources estimate up to 30% of world soil carbon" (Scurlock & Hall, 1998: 229; see also Anderson, 1991; Eswaran et al., 1993).

More recent sources also point to the neglect of the biogeochemical role of savanna vegetation (Campbell, 2017). For example, Zhou et al., 2022: 445) argue that although savannas cover 20% of the Earth's land surface, contribute over 30% of the terrestrial net primary production, and 75% of global area burned, and over 50% of global fire-driven carbon emissions, there is lack of details of the whole-ecosystem carbon storage for savannas, especially for those savannas under altered fire regimes. Using a case study of the African savanna, the findings of their study indicated that "fire suppression increased whole-ecosystem carbon storage by only 35.4 ± 12% (mean ± standard error), even though tree cover increased by 78.9 ± 29.3%, corresponding to total gains of 23.0 ± 6.1 Mg C ha^{-1} at an average of about 0.35 ± 0.09 Mg C ha^{-1} year^{-1}, more than an order of magnitude lower than previously assumed." It was also noted that savannas under frequent burning had large amounts of below-ground carbon, mostly in biomass and deeper soils, and that these subterranean carbon reservoirs "are not fully considered in afforestation or fire-suppression schemes but may mean that the decadal sequestration potential of savannas is negligible, especially weighed against concomitant losses of biodiversity and function" (Zhou et al., 2022: 445).

Williams et al. (2007) also discuss African savanna biogeochemistry in respect of carbon cycling, noting that Africa's primary net carbon release is land conversion, mostly through burning of forests, and these fires may return carbon to the atmosphere within several types of compounds, some of which may be chemically or radiatively active (e.g., methane, carbon monoxide, and aerosols), and may even be forerunners to radiatively active gases (e.g., ozone precursors)" (Williams et al., 2007; see also Crutzen & Andreae, 1990; Scholes & Andreae, 2000). Savanna fire emissions are large, but mostly short-term episodes offset by the following regrowth.

The contribution of savanna carbon fluctuations is mostly from savanna fire emissions, and Africa's carbon emissions and their interannual variability are likely to undergo substantial increases through the twenty-first century, due to the large carbon stocks, high vulnerability to climate and land use change, and increasing human populations and industrialization. Williams et al. (2007) maintain that carbon emissions in savanna fires are shorter-term carbon losses than forest fires, because the main combustible fuel is dead herbaceous vegetation, which may be only one or two years of growth, and hence, savanna fires may result in faster cycling of biomass carbon rather than net emissions. Where carbon emissions from savanna fires may be balanced over the long-term by growth in later years, the fires create intensive and small-scale contributions of carbon into the atmosphere, which can shift the seasonal or interannual distribution of CO_2 releases" (Williams et al., 2007).

Soils in temperate savanna may also contribute to carbon cycling through black carbon (BC) (Dai et al., 2005; Ansley et al., 2006). Liu et al. (2016, 1) define black carbon as "a continuum of incomplete combustion products of vegetation and fossil fuels" which "can be used as a proxy for wildfires and paleovegetation reconstruction." Dai et al. (2005: 1879) point out that "black carbon (BC) or charcoal is thought to represent an important component of the carbon cycle but has seldom been quantified in soils" and use a case study of a temperate mixed-grass savanna in the southern Great Plains using benzenecarboxylic acids as molecular markers for BC. The soil samples were collected from summer and winter fires, with the black carbon concentrations ranging from "50 to 130 g BC kg^{-1} of soil organic carbon (SOC), or from 0.55 to 1.07 g BC kg^{-1} of whole soil in this mixed grass savanna" (Dai et al., 2005: 1879). The findings indicated that the BC contribution to SOC increased significantly with soil depth ($P < 0.05$), and repeated fires increased BC slightly in comparison with the unburned controls. The effects of repeated fires on BC were not statistically significant in studied mixed-grass savanna. Dai et al. (2005: 1879) conclude that the findings reveal estimates of BC concentrations for uncultivated mixed-grass savanna, and also show that a few fires will have minimal impacts on the size of the soil BC pool in the studied ecosystem type and region.

Bork and Badiou (2017) also researched the importance of temperate grasslands in the global carbon cycle, recording that grasslands cover between 25% and 40% of the earth's land surface and may include as much as 30% of the Earth's soil organic carbon stocks. Additionally, the temperate grasslands of the world comprise an estimated 9.0–12.5 million km^2 or 7.0–9.7% of the Earth's land area and are extremely "diverse and productive terrestrial biomes" (see also Henwood, 1998; World Resources Institute (WRI), 2000; White et al., 2000; Braun et al., 2013; Hewins et al., 2015; Wang et al., 2014). Bork and Badiou (2017:8) conclude that temperate grasslands are key players in the global carbon cycle because of their high productivity rates, high carbon sequestration rates and broad geographical ranges. The authors however note that there can be significant losses of soil organic carbon due to the conversion or degradation of wetland ecosystems, which can result in these ecosystems converting to net radiative sources. Conversely, significant developments for carbon sequestration can result from the conversion of marginal croplands to grasslands and the management of grasslands/grazing lands (note: "Carbon sequestration is the process of capturing and storing atmospheric carbon dioxide," see United States Geological Survey, 2023).

Flooded grasslands and savannas are also important for global carbon change (Fidelis et al., 2013; Joyce et al., 2016). This study examined "above- and below-ground biomass and carbon dynamics in Brazilian Cerrado wet grasslands," with the main question being the importance of the Brazilian tropical wet grasslands as carbon sinks, and possible roles of fire frequency and season in affecting biomass and carbon allocation in Brazilian wet grasslands (Fidelis et al., 2013). In this study, the biomass was classified into two groups: "above-ground graminoids, forbs and dead matter, and below-ground roots and other below-ground organs" (Fidelis et al., 2013: 356). The findings indicated that over 40% of the biomass and carbon stocks were located below-ground, mostly in roots, but a high proportion of dead biomass was found above-ground, which the authors ascribe to possible low decomposition rates and annual accumulation. It was also argued that despite the lack of water stress in the studied type of grasslands, there was significant evidence of "resource re-allocation from below-ground organs to the above-ground biomass in the rainy season" (Fidelis et al., 2013: 356). The authors concluded that there was more dead biomass in the rainy season, perhaps the result of low decomposition rates, which may increase fire risk during the later dry season. Importantly, "these tropical wet grasslands stored high amounts of carbon... mostly in the roots. Thus, policymakers should consider tropical grasslands as potential carbon stocks since they are one of the most threatened and unprotected ecosystems in Brazil" (Fidelis et al., 2013: 356).

Montane grasslands and savannas are another important carbon sink (Rogger et al., 2022). For example, Knowles et al. (2020) conducted a study of "sky islands" "which are characteristic of sequential mountain-valley terrain where mountains form an island archipelago rising from surrounding valleys of desert sea." In the case study of mixed evergreen conifer forests on the Madrean sky islands of the southwestern United States and Mexico between 2009 and 2018, the findings indicated that throughout the year the forest vegetation functioned as net sink of carbon dioxide, resulting in higher levels of carbon sequestration than other studied montane forests in the continental western United States. The sustained forest activity resulted from mild winter temperatures and a bimodal precipitation regime, with moisture during the summer and winter. The forecasted climatic warming "could thus stimulate forest carbon gains during the winter and either increase or decrease respiratory carbon loss during summer as a function of moisture" due to the gross primary production being "temperature limited in winter and could become moisture limited during the dry early summer period depending on antecedent snowmelt moisture" (Knowles et al., 2020). Additionally, ecosystem respiration was found to be more sensitive than gross primary production to moisture availability during the nonwinter period. The authors concluded that the best predictor of the warm season carbon balance overall, was "a metric of snow aridity that included snow depth and potential evapotranspiration" and "the seasonally dissimilar impacts of warming and drying identified by this work inform current understanding of how climate change and/or variability may affect forest water and carbon cycling dynamics throughout the montane forest biome" (Knowles et al., 2020).

A case study of Peruvian montane grasslands found that montane tropical soils are a large carbon reservoir, functioning as both a source and a sink of CO_2 (Oliver et al., 2017). Increased CO_2 emissions may partially result from the decomposition and other changes of soil organic matter, sometimes after human actions. Changing soil organic matter may influence the long-term carbon storage and increasing atmospheric CO_2 concentrations, which is an indicator of SOM modification in response to land management practices. The study investigated the impacts of grazing and fire history on soil carbon in the Peruvian montane grassland case study. This involved using a density fractionation method to quantify the changeable and stable organic matter concentrations, and soil CO_2 flux and decomposition measurements. The findings indicated that grazing and burning increased soil CO_2 fluxes and decomposition rates and decreased the proportion of carbon in the free light fraction, mostly deeper (10–30 cm) in the soil (note that the "light fraction is a short-term reservoir of plant nutrients and the primary fraction for soil carbon formation and serves as a readily decomposable substrate for soil microorganisms. Its size is a balance between residue inputs and decomposition"; Oliver et al. (2017: 5633 see also Song et al., 2012). Oliver et al. (2017: 5633) found that in the control soils, about 20% of recovered material was in the free light fraction, which also contained 30% of the soil carbon content, while the burnt-grazed soil had the lowest recovery of the free light fraction and (10%) a lower carbon content (14%). Additionally, the burnt soils had higher proportions carbon in the occluded light fraction (12%) than the not-burnt soils (7%), with no significant difference for the treatments in the heavy fraction (F) (~ 70%). Finally, it is noted that burning and grazing changed the soil carbon dynamics, increased CO_2 fluxes, and "the dominant temperature driver was obscured by some other process, such as changes in plant C and N allocation" and the free light fraction was reduced when burning and grazing occurred at the same site, and there was no significant long term change in the total soil carbon stocks, and in the carbon storage in the in the occluded light fraction and the heavy fraction (Oliver et al. (2017).

Another study of mountain grassland in Switzerland by Rogger et al. (2022) used CO_2 flux measurements on a managed 1000 m elevation area, investigating "the potential of grasslands to mitigate climate change by carbon (C) sequestration in soils," which "depends on agricultural management and the response of the grassland C cycle to a warming and more variable climate." The study examined the response of CO_2 fluxes to environmental drivers and grazing and cutting events, as well as warming trends and inconsistent weather conditions on the annual C sequestration potential and estimated the carbon budgets including atmospheric and management-related carbon fluxes (fertilization, cutting/grazing). The findings indicated light, and temperature were the main environmental drivers of the CO_2 fluxes. Additionally, human-induced removal of the above-ground canopy reduced photosynthesis and also induced a saturation in the photosynthetic response to light at lower intensities compared with the time with human removal of vegetation (Rogger et al., 2022). Rogger et al. (2022) found that "with a net ecosystem exchange of −357 (±76) g C m^{-2} $year^{-1}$ and a net biome production of −154 (±80) g C m^{-2} $year^{-1}$ (including management C fluxes), the grassland on average acted as a net

C sink. There was also an anomalous increased net CO_2 during weather change negative for photosynthesis and regrowth following harvesting. During the period of study, the release anomalies led to an annual net C loss. Speculating into the future, the authors argue that under future climatic conditions, the C sequestration potential at the mountain site will be affected by the extent to which spring photosynthesis is offset by the climate-management interactions described and a climate warming-induced increase in ecosystem respiration (Rogger et al., 2022).

Tundra is an important component of carbon cycling (Belshe et al., 2013; Anthony et al., 2014). Schuur (2019) observed that the northern permafrost region soils contain 1460–1600 billion metric tons of organic carbon, which is double the quantity in the atmosphere. The organic carbon in the tundra region is climate sensitive, in that the warming conditions influence microbial conversion of permafrost carbon into greenhouse gases (carbon dioxide and methane) which are released into the atmosphere as a response to climate warming. Schuur (2019) also argues that the recent measurements of regional and winter season ecosystem carbon dioxide flux indicate permafrost region ecosystems released net carbon into the atmosphere. Schuur (2019) points out that "these observations signify that the feedback to accelerating climate change may already be underway."

Belshe et al. (2013: 1307) asked questions, and justified their questions: "Are tundra ecosystems currently a carbon source or sink? What is the future trajectory of tundra carbon fluxes in response to climate change? These questions are of global importance because of the vast quantities of organic carbon stored in permafrost soils." They conducted an incisive meta-analysis, based on data covering 40 years of CO_2 flux observations from the northern high latitudes. The objective was to determine whether seasonal or annual CO_2 fluxes underwent temporal change, and if spatial differences in mean annual temperature explain temporal changes in CO_2 flux" (Belshe et al., 2013: 1307). The results indicated that the growing season net CO_2 uptake increased after the 1990s, and some evidence that winter CO_2 emissions increased in the 1990s and the early 2000s. The authors conclude that "in spite of the uncertainty in the winter trend, we estimate that tundra sites were annual CO_2 sources from the mid-1980s until the 2000s, and data from the last 7 years show that tundra continue to emit CO_2 annually" (Belshe et al., 2013: 1307). Also, the CO_2 emissions exceeded CO_2 uptake across the different temperatures in the tundra, this suggesting that "despite increases in growing season uptake, tundra ecosystems are currently CO_2 sources on an annual basis" (Belshe et al., 2013: 1307).

Desert is another source of carbon (Wang et al., 2007; Thomas & Hoon, 2010; Zhao et al., 2017). For example, Throop et al. (2020, 423–443) studied soil respiration (Rs) as an important factor for carbon release to the atmosphere, taking the case study of the Namib desert in Southern Africa. The main topic was the variability in soil respiration response to "rainfall pulses at multiple levels of spatial heterogeneity: (1) along a hyper-arid to arid rainfall gradient, (2) across soil surfaces that differ in stability and composition, and (3) among different geomorphic and vegetation patch types." The case study included comparisons between two soil surfaces, with two vegetation/geomorphic patch types, differing in organic matter sources and transport processes, with one "characterized by depositional inputs" and the other "characterized by erosional losses" (Throop et al., 2020: 423). The findings showed that soil respiration was responsive to rainfall pulses, but in many cases soil surfaces and patch types had more effect on soil respiration than rainfall pulses did. Higher rainfall had more impact than lighter rainfall on soil respiration. Additionally, greater soil respiration occurred on stable soil surfaces than unstable soil surfaces, and in greater soil respiration occurred on depositional than erosional patch types. The authors conclude that "large differences in Rs among rainfall zones, soil surfaces, and patch types point to the need to carefully consider multiple scales of spatial heterogeneity when interpreting dryland biogeochemical fluxes" (Throop et al., 2020: 423).

De Graaff et al. (2014: 931), researching on biogeochemical cycling in desert and xeric grassland, argue that semiarid and arid ecosystems dominated by shrubs ("dry shrublands") are important for the global C cycle, but there has been little assessment of the effects of climate change and increased atmospheric CO_2 on biogeochemical cycling in these arid ecosystems. De Graaff et al. (2014: 931) examined data from manipulative studies (note: a manipulative study is where deliberate changes are made to determine effects), and comparative studies contrasting ecosystem processes in different vegetation types (e.g., shrub or herbaceous canopy versus intercanopy vegetation, microsites), to determine the impact of climatic and atmospheric CO_2 changes on biogeochemical cycles, through the alteration of plant and microbial physiology and ecosystem structures, and how ecosystem structures impacts on biogeochemical cycles across a climate gradient.

The findings indicated that there were insufficient numbers of manipulative studies and measurements in those studies, hence limited knowledge of explanations of biogeochemical changes (De Graaff et al., 2014: 931). Additionally, it was found that changes in ecosystem structure affect biogeochemical cycling, with decreasing pools and fluxes of C and N if the vegetation canopy microsites declined and "differences in biogeochemical cycling between microsites are predictable with a simple aridity index…, where the relative difference in pools and fluxes of C and N between vegetation canopy and inter-canopy microsites is positively correlated with aridity" (De Graaff et al., 2014: 931). The authors concluded that with climate change sufficient to alter ecosystem structure, there will be strong impacts on biogeochemical cycles, with increasing aridity resulting in greater heterogeneity in biogeochemical cycling among the microsites. The authors conclude that "more studies, especially long-term manipulative experiments in dry shrubland ecosystems, will be needed for more effective prediction of climate change impacts on desert biogeochemical cycling" (De Graaff et al., 2014: 931).

For Mediterranean ecosystems (shrubland), there is evidence that shrublands are important for the reduction of atmospheric CO_2 and contribute to the amelioration of the impacts of climate change, because of their role as carbon sinks and their large area extents (Carrión-Prieto et al., 2017: 67). For example, these authors selected for study "two of the most representative shrub species in the Iberian Peninsula, *Cistus ladanifer* L. and *Erica arborea* L." which "were studied in terms of biomass distribution and carbon and nitrogen contents in the different fractions" (Carrión-Prieto et al., 2017: 67). The objective of the study was an effective estimation method for biomass, using a new method for simple root-to-shoot calculation, which would be based on vibrational data. Using this technique, carbon sequestration, was determined, which for these two species was higher than for most other shrubs and justified the use of these two species for restoration and carbon offsetting.

Gratani et al. (2013: 1167) nevertheless point out that "to date, only a few attempts have been done to estimate the contribution of Mediterranean ecosystems to the global carbon cycle." Their case study was of the shrub species, composition, and structure of the Mediterranean shrublands (Maquis or macchia shrubland biome) on the Latium coast of Italy, which were studied to determine contribution to carbon sequestration. The methodology classified the shrubs into small, medium, and large, based on volume and leaf area index, and the annual carbon dioxide sequestration per species was calculated by multiplying the photosynthetic leaf surface area of each species by the mean yearly photosynthetic rate and the total yearly photosynthetic activity time. The findings indicated that for sequestration per species, evergreen oak (*Quercus ilex* L.) and the strawberry tree (*Arbutus unedo*, L. 1753) had the highest levels, followed by the green olive tree (*Phillyrea latifolia*, L.), tree heath (*Erica arborea*, L.), Mediterranean Heath (*Erica multiflora*, L.), rockrose (*Cistus incanus*, L.), Mastic tree (*Pistacia lentiscus*, L.), Rosemary (*Rosmarinus officinalis*, L.), and common smilax (*Smilax aspera*, L.), the interspecies carbon sequestration varying from 46.2 ± 15.8 kg CO_2 year^{-1}, mean value), to 6.8 ± 4.2 kg CO_2 year^{-1}, mean value). The authors conclude that "the total yearly CO_2 sequestration per shrub (SCshy) was 149 ± 5 kg CO_2 year^{-1} in L (large shrubs) decreasing 30% in M (medium shrubs) and 80% in S (small shrubs)" (Gratani et al., 2013: 1167) Using this information for the different sized shrubs and their sequestration, the authors estimated the total CO_2 sequestration of the Mediterranean maquis vegetation (Maquis or macchia shrubland biome) as quantified as 80 Mg CO2 ha ^{1} year ^{1}, or 22 Mg C ha ^{1} year ^{1} (Gratani et al., 2013: 1167).

Beier et al. (2009) also support the vital ecological importance of shrublands, and the relative neglect of biogeochemical studies for these ecosystems relative to forests or grasslands. Arguably, "as climate change progresses, the potential feedback from terrestrial ecosystems to the atmosphere through changes in carbon stocks, carbon sequestration, and general knowledge on biogeochemical cycles becomes increasingly important" (Beier et al., 2009:1). Their study examines the carbon and nitrogen balances of six shrublands on a climatic gradient across the European continent, with the objective of evaluating the form of carbon storage in European shrublands. The findings indicated that across the sites "the net carbon storage in the systems ranged from 1163 g C m^{-2} to 18,546 g C m^{-2}, and the systems ranged from being net sinks (126 g C m^{-2} a^{-1}) to being net sources (−536 g C m^{-2} a^{-1}) of carbon with the largest storage and sink of carbon at wet and cold climatic conditions" (Beier et al., 2009).

Other findings were that the soil carbon dominated the carbon budget, especially in cold and wet climates (here soil carbon was 95% of the total ecosystem carbon); carbon respiration from soil organic matter pool was the dominant carbon loss, over carbon loss from above-ground litter decomposition; and below-ground carbon allocation was over 5 times above-ground litterfall carbon ("greater than the factor of 2 reported in a global analysis of forest data") (Beier et al., 2009). The authors conclude and speculate that a future, climate-driven land cover change between grasslands and shrublands in Europe may result in increased ecosystem C "where shrublands are promoted and less where grasses are promoted" and ":if feedbacks on the global carbon cycle are to be predicted it is critically important to quantify and understand below-ground carbon allocation and processes as well as soil carbon pools, particularly on wet organic soils, rather than plant functional change as the soil stores dominate the overall budget and fluxes of carbon" (Beier et al., 2009).

The Tools of Geomatics

All the environmental dynamics described earlier may be measured by or related to measurable parameters, where a parameter is a numerical or other measurable factor that is part of system and may affect the operation of the system, in this case the attributes and components of micro, local, regional or global ecosystems. These parameters include the components of soil, vegetation, climate and/or topography (Campbell, 2017, 2020). These components may be measured, and the measurements quantified using software of increasingly sophisticated geomatics based technology (Campbell, 2018). Therefore, the tools of geomatics, including those of remote sensing, geodesy, geomatics engineering, and other related survey methods, are highly relevant to both biogeochemistry and biogeography, as space, place, location, patterns, temperature, moisture, and movement are vital parameters for the measurement of environmental dynamics, including those that determine biological and geochemical features (Campbell, 2017, 2018). For example, Pagiatakis et al. (2009: 25) note that "geomatics engineering is revolutionizing science, natural resource exploration, space technology, mapping, economic growth and safety-of-life solutions by means of new theories and methodologies, and emerging technological developments." Plag (2000) adds that "integrated global monitoring is a prerequisite for global environmental management of a sustainable Earth. Specialised global observing systems have already been initiated and partly are implemented (e.g., the G3OS, the Global Terrestrial, Climate and Ocean Observing Systems: GTOS, GCOS, GOOS)." It is contended that it is "of fundamental value to fully integrate geodetic monitoring techniques into a global integrated Earth monitoring system, which is a significant step towards an Earth information system for sustainability. This integration should follow the Global Integrated Observing Strategy." This section considers three components of geomatics-based remote sensing (geodesy, radio detection and ranging – RADAR, and light detection and ranging – LiDAR) and how they are connected to biogeochemistry and biogeographical research.

Geodesy

Remote sensing and GIS techniques are all based on the theoretical and practical analyses of the science of geodesy. Geodesy is fundamental for the geosciences that consider locations on the earth's surface for technical, environmental, and social applications. The National Oceanic and Atmospheric Administration (1977) notes that "although geodesy is the oldest among earth sciences, dating back to the Sumerians of 5000 years ago, its subject and even its name have been widely unknown." This organization notes that geodesy developed as a science and applied technology from practical needs, these including applications to land and cadastral surveys, civil engineering, resource location and navigation. The National Oceanic and Atmospheric Administration (2021a) defines geodesy as "the science of accurately measuring and understanding three fundamental properties of the Earth: its geometric shape, its orientation in space, and its gravity field – as well as the changes of these properties with time." Meyer (2006, 3) defines geodesy as "the science that studies the macroscopic shape of the Earth, including its gravitational and magnetic fields. A basic function of geodesy is to create meaningful coordinate systems for the purpose of surveying, mapping, navigation, and any other application of spatial data. In particular, geodesy is what gives meaning to latitude, longitude, and height. Geodesy is the cornerstone of all spatial sciences and has applications in geology, geophysics, and astronomy, to name a few." Meyer (2018, 3) also cites the Geodetic Glossary (National Geodetic Survey, 2009), where geodesy is defined as "the science concerned with determining the size and shape of the Earth" or "The science that locates positions on the Earth and determines the Earth's gravity field." Hay (2000, 1) notes that geodesy is the "discipline concerned with the measurement of the size and shape of the Earth and positions on it. It is important to remote sensing and geographical information systems (GIS) as it underpins the essential process of georeferencing images and associated vector coverages to Earth surface locations" … including the concepts of "geoids. reference ellipsoids, datums, projections and grid referencing systems" (see also Burkard, 1968; Snyder, 1987; Smith et al., 1988).

GIS Resources (2021) defines geodesy as primarily "the science which deals with the methods of precise measurements of elements of the surface of the earth and their treatment for the determination of geographic positions on the surface of the earth." It also deals with the theory of the size and shape of the earth. The fundamental status of geodesy is revealed by the fact that it has dispersed into supporting roles for numerous fields, such as geomatics, GIS, remote sensing, cartography and even epidemiology and landuse planning. Geodetic techniques in have even been applied to application to animal ecology, which may have unique, mostly scale-based requirements, including the georeferencing of spatial, habitat surveys, elevation studies, the development of increased precision in animal and physical feature location, and more detailed landscape and land feature change (Campbell, 2022a). The National Aeronautics and Space Administration (2018) describes several geodetic techniques, which are used to observe the geodetic properties of the Earth including the space-geodetic techniques of Very Long Baseline Interferometry

(VLBI), Satellite Laser Ranging (SLR), Global Navigation Satellite Systems (GNSS) like the US Global Positioning System (GPS), and the French Doppler Orbitography and Radio-Positioning by Integrated Satellite (DORIS) system.

Geodesy has also become linked to geophysics, which is a major supporting discipline for biogeochemistry (Campbell, 2020). At the institutional and organizational levels, this link is shown by the organizations (e.g., the International Union of Geodesy and Geophysics (IUGG), Geodesy and Geophysics Laboratory at the National Aeronautics and Space Administration Goddard Flight Space Center) that work with an integration of geodesy and geophysics. For the latter organization, work "areas of Earth time variable and static geopotential and geomagnetic fields, Earth orientation, surface deformation, characterization and change, tides, land ice mass evolution, global and regional sea level, and airborne and spaceborne laser altimetry" (National Aeronautics and Space Administration Goddard Flight Space Center, 2021). The American Geophysical Union (2023) actually defines geodesy as a branch of geophysics, "that studies the geometrical, structural, and gravitational properties of the Earth, their time evolution, and the dynamic interactions of the solid Earth with other physical components of the Earth system (atmosphere, hydrosphere, cryosphere, and the core), at a wide range of temporal and spatial scales."

Geodesy has made an impact through geomatics engineering, where there are possibilities for higher precision in earth feature recording, measurement and observation, and the design and use of terrestrial, marine and space sensors, which can be used for applications such agriculture, forestry, transportation, navigation, climate change, planning and health (Campbell, 2022a). Bowditch (2002: 18) describes vertical surveying as one of the most important applications, principally "the process of determining elevations above mean sea-level. In geodetic surveys executed primarily for mapping, geodetic positions are referred to an ellipsoid, and the elevations of the positions are referred to the geoid." Additionally, "the geoidal heights must be considered to establish the correct height above the geoid. Precise geodetic leveling is used to establish a basic network of vertical control points. From these, the height of other positions in the survey can be determined by supplementary methods" (National Geodetic Survey, 2017). With such information, geodesy "has revolutionized the study of solid Earth processes through its ability to measure the deformation of the Earth's surface and the Earth's gravity field with extraordinary accuracy" and such may "provide information about the motion of the Earth's tectonic plates, insight into the cause and timing of earthquakes and volcanic eruptions, and constraints on the internal forces that drive them" comprising "integrated and geo-referenced sets of observations on global to regional spatial scales and with high spatial and temporal resolution" (National Aeronautics and Space Administration, 2018).

Geodetic derived data, including that from satellites, includes inputs for geophysics, hydrography, and oceanography (Seeber, 2003; Campbell, 2022a). Seeber (2003) describes some of the observation methods of satellite applications as satellite laser ranging, satellite altimetry, gravity field missions, baseline interferometry, Doppler methods, and Global Navigation Satellite Systems (GNSS) and strong focus on positioning techniques, such as the NAVSTAR Global Positioning System

(GPS), and to applications. Some developments include global and regional control networks, gravity field modeling, Earth rotation and global reference frames, crustal motion monitoring, cadastral and engineering surveying, geoinformation systems, navigation; marine and glacial geodesy, and photogrammetry and remote sensing (Seeber, 2003). Satellite images are generally more effective than aerial photographs, as they are taken from much higher altitudes and have much larger coverage (Jensen, 2007; Campbell, 2018). Image enhancement techniques were developed, with the advancement of satellite technology, with improved resolutions of on meter for panchromatic images and 4 m for multispectral images (Campbell, 2018). Here, panchromatic refers to an emulsion produces a more realistic scene. Panchromatic imagery has a pixel equivalent to an area 0.6 m × 0.6 m, but the multispectral pixels represent an area of 2.4 m × 2.4 m (Livingston, 2015). Table 4.6 lists some of the recent satellites.

Other developments included increased use of multispectral images, which are produced in bands that represent a wider range of the electromagnetic spectrum, i.e., including the ultraviolet, visible, and infrared bands. The number of bands may vary between three to ten bands. For example, Landsat 5, produces 7 band images, between wavelengths of 450 and 1250 nm. Global positioning systems (GPS) were also developed, "a U.S.-owned utility that provides users with positioning, navigation, and timing (PNT) services. This system consists of three segments: the space segment, the control segment, and the user segment. The U.S. Space Force develops, maintains, and operates the space and control segments" (National Institute of Standards and Technology, 2023). GPS is based on 24 orbiting satellites, with a control segment that monitors the satellites, and a user segment that receives user location information. Beutler (1998) argues that "GPS may be considered a fully mature space geodetic technique in 1997. Some people even would say that GPS is the dominant technique in geodesy and geodynamics." GPS applied to environmental sciences and ecology generally locates features, animals, and habitat parameters, for more effective ecological analyses (Campbell, 2018). These techniques all emphasize the measurement of the Earth's form, including physical and biological features which may be measured, at local, regional, and global scales, additionally through time series monitoring (Campbell & Jha, 2021).

A particularly relevant application of geodetic methods to biogeography and the environmental parameters of biogeochemical change is the measurement of the physical features of animal habitats, including that of movements such as migration, including topography, elevation, climatic and geomorphological events, such as storms and hurricanes, sudden earth movements (landslides, volcanism, earthquakes), and ocean events (Campbell, 2018). Such events and parameters may affect vegetation, water bodies and elevated features upon which animals may depend for foraging, nesting, and migration (Campbell, 2015).

Such modern geodetic measurements used in the study of the earth and environmental sciences has contributed to monitoring and analysis of geohazards, including weather events, earthquakes, volcanic eruptions, debris flows, landslides and subsidence and sea level change, all relevant for geophysical and geochemical studies. The National Aeronautics and Space Administration (2018) adds that "geodesy

Table 4.6 Recent satellite launched by nations including the United States

Name	Nation	Dates (launch to retirement)	Spatial resolution (m)
EROS-A	Israel	Dec. 5, 2000–May 2016	1.8 m
QuickBird	USA	Oct. 18, 2001–Dec. 17, 2014	60 cm PAN 2.4 m MS
Spot 5	France	May 4, 2002–Mar. 31, 2015.	5 m PAN 10 m MS
OrbView-3	USA	June 2003–March 13, 2011	1 m PAN 4 m MS
Carosat-1	India	May 5, 2005–2018	2.5 m PAN
EROS-B	Israel	Apr. 25, 2006–2018	70 cm
EROS-B	Israel	Apr. 25, 2006–2018	70 cm
SAR-Lupe 1–5	Germany	Dec. 19, 2006–July 22, 2008 +	<1 m, SAR
Cartosat-2	India	Jan. 10, 2007–2018	<1 m
Cartosat-2	India	Jan. 10, 2007–2018	<1 m
Cosmo SkyMed-1	Italy	June 8, 2007–2018	SAR
TerraSar-X	Germany	June 15, 2007	Variable, to 1 m SAR
Cosmo SkyMed-2	Italy	Dec. 9, 2007–2018	SAR
RADARSAT-2	Canada	Dec. 14, 2007–2018	3–100, <1 m SAR
WorldView-1	USA	Sept. 18, 2007–2018	0.5 m PAN
Cartosat-2A	India	April 28, 2008	<1 m PAN
Cosmo SkyMed-3	Italy	Oct. 25, 2008–2018	SAR
GeoEye-1	USA	Sep. 6, 2008–2018	0.41–46 m PAN, 1.65 m MS
WorldView-2	USA	Oct. 8, 2009–2018	46 cm PAN 1.84 m MS
TanDEM-X	Germany	June 21, 2010–2018	2–4 m
Cartosat-2B	India	July 12, 2010–2018	<1 m PAN
Cosmo SkyMed-4	Italy	Nov. 6, 2010–2018	SAR
NigComSat-1R	Nigeria	Dec. 19, 2011–2018	2.5 m PAN 5 m MS
Pleiades-1A	France	Dec. 16, 2011–2018	PAN 0.5 m, 2 m MS
Pleiades-1B	France	Dec. 2, 2012–2018	PAN 0.5 m 2-m MS
Spot 6	France	Sept. 9, 2012–2018	PAN – 1.5 m, MS – 6 m, (BGR, NIR)
Spot 7	France	June 30, 2014–2018	PAN 1.5 m; MS 6 m (BGR, NIR)
WorldView-3	USA	Aug. 13, 2014–2018	31 cm PAN, 1.24 m MS, 3.7 m SWIR, 30 m CAVIS

NB: PAN refers to panchromatic, MS refers to multispectral and SAR to synthetic aperture radar. SW refers to short wave, IR to infrared and SWIR to shortwave infrared imagery. BGR refers to blue, green and red. Sources: Livingston (2015, 15–16), Indian Space Research Organization (2018); NASA (2018d). NB: PAN refers to panchromatic, MS refers to multispectral and SAR to synthetic aperture radar. CAVIS refers to Clouds, Aerosols, Vapors, Ice, and Snow. SW refers to short wave, IR to infrared, NIR to near infrared and SWIR to shortwave infrared imagery

is at the heart of all present-day ocean studies. It is how the orbit reference is computed for analysis of all satellite altimeter data, and the accuracy with which that data can be tied to tide gauge and other data. Knowledge of the reference frame geo-center and its change are needed to study regional and global sea level change and ocean-climate cycles like El Niño, the North Atlantic Oscillation, and the Pacific Decadal Oscillation" (National Aeronautics and Space Administration, 2018).

Atmospheric events are an important area of application for geodetic methods, and the results of which can form the basis for studies of atmospheric, biogeochemical cycling (Campbell, 2020). For example, the National Aeronautics and Space Administration (2018) maintains that one of the main factors for geodesy applications concerns the georeferencing of the events and associated data to a common frame, with an example of weather events, for which geodetic methods may support the geo-referencing of meteorological observing data, the provision of atmospheric weather models with space- and time-varying gravity fields, and tracking of global changes in stratospheric mass and lower tropospheric water vapor fields (National Aeronautics and Space Administration, 2018). Other relevant measurements include a reference system may be for the study of "sub-seasonal, seasonal and secular movements of continental and basin-scale water masses; the wet part of the troposphere from atmospheric sounding; loading and un-loading of the land surface due to seasonal changes of groundwater; local measurement of the integral variation of ground water from permanent gravimetric tidal stations; measurement of water level of major lakes and rivers by satellite altimetry; improved digital terrain models, as basis for flux modeling of surface water and flood modeling; geographic information systems for the establishment of comprehensive referenced data modeling" (National Aeronautics and Space Administration, 2018).

Geodetic measurements may also measure the Earth's gravity field, which affects Earth surface systems and features (Campbell, 2018). For example, Jekeli (2011) points out that the Earth's gravitational field is the field generated by the masses of the terrestrial body, for which measurements made on the surface include "tidal components due to the sun and moon (and theoretically other planets) and the atmosphere, as well as the centrifugal acceleration due to Earth's rotation. Therefore, geodesists distinguish between terrestrial gravitation (mass attraction of the Earth) and gravity (gravitation plus centrifugal acceleration), where tidal and atmospheric effects are treated as corrections" (Jekeli, 2011). The National Oceanic and Atmospheric Administration (2021b) writes that measurements of the Earth's gravity enable the determination of mean sea level, and the elevation of locations on the Earth's surface is based on the mean sea level. The National Oceanic and Atmospheric Administration (2021b) argues that "knowing how gravity changes sea level helps geodesists make more accurate measurements. In general, in areas of the planet where gravitational forces are stronger, the mean sea level will be higher. In areas where the Earth's gravitational forces are weaker, the mean sea level will be lower." Measurements of the Earth's gravity field may be conducted from the Earth surface and from satellites (Rapp, 1975; Campbell, 2018). Naranjo (2005) cite the Gravity Recovery and Climate Experiment (GRACE) mission, launched in 2002 used to map the Earth's gravity field, with a high resolution (see also Dehne et al., 2009). This system uses two identical satellites, flying about 220 km apart, with a microwave ranging system monitoring the distance between the two satellites at the resolution of one micron (one centimeter (cm) is equal to 10,000 micrometers (microns).

The Earth's gravity may be calculated at any location, using the changes in distance between the two satellites, which is due to (Naranjo, 2005).

Radio Detection and Ranging (RADAR)

Another important technological development in the field of remote sensing is Radio Detection and Ranging (RADAR), which originally a military application for the detection of aircraft, spacecraft, surface vessels, land vehicles, and even missiles, and later civilian applications such as animal (e.g., radar ornithology), terrain, and weather formations. A radar unit is composed of a transmitter that produces electromagnetic waves with the radio or microwave range, and one or two antennas: either one for transmitting the signals and the other for receiving the reflected waves, or one for both emitting and receiving. A processor analyzes the received data to derive the required information about the monitored object or objects (size, form, movement, speed, and pattern) (Campbell, 2018). Skolnik (2022) notes that "what distinguishes radar from optical and infrared sensing devices is its ability to detect faraway objects under adverse weather conditions and to determine their range, or distance, with precision."

Modern radar uses a pulse system, in which the radar signal consists of rapid, repetitive pulses, with the range accuracy of the pulse radar dependent on the width of the pulse. Shorter pulses give more accurate range measurements, but require wider bandwidths in the receiver and transmitter, as the bandwidth is equal to the reciprocal of the pulse width (Skolnik, 2022). The accuracy of radar technology varies from a few centimeters to tens of meters. Skolnik (2022) notes radar performance is based on 6 attributes; its maximum range for target identification and measurement; the accuracy of range and angle measurement of target; the accuracy of differentiation of targets; the detection of the target echo among noises; the recognition of target type; and the logistics of operation.

Recent radar types include Synthetic Aperture Radar (SAR) and Doppler Radar. SAR is "is a type of active data collection where a sensor produces its own energy and then records the amount of that energy reflected back after interacting with the Earth. While optical imagery is similar to interpreting a photograph, SAR data require a different way of thinking in that the signal is instead responsive to surface characteristics like structure and moisture" (Herndon et al., 2020). SAR is as the "spatial resolution of radar data is directly related to the ratio of the sensor wavelength to the length of the sensor's antenna. For a given wavelength, the longer the antenna, the higher the spatial resolution. From a satellite in space operating at a wavelength of about 5 cm (C-band radar), in order to get a spatial resolution of 10 m, you would need a radar antenna about 4250 m long" (Herndon et al., 2020). This, an impossibility is simulated by a sequence of acquisitions from short antennas, which are combined to simulate a larger antenna, enabling higher resolution data (Herndon et al., 2020). SAR data can be used for Interferometry (InSAR), which uses the sensor information to record distances between the sensor and the target. With two or more observations of the target, the distance, with sensor geometric information, land surface topographical change may be recorded, at high resolutions (Herndon et al., 2020).

Doppler radar is used to ascertain the speed of an object, basically it "bounces a pulse of radio waves off the target object. Once the wave bounces, it acts just like the sound waves from the police siren. And by measuring how much the frequency of the radio waves changes after bouncing off the object, you can calculate the speed of that object" (Wood, 2023). Doppler radar uses the Doppler effect, which is the change in frequency of a wave in relation to an observer moving in relation to the wave source, or "the apparent difference between the frequency at which sound or light waves leave a source and that at which they reach an observer, caused by relative motion of the observer and the wave source" (Encyclopedia Britannica, 2023). Petrescu (2012) describes the Doppler effect: "the change in frequency of a wave for an observer moving relative to the source of the wave… The total Doppler Effect may therefore result from motion of the source, motion of the observer, or motion of the medium." Fessenden (2014) points out that "the radio waves sent out by doppler radar bounce off raindrops and birds alike and return a signal to the receiver."

Numerous applications have emerged from radar technology (Campbell, 2018). Radar has increasingly been applied to nonmilitary activities, such as air traffic control for civilian airports, civilian aircraft and ship navigation, weather observation, planetary observation and space surveillance, remote sensing-based investigations of the environment, measurements for industrial applications and for law enforcement (Skolnik, 2022). From the perspective of biogeochemistry, the most important applications of radar are for animal and habitat structure monitoring (Campbell, 2018). The Canada Center for Remote Sensing and the Canadian Space Agency developed RADARSAT 1, RADARSAT 2 and RADARSAT Constellation satellite missions, enabled several data applications: "Surface water (incl. flooding) and wetlands; Polarimetric RADARSAT-2 for Monitoring Canadian Wetlands; Monitoring Open Fresh Water in Northern Environment, using SAR imagery and FnFCE (Forest non-Forest Class Extraction); Freshwater lake and river ice; Mapping and Monitoring Lake Ice using SAR Satellites; River Ice Mapping and Monitoring using SAR Satellites; Surface deformation and slope stability; InSAR Measurements of Terrain Stability in Canada's North; InSAR monitoring of permafrost activity, NWT, Canada" (Government of Canada, 2021). Liverman et al. (2006) describes the use shuttle radar topography data to create a digital elevation model landcover analysis for Newfoundland, Canada, including representing topographical features such as glacial features landforms, and vegetation forms. Liverman et al. (2006) "Shuttle Radar Topography Mission (SRTM) data offers a new means of examining landscape and topography. This is best done through production of digital elevation models and false shading. The data is limited in its resolution but is effective in delineating landscape features, particularly in heavily vegetated areas."

Modern radar applications include a 3D Bird Radar System, for detecting and tracking birds in flight and monitoring birds near airports to avoid collision hazards and provide a total Bird-Aircraft Strike Hazard solution (Miltronix Limited, 2021). There is also Foliage Penetration Radar, with "the capability to see, remotely from the air, what is hiding under the forest cover" (Miltronix Limited (2021). Vegetation and habitat patterns are measurable: for SAR, "due to its cloud free capability, Synthetic Aperture Radar (SAR) is an ideal source of information on forest dynamics in countries with near-constant cloud-cover. However, few studies have investigated the use of SAR for forest cover estimation in landscapes with highly sparse and fragmented forest cover" (Devaney et al., 2015). For the latter study, the authors concluded that the "findings indicate that spaceborne radar could aid inventories in regions with low levels of forest cover in fragmented landscapes. The reduced accuracies observed for the global and pan-continental forest cover maps in comparison to national and SAR-derived forest maps indicate that caution should be exercised when applying these datasets for national reporting" (Devaney et al., 2015).

Hansen et al. (2020) examined study sites of forest biomes in Alaska, Florida, Finland, UK, Indonesia, and Colombia, using Synthetic Aperture Radar, with improvements over similar studies (longer wavelengths, such as L-band, more study sites and forest types, longer time series and dual-polarimetric radar that transmits and receives pulses in both a horizontal and vertical orientation) and information from Sentinel-1 Single Look Complex (SLC) data (Hansen et al., 2020). The objective was to test the effectiveness of C-band Sentinel-1 data in distinguishing forest from nonforest in the study sites, as "using the time series for a full year significantly increases the classification accuracy compared to a single scene (a mean of 85% compared to 77% across the study sites for the best classifier)." Ground Range Detected (GRD) products were compared, and contrasted with SLC products, which contain both phase information and backscatter amplitude, which are tested for additional information content. The findings indicated increases over the mean accuracy of 87% (there were variable accuracies, with accuracies of up to 93% for the Finland case, and 80% for the Alaska case). Additionally, the findings indicated that the best performance occurred when a Support Vector Machine classifier was used, this outperforming random forest, k-Nearest-Neighbors, and Quadratic Discriminant Analysis. The authors conclude that "for the purposes of forest mapping the smaller file size and easier to process Ground Range Detected (GRD) products are sufficient, unless the Single Look Complex (SLC) products are used to compute the temporal coherence which was not tested in this study" (Hansen et al., 2020).

Radar has also been used in the study of peatlands, which are complex due to the combination of water, soils, and vegetation (Romshoo, 2003). Synthetic Aperture Radar is an effective tool for such studies. Herndon et al. (2020) define Synthetic Aperture Radar as "a type of active data collection where a sensor produces its own energy and then records the amount of that energy reflected back after interacting with the Earth. While optical imagery is similar to interpreting a photograph, SAR data require a different way of thinking in that the signal is instead responsive to surface characteristics like structure and moisture." For example, Romshoo (2003: 63, 80) argue that peatlands are important multipurpose ecosystems, which support large amounts of biomass and peat soil carbon and are a "repository of live forest and soil carbon and other biogeochemicals." Their study uses a "time series of Japanese Earth Resource Satellite-1 (JERS-1) L-band Synthetic Aperture Radar

(SAR) data," the objective being the monitoring of deforestation and inundation patterns: "two important dynamic processes in the tropical peatland forest ecosystems, i.e., deforestation due to anthropogenic activities and fires; and the hydrological characterization and the inundation patterns" (Romshoo (2003: 63). The study also used a radiative transfer based radar backscattering model, MIMICS to define the scattering mechanisms under ecological and hydrological conditions. The accumulation of ground biomass stocks after anthropogenic disturbances was simulated using an ecosystem dynamics model, CENTURY (see also Parton et al., 1993).

The justification for the study was that peatlands are vital links in water and carbon cycling, and their "total carbon pool exceeds that of the worlds forests and equals that of the atmosphere" and store 12–30% of the global soil carbon, on 3% of the Earth's land surface, and Southeast Asian peatlands have high rates of accumulated carbon compared to northern peatlands (Romshoo, 2003: 63). Regarding the Indonesian peatlands which are the object of the study, "the Central Kalimantan peatlands… are a repository of carbon stocks and are very often more than 3 m deep and sometimes up to 20 m deep" (Romshoo, 2003: 63). The findings indicated that anthropogenic and natural factors have contributed to deforestation (areas with low biomass), as the L-band SAR detected the pixel-wise seasonal and spatial inundation information for forest types where there was significant transmissivity of the L-band SAR signals. It also demonstrated the utility of L-band SAR for monitoring deforestation and inundation patterns. These initiated the loss of carbon rich forests and a net release of carbon to the atmosphere, and "the high-resolution radar remote sensing, when used operationally, would reduce the uncertainty in the present estimates" (Romshoo (2003: 63). Additionally, the model simulations could be quantitatively verified using the optical remote sensing data. The seasonal and spatial variations of inundation patterns can be mapped using SAR observations and modeled by a radiative transfer approach (Romshoo, 2003: 63).

Torbick et al. (2017: 99) used multiscale synthetic aperture radar and optical imagery, and a processed-based biogeochemical model to map rice greenhouse gas emissions in the Red River Delta, Vietnam. The justification for the study was that agricultural land use contributes to approximately 50% of global atmospheric inputs of methane (CH_4), and agricultural soils emit 75% of global nitrous oxide. In the case study region, agricultural land covers nearly 10 million ha in Vietnam, mostly under rice cultivation. Therefore, rice land use represents a significant factor for greenhouse gases (GHG) mitigation by possible reductions of CH_4 and N_2O emissions and soil carbon sequestration (Torbick et al., 2017: 99). Additionally, Vietnam's Second National Communication to the United Nations Framework Convention on Climate Change (UNFCCC) reported total emissions of 151 million tons of GHG in carbon dioxide equivalent (CO_2e), with the largest (58%) emitter in the agricultural sector (43% of total). The authors contend that "improved biogeochemical modeling with high-resolution remote sensing parameterization is the number one need to reduce uncertainty in CH_4 emissions assessment" (Torbick et al., 2017: 100; see also Zhang et al., 2002).

In this study by Torbick et al. (2017), the synthetic aperture radar and optical imagery data were entered into a random forest classifier, with field materials as training data for mapping the range of rice cultivation. Time series analysis generated information on "crop calendar, hydroperiod and cropping intensity to use as parameters for the denitrification–decomposition (DNDC) model to estimate emissions" (Torbick et al. (2017: 99). The findings indicated in 2015 that the rice extent was 583,470 ha, the total harvested area was 1,078,783 ha, and total methane emissions for the delta were 345.4 million kg CH_4-C, which was equivalent to 11.5 million tonnes (carbon dioxide equivalent CO_2e). The contribution of satellite remote sensing was the accurate assessment of water management, modeling of the impacts of decisions such as irrigation practices, changes in GHG emissions, mitigation initiatives, and for rice applications, the backscatter microwave observations sensitive to growth stages (biomass development, plant height, leaf–ground double bounce, and inundation frequency and duration).

The authors noted that "during rice transplanting periods, the surface contribution of a rice paddy causes low backscatter. As plant tillering, biomass and haulm develop the backscatter response increases with more interaction, and volume scattering tends to cause a decrease in backscatter as the crop peaks and approaches harvest" (Torbick et al., 2017: 100). The combined use of Landsat-8 Operational Land Imager (OLI), Sentinel-2, Sentinel-1, and Phased Array L-band Synthetic Aperture Radar-2 (PALSAR-2) was particularly effective. Torbick et al. (2017: 106–107) conclude that the use of SAR was a "true game changer," and "fusion of SAR and optical satellite data provides very high accuracy of crop-type maps... New, open-access (free), high temporal frequency Sentinel-1 radar (C-band) data is a substantial advance in operational assessment and monitoring for agriculture, especially rice crops and regions with high cloud cover." The measurement of regional estimates of GHG emissions would be improved, as SAR-derived maps of rice extent, cropping intensity, growing season length and hydroperiod create more accurate process models.

Light Detection and Ranging (LiDAR) Technologies

Another important remote sensing tool is Light Detection and Ranging (LiDAR), which uses a pulsed laser to measure ranges to the Earth (Campbell, 2018, 2022c; National Oceanic and Atmospheric Administration, 2023). Wehr and Lohr (1999, 68) define LiDAR as airborne laser scanning and as a collector of "high-resolution data for generation of a digital surface or digital terrain model," a "rapid, highly accurate and efficient method of capturing 3D data of large areas, such as agricultural or forestry sites, urban areas, industrial plants, etc." and "currently the most detailed and accurate method of creating digital elevation models, ground surfaces such as rivers, paths, cultural heritage sites, etc., which are concealed by trees." These applications may be classified as topographic (to measures the Earth's surface) or bathymetric (to measure the sea floor and riverbeds) (Campbell, 2018). Eagle Mapping (2022) notes that LiDAR uses laser light to measure distance in a similar manner to RADAR but unlike the latter, LiDAR uses light rather than radio waves. LiDAR produces accurate, detailed surface models generally faster and often more cheaply than conventional photogrammetric mapping. LiDAR also enables much faster surveys than ground surveying. Hence, as Environmental Systems Research Institute (ESRI) (2019) notes, "lidar, primarily used in airborne laser mapping applications, is emerging as a cost-effective alternative to traditional surveying techniques such as photogrammetry."

The LiDAR system is mounted on a platform such as an aircraft, including aeroplanes and helicopters, and comprises a laser, scanner, and a global positioning system (GPS) (National Oceanic and Atmospheric Administration, 2023). The light pulses emitted, and the information represented by their contact with the Earth's surface, assisted by integrated, supporting data and information, may generate precise, three-dimensional information on the form of the Earth's surface form. (Campbell, 2018). Applications in the geosciences include high-resolution digital elevation maps and the detection of geological features such as fault lines, and atmospheric applications including the recording of clouds, winds, aerosols, and larger airborne objects. Applications for surveying, include the creation of systems termed Digital Terrain Models (DTMs) or Digital Elevation Modes (DEMs), and for forestry the calculation of parameters such as tree height ranges, crown width and diameter, canopy volumes and vegetation cover. These applications may also be useful for animal ecological and zoological research such as the study of animal habitats, migrations, and foraging movements (Campbell, 2018).

Wang and Menenti (2021) note that LiDAR developed after the invention of laser in 1960 (see also Maiman, 1960; Woodbury et al., 1961; Smullin & Fiocco, 1962; Schotland, 1966; Cooney, 1968; Melfi et al., 1969). Here, laser refers to light amplification by stimulated emission of radiation, which is a device that stimulates atoms or molecules to emit light at specific wavelengths, the amplification of the light creating a narrow beam of light (Taylor, 2000). More advanced LIDAR systems have been developed (Fiocco & Smullin, 1963; Weitkamp, 2005; Kashani et al., 2015; Wang & Menenti, 2021). The LiDAR system records the time for the emitted light to reach the ground and return, and the time record is used to calculate distance, and the distance is converted to elevation. Global Positioning Systems are used to identify the X, Y, and Z locations of the light energy (Wasser, 2020).

The two main LiDAR types are (1) discrete-return LiDAR devices, which measure either one (single-return systems) or a few (multiple-return systems) of heights by identifying, in the return signal, the main peaks that represent discrete objects in the path of the laser illumination and (2) waveform recording devices which measure the time-differentiated intensity of the returned energy from the laser pulses, thus providing a definite record of the height distribution of the surfaces illuminated by the laser pulse (Lefsky et al., 2002, 20). Lefsky et al. (2002, 20) compares the two systems: "by analogy to chromatography, the discrete-return systems identify, while receiving the return signal, the retention times and heights of major peaks; the waveform-recording systems capture the entire signal trace for later processing."

Recent, advanced applications of LiDAR include terrain digital models, cryosphere studies, terrestrial ecology, atmospheric science, and hydrology/oceanography. Recent platforms include NASA's Cloud-Aerosol Lidar and Infrared Pathfinder

Satellite Observations (CALIPSO) (Winker et al., 2010), the Cloud, and land Elevation Satellite (ICESat) (Markus et al., 2017) and ESA's Aeolus wind satellite (Kanitz et al., 2020). There are also unmanned aerial vehicle (UAV or "drone") platforms (González-Jorge et al., 2017; Wang & Menenti, 2021). The data may be analyzed as point cloud data. ESRI (2019) notes that "post-processed spatially organized LiDAR data is known as point cloud data." These initial point clouds are groups of 3D elevation points, including *x*, *y*, and *z*, with additional attributes such as GPS time stamps. Additionally, "elevations for the ground, buildings, forest canopy, highway overpasses, and anything else that the laser beam encounters during the survey constitutes point cloud data (ESRI, 2019).

Doppler LiDAR is a new method that can be applied to model air movements and related bird flight (Campbell, 2018, 2022c). Newsom and Krishnamurthy (2020) note that "the Doppler lidar (DL) is an active remote-sensing instrument that provides range- and time-resolved measurements of radial velocity, attenuated backscatter, and signal-to-noise ratio (SNR)"; additionally "the principle of operation is similar to radar in that pulses of electromagnetic energy (infrared in this case) are transmitted into the atmosphere; the energy scattered back to the transceiver is collected and measured as a time-resolved signal." In terms of applications, Hardesty (2015) defines Doppler LiDAR as a technology that "can be designed primarily to measure winds from aerosol-scattered radiation, or from molecule-scattered radiation, or from both… A Doppler LiDAR is typically made up of a laser transmitter to produce pulses of energy that irradiate the atmospheric volume of interest; a receiver that collects the backscattered energy and estimates the backscattered energy and Doppler shift of the return; and a beam-pointing mechanism that directs the transmitter and receiver together in various directions to probe different atmospheric volumes and measure different components of the wind" (Hardesty, 2015).

Koch (2006: 315) compares the similarities between Doppler LiDAR and Doppler radar, maintaining that Doppler LiDAR is analogous to Doppler radar, but with radar using long radio wavelengths and LiDAR using short infrared wavelengths. Koch (2006: 315) also remark on similar points to those of Hardesty (2015), that "doppler LiDAR allows imaging of wind fields to reveal the structure of updrafts and downdrafts in a thermal… imaging wind fields with doppler lidar can be a useful tool for studying thermals and how they are used by soaring birds. An effective combination for further study of bird flight interaction with wind phenomena would be to add lidar measurements to an established means of tracking bird flight by radio or GPS transmitters, aircraft tracking, or radar." Doppler LiDAR transmits pulses of light from a laser, which are reflected by the suspended atmospheric aerosols, and the reflected laser light is collected by a telescope and focused into an optical fiber. An optical mixing process, termed heterodyning, measures shifts in the frequency of the laser light. The reflection point's distance is calculated by comparing the speed of light to the timing of the transmitted and received pulses. The light pulses are reflected by mobile small particles moving with the wind, and the measured Doppler shift from these moving particles shows the wind direction and speed (Koch, 2006).

Applications of LiDAR are numerous, but habitat and vegetation for a prominent application (Campbell, 2018, 2022a, b, c). For the study of biogeochemical agents such as vegetation cover and related chemical cycling, and animal habitats based on the vegetation variables, a primary application of LiDAR is the measurement of the vegetation structures, and associated variables such as water, soils, and gases, that are relevant to biogeochemical cycling and biodiversity (Rustad et al., 2020; Wasser, 2020). Wasser (2020) notes that "scientists often need to characterize vegetation over large regions to answer research questions at the ecosystem or regional scale… LiDAR directly measures the height and density of vegetation on the ground making it an ideal tool for scientists studying vegetation over large areas." Goetz et al. (2010: 254) contend that the findings of their study using LiDAR for avian habitat ecology, is that LiDAR metrics (canopy structural diversity, including canopy height, topography, and the vertical distribution of canopy elements) were "consistently better predictors than traditional remotely sensed variables such as canopy cover, indicating that LiDAR provides a valuable resource for biodiversity research applications." This was because "the canopy vertical distribution information was consistently found to be the strongest predictor of species richness, and this was predicted best when stratified into guilds dominated by forest, scrub, suburban and wetland species" (Goetz et al., 2010: 254).

LiDAR based techniques are commonly applied to the study of closed or semi-closed canopy forest, and the soil ecosystems associated with these ecological regions (Rustad et al., 2020). For vegetation heights, one of the more basic LiDAR applications, the United States Geological Survey (2020) describes high-resolution LiDAR that can measure surface features with a vertical accuracy of 10 cm. Lefsky et al. (2002: 19–20) also argue that when the vertical distance between a sensor in a level-flying aircraft and the Earth's surface is measured along a transect, an outline of both the ground surface and any covering vegetation is obtained. Even with high vegetation cover, where many measurements are returned from plant canopies, some of the measurements may be returned from the underlying ground surface, which may enable an accurate map of the canopy height.

For more complex structures, LiDAR is also useful (Campbell, 2018). For example, Lefsky et al. (2002, 19) write that LiDAR sensors are capable of measuring the three-dimensional distribution of plant canopies and crucially the subcanopy topography, which provide the materials for the production of high-resolution topographic maps and accurate estimates of vegetation height, cover, and canopy structure. LiDAR may also be used to accurately estimate the leaf area index (LAI) and above-ground biomass "even in those high-biomass ecosystems where passive optical and active radar sensors typically fail to do so" (Lefsky et al., 2002, 19). LiDAR-based data can therefore measure vegetation variables which indicate possibilities for chemical element movements such a carbon cycling, nitrogen, and phosphorus accumulation and water evaporation, important vegetation variables including canopy height and cover, leaf area index, vertical forest structure. Species identification can be possible when using high quality LiDAR technology (Świątek & Pietrzykowski, 2022).

LiDAR can also be used for topographical analyses, and "can produce topographic maps of amazing detail and accuracy, even where the ground is obscured by forest canopy" (Krogstad & Schiess, 2004). For topographic structure and included vegetation, advanced systems such as discrete-return and waveform-recording lidar sensors have strong points (Campbell, 2022c). Discrete-return LiDAR sensors may have high spatial resolutions, which is due to "the small diameter of their footprint and the high repetition rates of these systems (as high as 33,000 points per second), which together can yield dense distributions of sampled points" (Lefsky et al., 2002: 20–21; see also Campbell, 2022c). This technology is useful for detailed ground mapping and canopy surfaces. Another advantage, due to the high resolution, is the possibility that data can be aggregated at various scales, allowing precise locations such as tree canopies to be recorded. The discrete-return systems are commonly available, and under continual development, for surveying, digital terrain mapping and modeling, and higher resolution topographical mapping and photogrammetry (Campbell, 2022c).

Waveform-recording LiDAR is also useful, due to their enhanced ability to record variations in canopy structure over increasingly large areas, and global data set availability (Lefsky et al., 2002). Waveform-recording LiDAR systems "record the entire time-varying power of the return signal from all illuminated surfaces and are therefore capable of collecting more information on canopy structure than all but the most spatially dense collections of small-footprint LiDAR" (Lefsky et al., 2002, 20). Waveform-recording LiDAR also cover greater spatial extents and can provide broader information on these areas, with more efficient storage and analysis. Waveform-recording LiDAR systems may be based on space-borne platforms, with examples being Shuttle Laser Altimeter missions, with footprints of about 100 m, and the Ice, Cloud, and Land Elevation Satellite (ICESat) mission, carrying the Geoscience Laser Altimeter System, with 70-m diameter footprints (Harding et al., 2001). Waveform-recording lasers have also been developed by NASA's Goddard Space Flight Center (Lefsky et al., 2002).

LiDAR is also important for the recording and analysis of forest understory, which characterizes vegetation systems as much as canopy and topography (Campbell, 2022c). Understory studies are vital for biogeochemical and biogeographical variables, as chemical elements and compounds may vary across these strata (Campbell, 2022b). For example, Wang et al. (2021) present a study titled "short-term canopy and understory nitrogen addition differ in their effects on seedlings of dominant woody species in a subtropical evergreen broadleaved forest," which maintains that atmospheric nitrogen (N) deposition is increasing, and information on the impact of nitrogen deposition on seedling performance of tree species is important for the prediction of forest regeneration. The findings indicated that nitrogen deposition could strongly affect the woody species seedling growth. Additionally, the understory addition of nitrogen was more important than the canopy addition of nitrogen, regarding impacts on seedling height, basal diameter and biomass. The authors argue that consequently, "the traditional use of understory addition of N to simulate atmospheric N deposition may overestimate the effects of N deposition on seedling performance" (Wang et al., 2021).

Although most LiDAR systems applications to vegetation structure examine canopy structures, vegetation layers may obscure the forest understory. For example, Hamraz et al. (2017) argue that LiDAR) point clouds covering large, forested areas nay be developed to segment individual trees and to extract tree-level information. The authors contend although current segmentation procedures may generally detect over 90% of overstory trees, they may detect only 60% of understory trees due to the occlusion effect of higher canopy layers. The understory trees are however an essential part of the forest ecosystem, especially as habitat for numerous wildlife species and a factor for stand development. Hamraz et al. (2017) modeled the occlusion effects and examined the impacts of several canopy layers, such as one over story, and several understory layers. They argue that "at a density of ~170 pt/m^2 understory trees can likely be segmented as accurately as overstory trees" using advanced LiDAR methods and create a more accurate assessment of forest structures (Hamraz et al., 2017). The findings indicated that to offset the occlusion effect a point cloud (PCD) of 170 pt/m^2 is needed to segment understory trees down to the third canopy layer. The study modeled the LiDAR occlusion effect of higher canopy layers, for estimation of the PCD required for the segmentation of trees of the lower canopy layers (Hamraz et al., 2017).

Venier et al. (2019) also examined the measurement of forest understory vegetation, explaining the value of predicting and mapping understory. Four understory metrics (fractional cover, leaf area density, voxel cover (value on a regular grid in three-dimensional space) and normalized cover) were compared, and important explanatory variables related to occlusion were included, and there was comparison of the mixed effects vs. Random Forest approach for generating predictions, both being used for modeling understory vegetation using LiDAR data (Venier et al., 2019). The random forest approach is used to model forest inventory variables, and LiDAR derived predictors are used to create the best prediction of the forest inventory variables. Problems with the random forest approach include the production of estimates of variable importance, rather "an ecologically interpretable relationship," it does not make assumptions about data structure, it is better for the prediction of nonlinear relationships, it is not sensitive to correlations between variables, and it may be biased when incomplete ranges of conditions are being sampled (Venier et al., 2019). The linear mixed-effects regression models, make assumptions of homoscedasticity (the variance of the residual, or error term, in a regression model is constant) but may create simpler, more interpretable models. Venier et al. (2019) argue that "in random forest models, large suites of variables are usually included to achieve the best predictive capacity. In the regression models, it is more important to limit the number of variables included to avoid overfitting and strong correlations between explanatory variables" (Venier et al., 2019).

The model used by Venier et al. (2019) included three vertical strata for the forest understory (0.5 m to <1.5 m, 1.5 m to <2.5 m, 2.5 m to <3.5 m), and four understory LiDAR metrics designed to control for the spatial heterogeneity of sampling density, to capture the vegetation understory density (Venier et al. (2019). The findings indicated occlusion of LiDAR pulses in the lowest stratum in the forest model, but no evidence that the occlusion affected the predictability of understory structure.

The mixed-effects regression models were favored over the Random Forest model, despite similar results: "the best approach to predict understory structure is using the mixed-effects model with the voxel-based understory LiDAR metric along with vertical stratum, because it yielded the highest explained variance with the fewest number of variables." The findings also indicated that the other understory LiDAR metrics (fractional cover, normalized cover, and leaf area density) could be effective in the mixed-effects and random forest modeling approaches, and both produced predictions usable for ecological and forest planning applications, and increased accuracy for spatial mapping of wildlife habitat, fire behavior, or forest ecosystem dynamics; "our results fully support the indirect evidence provided from wildlife studies that LiDAR can predict understory vegetation structure even in the presence of a mature tree canopy" (Venier et al., 2019).

Estimations of canopy height and canopy cover using LiDAR are also vital (Campbell, 2018; Campbell & Jha, 2021). For example, Wasser et al. (2013) studied the reliability of LiDAR by comparing the findings of LiDAR data with field assessments, to "determine if existing LiDAR data flown in leaf-off conditions for applications such as terrain mapping can effectively estimate forested riparian buffer H (canopy height) and FC (fractional canopy cover) within a range of riparian vegetation types" (Wasser et al., 2013). The findings indicated that canopy height models (CHMs) underestimated the canopy height, H more than the LiDAR percentile methods, partly due to the vegetation tree leaf type (conifer needle, deciduous simple leaf, or deciduous compound leaf) and the variation in canopy height. The LiDAR estimates of fractional canopy cover were within 10% of the findings of plot measurements during the leaf-on periods, but for the leaf-off periods the LiDAR estimates were mostly under-estimated except for mixed and conifer plots. The depth of the laser pulse penetration lower in the canopy varied more than for the upper canopy layer, possibly affecting the within canopy vegetation structure estimates. The conclusion indicated the study "demonstrates that leaf-off LiDAR data can be used to estimate forested riparian buffer canopy height within diverse vegetation conditions and fractional canopy cover within mixed and conifer forests when leaf-on lidar data are not available" (Wasser et al., 2013).

Another important issue is the measurement of vegetation indices in heterogeneous landcover and the links with wildlife (Campbell, 2018). Carrasco et al. (2019) examined a counter to existing structure–animal diversity work, which is generally focused on structural metrics derived from LiDAR returns from only canopy and terrain features, stating that "despite the ability of LiDAR pulses to penetrate the top layers of the forest, most animal diversity studies hitherto have focused on building LiDAR-derived structural metrics using only canopy cover, or the variation in canopy height," and that structural metrics that can capture the whole 3D forest structure may create higher predictive power for the assessment and study of multiple taxa biodiversity (Carrasco et al., 2019). Reality indicates local animal diversity patterns are related to vegetation structural heterogeneity, and LiDAR developments have created important 3D forest structure studies. The study created new LiDAR structural metrics, based on the Leaf Area Density (LAD) at the vegetation height layers, and the links of these to bird species presence and density. Investigations queried if LAD-based metrics better explained bird species richness compared to top of the canopy metrics, and the different aspects of vegetation structural heterogeneity impacted differently bird species richness (Carrasco et al., 2019).

The study, based on LiDAR full waveform and discrete data, used two sets of metrics, Canopy Height Metrics, and the LAD-Based Metrics. The Canopy Height Metrics, which used LiDAR's first returns, represented top-of-the-canopy points (from which the maximum height, mean height and standard deviation were calculated) and canopy models (where digital surface models were calculated for each plot using a points to raster (p2r) algorithm, creating an output raster based on the height of the highest LiDAR returns at each pixel, with mean and standard deviation calculated from the canopy models for each plot) (Carrasco et al., 2019). The LAD-Based Metrics were based on all the LiDAR returns for estimation of vegetation density at different height layers. The leaf area density (LAD) was measured from LiDAR point-cloud data, and the LAD estimation used leaf area index (LAI, the projected total surface area of leaves per unit ground surface area) and the gap fraction (GF, the light transmittance across a forest layer if a random leaf distribution is assumed), based on the assumption of a random leaf distribution (Carrasco et al., 2019). The findings were that the metrics based on LAD measurements had stronger explanatory power (43% of variance explained) than the Canopy Height Metrics based on the variation of canopy heights (32% of variance explained) (Carrasco et al., 2019). Bird species richness varied with horizontal heterogeneity but not vertical heterogeneity. Hence, the LAD-based metrics generally showed "higher explanatory power with regard to bird richness than metrics based on top of the canopy measurements" and models based on only LAD metrics were more accurate than canopy-based metrics, and LAD metrics were more important than canopy-based metrics in the mixed-metrics models (Carrasco et al., 2019).

Regionalization of vegetation variation is another important vegetation issue to which LiDAR can be applied for biogeographical, biogeochemical, wildlife habitat mapping and species distribution modeling (Campbell, 2018). Guo et al. (2017: 50) justified such topics for study, as "large – area structural inventories using LiDAR-derived variables that characterize generic habitat structure have rarely been done yet would be helpful for guiding biodiversity monitoring and conservation assessments of species at regional levels." Their study processed regional-scale LiDAR into classes of vegetation, using cluster analysis on six LiDAR-derived habitat-related variables, with vegetation structure in eight classes within ten natural, forested subregions in forests in Alberta, Canada. The findings indicated that the vegetation structure of the Lower Foothills Natural Subregions was comparatively complex, with wildfire the most important disturbance factor for all vegetation classes. Only the rarest class of vegetation, had low canopy cover partly due to timber harvesting. LiDAR technology, with a spatial resolution (30 m) was finer than polygon-based forest inventories (Guo et al., 2017: 50).

Guo et al. (2018) applied LiDAR to large-scale landcover studies, on the premise that "the application of regional-level airborne LiDAR (light detection and ranging) data to characterize habitat patches and model habitat connectivity over large landscapes has not been well explored" and "maintaining a connected network of habitat in the presence of anthropogenic disturbances is essential for regional-level conservation planning and the maintenance of biodiversity values." The vegetation was classified using LiDAR-identified vegetation structural attributes, creating habitat patches, and habitat networks were created for forest categories and analyzed using network connectivity metrics. Landcover change scenarios were simulated by habitat patch removals, which simulated the landcover change and management impacts. Findings indicated simulated constructions could represent habitat networks, and spatial configuration of habitat landcover is important for ecological studies. Hence, "the scheme prioritizing only habitat area resulted in immediate near-term losses in connectivity, whereas the scheme considering both habitat area and their spatial configurations maintained the overall connectivity most effectively" (Guo et al., 2018).

Doppler LiDAR is also used for habitat analysis, especially in relation to wind movements in vegetation systems. Träumner et al. (2012: 399) explain the use of Doppler lidar to provide data for the analysis of turbulence structure above forest edges, where "comparison with state-of-the-art laser Doppler anemometry measurements on a physical model in an atmospheric boundary layer wind tunnel shows that the lidar is able to resolve details of the wind field at a change of roughness in an appropriate order of magnitude" (Träumner et al., 2012: 399). Wind movements at the forest edge are important, because such turbulence may affect bird flight during foraging (Campbell, 2015). Träumner et al. (2012: 399) point out that measurements of real wind situations with moderate wind speed directed toward and away from forest edges, and the examination of a storm event with wind velocities below 45 m s^{-1}, represent common features of the isolines of average wind speed and its standard deviation and agree with results in previous studies. LiDAR with wide measuring zones can cover a dimension of air with a vertical measurement of over 6 tree heights by a horizontal measurement of 30 tree heights, covering a wide area for the measurement of changing air perturbation, illustrated by a change in roughness (Träumner et al., 2012).

Vegetation and hence habitat analysis using LiDAR may record the presence of frequencies and habitats of prey species, and related predator foraging (Campbell, 2022b, c). Olsoy (2015: 74) notes that "heterogeneous vegetation structure can create a variable landscape of predation risk – a fearscape – that influences the use and selection of habitat by animals." The author notes that this mapping of the vegetation properties that contribute to predation risk, such as concealment and visibility opportunities, can be difficult across wide landscapes, especially as the traditional ground-based measures of predation risk are usually location specific, with limited spatial resolution. Terrestrial laser scanning (TLS) may be used for the mapping of the vegetation properties that shape fearscapes (Olsoy et al., 2015: 74). In this study, terrestrial laser scanning was used for estimation of prey concealment from several vantage points, which represented predator sightlines and the visibility of potential predators from the prey locations. The author concludes that TLS facilitates a comprehensive data set enabling an examination of how habitat changes may affect prey and predators. These TLS methods may be combined with other remotely sensed imagery, to allow the use of fearscape analyses with the management and restoration of landscapes (Olsoy et al., 2015: 74).

Savanna vegetation, a major global ecosystem composed of mixed grass, shrub, and tree vegetation, may be assessed using LiDAR, either as part of a mixed forest/savanna vegetation or savanna study (Campbell, 2022b). As noted earlier in this chapter, savanna and other open vegetation associations are important ecosystems and vital for global biogeochemistry and biogeography. For example, Gwenzi (2017: 235) contain that "despite occupying almost a fifth of the global terrestrial vegetation system, savanna ecosystems are relatively understudied in the Earth observation field. As a result, their contribution to global socioecological functions, such as carbon sequestration, habitat provision, watershed protection, biodiversity, and communal supply of timber and nontimber products, is inadequately accounted for." LiDAR applications to savanna-based studies may differ from those required for forest studies (Campbell, 2022b). For example, Gwenzi and Lefsky (2014: 338) maintain that savannas require different models from those developed for studying forests and it is vital to record the high canopy heterogeneity associated with savannas. However, "studies using Light Detection and Ranging (LIDAR) have demonstrated the technology's ability to measure canopy height and the strong relationship between canopy height and structural attributes such as above-ground biomass, but most of this work has focused on closed canopy forests." Gwenzi and Lefsky (2014: 338) examined the use of space-borne LiDAR in the estimation of canopy height for the assessment of biomass in savannas, and their findings indicated that forest-based methods were inadequate, and their methods "demonstrated the capability of waveform LiDAR to assess vegetation structural attributes in savannas. The challenge in canopy height modeling using this technique in such ecosystems not only is limited to terrain slope but also includes the interacting influence of low canopy cover and short height."

Savanna vegetation may grade into forest of varying density, usually dry sparse forest to denser more moist forest, and such gradients may be assessed with LiDAR applications. Canopies may vary in frequency density, especially where there is derived savanna, which is savanna that exists due to human action such as tree removal for agriculture or fuel wood extraction. When human action ceases or intensifies, variable mixed savannas result (Campbell, 2022b). Marselis et al. (2018: 626) also studied LiDAR applications to vegetation gradients between savanna and forest, while noting that "tropical forest vegetation structure is highly variable, both vertically and horizontally, and provides habitat to a large diversity of species," their research intended "to assess whether large footprint full-waveform LiDAR data can be used to distinguish successional vegetation types based on their vertical structure." The vegetation metrics were derived from the LiDAR waveforms: canopy height, canopy fractional cover, total Plant Area Index (PAI) and vertical profile of PAI. The findings indicated that airborne waveform LiDAR produced similar results to Terrestrial Laser Scanning, that waveforms are an effective tool to distinguish between the vegetation types in a gradient format, and that LiDAR derived vegetation profiles provide strong ecological information (Marselis et al., 2018: 626).

Grassland, defined as vegetation dominated by grassy species, with fewer trees and shrubs than savanna may also be studied with LiDAR technologies (Campbell, 2018). For example, Zhang et al. (2021) note that unmanned aerial vehicle (UAV) light detection and ranging (LiDAR) sensors with "a single laser beam emission capability can rapidly detect grassland vegetation parameters, such as canopy height, fractional vegetation coverage (FVC) and aboveground biomass (AGB)." Their findings showed that the LiDAR method was successful, as compared with other prediction maps, these maps had higher accuracy, wider spatial coverage cover and a higher spatial resolution. Grassland under variable grazing intensities, can also be assessed using advanced LiDAR techniques. Marcinkowska-Ochtyra et al. (2018) examined two grass species (*Molinia caerulea*, *Calamagrostis epigejos*), using Airborne 1-m resolution HySpex images and LiDAR data, and vegetation structures from discrete and full-waveform data and topographic indexes were created based on a LiDAR point cloud generated Canopy Height Model. The classifications were performed using a Random Forest algorithm. Marcinkowska-Ochtyra et al. (2018) concluded that hyperspectral (HS) and LiDAR airborne data is applicable to the study of grasslands, including assessments and monitoring of grassland species encroaching into habitats.

Conclusions

This chapter has examined some topics covering key issues in the relationship between biogeochemistry and biogeography, the latter science examined principally as a terrestrial science, as marine biogeography and its relationship with biogeochemistry will be examined in the next chapter. The development of the tools of geomatics is a recent, crucial application for environmental research, lending analyses and precision to case studies, and superior spatial perspectives to regional and global studies. The main links between biogeochemistry and terrestrial biogeography concern the physical features of the Earth's surface, especially vegetation structures, for which exact measurements may allow a more accurate assessment of their role in biogeochemical cycling. Geodesy, radar, and LiDAR are still developing; therefore, the application of these techniques to the environmental sciences is an ongoing exercise, with the main issues being increasing precision, resolution, and detail and barrier penetration (such as through canopies). Some of the biogeographical and geomatics-based studies hint at the contributions to predictive models, which may become increasingly important, given the current sometimes extreme acknowledgement of rapid, trend-based environmental change. This development will contribute to the development of the environmental sciences and strengthens the status of biogeochemistry as a key science of the study of the Earth.

References

Alongi, D. M. (2014). Carbon cycling and storage in mangrove forests. *Annual Review of Marine Science, 6*, 195–219. https://doi.org/10.1146/annurev-marine-010213-135020

American Geophysical Union (AGU). (2023). *About Geodesy*. Retrieved from https://connect.agu.org/geodesy/about/about-us

Anderson, J. M. (1991). The effects of climate change on decomposition processes in grassland and coniferous forests. *Ecological Applications, 1*, 326–347.

Ansley, R. J., Boutton, T. J., & Skjemstad, J. O. (2006). Soil organic carbon and black carbon storage and dynamics under different fire regimes in temperate mixed-grass savanna. *Global Biogeochemical Cycles, 20*(3), GB3006. https://doi.org/10.1029/2005GB002670

Anthony, K. M., Zimov, S. A., Grosse, G., Jones, M. C., Anthony, P. M., Chapin, F. S., 3rd, Finlay, J. C., Mack, M. C., Davydov, S., Frenzel, P., & Frolking, S. (2014). A shift of thermokarst lakes from carbon sources to sinks during the Holocene epoch. *Nature, 511*(7510), 452–456. https://doi.org/10.1038/nature13560

Baudena, M., Dekker, S. C., van Bodegom, P. M., Cuesta, B., Higgins, S. I., Lehsten, V., Reick, C. H., Rietkerk, M., Scheiter, S., Yin, Z., Zavala, M. A., & Brovkin, V. (2015). Forests, savannas, and grasslands: Bridging the knowledge gap between ecology and Dynamic Global Vegetation Models. *Biogeosciences, 12*, 1833–1848. https://doi.org/10.5194/bg-12-1833-2015

Beaumont, L. J., & Duursma, D. (2012). Global projections of 21st century land-use changes in regions adjacent to Protected Areas. *PLoS One, 7*(8), e43714. https://doi.org/10.1371/journal.pone.0043714

Beeby, A. (1993). *Applying ecology*. Chapman and Hall.

Begon, M., Harper, J. L., & Townsend, C. R. (2006). *Ecology: Individuals, populations and communities*. Blackwell Science Ltd.

Beier, C., Emmett, B., Tietema, A., Schmidt, I., Penuelas, J., Lang, E., Duce, P., Angelis, P., Gorissen, A., Estiarte, M., De Dato, G., Sowerby, A., Kroel-Dulay, G., Lellei-Kovács, E., Kull, O., Mänd, P., Petersen, H., Gjelstrup, P., & Spano, D. (2009). Carbon and nitrogen balances for six shrublands across Europe. *Global Biogeochemical Cycles, 23*(4), 1–13. https://doi.org/10.1029/2008GB003381

Belshe, F. E., Schuur, A. G., & Bolker, B. M. (2013). Tundra ecosystems observed to be carbon dioxide sources due to differential amplification of the carbon cycle. *Ecology Letters, 16*(10), 1307–1315. https://doi.org/10.1111/ele.12164

Berlanga-Robles, C. A., & Ruiz-Luna, A. (2020). Assessing seasonal and long-term mangrove canopy variations in Sinaloa, northwest Mexico, based on time series of enhanced vegetation index (EVI) data. *Wetlands Ecology and Management, 28*, 229–249. https://doi.org/10.1007/s11273-020-09709-0

Beutler, G. (1998). The role of GPS in space geodesy. In P. J. G. Teunissen & A. Kleusberg (Eds.), *GPS for geodesy* (pp. 625–650). Springer-Verlag. https://doi.org/10.1007/978-3-642-72011-6_16

Bianchi, T. S. (2021). The evolution of biogeochemistry: Revisited. *Biogeochemistry, 154*, 141–181. https://doi.org/10.1007/s10533-020-00708-0

Biber, M. F., Voskamp, A., Niamir, A., Hickler, T., & Hof, C. (2019). A comparison of macroecological and stacked species distribution models to predict future global terrestrial vertebrate richness. *Journal of Biogeography, 47*(1), 114–129. https://doi.org/10.1111/jbi.13696

Bork, E., & Badiou, P. (2017). *The importance of temperate grasslands in the global carbon cycle*. Retrieved from https://abnawmp.ca/wp-content/uploads/2020/09/The-Importance-of-Temperate-Grasslands-in-the-Global-Carbon-Cycle_April-27-2017-002_web.pdf

Bowditch, N. (2002). Geodesy and Datums in navigation. In N. Bowditch (Ed.), *The American practical navigator: An epitome of navigation* (pp. 15–22). National Imagery and Mapping Agency.

Braun, M., Bai, Y., McConkey, B., Farrell, R., Romo, J. T., & Pennock, D. (2013). Greenhouse gas flux in a temperate grassland as affected by landform and disturbance. *Landscape Ecology, 28*(4), 709–723.

Britannica, T. Editors of Encyclopaedia. (2020). *Biogeography. Encyclopedia Britannica*. https://www.britannica.com/science/biogeography

Britannica, T. Editors of Encyclopaedia. (2023). *Doppler effect. Encyclopedia Britannica*. https://www.britannica.com/science/Doppler-effect

Burkard, R. K. (1968). *Geodesy for the Layman*. Aeronautical Chart And Information Center.

Burrows, C. J. (1990). *Processes of vegetation change*. Unwin Hyman.

Cameron, C., Hutley, L. B., & Friess, D. A. (2019). Estimating the full greenhouse gas emissions offset potential and profile between rehabilitating and established mangroves. *Science of the Total Environment, 665*, 419–431. https://doi.org/10.1016/j.scitotenv.2019.02.104

Campbell, M. (1998). *Interactions between biogeography and rural livelihoods in the coastal savanna of Ghana*. Unpublished PhD Thesis, University of London (Imperial College).

Campbell, M. (2015). *Vultures: Their evolution, ecology and conservation*. Taylor and Francis (CRC Press).

Campbell, M. (2017). *Biological conservation for the 21st century: A conservation biology of large wildlife*. Nova Science Publishers.

Campbell, M. (2018). *Geomatics and conservation biology*. Nova Science Publishers.

Campbell, M. (2020). *The physical geography of El Salvador: A geophysical and ecological approach*. Nova Science Publishers.

Campbell, M. (2022a). Geodesy and geophysical applications for large raptor ecological research. In M. Campbell & K. K. Jha (Eds.), *Dynamics and interrelations between nature, science, and society* (pp. 160–219). Nova Science Publishers.

Campbell, M. (2022b). *The great eagles: Their evolution, ecology and conservation*. Taylor and Francis (CRC Press).

Campbell, M. (2022c). Geophysical techniques and applications for animal ecological research: The case of light detection and ranging (LiDAR). In M. Campbell & K. K. Jha (Eds.), *Dynamics and interrelations between nature, science, and society* (pp. 247–292). Nova Science Publishers.

Campbell, M., & Jha, K. K. (2021). *Critical research techniques in animal and habitat ecology: International examples*. Nova Science Publishers.

Carrasco, L., Giam, X., Papeş, M., & Sheldon, K. (2019). Metrics of Lidar-derived 3D vegetation structure reveal contrasting effects of horizontal and vertical forest heterogeneity on bird species richness. *Remote Sensing, 11*(7), 743. https://doi.org/10.3390/rs11070743. MDPI AG.

Carrión-Prieto, P., Hernández-Navarro, S., Martín-Ramos, P., Sánchez-Sastre, L. F., Garrido-Laurnaga, F., Marcos-Robles, J. L., & Martín-Gil, J. (2017). Mediterranean shrublands as carbon sinks for climate change mitigation: New root-to-shoot ratios. *Carbon Management, 8*(1), 67–77. https://doi.org/10.1080/17583004.2017.1285178

Cary Institute of Ecosystem Studies. (2020). *Definition of ecology*. Cary Institute of Ecosystem Studies. Retrieved from https://www.caryinstitute.org/news-insights/2-minute-science/definition-ecology

Chapin, F. S., Walker, B. H., Hobbs, R. J., Hooper, D. U., Lawton, J. H., Sala, O. E., & Tilman, D. (1997). Biotic control over the functioning of ecosystems. *Science, 277*(5325), 500–504. https://doi.org/10.1126/science.277.5325.500

Charles, S. P., Kominoski, J. S., Armitage, A. R., Guo, H., Weaver, C. A., & Pennings, S. C. (2020). Quantifying how changing mangrove cover affects ecosystem carbon storage in coastal wetlands. *Ecology, 101*(2), e02916. https://doi.org/10.1002/ecy.2916

Chiesi, M., Moriondo, M., Maselli, F., Gardin, L., Fibbi, L., Bindi, M., & Running, S. W. (2010). Simulation of Mediterranean forest carbon pools under expected environmental scenarios. *Canadian Journal of Forest Research, 40*(5), 850–860. https://doi.org/10.1139/X10-037

College of Earth, Ocean, and Atmospheric Sciences. (2022). *Ocean Ecology and Biogeochemistry*. College of Earth, Ocean, and Atmospheric Sciences.

Cooney, J. A. (1968). Measurements on the Raman component of laser atmospheric backscatter. *Applied Physics Letters, 12*(2), 40–42.

Cramer, W., Bondeau, A., Woodward, F. I., Prentice, I. C., Betts, R. A., Brovkin, V. A., Cox, P. M., Fisher, V. A., Foley, J. A., Friend, A. D., Kucharik, C. J., Lomas, M. R., Ramankutty, N., Sitch, S. A., Smith, B., White, A., & Young-Molling, C. (2001). Global response of terrestrial ecosystem structure and function to CO2 and climate change: Results from six dynamic global vegetation models. *Global Change Biology, 7*. https://doi.org/10.1046/j.1365-2486.2001.00383.x

Crisci, J., Katinas, L., & Posadas, P. (2003). *Historical biogeography: An introduction*. Harvard University Press.

Crutzen, P. J., & Andreae, M. O. (1990). Biomass burning in the tropics: Impact on atmospheric chemistry and biogeochemical cycles. *Science (New York, N.Y.), 250*(4988), 1669–1678. https://doi.org/10.1126/science.250.4988.1669

Dai, X., Boutton, T., Glaser, B., Ansley, R., & Zech, W. (2005). Black carbon in a temperate mixed-grass savanna. *Soil Biology & Biochemistry, 37*(10), 1879–1881. https://doi.org/10.1016/j.soilbio.2005.02.021

Dansereau, P. (2015). *Biogeography. The Canadian Encyclopedia*. Retrieved from https://www.thecanadianencyclopedia.ca/en/article/biogeography

De Graaff, M. A., Throop, H. L., Verburg, P. S. J., Arnone, J. A., III, & Campos, X. (2014). A synthesis of climate and vegetation cover effects on biogeochemical cycling in shrub-dominated drylands. *Ecosystems, 17*, 931–945. https://doi.org/10.1007/s10021-014-9764-6

Dehne, M., Cervantes, F.G., Sheard, B.S., Heinzel, G., & Danzmann, K. (2009). Laser interferometer for spaceborne mapping of the Earth's gravity field. *Journal of Physics: Conference Series, 154*, 7th International Lisa Symposium 16–20 June 2008, Barcelona, Spain. https://doi.org/10.1088/1742-6596/154/1/012023

Devaney, J., Barrett, B., Barrett, F., Redmond, J., & Halloran, O. J. (2015). Forest cover estimation in Ireland using radar remote sensing: A comparative analysis of forest cover assessment methodologies. *PLoS One, 10*(8), e0133583. https://doi.org/10.1371/journal.pone.0133583

Donato, D. C., Kauffman, J. B., Mackenzie, R. A., Ainsworth, A., & Pfleeger, A. Z. (2012). Whole-Island carbon stocks in the tropical Pacific: Implications for mangrove conservation and upland restoration. *Journal of Environmental Management, 97*, 89–96. https://doi.org/10.1016/j.jenvman.2011.12.004

Eagle Mapping. (2022). *LiDAR*. Retrieved from https://www.eaglemapping.com/acquisition-lidar

Earth Observatory. (2022). *Coniferous forest*. Retrieved from https://earthobservatory.nasa.gov/biome/bioconiferous.php#:~:text=Precipitation%20in%20coniferous%20forests%20varies,depends%20on%20the%20forest%20location

Ebach, M. C. (2015). A history of biogeography for the twenty-first century biogeographer. In *Origins of biogeography. History, philosophy and theory of the life sciences* (Vol. 13). Springer. https://doi.org/10.1007/978-94-017-9999-7_1

Engler, J., Secondi, J., Dawson, D., Elle, O., & Hochkirch, A. (2015). Range expansion and retraction along a moving contact zone has no effect on the genetic diversity of two passerine birds. *Ecography, 39*(9), 815–911. https://doi.org/10.1111/ecog.01520

Environmental Systems Research Institute (ESRI). (2019). *What is Lidar data*? Retrieved from https://desktop.arcgis.com/en/arcmap/10.3/manage-data/las-dataset/what-is-lidar-data-.htm

Eswaran, H., van den Berg, E., & Reich, P. (1993). Organic carbon in soils of the world. *Soil Science Society of America Journal, 57*, 192–194.

Faber-Langendoen, D., Keeler-Wolf, T., Meidinger, D., Josse, C., Weakley, A., Tart, D., Navarro, G., Hoagland, B., Ponomarenko, S., Saucier, J.-P., Fults, G., & Helmer, E. (2016). *Classification and description of world formation types. General technical report RMRS-GTR-346*. USDA Forest Service, Rocky Mountain Research Station.

Fessenden, M. (2014). How bird migrations show up beautifully on Doppler radar. *Smithsonian Magazine*. Retrieved from https://www.smithsonianmag.com/smart-news/how-doppler-radar-can-track-bird-migrations-180952834/

Fidelis, A., Lyra, M. F. S., & Pivello, V. R. (2013). Above- and below-ground biomass and carbon dynamics in Brazilian Cerrado wet grasslands. *Journal of Vegetation Science, 24*(2), 356–364. https://doi.org/10.1111/j.1654-1103.2012.01465.x

Filho, W. L. (2017). *Climate change impacts and adaptation strategies for coastal communities*. Springer.

Fiocco, G., & Smullin, L. (1963). Detection of scattering layers in the upper atmosphere (60–140 km) by optical radar. *Nature, 199*, 1275–1276. https://doi.org/10.1038/1991275a0

Fung, I. Y., Doney, S. C., Lindsay, K., & John, J. (2005). Evolution of carbon sinks in a changing climate. *Proceedings of the National Academy of Sciences of the United States of America, 102*(32), 11201–11206. https://doi.org/10.1073/pnas.0504949102

Geist, H. (2005). *The causes and progression of desertification*. Ashgate.

Geist, H. J., & Lambin, E. F. (2004). Dynamic causal patterns of desertification. *Bioscience, 54*(9), 817–829. https://doi.org/10.1641/0006-3568(2004)054[0817:DCPOD]2.0.CO;2

Gil-Tena, A., Aquilué, N., Duane, A., De Cáceres, M., & Brotons, L. (2016). Mediterranean fire regime effects on pine-oak forest landscape mosaics under global change in NE Spain. *European Journal of Forest Research, 135*, 403–416. https://doi.org/10.1007/s10342-016-0943-1

GIS Resources. (2021). *What is Geodesy*? https://gisresources.com/what-is-geodesy/#:~:text=Geodesy%20is%20the%20science%20which,the%20surface%20of%20the%20earth

Goetz, S. J., Steinberg, D. C., Betts, M. G., Holmes, R. T., Doran, P. J., Dubayah, R. O., & Hofton, M. A. (2010). Lidar remote sensing variables predict breeding habitat of a Neotropical migrant bird. *Ecology, 91*(6), 1569–1576. https://doi.org/10.1890/09-1670.1

González-Jorge, H., Martínez-Sánchez, J., Bueno, M., & Arias, A. P. (2017). Unmanned aerial systems for civil applications: A review. *Drones, 1*(1), 2. MDPI AG. Retrieved from https://doi.org/10.3390/drones1010002

Government of Canada. (2021). *Synthetic Aperture Radar*. Retrieved from https://natural-resources.canada.ca/earth-sciences/geomatics/satellite-imagery-and-air-photos/sensors-and-methods/synthetic-aperture-radar/10968

Gratani, L., Varone, L., Ricotta, C., & Catoni, R. (2013). Mediterranean shrublands carbon sequestration: Environmental and economic benefits. *Mitigation and Adaptation Strategies for Global Change, 18*, 1167–1182. https://doi.org/10.1007/s11027-012-9415-1

Guo, X., Coops, N. C., Tompalski, P., Nielsen, S. E., Bater, C. W., & Stadt, J. J. (2017). Regional mapping of vegetation structure for biodiversity monitoring using airborne lidar data. *Ecological Informatics, 38*, 50–61. https://doi.org/10.1016/j.ecoinf.2017.01.005

Guo, X., Coops, N. C., Gergel, S. E., Bater, C. W., Nielsen, S. E., Stadt, J. J., & Drever, M. (2018). Integrating airborne lidar and satellite imagery to model habitat connectivity dynamics for spatial conservation prioritization. *Landscape Ecology, 33*(3), 491–511. https://doi.org/10.1007/s10980-018-0609-0

Guo, F., Guénard, B., Economo, E. P., Deutsch, C. A., & Bonebrake, T. C. (2020). Activity niches outperform thermal physiological limits in predicting global ant distributions. *Journal of Biogeography, 47*(4), 829–842. https://doi.org/10.1111/jbi.13799

Gwenzi, D. (2017). Lidar remote sensing of savanna biophysical attributes: Opportunities, progress, and challenges. *International Journal of Remote Sensing, 38*(1), 235–257. https://doi.org/10.1080/01431161.2016.1259683

Gwenzi, D., & Lefsky, M. A. (2014). Modeling canopy height in a savanna ecosystem using spacebome lidar waveforms. *Remote Sensing of Environment, 154*, 338–344. https://doi.org/10.1016/j.rse.2013.11.024

Hamraz, H., Contreras, M. A., & Zhang, J. (2017). Forest understory trees can be segmented accurately within sufficiently dense airborne laser scanning point clouds. *Scientific Reports, 7*(1), 6770. https://doi.org/10.1038/s41598-017-07200-0

Hansen, J. N., Mitchard, E. T. A., & King, S. (2020). Assessing forest/non-forest separability using Sentinel-1 C-band synthetic aperture radar. *Remote Sensing, 12*(11), 1899. MDPI AG. Retrieved from https://doi.org/10.3390/rs12111899

Hardesty, R. M. (2015). LIDAR | Doppler. In G. R. North, J. Pyle, & F. Zhang (Eds.), *Encyclopedia of atmospheric sciences* (2nd ed., pp. 289–295). https://doi.org/10.1016/B978-0-12-382225-3.00207-3

Harding, D. J., Lefsky, M. A., Parker, G. G., & Blair, J. B. (2001). Laser altimeter canopy height profiles: Methods and validation for closed-canopy, broadleaf forests. *Remote Sensing of Environment, 76*(3), 283–297.

Harris, N., & Gibbs, D. (2021). *Forests absorb twice as much carbon as they emit each year*. Retrieved from https://www.wri.org/insights/forests-absorb-twice-much-carbon-they-emit-each-year

Hartl, D. L., & Clark, A. G. (2007). *Principles of population genetics* (4th ed.). Sinauer Associates.

Hay, S. I. (2000). An overview of remote sensing and geodesy for epidemiology and public health application. *Advances in Parasitology, 47*, 1–35. https://doi.org/10.1016/s0065-308x(00)47005-3

Heads, M. (2015). The relationship between biogeography and ecology: Envelopes, models, predictions. *Biological Journal of the Linnean Society, 115*(2), 456–468. https://doi.org/10.1111/bij.12486

Henwood, W. D. (1998). An overview of protected areas in the temperate grasslands biome. *Parks, 8*(3), 3–8.

Herndon, K., Meyer, F., Flores, A., Cherrington, E., & Kucera, L. (2020). *What is Synthetic Aperture Radar*? Retrieved from https://earthdata.nasa.gov/learn/backgrounders/what-is-sar

Hewins, D. B., Fatemi, F., Adams, B., Carlyle, C. N., Chang, S. X., & Bork, E. W. (2015). Grazing, regional climate and soil biophysical impacts on microbial enzyme activities in grassland soil of western Canada. *Pedogiologia – Journal of Soil Ecology, 58*, 201–209.

Higuera, P. E., Chipman, M. L., Barnes, J. L., Urban, M. A., & Hu, F. S. (2011). Variability of tundra fire regimes in Arctic Alaska: Millennial-scale patterns and ecological implications. *Ecological Applications, 21*(8), 3211–3226. https://doi.org/10.1890/11-0387.1

Indian Space Research Organization. (2018). *Indian Spacecraft*. Government of India.

Jekeli, C. (2011). Gravity field of the earth. In H. K. Gupta (Ed.), *Encyclopedia of solid earth geophysics. Encyclopedia of earth sciences series* (pp. 471–484). Springer. https://doi.org/10.1007/978-90-481-8702-7_91

Jenkins, D. G., & Ricklefs, R. E. (2011). Biogeography and ecology: Two views of one world. *Philosophical Transactions of the Royal Society of London. Series B, Biological Sciences, 366*(1576), 2331–2335. https://doi.org/10.1098/rstb.2011.0064

Jensen, J. R. (2007). *Remote sensing of the environment: An earth resource perspective* (2nd ed.). Pearson Prentice Hall.

Johnson, D. L. (1992). Ecology, biology, field experimentation and environmental impact. In C. Lomer & C. Prior (Eds.), *Biological control of locusts and grasshoppers* (pp. 267–278). CAB International.

Joyce, C. B., Simpson, M., & Casanova, M. (2016). Future wet grasslands: Ecological implications of climate change. *Ecosystem Health, and Sustainability, 2*(9), e01240. https://doi.org/10.1002/ehs2.1240

Kanitz, T., Ciapponi, A., Mondello, A., D'Ottavi, A., Baselga Mateo, A. B., Straume, A.-G., Voland, C., Bon, D., Checa, E., Alvarez, E., Bellucci, I., Pereira Do Carmo, J., Brewster, J., Marshall, J., Schillinger, M., Hannington, M., Rennie, M., Reitebuch, O., Lecrenier, O., Bravetti, P., Sacchieri, V., De Sanctis, V., Lefebvre, A., Parrinello, T., & Wernham, D. (2020). ESA's Lidar missions Aeolus and EarthCARE. *EPJ Web Conference, 237*, 01006. https://doi.org/10.1051/epjconf/202023701006

Karl, D. M. & Schlesinger, W. H. (2014). Volume Editors' Introduction. In Holland, H. D. & Turekian, K. K. (Eds.), *Treatise on geochemistry* (2nd Ed.) 10: xxiii–xxvi). Amsterdam: Elsevier Science. https://doi.org/10.1016/B978-0-08-095975-7.09983-6

Kashani, F., Kashani, P., Moghimian, M., & Shakour, M. (2015). Effect of stress inoculation training on the levels of stress, anxiety, and depression in cancer patients. *Iranian Journal of Nursing and Midwifery Research, 20*(3), 359–364.

Katinas, L., & Crisci, J. (2018). Agriculture biogeography: An emerging discipline in search of a conceptual framework. *Progress in Physical Geography: Earth and Environment, 42*(4). https://doi.org/10.1177/0309133318776493

Knowles, J., Scott, R., Minor, R., & Barron-Gafford, G. (2020). Ecosystem carbon and water cycling from a sky Island montane forest. *Agricultural and Forest Meteorology, 281*, 107835. https://doi.org/10.1016/j.agrformet.2019.107835

Koch, G. J. (2006). Using a Doppler light detection and ranging (lidar) system to characterize an atmospheric thermal providing lift for soaring raptors. *Journal of Field Ornithology, 77*(3), 315–318. https://doi.org/10.1111/j.1557-9263.2006.00058.x

Krinner, G., Viovy, N., de Noblet-Ducoudré, N., Ogée, J., Polcher, J., Friedlingstein, P., Ciais, P., Sitch, S., & Prentice, I. C. (2005). A dynamic global vegetation model for studies of the coupled atmosphere-biosphere system. *Global Biogeochemical Cycles, 19*(1). https://doi.org/10.1029/2003GB002199

Krogstad, F., & Schiess, P. (2004). The allure and pitfalls of using LiDAR topography in harvest and road design. In J. Nelson, & M. Clark (Eds.). *Proceedings of the Joint Conference of IUFRO 3.06 Forest Operations in Mountainous Conditions and the 12th International Mountain Logging Conference*, Vancouver, June 13–16, pp. 1–10.

Krupa, J. J., Hopper, K. R., Gruber, S. B., Schmidt, J. M., & Harwood, J. D. (2020). Plant-animal interactions between carnivorous plants, sheet-web spiders, and ground-running spiders as guild predators in a wet meadow community. *Ecology and Evolution, 10*(11), 4762–4772. https://doi.org/10.1002/ece3.6230

Kucuksezgin, F., Pazi, I., Kocak, F., Gonul, T., Duman, M., & Eronat, H. (2019). Chapter 12 – The coasts of Turkey. In C. Sheppard (Ed.), *World seas: An environmental evaluation. Volume I: Europe, the Americas, and West Africa* (2nd ed., pp. 307–332). Elsevier.

Lahdenperä, M., Jackson, J., Htut, W., & Lummaa, V. (2019). Capture from the wild has long-term costs on reproductive success in Asian elephants. *Proceedings. Biological sciences, 286*(1912), 20191584. https://doi.org/10.1098/rspb.2019.1584

Lal, R., & Lorenz, K. (2012). Carbon sequestration in temperate forests. In R. Lal, K. Lorenz, R. Hüttl, B. Schneider, & J. von Braun (Eds.), *Recarbonization of the biosphere* (pp. 187–201). Springer. https://doi.org/10.1007/978-94-007-4159-1_9

Le Maitre, D. C., & Midgley, J. J. (1992). Plant reproductive ecology. In R. M. Cowling (Ed.), *The ecology of fynbos. Nutrients, fire and diversity* (pp. 135–174). Oxford University Press.

Lefsky, M. A., Cohen, W. B., Harding, D. J., Parker, G. G., Acker, S. A., & Gower, S. T. (2002). Lidar remote sensing of above ground biomass in three biomes. *Global Ecology and Biogeography, 11*(5), 393–399. https://doi.org/10.1046/j.1466-822x.2002.00303.x

Lehmann, C. E., & Parr, C. L. (2016). Tropical grassy biomes: Linking ecology, human use and conservation. *Philosophical Transactions of the Royal Society of London. Series B, Biological Sciences, 371*(1703), 20160329. https://doi.org/10.1098/rstb.2016.0329

Leone, A., Puncher, G. N., Ferretti, F., et al. (2020). Pliocene colonization of the Mediterranean by Great White Shark inferred from fossil records, historical jaws, phylogeographic and divergence time analyses. *Journal of Biogeography, 47*(5), 1119–1129. https://doi.org/10.1111/jbi.13794

Linder, H. P. (2005). Evolution of diversity: The Cape flora. *Trends in Plant Science, 10*(11), 536–541. https://doi.org/10.1016/j.tplants.2005.09.006

Liu, L., Qiao, Y., & Ha, Z. (2016). Black carbon concentration and isotopic composition of surface sand from deserts and dune fields in Northern China. *Palaeogeography, Palaeoclimatology, Palaeoecology, 445*, 1–7. https://doi.org/10.1016/j.palaeo.2015.12.026

Liverman, D. G., Batterson, M. J., Bell, T. J., Nolan, L. W., Marich, A. S., & Putt, M. (2006). *Digital elevation models from shuttle radar topography mission data – New insights into the quaternary history of Newfoundland* (pp. 177–189). Current Research. Newfoundland and Labrador Department of Natural Resources Geological Survey, Report 06-1.

Livingston, S. (2015). *Commercial remote sensing satellites and the regulation of violence in areas of limited statehood* (CGCS Occasional Paper Series on ICTs, State building, and Peacebuilding in Africa No. 5). University of Pennsylvania.

Lorenz, K., & Lal, R. (2010). *Carbon sequestration in forest ecosystems*. Springer.

Maiman, T. H. (1960). Stimulated optical radiation in ruby. *Nature, 187*, 493–494.

Manca, G., Tirone, G., Matteucci, G., Tonon, G., Cherubini, P., Ballarin, I., Duerr, M., Matteucci, M., & Seufert, G. (2009). The effect of climate variability on the carbon cycle of a Mediterranean forest. Geophysical Research Abstracts 11, EGU2009-11475-1.

Marcinkowska-Ochtyra, A., Jarocińska, A., Bzdęga, K., & Tokarska-Guzik, B. (2018). Classification of expansive grassland species in different growth stages based on hyperspectral and LiDAR data. *Remote Sensing, 10*(12), 2019. https://doi.org/10.3390/rs10122019. MDPI AG.

Marin, J., & Hedges, S. B. (2016). Time best explains global variation in species richness of amphibians, birds, and mammals. *Journal of Biogeography, 43*(6), 1069–1079.

Marin-Spiotta Biogeography and Biogeochemistry (BiogeoLab) Research Group. (2022). *Marin-Spiotta BiogeoLab: Biogeochemistry – Biogeography – Global Change*. Retrieved from https://marinspiotta.wordpress.com/

Markus, T., Neumann, T., Martino, A., Abdalati, W., Brunt, K., Csatho, B., Farrell, S., Fricker, H., Gardner, A., Harding, D., Jasinski, M., Kwok, R., Magruder, L., Lubin, D., Luthcke, S., Morison, J., Nelson, R., Neuenschwander, A., Palm, S., Popescu, S., Shum, C. K., Schutz, B. E., Smith, B., Yang, Y., & Zwally, J. (2017). The Ice, Cloud, and land Elevation

Satellite-2 (ICESat-2): Science requirements, concept, and implementation. *Remote Sensing of Environment, 190*, 260–273. https://doi.org/10.1016/j.rse.2016.12.029
Marselisa, S. M., Tang, H., Armston, J. D., Calders, K., Labrière, N., & Dubayah, R. (2018). Distinguishing vegetation types with airborne waveform lidar data in a tropical forest-savanna mosaic: A case study in Lopé National Park, Gabon. *Remote Sensing of Environment, 216*, 626–634. https://doi.org/10.1016/j.rse.2018.07.023
Marti, O., Braconnot, P., Bellier, J., Benshila, R., Bony, S., Brockmann, P., Cadule, P., Caubel, A., Denvil, S., Dufresne, J. L., Fairhead, L., Filiberti, M.-A., Foujols, M.-A., Fichefet, T., Friedlingstein, P., Goosse, H., Grandpeix, J. Y., Hourdin, F., Krinner, G., Lévy, C., Madec, G., Musat, I., de Noblet, N., Polcher, J., & Talandier, C. (2005). *The new IPSL climate system model: IPSL-CM4, Note du Pôle de Modélisation 26*. Institut Pierre Simon Laplace.
Melfi, S. H., Lawrence, J. D., Jr., & McCormick, M. (1969). Observation of Raman scattering by water vapor in the atmosphere. *Applied Physics Letters, 15*(9), 295–297.
Melillo, J. M., Borchers, J., Chaney, J., Fisher, H., Fox, S., Haxeltine, A., Janetos, A., Kicklighter, D. W., Kittel, T. G. F., McGuire, A. D., McKeown, R., Neilson, R., Nemani, R., Ojima, D. S., & Painter, T. (1995). Vegetation/Ecosystem Modeling and Analysis Project: Comparing biogeography and biogeochemistry models in a continental-scale study of terrestrial ecosystem responses to climate change and CO2 doubling. *Global Biochemical Cycles, 9*(4), 407–437. https://doi.org/10.1029/95GB02746
Meyer, T. (2018). *Introduction to geometrical and physical geodesy: Foundations of geomatics*. ESRI Press.
Meyer, T., Roman, D., & Zilkoski, D. (2006). What does height really mean? Part I: Introduction. *Surveying and Land Information Science, 66*, 127–137.
Miltronix Limited. (2021). *A real breakthrough in foliage penetration radar*. Retrieved from https://miltronix.co.uk/portfolio/folpen-middrs-2/
Monaco, M. E., Kendall, M. S., Higgins, J. L., Alexander, C. E., & Tartt, M. S. (2003). Biogeographic assessments of NOAA National Marine Sanctuaries: The integration of ecology and GIS technology. In *Oceans 2003. Celebrating the Past ... Teaming Toward the Future (IEEE Cat. No.03CH37492)* (Vol. 3, pp. 1489–1498), San Diego, CA. https://doi.org/10.1109/OCEANS.2003.178086
Mortimore, M. J. (1989). *Adopting to drought: Farmers, famines and desertification in West Africa*. Cambridge University Press.
Mueller-Dombois, D. (2001). Island biogeography. In S. A. Levin (Ed.), *Encyclopedia of biodiversity* (pp. 565–580). Elsevier B.V. https://doi.org/10.1016/B0-12-226865-2/00166-8
Mueller-Hohenstein, K. (2001). Biogeography. In N. J. Smelser & P. B. Baltes (Eds.), *International encyclopedia of the social & behavioral sciences* (pp. 1201–1206). Elsevier B.V. https://doi.org/10.1016/B0-08-043076-7/02599-7
Murphy, P. G., & Lugo, A. E. (1986). Ecology of tropical dry forest. *Annual Review of Ecology and Systematics, 17*, 67–88. http://www.jstor.org/stable/2096989
Naranjo, L. (2005). *Matter in motion: Earth's changing gravity*. Retrieved from https://www.earthdata.nasa.gov/learn/sensing-our-planet/matter-in-motion-earths-changing-gravity
National Aeronautics and Space Administration. (2018). *Science and applications*. Retrieved from https://space-geodesy.nasa.gov/science_and_applications/science_and_applications.html
National Geodetic Survey. (2009). *Geodetic glossary*. Retrieved from http://www.ngs.noaa.gov/CORSProxy/Glossary/xml/NGS_Glossary.xml
National Geographic Society. (2022). *The five major types of biomes*. Retrieved from https://education.nationalgeographic.org/resource/five-major-types-biomes
National Institute of Standards and Technology. (2023). *GPS*. Retrieved from https://csrc.nist.gov/glossary/term/gps#:~:text=The%20Global%20Positioning%20System%20(GPS,segment%2C%20and%20the%20user%20segment
National Oceanic and Atmospheric Administration. (1977). *Basic geodesy*. National Oceanic and Atmospheric Administration.
National Oceanic and Atmospheric Administration. (2021a). *What is Geodesy?* Retrieved from https://oceanservice.noaa.gov/facts/geodesy.html

National Oceanic and Atmospheric Administration. (2021b). *Gravity*. Retrieved from https://oceanservice.noaa.gov/education/tutorial_geodesy/geo07_gravity.html
National Oceanic and Atmospheric Administration. (2023). *What is Lidar*? Retrieved from https://oceanservice.noaa.gov/facts/lidar.html
Natural Resources Canada. (2022). *Forest carbon*. Retrieved from https://www.nrcan.gc.ca/climate-change-adapting-impacts-and-reducing-emissions/climate-change-impacts-forests/forest-carbon/13085
Newsom, R. K., & Krishnamurthy, R. (2020). *Doppler Lidar (DL) Instrument Handbook*. Retrieved from https://www.arm.gov/publications/tech_reports/handbooks/dl_handbook.pdf
Odum, E. P. (1971). *Fundamentals of ecology* (3rd ed.). W.B. Saunders Company.
Ohte, N., & Tokuchi, N. (2011). Hydrology and biogeochemistry of temperate forests. In D. Levia, D. Carlyle-Moses, & T. Tanaka (Eds.), *Forest hydrology and biogeochemistry. Ecological studies 216*. Springer. https://doi.org/10.1007/978-94-007-1363-5_12
Oliver, J. E. (2008). *Encyclopedia of world climatology*. Springer Science & Business Media.
Oliver, V., Oliveras, I., Kala, J., Lever, R., & Teh, Y. A. (2017). The effects of burning and grazing on soil carbon dynamics in managed Peruvian tropical montane grasslands. *Biogeosciences, 14*, 5633–5646. https://doi.org/10.5194/bg-14-5633-2017
Olson, D. M., Dinerstein, E., Wikramanayake, E. D., Burgess, N. D., Powell, G. V. N., Underwood, E. C., D'Amico, J. A., Itoua, I., Strand, H. E., Morrison, J. C., Loucks, C. J., Allnutt, T. F., Ricketts, T. H., Kura, Y., Lamoreux, J. F., Wettengel, W. W., Hedao, P., & Kassem, K. R. (2001). Terrestrial ecoregions of the world: A new map of life on Earth. *Bioscience, 51*(11), 933–938.
Olsoy, P. J., Forbey, J. S., Rachlow, J. L., Nobler, J. D., Glenn, N. F., & Shipley, L. A. (2015). Fearscapes: Mapping functional properties of cover for prey with terrestrial LiDAR. *Bioscience, 65*(1), 74–80. https://doi.org/10.1093/biosci/biu189
Pagiatakis, S., Bisnath, S., Armenakis, C., & Wang, J. (2009). The establishment of a geomatics engineering program and its challenges: The York University Case. *Geomatica, 63*(2), 25–36.
Parton, W. J., Scurlock, M. O., Ojima, D. S., Gilmanov, T. G., Scholes, R. J., Schimel, D. S., Kirchner, T., Menaut, J. C., Seastedt, T., Moya, E. G., Kamnalrut, A., & Kinyamario, J. I. (1993). Observations and modelling of biomass and soil organic matter dynamics for the grassland biome worldwide. *Global Biogeochemical Cycles, 7*(4), 785–809.
Petrescu, F. I. (2012). *A new Doppler effect*. Books on Demand GmbH. https://doi.org/10.13140/RG.2.1.1142.8560
Petruzzello, M. (2022). *Mediterranean vegetation. Encyclopedia Britannica*. https://www.britannica.com/plant/Mediterranean-vegetation
Pickett, S. T. A., Parker, V. T., & Fiedler, P. L. (1992). The new paradigm in ecology: Implications for conservation biology above the species level. In P. L. Fiedler & K. Jain (Eds.), *Conservation biology: The theory and practice of nature conservation, preservation and management* (pp. 66–87). Chapman and Hall.
Pimm, S. (1991). *The balance of nature? Ecological issues in the conservation of species and communities*. University of Chicago Press.
Plag, H. P. (2000). Integration of geodetic techniques into a global Earth monitoring system and its implication for Earth system sciences. In R. Rummel, H. Drewes, W. Bosch, & H. Hornik (Eds.), *Towards an Integrated Global Geodetic Observing System (IGGOS). International Association of Geodesy Symposia, 120*. Springer. https://doi.org/10.1007/978-3-642-59745-9_11
Quinn, J. A. (2008). *Arctic and alpine biomes*. Greenwood Press.
Rapp, R. H. (1975). Status of the geopotential. *Reviews of Geophysics, 13*(3), 267–268. https://doi.org/10.1029/RG013i003p00267
Reed, S., Cavaleri, M. A., Alonso-Rodríguez, A. M., Kimball, B. A., & Wood, T. E. (2016). *Tropical rain forest biogeochemistry in a warmer world: Initial results from a novel warming experiment in a Puerto Rico tropical forest*. American Geophysical Union, Fall Meeting 2016.
Riddle, B. R. (2017). Biogeography: History. In *International encyclopedia of geography: People, the earth, environment and technology*. Wiley. https://doi.org/10.1002/9781118786352.wbieg1063

Rogger, J., Hörtnagl, L., Buchmann, N., & Eugster, W. (2022). Carbon dioxide fluxes of a mountain grassland: Drivers, anomalies, and annual budgets. *Agricultural and Forest Meteorology, 314*, 108801. https://doi.org/10.1016/j.agrformet.2021.108801

Romshoo, S. A. (2003). Radar remote sensing for monitoring of dynamic ecosystem processes related to biogeochemical exchanges in tropical peatlands. *Visual Geosciences, 8*, 63–82. https://doi.org/10.1007/s10069-003-0015-9

Rustad, L. E., Knoepp, J., Richter, D. D., & Scott, D. A. (2020). Biogeochemical cycling in forest and rangeland soils of the United States. In R. Pouyat, D. Page-Dumroese, T. Patel-Weynand, & L. Geiser (Eds.), *Forest and rangeland soils of the United States under changing conditions*. Springer. https://doi.org/10.1007/978-3-030-45216-2_4

Sampson, G., & Hamilton, E. (2014). *Tropical and sub-tropical broadleaf forests*. Retrieved from https://sites.google.com/site/ms12biomes/tropical-and-sub-tropical-broadleaf-forests

Sanmartín, I. (2012). Historical biogeography: Evolution in time and space. *Evolution: Education and Outreach, 5*, 555–568. https://doi.org/10.1007/s12052-012-0421-2

Scholes, M., & Andreae, M. O. (2000). Biogenic and pyrogenic emissions from Africa and their impact on the global atmosphere. *Ambio, 29*(1), 23–29. https://doi.org/10.1579/0044-7447-29.1.23

Scholes, R., & Walker, B. (1993). *An African Savanna: Synthesis of the Nylsvley Study (Cambridge studies in applied ecology and resource management)*. Cambridge University Press. https://doi.org/10.1017/CBO9780511565472

Schotland, R. M. (1966). Some observations of the vertical profile of water vapor by a laser optical radar. In *Proc. 4th symposium on remote sensing of environment*. Ann Arbor, University of Michigan 12–14, pp. 273–283.

Schuur, T. (2019). *Permafrost and the global carbon cycle*. Retrieved from https://arctic.noaa.gov/Report-Card/Report-Card-2019/ArtMID/7916/ArticleID/844/Permafrost-and-the-Global-Carbon-Cycle#:~:text=Northern%20tundra%20and%20boreal%20ecosystems,soil%20back%20to%20the%20atmosphere

Scurlock, J. M. O., & Hall, D. O. (1998). The global carbon sink: A grassland perspective. *Global Change Biology, 4*(2), 229–233. https://doi.org/10.1046/j.1365-2486.1998.00151.x

Seeber, G. (2003). *Satellite geodesy: Foundations, methods, and applications*. De Gruyter. https://doi.org/10.1515/9783110200089

Sharma, S., Sharma, H. P., Chaulagain, C., Katuwal, H. B., & Belant, J. L. (2020). Estimating occupancy of Chinese pangolin (Manis pentadactyla) in a protected and non-protected area of Nepal. *Ecology and Evolution, 10*(10), 4303–4313. https://doi.org/10.1002/ece3.6198

Sheppard, C. (2013). *Coral reefs of the United Kingdom Overseas Territories*. Springer Science & Business Media.

Simmons, I. G. (1993). *Environmental history: A concise introduction*. Basil Blackwell.

Skolnik, M. I. (2022). *Radar. Encyclopedia Britannica*. https://www.britannica.com/technology/radar

Smith, G. R., Levin, R. H., Abel, P., & Jacobowitz, H. (1988). Calibration of the solar channels of the NOAA-9 AVHRR using high altitude aircraft measurements. *Journal of Atmospheric and Oceanic Technology, 5*(5), 631–639. https://doi.org/10.1175/1520-0426(1988)005<0631:COTSCO>2.0.CO;2

Smithsonian Tropical Research Institute. (2020). *Biogeography*. Mithsonian Tropical Research Institute. https://stri.si.edu/discipline/biogeography

Smullin, L. D., & Fiocco, G. (1962). Optical echoes from the moon. *Nature, 194*, 1267–1267. https://doi.org/10.1038/1941267a0

Snyder, J. P. (1987). *Map projections—A working manual. U.S. Geological Survey Professional Paper 1395*. United States Government Printing Office.

Song, B., Niu, S., Zhang, Z., Yang, H., Li, L., & Wan, S. (2012). Light and heavy fractions of soil organic matter in response to climate warming and increased precipitation in a temperate steppe. *PLoS One, 7*(3), e33217. https://doi.org/10.1371/journal.pone.0033217

Soper, F. M., MacKenzie, R. A., Sharma, S., Cole, T. G., Litton, C. M., & Sparks, J. P. (2019). Non-native mangroves support carbon storage, sediment carbon burial, and accretion of coastal ecosystems. *Global Change Biology, 25*(12), 4315–4326. https://doi.org/10.1111/gcb.14813

Spagopoulou, F. (2020). Transgenerational maternal age effects in nature: Lessons from Asian elephants. *Journal of Animal Ecology, 89*(4), 936–939. https://doi.org/10.1111/1365-2656.13218

Sternberg, T., Rueff, H., & Middleton, N. (2015). Contraction of the Gobi Desert, 2000–2012. *Remote Sensing, 7*(2), 1346–1358.

Stock, W. D., & Allsopp, N. (1992). Functional perspective of ecosystems. In R. M. Cowling (Ed.), *The ecology of fynbos. Nutrients, fire and diversity* (pp. 241–259). Oxford University Press.

Świątek, B., & Pietrzykowski, M. (2022). Impact of leaf litter and fine roots in the pool of carbon, nitrogen and phosphorus accumulated in soil in various scenarios of regeneration and reconstruction of forest ecosystems. *Forests, 13*(8), 1207. https://doi.org/10.3390/f13081207. MDPI AG.

Tallis, J. H. (1991). Forest and Moorland in the South Pennine Uplands in the Mid-Flandrian Period.: III. The spread of Moorland—Local, regional and national. *Journal of Ecology, 79*(2), 401–415. https://doi.org/10.2307/2260722

Taylor, N. (2000). *LASER: The inventor, the Nobel laureate, and the thirty-year patent war*. Simon & Schuster.

Thomas, A. D., & Hoon, S. R. (2010). Carbon dioxide fluxes from biologically crusted Kalahari Sands after simulated wetting. *Journal of Arid Environments, 74*, 131–139. https://doi.org/10.1016/j.jaridenv.2009.07.005

Throop, H. L., Archer, S. R., & McClaran, M. P. (2020). Soil organic carbon in drylands: Shrub encroachment and vegetation management effects dwarf those of livestock grazing. *Ecological Applications, 30*(7), 1–13. https://doi.org/10.1002/eap.2150

Tilman, D., Reich, P. B., & Isbell, F. (2012). Biodiversity impacts ecosystem productivity as much as resources, disturbance, or herbivory. *Proceedings of the National Academy of Sciences of the United States of America, 109*(26), 10394–10397. https://doi.org/10.1073/pnas.1208240109

Torbick, N., Salas, W., Chowdhury, D., Ingraham, P., & Trinh, M. (2017). Mapping rice greenhouse gas emissions in the Red River Delta, Vietnam. *Carbon Management, 8*(1), 99–108. https://doi.org/10.1080/17583004.2016.1275816

Träumner, K., Wieser, A., Ruck, B., Frank, C., Röhner, L., & Kottmeier, C. (2012). The suitability of Doppler lidar for characterizing the wind field above forest edges. *Forestry: An International Journal of Forest Research, 85*(3), 399–412. https://doi.org/10.1093/forestry/cps038

Tudge, C. (1991). *Global ecology*. The Natural History Museum.

United Nations Environment Programme. (1994). *United Nations Convention to Combat Desertification*. UNEP.

United States Geological Survey. (2020). *What is Lidar data and where can I download it?* Department of the Interior.

United States Geological Survey. (2023). *What is carbon sequestration?* United States Department of the Interior.

University of Arizona. (2022). *Eurasian flooded grassland*. Retrieved from https://wrangle.org/ecotype/eurasian-flooded-grassland

Van Oijen, M., Barcza, Z., Confalonieri, R., Korhonen, P., Kröel-Dulay, G., Lellei-Kovács, E., Louarn, G., Louault, F., Martin, R., Moulin, T., Movedi, E., Picon-Cochard, C., Rolinski, S., Viovy, N., Wirth, S. B., & Bellocchi, G. (2020). Incorporating biodiversity into biogeochemistry models to improve prediction of ecosystem services in temperate grasslands: Review and roadmap. *Agronomy, 10*(2), 259. https://doi.org/10.3390/agronomy10020259

Venier, L. A., Swystun, T., Mazerolle, M. J., Kreutzweiser, D. P., Wainio-Keizer, K. L., McIlwrick, K. A., Woods, M. E., & Wang, X. (2019). Modelling vegetation understory cover using LiDAR metrics. *PLoS One, 14*(11), e0220096. https://doi.org/10.1371/journal.pone.0220096

Vitt, L. J., & Caldwell, J. P. (2014). *Herpetology. An introductory biology of amphibians and reptiles* (4th ed.). Elsevier.

Wang, Z., & Menenti, M. (2021). Challenges and opportunities in lidar remote sensing. *Frontiers in Remote Sensing, 2*, 641723. https://doi.org/10.3389/frsen.2021.641723

Wang, L., D'Odorico, P., Ringrose, S., Coetzee, S., & Macko, S. A. (2007). Biogeochemistry of Kalahari sands. *Journal of Arid Environments, 71*, 259–279.

Wang, X., VandenBygaart, A. J., & McConkey, B. C. (2014). Land management history of Canadian grasslands and the impact on soil carbon storage. *Rangeland Ecology and Management, 67*(4), 333–343.

Wang, J., Hui, D., Ren, H., Liu, N., Sun, Z., Yang, L., & Lu, H. (2021). Short-term canopy and understory nitrogen addition differ in their effects on seedlings of dominant woody species in a subtropical evergreen broadleaved forest. *Global Ecology and Conservation, 31*, e01855. https://doi.org/10.1016/j.gecco.2021.e01855

Wasser, L. A. (2020). *The basics of LiDAR – Light detection and ranging – Remote sensing*. Retrieved from https://www.neonscience.org/resources/learning-hub/tutorials/lidar-basics

Wasser, L., Day, R., Chasmer, L., & Taylor, A. (2013). Influence of vegetation structure on lidar-derived canopy height and fractional cover in forested riparian buffers during leaf-off and leaf-on conditions. *PLoS One, 8*(1), e54776. https://doi.org/10.1371/journal.pone.0054776

Wehr, A., & Lohr, U. (1999). Airborne laser scanning – An introduction and overview. *ISPRS Journal of Photogrammetry and Remote Sensing, 54*(2–3), 68–82. https://doi.org/10.1016/S0924-2716(99)00011-8

Weigelt, P., König, C., & Kreft, H. (2020). GIFT – A global inventory of floras and traits for macroecology and biogeography. *Journal of Biogeography, 47*(1), 16–43. https://doi.org/10.1111/jbi.13623

Weitkamp, C. (2005). LiDAR: Range-resolved optical remote sensing of the atmosphere. In *Springer series in optical sciences (SSOS, volume 102)*. Springer. https://doi.org/10.1007/b106786

White, R., Murray, S., & Rohweder, M. (2000). *Pilot analysis of global ecosystems: Grassland ecosystem*. World Resources Institute.

Williams, C. A., Hanan, N. P., Neff, J. C., Scholes, R. J., Berry, J. A., Denning, A. S., & Baker, D. F. (2007). Africa and the global carbon cycle. *Carbon Balance and Management, 2*(3). https://doi.org/10.1186/1750-0680-2-3

Wilson, E. O. (2016). *Half-earth: Our planet's fight for life*. Liveright Publishing Corporation, a division of W. W. Norton & Company.

Winker, D. M., Pelon, J., Coakley, J. A., Jr., Ackerman, S. A., Charlson, R. J., Colarco, P. R., Flamant, P., Fu, Q., Hoff, R., Kittaka, C., & Kubar, T. L. (2010). The CALIPSO mission: A global 3D view of aerosols and clouds. *Bulletin of the American Meteorological Society, 91*(9), 1211–1229. https://doi.org/10.1175/2010BAMS3009.1

Wood, D. (2023). *What is Doppler radar? – Definition & uses*. Retrieved from https://study.com/academy/lesson/what-is-doppler-radar-definition-uses.html#:~:text=Doppler%20radar%20bounces%20a%20pulse,the%20speed%20of%20that%20object

Woodbury, E. J., Congleton, R. S., Morse, J. H., & Stitch, M. L. (1961). *Design and operation of an experimental Colidar*. IRE WESCON Convention.

World Resources Institute (WRI). (2000). *World resources 2000–2001: People and ecosystems: The fraying web of life*. World Resources Institute.

World Wildlife Fund. (2006). *Conservation Science > Terrestrial Ecoregions of the World*. Retrieved from http://omap.africanmarineatlas.org/BIOSPHERE/data/note_areas_sp/Ecoregions_Ecosystems/WWF_Ecoregions/WWFecoregions.htm

Zhang, Y., Li, C., Trettin, C. C., Li, H., & Sun, G. (2002). An integrated model of soil, hydrology, and vegetation for carbon dynamics in wetland ecosystems. *Global Biogeochemical Cycles, 16*(4): 9–1–9–17. https://doi.org/10.1029/2001GB001838

Zhang, Y., Meng, Q., Wang, Y., Zhang, X., & Wang, W. (2020). Climate change-induced migration patterns and extinction risks of Theaceae species in China. *Ecology and Evolution, 10*(10), 4352–4361. https://doi.org/10.1002/ece3.6202

Zhang, X., Bao, Y., Wang, D., Xin, X., Ding, L., Xu, D., Hou, L., & Shen, J. (2021). Using UAV LiDAR to extract vegetation parameters of inner Mongolian grassland. *Remote Sensing, 13*(4), 656. https://doi.org/10.3390/rs13040656

Zhao, Z., Peng, C., Yang, Q., Meng, F. R., Song, X., Chen, S., Epule, T. E., Li, P., & Zhu, Q. (2017). Model prediction of biome-specific global soil respiration from 1960 to 2012. *Earth's Future, 5*, 715–729. https://doi.org/10.1002/2016EF000480

Zhou, Y., Singh, J., Butnor, J. R., Coetsee, C., Boucher, P. B., Case, M. F., Hockridge, E. G., Davies, A. B., & Staver, A. C. (2022). Limited increases in savanna carbon stocks over decades of fire suppression. *Nature, 603*(7901), 445–449. https://doi.org/10.1038/s41586-022-04438-1

Chapter 5
Biogeochemistry and Oceanography

Michael O'Neal Campbell

Abstract The atmosphere and oceanic hydrosphere constitute the largest aspects of the Earth's surface, creating the coloration that from space led the Earth to be named the "blue planet." Predominantly air (nitrogen, oxygen, argon, carbon dioxide, etc.) and water (hydrogen, oxygen), both spheres are essential for all plant and animal life and form the links between the biological, biochemical, biophysical, geochemical, and geophysical worlds, and hence are the main components of biogeochemistry. The studied relationship between biogeochemistry, biogeography, and atmospheric and oceanic sciences is still evolving, and debates prevail concerning this complex relationship at the global level. This chapter examines the relations between the biogeochemistry of oceanic ecosystems and marine life and those parts of the atmosphere just above the oceans. The biogeography of the oceans plays an interacting role with oceanic biogeochemistry, as biogeochemistry enables the existence of marine plant and animal life, which also enable chemical cycling. Case studies are cited from the Earth's oceans, and the historical background of oceanography and associated sciences is also cited. It is concluded that biogeochemistry, biogeography, and oceanography comprise a growing scientific complex, and the understanding of the clusters of these sciences is important for the assessment of the expanding science of ocean biogeochemistry.

Introduction

Oceanography is the core discipline behind the development of a field of ocean biogeochemistry, which must also draw on biogeography, geochemistry, the marine sciences (geology, physics, chemistry, biology), and possible research methods such as Big Data and remote sensing. Oceanography, oceanology, or ocean science is clearly a vast discipline in that it studies over 70% of the Earth's surface, and the critical constituent of the Earth's materials, namely water, is increasingly defined by

M. O. Campbell (✉)
Simon Fraser University, Burnaby, BC, Canada
e-mail: evaz@torontomu.ca

M. O. Campbell, *Biogeochemistry and the Environment*,
https://doi.org/10.1007/978-3-031-47017-2_5

its breadth, depth and scientific inclusivity, and technological advancements (Dixon et al., 2021). The National Ocean Service of the United States notes that it is "hard to imagine, but about 97% of the Earth's water can be found in our ocean. Of the tiny percentage that is not in the ocean, about two percent is frozen up in glaciers and ice caps. Less than one percent of all the water on Earth is fresh. A tiny fraction of water exists as water vapor in our atmosphere" (National Oceanic and Atmospheric Administration, 2021a). The United States Geological Survey (2022) states that there are 1,338,000,000 cubic kilometers (321,000,000 cubic miles) of water in oceans, seas, and bays, or 96.54% of the water on Earth (see also Shiklomanov, 1993).

There are several recent definitions for oceanography from research universities. For example, Texas A&M University (2022) provides the definition: "Oceanography is an interdisciplinary science where math, physics, chemistry, biology, and geology intersect… Traditionally, we discuss oceanography in terms of four separate but related branches: physical oceanography, chemical oceanography, biological oceanography, and geological oceanography." Further, physical oceanography is the study of seawater's physical properties (temperature, density, etc.) and movements (waves, currents, tides) and the physical interactions between the ocean waters and the atmosphere. Chemical oceanography studies the chemical composition of seawater and the biogeochemical cycles that affect the ocean and may link with the atmosphere and the lithosphere and the rest of the hydrosphere (e.g., rainfall and rivers). Biological oceanography studies the marine biological organisms in the oceans (from microscopic bacteria, phytoplankton, zooplankton to fish and marine mammals) and some wider environmental issues than marine biology. Geological oceanography studies the geological structure, features, and evolution of the oceans (Texas A&M University, 2022).

The Massachusetts Institute of Technology (MIT) (2022) (with the Woods Hole Oceanographic Institution) offers a similar definition: Physical oceanography examines the physics and geography of the ocean currents and water properties. Chemical oceanography studies the movements of chemical elements and compounds through the oceans, across ranges of time and space scales. Biological oceanography examines oceanic organisms, including their distribution, abundance, and production, and the factors and processes that determine and influence the spread and development of species. Also added is Marine Geology and Geophysics, defined as the study of "the structure of the earth beneath the oceans, the processes that shape the seafloor, the interactions between geological and biological systems, the history of ocean circulation patterns and climate change preserved in seafloor and lake sediments, corals, ice sheets, and other natural archives, and climate variability and impacts." Environmental Science.Org. (2022a) adds that chemical oceanographers "monitor the chemical composition of the ocean water to better understand how they shape the planet. They may study pollution or help find naturally occurring resources on the seafloor," physical oceanographers study "the movements of the oceans, in the waves and currents and tides that move the water itself" and geological oceanographers examine "ocean's floor…." and "may study undersea volcanic activity and its relation to the movement of tectonic plates or the deep oceanic trenches that plunge thousands of feet."

Another incisive definition is that of the University of British Columbia (2022), which holds that "oceanographers study the physical and chemical properties of ocean water, the life within it, the sea floor, and the ocean's connection to the rest of the planet. Oceanography draws from biology, physics, chemistry, and geology, making it a truly multidisciplinary science. Oceanographers work in diverse areas, including biological oceanography; the physics of waves, tides, and currents; the creation, evolution, and structure of sea floors; and chemical oceanography." This is similar to the definition of Carroll University (2022), which holds that oceanography is the "interdisciplinary study of the global oceans – from shallow coastal areas to the deepest ocean trenches, from the tropics to the polar seas." Four subdisciplines are mentioned, these being "physical oceanography (the study of waves, currents, tides, and ocean energy); geological oceanography (the study of the sediments, rocks, and structure of the seafloor and coastal margins); chemical oceanography (the study of the composition and properties of seawater and how it is affected by physical mixing processes and interactions with the seafloor, the atmosphere and ocean life); and biological oceanography (the study of marine organisms and their interactions with the ocean environment."

National Geographic (2022) defines oceanography as "the study of the physical, chemical, and biological features of the ocean, including the ocean's ancient history, its current condition, and its future" and that "oceanography applies chemistry, geology, meteorology, biology, and other branches of science to the study of the ocean. It is especially important today as climate change, pollution, and other factors are threatening the ocean and its marine life." National Geographic (2022) links these definitions with current environmental issues and contends that "in a time when the ocean is threatened by climate change and pollution, coastlines are eroding, and entire species of marine life are at risk of extinction, the role of oceanographers may be more important now than it has ever been." Based on these developments, "one of the most critical branches of oceanography today is known as biological oceanography. It is the study of the ocean's plants and animals and their interactions with the marine environment" (National Geographic, 2022).

The rest of this chapter examines the branches of oceanography, and their links with associated sciences (i.e., biological oceanography and marine biology and marine biogeography; geological oceanography and marine geology; physical oceanography and marine physics; and chemical oceanography and marine chemistry and biochemistry), how the oceanographic sciences relate to the related science (the oceanography branch usually has a broader, environmental focus, while the subdiscipline with the prefix "marine" is more specialized, more linked to the basic science), and links with technology, such as LiDAR. These issues are then related to biogeochemistry.

Links Between Biogeochemistry, Oceanography, and Biogeography

Biological Oceanography

Biological oceanography as a science may be confused with marine biology, but distinctions are discernable based on the focus and breadth of study, as biological oceanography looks at the broader, food web, and physical and chemical relations of oceanic life, while marine biology focuses more on biological, species-oriented studies (Lalli & Parsons, 1993; Miller & Wheeler, 2012). Environmental Science. Org. (2022a) mentions links between biological oceanography and marine biology, as marine biologists are defined as oceanographers who study marine ecosystems and their inhabitants, which includes working with research animals, field experimentation in the ocean, with data collection and tracking studies. Ellis (2007) argues that "marine biology and biological oceanography are two disciplinary subfields that have long struggled with their definitions. Should marine biology simply be considered a part of biology that takes place in the marine environment or is it a distinct entity, with conceptual problems and methodological approaches all its own? Similarly, biological oceanography could be seen as a necessary adjunct to physical and chemical oceanography, or it could be defined more as an extension of biology into the marine realm." Despite the overlaps between the two disciplines of biological oceanography and marine biology (Environmental Science.Org. (2022b), compared with marine biology, biological oceanography will to a greater degree examine wider factors that impact life in the oceans; the "primary concern is *relationships* in the ocean" including for example, the processes that lead to the ecology of life around undersea thermal vents, the impact that subaquatic earthquakes on an oceanic ecology and related physical attributes, and the chemical changes that can impact on the life forms (Environmental Science.Org., 2022b).

The Florida Institute of Technology (2022) also attempts to distinguish between biological topics in oceanography and the coverage of marine biology. Referring to university studies as a guide, the comment is that "oceanography and marine biology may seem like basically the same thing, but these two majors lead to very different career paths and jobs.... While oceanographers study the oceans themselves – the chemistry, physics, and geology of ocean systems and how organisms shape these systems, marine biologists study marine organisms – their characteristics, physiology, and life history." Basically, oceanographers study the role of oceans in the "big picture of Earth's ecosystems and the ways in which oceans impact life on Earth and vice versa" and study the oceans from a variety of perspectives including those of mathematics, chemistry, physics, geology, and meteorology (Florida Institute of Technology, 2022). Oceanographers also study marine life, but from the viewpoint of how marine life (linked to chemicals, temperature, weather, and other environmental factors) interrelates with ocean conditions, contrasting with marine biologists, who "might study algae to catalog it, discover where it can survive, what it eats, and what eats it," while an oceanographer "would study the impact of algae on the ocean water itself, such as when algae release toxic substances into the ocean and cause the water to change color, or cause other forms of life to die" (Florida Institute of Technology, 2022).

Vaillancourt (2016) concurs with these distinctions, namely that marine biology, while broad in focus may focus more on biology and ecological relations with the environment, while biological oceanography may include more perspectives from environmental (chemistry, geology, physics) sciences. Basically, the two disciplines have overlapped, but marine biologists "focus more on the physiological ecology, biology, and taxonomy of the organism or taxonomic group, within the context of its environment" with the emphasis on studies mostly at the organismal level (Vaillancourt (2016). By contrast, the biological oceanographer looks more at the organisms at the bottom of the food chain, "the smaller plankton, whose biomass, productivity and behavior are determined more by advection, turbulence, and mixing of the water column and so is more intimately linked to the physical and chemical environment" and possibly the biological oceanographer may be seen as "treating the ocean as an organism itself… with energy, mass and gas flows mediated by the activity of the organisms but also the physical processes of the ocean/atmosphere environment" (Vaillancourt, 2016).

There are also overlaps in the coverage of the biological oceanography and marine biology journals. For example, an old study by Wood et al. (1967), published online by the Journal of the Marine Biological Association of the United Kingdom as the most cited, was based on a mostly chemical oceanography topic: "An accurate, dependable determination of 0–60 µg-at./l. of $NO-_3$-N in seawater has been developed. The sample is treated with tetrasodium ethylenediaminetetraacetate solution and passed through a column of copperized cadmium filings. A nearly quantitative reduction of nitrate to nitrite results. Nitrite is then determined by a diazotization method. Neither sulphide nor high nitrite concentrations interferes" (Wood et al. (1967: 23).

Another study by Sun et al. (2022: 1578) in the *Journal of Oceanography and Limnology* could be classified as under the umbrella of marine biology: titled "Diverse habitat preferences of two sea cucumber species and the seasonal change in a coral reef area." It examines "two common tropical sea cucumbers (*Holothuria edulis* and *Stichopus chloronotus*) to explore the reasons that affect their habitat selection and seasonal changes, so as to provide support for the spatial planning of sea cucumbers reserve in coral reef area" (Sun et al., 2022: 1578). The findings of the research indicated that the shift from dry to rainy season is important for habitat preferences with pronounced variations in habitat preference between the seasons for both species.

Both marine biology and biological oceanography may focus on the study of plankton, as this species group is basic to many marine ecosystems (Catlett et al., 2022; Journal of Experimental Marine Biology and Ecology, 2022; Journal of the Marine Biological Association of the United Kingdom, 2022; Marine Ecology Progress Series (MEPS), 2022; Marine Biology, 2022; Takahashi et al., 2022). For example, Hostetler et al. (2018: 127) remark that marine ecosystems are complex entities including huge numbers of species over a vast range of spatial and temporal scales. Phytoplankton constitutes the base of most marine ecosystems, with an annual net photosynthetic carbon equivalent to all terrestrial vegetation (Field et al., 1998; Behrenfeld et al., 2001). The basic production system of oceanic plankton is linked to the CO_2 and other exchanges between the atmosphere and ocean, deep ocean carbon sequestration, and ocean food webs, including larger organisms (see also Falkowski et al., 1992; Antoine et al., 1996; Behrenfeld & Falkowski, 1997; Falkowski et al., 1998; Meskhidze & Nenes, 2006; DeVries et al., 2012; Gantt & Meskhidze, 2013).

Tables 5.1, 5.2, 5.3 and 5.4 and Figs. 5.1 and 5.2 illustrate some differences between biological oceanography and marine biology, referring to relevant journals. From the many reputable journals covering marine biology and biological oceanography, four journals each are selected to illustrate some of the different topics (Journal of Oceanography, Annual Review of Marine Science, Journal of Oceanology and Limnology, Limnology and Oceanography for biological oceanography; Marine Biology, Journal of Experimental Marine Biology and Ecology, Journal of the Marine Biological Association of the United Kingdom. Marine Ecology Progress Series for marine biology). The aims and scopes of the journals indicate a broader, more environmental, focus for the oceanographic journals and a more organism/ecology focus for the marine biology journals.

The main relationship between biological oceanography and biogeochemistry concerns the biogeochemical cycling that affects the ecosystems and related environmental components and patterns that are the focus of biological oceanography, and any possible role that the latter have in affecting biogeochemical cycling. These relations are also relevant to marine biogeography (Fig. 5.2), which is the spatial element of marine biology. The environmental components of biological oceanography may affect species distributions (the biogeography of marine life and ecosystems) (National Oceanic and Atmospheric Administration, 2021b). These relations are depicted in the tables below. In this regard, marine biogeography may be defined

Table 5.1 Biological oceanography journals' aims and scopes

Journal title	Aims and scopes
Journal of Oceanography	Promote "understandings of ocean systems from various aspects including physical, chemical, biological, geological oceanography as well as paleoceanography, etc." The favored region is the western North Pacific and Asian coastal waters, "but the study region is not limited to the Asian Pacific"
Annual Review of Marine Science	Based on "diverse topics within the major disciplines of coastal and blue water oceanography (biological, chemical, geological and physical) as well as subjects in ecology, conservation, and technological developments with the marine environment as the unifying theme"
Journal of Oceanology and Limnology	Considers "aspects of oceanology (oceanography) and limnology from all over the world, including those in academic coverage: biology (bioinformatics, bioengineering), physics (mathematical and computational physics), geology (sedimentology, geophysics, paleontology, geochemistry), chemistry (geochemistry, biochemistry), hydrology, meteorology, and geography; and those in practical coverage: aquaculture, marine resource exploration, remote-sensing, environmental protection, marine engineering, pharmacology, and instrumentation"
Limnology and Oceanography	All aspects of the sciences of limnology and oceanography… understanding of aquatic systems… "the fields of limnology and oceanography, whether physical, chemical, or biological in nature, empirical or theoretical in method, and from elemental to geological, ecological to evolutionary, species to ecosystem, or system to global in scale." Topics mostly within the "scientific philosophy of novel, hypothesis-driven or observational research that contributes significantly to a broader understanding of aquatic sciences"

Source: Journal of Oceanography (2022); Annual Review of Marine Science (2022); Journal of Oceanology and Limnology (2022); Limnology and Oceanography (2022)

Table 5.2 Marine biology journals' aims and scopes

Journal title	Aims and scopes
Marine Biology	All fields of marine biology. Special emphasis on "the understanding of life in the sea, organism-environment interactions, interactions between organisms, and the functioning of the marine biosphere." Particular interest on "marine biogeochemistry, ecology of marine populations and communities, marine ecophysiology, behavioural biology of marine organisms, development and life cycles of marine organisms, biochemistry and physiology of marine organisms, evolution of marine organisms, population genetics of marine organisms, marine conservation, technology developments in marine biology, global change ecology of marine organisms, marine invasion ecology, marine biodiversity"
Journal of Experimental Marine Biology and Ecology. From Jan 1, 2022, Marine Biodiversity Records added	Experimental ecological research on marine organisms and their environmental relations. Especially "biochemistry, physiology, behavior, genetics, and ecological theory." Includes "hypothesis driven experimental work, both from the laboratory and the field" and "natural experiments or descriptive studies that elucidate fundamental ecological processes" and topics "should have a broad ecological framework beyond the specific study organism or geographic region." With the merger of Marine Biodiversity Records, additional topics include "geographical ranges of marine species, including the effects of the introduction of novel or alien species to marine ecosystems, and of taxonomic studies relevant to these changes"
Journal of the Marine Biological Association of the United Kingdom	All aspects of marine biology, currently especially "marine ecology, behaviour and fisheries; Biodiversity and population studies of marine ecosystems, especially potential impacts of global warming, ocean acidification and climate change on ecosystem resilience; Physiology, biochemistry, genetics and molecular biology of marine organisms, including microbiology, particularly related to the integrative ecophysiology of marine organisms; Taxonomic syntheses, including molecular phylogenies; Morphology, life history and developmental biology of marine organisms, especially model systems; Chemical and physical oceanography, as directly relevant to marine biology; Marine environmental health, and in particular its relationship to human health and well-being"
Marine Ecology Progress Series	All aspects of "marine, coastal, and estuarine ecology. MEPS coverage includes the whole spectrum of species, habitats, biological organisation (cells to ecosystems) and research (fundamental and applied)" with priority "given to outstanding research that advances our ecological understanding." Topics covered include "microbiology, botany, zoology, ecosystem research, biological oceanography, ecological aspects of fisheries and aquaculture, pollution, environmental protection, conservation, resource management." Examples cited include environmental factors (including "tolerances and responses of marine organisms (microorganisms, autotrophic plants and animals) to variations in abiotic and biotic components of their environment; radioecology"), physiological-ecological mechanisms, Molecular marine ecology, cultivation/aquaculture, dynamics (this refers to "energy, matter flows, in organisms, populations and ecosystems; biodiversity; trophic interrelations; production, transformation, decomposition of organic matter; plankton, benthos, estuarine, coastal ecology; Wadden sea, coral reef, deep-sea, open-ocean, polar, theoretical ecology; ecological methodology, technology; modeling and computer simulation), ocean use management, and Ecoethics ("relevant to human thought and conduct")"

Sources: Journal of Experimental Marine Biology and Ecology (2022); Journal of the Marine Biological Association of the United Kingdom (2022); Marine Ecology Progress Series (MEPS), 2022; Marine Biology (2022)

Table 5.3 Biological oceanography research examples

	Title	Topic and description
1	Occurrence patterns of larval mesopelagic fishes in the mouth of highly eutrophic Tokyo Bay, central Japan	Broad ecological study, examining the "occurrence patterns of larval mesopelagic fish species in the surface layers of the highly eutrophic Tokyo Bay." Focuses on mesopelagic fishes, with sample of 2276 individuals of 78 species
2	Phytoplankton and ice-algal communities in the seasonal ice zone during January (Southern Ocean, Indian sector)	Broad ecological study, examining relations between phytoplankton species (especially *Fragilariopsis*, F. Hustedt, 1913, *Pseudo-nitzschia* H. Perag), composition and sea-ice algae in seawater after melting from sea ice
3	Evolution, Microbes, and Changing Ocean Conditions	Broad ecological study: Evolutionary tools for four areas of ocean change biology: plastic and evolutionary trait changes, environmental variability and trait values, multiple environmental drivers in trait change, and population tolerance limits
4	New Microbial Biodiversity in Marine Sediments	Broad ecological study of microbes in marine sediments. Shows how advancements in the computational analysis of large genomic data sets have enabled the reconstruction of individual genomes from complex microbial communities
5	The Nutritional Ecology of Marine Apex Predators	Broad ecological study of marine apex predators in environmental contexts. Multidimensional approach, the nutritional geometry framework, nutritional geometry and marine prey, ecological niche concept, predator research
6	Diverse transformations of sulfur in seabird-affected sediments revealed by microbial and stable isotope analyses	Broad ecological study, microbial ecology of sulfides and sulfates, in sediments, in a penguin-affected lake and a penguin free lake in Antarctica Peninsula. Penguin organic inputs affected sulfur transformations, microbe ecology
7	Effects of climate change on the potential habitat distribution of swimming crab *Portunus trituberculatus* under the species distribution model	Broad ecological study, using predictive species distribution model (SDM), on climate change impact on distribution of the swimming crab *Portunus trituberculatus*. Predicted habitat loss, China Sea, Yellow Sea, not Bohai Sea
8	Integrating phytoplankton pigment and DNA meta-barcoding observations to determine phytoplankton composition in the coastal ocean	Broad ecological study, comparing DNA meta-barcoding and high-performance liquid chromatography pigment analysis for phytoplankton composition. Both identify eukaryotic phytoplankton taxa (diatoms, dinoflagellates, chlorophytes, prymnesiophytes) differ on biomarker pigmentation

1. Nagaiwa et al. (2022), Journal of Oceanography; 2. Takahashi et al. (2022) Journal of Oceanography; 3. Collins et al. (2020), Annual Review of Marine Science; 4. Baker et al. (2021), Annual Review of Marine Science; 5. Machovsky-Capuska and Raubenheimer (2020), Annual Review of Marine Science; 6. Shen et al. (2022), Journal of Oceanology and Limnology; 7. Liu et al. (2022), Journal of Oceanology and Limnology; 8. Catlett et al. (2022), Limnology and Oceanography

Table 5.4 Marine biology research examples

	Title	Topic and description
1	Low population genetic structure is consistent with high habitat connectivity in a commercially important fish species (*Lutjanus jocu*)	Single species ecological and genetic study of the dog snapper (*Lutjanus jocu*, Bloch & J. G. Schneider, 1801). Hypothesis that habitat access affects genetic divergence among populations and genetic diversity within populations. Findings indicated habitat change did not change genetic divergence between populations
2	Intraspecific trophic variation in brown booby (*Sula leucogaster*) from the Southwestern Atlantic	Single species ecological study on the brown booby (*Sula leucogaster*), hypothesis that sexual size dimorphism reduces intraspecific food competition. The pelagic (sea surface feeding) dogtooth herring (*Chirocentrodon bleekerianus*; Poey, 1867) was mostly taken by females, the demersal (seafloor feeding) Stripped weakfish (*Cynoscion guatucupa*, Cuvier, 1830) was taken by male boobies, a sex trophic segregation
3	Similar metabolic responses of co-occurring postsettlement mussels to temperature change despite distinct geographical distributions	Single species ecological study of the Mediterranean mussel (*Mytilus galloprovincialis*, Lamarck, 1819), compared with the Perna perna mussel (*Perna perna*, L.1758), the hypothesis being "physiological constraints on early stages of development could explain adult distributions and life histories." For adults, there were interspecies differences in metabolic regulation capacities in response to temperature, but none in young specimens
4	Differential hypoxia tolerance of eastern oysters from the northern Gulf of Mexico at elevated temperature	Single species ecological study of eastern oysters (*Crassostrea virginica*; Gmelin, 1791), and responses to hypoxia in Gulf of Mexico estuaries. Findings indicated differences in hypoxia tolerance between populations, and some oysters were better able to protect tissues against acidosis and oxidative damage during hypoxia and high temperature stress, with variations in response to hypoxia and high temperature stress
5	Chemical fingerprints suggest direct familiarization rather than phenotype matching during olfactory recognition in Australian sea lions (*Neophoca cinerea*)	Single species ecological study of Australian sea lions (*Neophoca cinerea*, Péron, 1816), using gas chromatography–mass spectrometry (GC–MS), for chemical fingerprints in mother-pup pairs. No GC-detectible mother-offspring similarity in chemical fingerprints detected, rather direct familiarization between mothers and young
6	Contrasting decadal recruitment patterns in the sea urchin *Strongylocentrotus droebachiensis* in the Gulf of Maine	Single species ecological study of green sea urchin (*Strongylocentrotus droebachiensis*, Müller, 1776), with declining populations from pressures of overfishing, anthropogenic climate change, predation by Jonah crabs (*Cancer borealis*, Stimpson, 1859)

(continued)

Table 5.4 (continued)

	Title	Topic and description
7	Differences in the morphological body condition index of sea turtles between species and size classes	Single species ecological study of sea turtles: green turtles (*Chelonia mydas*), loggerhead turtles (*Caretta caretta*) and hawksbill turtles (*Eretmochelys imbricata*). Fulton's condition index (bodyweight divided by the cube of straight carapace length (SCL)) indicates good body condition, problematic in not considering the mass-length relationship. With interspecies differences, high and low index values were found
8	Varying reef shark abundance trends inside a marine reserve: evidence of a Caribbean reef shark decline	Single species ecological study of Caribbean reef shark (*Carcharhinus perezi*, Poey, 1876) and nurse shark (*Ginglymostoma cirratum*, Bonnaterre, 1788) inside and outside marine-protected areas (MPAs) in Belize, Findings indicated a reef shark decline in MPAs, but stable nurse shark populations. Reef shark decline perhaps due to fishing and nurse shark stability to behavior and nationwide protection

1. Tovar Verba et al. (2023), Marine Biology; 2. Mancini et al. (2023), Marine Biology; 3. Nel et al. (2023), Marine Biology; 4. Coxe et al. (2023), Journal of Experimental Marine Biology and Ecology; 5. Wierucka et al. (2019), Journal of Experimental Marine Biology and Ecology; 6. Harris and Gibson (2023), Journal of Experimental Marine Biology and Ecology; 7. Nishizawa and Joseph (2022), Journal of the Marine Biological Association of the United Kingdom; 8. Flowers et al. (2022), Marine Ecology Progress Series

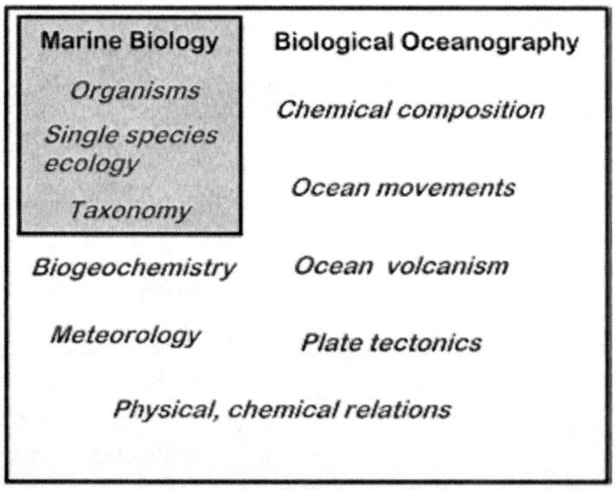

Fig. 5.1 Diagrammatic relations between biological oceanography and marine biology

as "the study of marine species, the geographic distribution of their habitats, and the relationships between living organisms and the environment" (National Oceanic and Atmospheric Administration, 2021b). Rivadeneira (2017) adds that "marine biogeography is a subfield of biogeography aimed at understanding the patterns and processes governing the distribution of marine taxa at geographic scales."

Marine biogeography is related to several disciplines and subdisciplines, including marine biology and ecology, physical and biological oceanography, ecophysiology, genetics, geography, geology, paleontology, and macroecology. Several subdisciplines have been proposed from the intersection with these branches,

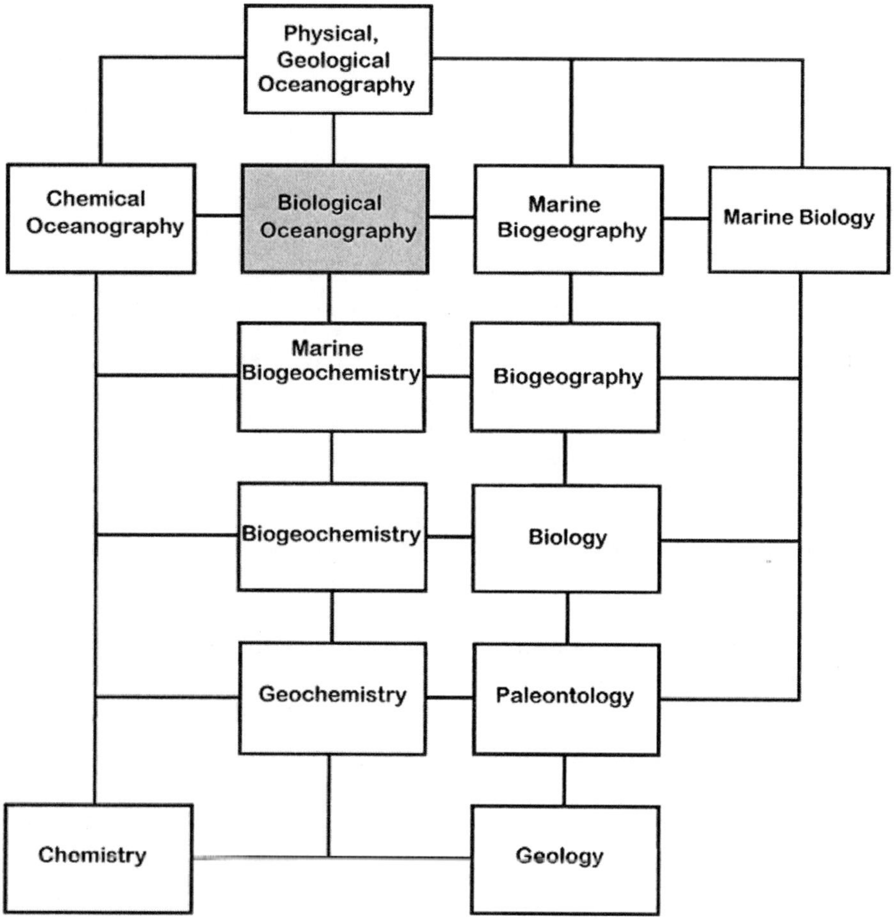

Fig. 5.2 Links between biological oceanography and related disciplines

alluding to the subject of study (e.g., phytogeography and zoogeography), the temporal scale of driving processes (e.g., ecological biogeography and historical biogeography), the use of phylogenetic and phylogeographic tools (e.g., comparative phylogeography), paleontological data (paleobiogeography), or the combined use of multiple approaches (e.g., integrative biogeography).

Older texts covering the development of marine biogeography include Ekman (1953), Hedgpeth (1957), and Briggs (1974). Examining these texts, Rivadeneira (2017) notes that Ekman's work was the "first biogeographic classification of the global ocean," Hedpeth's book was "the first major synthesis of marine biogeography, summarizing almost a century of scattered studies, including major biogeographic units recognized to that date, a lucid discussion of the role of sea temperature on species distribution, and the causes of the bipolarity phenomenon," and "Briggs provided the first detailed global biogeographic classification scheme, including more than fifty-three provinces worldwide."

More recent texts covering the topics of marine biogeography include the work of Longhurst (2010). Arctic Biodiversity Data Service (2016) notes that "at the first level of reduction, Longhurst recognized four principal biomes (also referred to as domains in earlier publications): the Polar Biome, the Westerlies Biome, the Trade-Winds Biome, and the Coastal Boundary Zone Biome." The next level of reduction classifies the ocean basin into provinces, averaging about ten for each basin and "these partitions provide a template for data analysis or for making parameter assignments on a global scale" (Arctic Biodiversity Data Service, 2016).

In the "biogeographic partition of the ocean," Longhurst (2007a) notes that "oceanic biogeography has inherent difficulties compared with terrestrial biogeography, such as high cost of collecting samples at sea, the high levels of expatriation of plankton species, the relative lack of isolation among natural regions, and the problem of observing three-dimensional distributions that vary in both space and time." Longhurst (2007b) also defines biomes as "the primary partition of the ocean" and mentions "four primary biomes of the upper ocean. They are (1) polar biome, (2) westerlies biome, (3) trades biome, (4) and coastal biome." Differences between these biomes are complex: the polar biome has "low taxonomic diversity at all trophic levels," while for the westerlies biome, "herbivore ecology is complex… because both the pattern of seasonal ontogenetic migration, typical of the polar seas, and the diel vertical migration pattern, typical of trade-wind seas, occur here"; for the trades biome, "the pelagic ecosystem is at its most taxonomically diverse and represents the climax community of the pelagos," and the "plankton, and especially the zooplankton, of the coastal biome at all latitudes differs consistently from the plankton of the open ocean because it comprises a higher proportion of meroplankton – the planktonic larvae of benthic and littoral invertebrates" (Longhurst, 2007b). Provinces are the next level of oceanic partition (Longhurst, 2007c). The creation of provinces is argued to be problematic, the author arguing that creating a partition of the ocean's surface is a fractal problem, as the physical phenomena occur across a great range of dimension and possibly "calculations of primary production should be partitioned between biogeochemical provinces (BGCPs) within which photosynthetic parameters, and the form of the chlorophyll profile, might be seasonably predictable" (Longhurst, 2007c).

Marine biogeochemistry is another field interstitial to biological oceanography and biogeochemistry (Fig. 5.2). The University of Southampton (2022) describes this discipline as the "biogeochemistry of marine systems, both in the open ocean and in shelf seas" and the study of "the ocean's role in global carbon, nutrient and other chemical cycles." Mote Marine Laboratory and Aquarium (2022) defines marine biogeochemistry as "the study of the physical, biological, geological, and chemical processes that distribute and transform nutrients and other chemical species in the marine environment. Observations of physical/biological/ecological coupling give insight into how these processes contribute to global biogeochemical cycles." Mlakar (2022) describes marine biogeochemistry as being focused on studying biogeochemical processes in marine systems, and the interactions with the atmosphere, coastal and terrestrial systems, and research ranging across the biogeochemical cycling of micronutrients, such as carbon, nitrogen, phosphorus, sulfur, and iron, and organic matter. These are important for the understanding of climate change, based on their role in the carbon cycle interactions with the climate systems. Aerosols are also significant factors for climate, pollution, biogeochemical cycles, and dust deposition levels in the ocean, and such aeolian sources of elements may be important for the carbon and nutrient cycling in the oceans. In the study of such dynamics, the research should "explore all processes between the atmosphere, oceans, lakes, rivers, and groundwater using an interdisciplinary approach following physical, chemical, biological, and geological variables" (Mlakar, 2022).

Stable isotope analysis has emerged as an important tool for marine animal research, including marine biochemical and biogeographic studies (MacKenzie et al., 2011). For example, Newsome et al. (2010: 509) contend that stable isotope analysis (SIA) is an established, important tool to quantify the flow of energy within and among ecosystems, and for the estimation of habitat use and movement patterns qualitatively, and the exploration of physiological processes from the molecular to the organismal level. The stable isotope composition is mostly assessed by the isotopic composition of the materials entering its body, to create tissues and biological minerals, the processes termed called isotopic fractionation. This technique "has proven especially useful in the study of animal diet, habitat use, movement, and physiology" (Newsome et al. (2010: 509). Newsome et al. (2010) note SIA is useful for recording diet and trophic level for marine mammals, as these species live in vast, spatial habitats, with great mobility during foraging ecology, habitat use, and/or migratory patterns, and most species undergo large migrations/movements (Newsome et al. (2010: 509).

Studies using SIA include those focused on contaminant concentrations (e.g., organochlorides and heavy metals in marine ecosystems), physiological issues including isotopic turnover or impacts of diet, physiological or reproductive conditions on isotopic fractionation, and the examination of marine mammal ecology on varying times scales (historic, archaeological, paleoecological) (Newsome et al., 2010). Marine studies using this technology include the areas of physiology and fractionation, foraging ecology and habitat use, ecotoxicology, and historic ecology and paleoecology (Newsome et al. (2010: 509). Newsome et al. (2010: 509) caution that the inability to conduct controlled laboratory experiments on many marine mammal species may compromise the research on these species, hence future studies may use isotopic data collected from other organisms and be aware of the main assumptions often made in the application of SIA to the study of animal ecology. Stable isotope values of carbon, nitrogen, hydrogen, and oxygen are however now commonly used for the study of foraging ecology and trophic status, habitat use, migration, population connectivity, and physiology of numerous species. Isotopes of other elements, including sulfur, lead, and strontium, have been used for ecological information, but less often (Newsome et al. (2010: 510).

Smith et al. (2021) also present a study of stable isotope analysis as "a powerful indirect tool that can be used to infer foraging behavior and habitat use retrospectively from archived specimens" and take the case study of beaked whales, "that are challenging to study in situ, and although Sowerby's beaked whale (*Mesoplodon bidens*) was discovered >200 years ago, little is known about its biology." The study measured δ13C and δ15N stable isotope composition in bone, muscle, and skin tissue from whale specimens, collected from the North Atlantic Ocean, to assess the whales' movement ecology and spatial population structure. The findings indicated that "median δ13C and δ15N values in Sowerby's beaked whale bone, muscle, and skin tissues significantly differed between whales sampled from the east and west North Atlantic Ocean" (Smith et al. (2021). Smith et al. (2021) state that these findings indicate that Sowerby's beaked whale has both short- and long-term site fidelity to the area where the specimens were obtained, this evidence suggesting that the whales' presence is composed of two or more populations or has a metapopulation structure, which may have implications for conservation and management policy. This also indicates that stable isotope analysis is a versatile tool for generating spatial data and may be applicable to other elusive species. The authors conclude that by "using a combination of tissues synthesized at different times or at different rates, researchers can infer spatial origin and diet across time" and "stable isotope analysis… has been used to infer animal migrations, the spatial origin of wildlife products, and even historical trophic structures. Two of the most commonly used isotopes for wildlife studies are carbon (expressed as δ13C values), used to evaluate habitat range and latitudinal shifts, and nitrogen (expressed as δ15N values), used for obtaining foraging and trophic information)" (Smith et al., 2021; see also Phillips & Eldridge, 2006; Ben-David & Flaherty, 2012; Hopkins & Ferguson, 2012; Vander Zanden et al., 2015).

Chemical Oceanography or Marine Chemistry

Definitions of chemical oceanography center on the chemical composition of the ocean, and associated relations; "chemical oceanography has to do with the composition of seawater and the biogeochemical cycles that affect it" (Britannica, 2019). The Department of Marine Sciences, University of Georgia (2022) documents that "chemical oceanography encompasses the study of the chemical components of the oceans, their reactions, and their pathways of transformation" considering "both organic and inorganic compounds, particulate and dissolved material, and the ocean sediments." Cochran (2014) comments about topics of chemical oceanography, in that they concern "distributions of the chemical elements in the oceans, how they are added (their sources) and removed (their sinks) and the processes that affect their concentrations" (Cochran, 2014). Some articles place marine chemistry within chemical oceanography, among other subdisciplines. For example, Hoyle and Lerner (2023) state that "chemical oceanography can be further divided into focused areas of study. For example, marine chemistry is concerned with the composition of seawater. Marine geochemistry is additionally concerned with the chemistry of the precipitated rocks and sediment found on the ocean floor. Additionally, marine biogeochemistry is concerned with the role of organisms (particularly microorganisms) in the alteration or formation of geological features in the oceans" (see also Donat & Bruland, 1995; Libes, 2009).

Boyle (2007: 305) notes that "chemical oceanography plays a central role in understanding the distribution and reactivity of chemical components within the ocean and at the earth–ocean, sediment–ocean, and atmosphere–ocean interfaces." Additionally, "the origin of the study of the chemistry of the oceans, frequently termed chemical oceanography or marine chemistry" began during the nineteenth century, which found the "constancy in the ratio of the major ions (sodium, calcium, magnesium, chloride, and sulfate) to each other throughout the oceanic water column" (Boyle, 2007: 305). It is also argued that since the development of this science, analytical chemistry has been fundamental to chemical oceanography, as a method to study trace elements at micromolar, nanomolar, and picomolar concentrations, which is needed to understand the distribution and reactivity of oceanic chemical components, and to study "chemical transformations via reaction mechanisms at the molecular level, chemical oceanography requires studying physical, geological, atmospheric, and biological processes to demonstrate how the elements and their compounds, both natural and anthropogenic, vary over a range of spatial and temporal scales" (Boyle, 2007: 305).

Doney (2010: 1512) argues that such studies are vital due to the increasingly dynamic ocean composition, changes, and ecosystems, as currently there is evidence of climate change, increasing levels of atmospheric carbon dioxide and nutrient inputs, and many types of pollution that may be changing the chemistry of the global ocean, sometimes at far greater rates than the changes documented during the historical and recent geological record. Currently, observed trends include a shifts in the acid-base chemistry of the oceans, reduced subsurface oxygen in coastal waters to the deep ocean, increasing coastal nitrogen levels and in mercury and persistent organic pollutants, "with most changes related to human uses of fossil fuels for combustion, fertilizer applications, and industrial development" (Doney, 2010: 1512).

Scott (2019) writes that research on the links between biogeochemistry and oceanography largely focuses on the chemistry and circulation of carbon, as this element is the basis of all life on Earth, and a factor for greenhouse gases and climate change (see also Thurman & Burton, 2001; Libes, 2009). Scott (2019) cites three American government projects which exemplify the recognition of the importance of carbon. These are as follows: (1) the United States Global Change Research Program (USGCRP), which aims to design a carbon cycle research program, with the involvement of the Department of Energy, the National Aeronautic and Space Administration, the National Oceanic and Atmospheric Administration, the National Science Foundation, and the U.S. Geological Survey; (2) the Global Ocean Ecosystems Dynamics (GLOBEC), which is supported by the National Science Foundation, aiming at understanding interactions and feedbacks and interactions between global change, marine ecosystem and physical climate systems; and (3) the Global Carbon Program (GCP), supported by the National Oceanic and Atmospheric Administration, aiming to increase prediction of human induced carbon dioxide additions to the atmosphere (Scott, 2019). Three relevant topics are also mentioned for the biogeochemistry of the oceans: (1) the use of carbon in the process of growth of diatoms and the consumption of diatoms by zooplankton, where "the transfer of this carbon toward the ocean floor and its partial burial in the sediments is often referred to as the carbon pump; it is one of the processes that slow the accumulation of CO_2 in the atmosphere"; (2) the silicon (mostly entering the oceans) from rivers used in the diatom shells, which remove silicon from the surface ocean water; and (3) the declining distribution of O_2 (oxygen) with depth, due to its sourcing from air at the ocean surface and consumption by bacteria in the ocean depths, with cold, dense oxygen rich water from the polar regions also flowing on the ocean floor (Scott, 2019).

Global carbon (including carbon dioxide) cycling is an important aspect of chemical oceanographic studies. For example, Boyle (2007: 305) refers to recent studies of chemical oceanography and carbon cycling. These include the work of Millero (2007), focusing on the Marine Inorganic Carbon Cycle. Boyle (2007: 305) argues that the chemistry of the surface ocean is largely dominated by the uptake of CO_2 into phytoplankton and the ocean CO_2 cycle has been unsettled by the burning of fossil fuels, as also documented by Millero (2007), which gives an analytical study of this phenomenon. The study by Morse et al. (2007) examined thermodynamics and kinetics of calcium carbonate ($CaCO_3$) formation and dissolution, and like the study by Millero (2007) studied oceanic change, especially increased acidification. Mopper et al. (2007) studied dissolved organic matter, which is described as comprising up to 90% of the organic carbon in the ocean and may be surveyed by mass spectrometry and NMR spectrometry for chemical structural information. Other studies (e.g., McNichol & Aluwihare, 2007) examine the use of radiocarbon methods to track oceanic organic matter, for which acceleratory mass spectrometry may be used to measure ^{14}C in water, the links between oceanic CO_2 and organic matter changes, the transfer of organic carbon between dissolved organic matter and particulate organic matter, and the transfer of particulate organic matter to sediments (McNichol & Aluwihare, 2007). Boyle (2007: 305) also adds that "although POM decomposition to bicarbonate/CO_2 occurs in sediments," a study by Burdige (2007) "reviews organic matter preservation in marine sediments, which helps sequester CO_2 from the atmosphere. The importance of oxygen concentrations and redox oscillations in sediments has a profound effect on OM preservation."

Neuer et al. (2017) report that from the preindustrial era, the oceans have absorbed at least a quarter of the carbon dioxide (CO_2) that people have emitted into the atmosphere, which has resulted in increased ocean acidity. Levels of oxygen dissolved in ocean water have also changed, due to warming and changing circulation patterns. Complications occur in coastal regions, as marine ecosystems are affected by human-induced events such as pollution, relevant land use changes, and unsustainable overfishing (see also Sabine et al., 2004; Bates et al., 2014; Tanhua et al., 2015). Sabine et al. (2004: 367) estimated a global oceanic anthropogenic carbon dioxide (CO_2) sink for nearly 200 years (from 1800 to 1994) of 118 ± 19 petagrams of carbon (note a petagram is a unit of mass equal to 1,000,000,000,000,000 (10^{15}) grams, or 1,000,000,000,000 kg). As the oceanic sink accounts for nearly 50% of the total fossil-fuel and cement-manufacturing emissions, the terrestrial biosphere was a net source of CO_2 to the atmosphere of about 39 ± 28 petagrams of carbon for the period documented, and "the current fraction of total anthropogenic CO_2 emissions stored in the ocean appears to be about one-third of the long-term potential" (Sabine et al., 2004: 367)

Tanhua et al. (2015) notes atmospheric carbon dioxide (CO_2) concentrations have increased by 42% since the onset of the industrial revolution from human created emissions from and in 2010, the oceans had absorbed an estimated 155 ± 30 petagrams of anthropogenic CO_2 the equivalent of 28% of the total CO_2 emissions during the same time. This ocean absorption limited the increases of CO_2 in the atmosphere, and reduced climate change, but created consequences for ocean chemistry and biology (see also Khatiwala et al., 2013). Tanhua et al. (2015) points out that increased enhanced ocean CO_2 uptake changed the marine carbonate system, affecting seawater acidity. When CO_2 dissolves in seawater, it forms carbonic acid (H_2CO_3), which is a weak acid that dissociates into bicarbonate (HCO_3^-) and hydrogen ions (H^+). The increased H^+ means increased acidity (lower pH). Ocean acidification rates are slowed by CO_3^{2-}, which binds up much of the H^+, and forming bicarbonate. This buffering reaction consumes CO_3^{2-}, which reduces the chemical capacity of the surface waters to take more CO_2. Hence, "currently, that capacity is only 70% of what it was at the beginning of the industrial era, it may well be reduced to only 20% by the end of the century" (Tanhua et al., 2015).

Godbold and Calosi (2013) concur with the evidence of increased atmospheric CO_2 concentrations, "from a pre-industrial level of approximately 280 ppm to approximately 385 ppm, with further increases (700–1000 ppm) anticipated by the end of the twenty-first century." This has been manifested by temperature changes over the past three decades, as changes in CO_2 have increased global average temperatures by about 0.2 °C per decade, and as much of the additional energy was absorbed by the oceans, these water bodies experienced a 0.8 °C rise over the past century. The uptake of heat energy and CO_2 by the global oceans resulted in changes in ocean carbonate chemistry, which included reductions in pH and carbonate saturation states, and increases in dissolved CO_2 and bicarbonate ions, which is termed "ocean acidification" (see also Hansen et al., 2006; Solomon et al., 2007; Doney, 2010).

Ocean acidification affects marine organisms and their relationship ecologies (Godbold & Calosi, 2013). Tanhua et al. (2015) notes the exponential growth of studies of increased CO_2 on marine organisms, and "in 2013, Thompson-Reuters labelled ocean acidification the number one research front in Ecology and Environmental Sciences." Godbold and Calosi (2013) point to key questions concerning marine biology and ocean acidification: "What are the effects of ocean acidification and climate change on species interactions?" "Are communities and their ability to maintain ecosystem functioning resilient to ongoing global change?" "Can organisms' capacity for phenotypic buffering and adaptation offset the consequences of environmental change?" Impacts and interactions with species interactions and community dynamics are important areas of enquiry (Godbold & Calosi, 2013). Godbold and Calosi (2013) make the point that "marine microorganisms that drive many global ocean processes (e.g., oxygen production, primary productivity and biogeochemical cycling) are able to adapt to ocean acidification" but relevant information of whether these organisms are also able to adapt to the "combined changes in pCO_2 and temperature is still lacking."

One impact concerns the development of organism shells through calcification. Some organisms, such as mollusks, corals, coralline algae, and some plankton, may be affected, as the development of shell or skeletal material through calcification required CO_3^{2-} (Tanhua et al., 2015). Lohbeck et al. (2012: 346) maintain that marine calcifying organisms may be negatively affected by ocean acidification, but "our present understanding of the sensitivity of marine life to ocean acidification is based primarily on short-term experiments, in which organisms are exposed to increased concentrations of CO_2" but there is the possibility that "phytoplankton species with short generation times, in particular, may be able to respond to environmental alterations through adaptive evolution" (Lohbeck et al., 2012: 346). The study is based on the "world's single most important calcifying organism, the coccolithophore" (*Emiliania huxleyi* (Lohm.) Hay and Mohler) and explores the ability of this species to evolve in response to ocean acidification, using two 500-generation selection experiments which involved exposure to increased CO_2 levels (Lohbeck et al., 2012: 346). The findings indicated that the populations kept at increased partial CO_2 pressure exhibited higher growth rates than those at ambient (local) CO_2 partial pressure, those selected at increased partial pressure exhibited higher growth rates, when both were tested under ocean acidification conditions. The authors conclude that "calcification was partly restored: rates were lower under increased CO_2 conditions in all cultures but were up to 50% higher in adapted compared with non-adapted cultures" and that with such adaptation "contemporary evolution could help to maintain the functionality of microbial processes at the base of marine food webs in the face of global change" (Lohbeck et al., 2012: 346).

Recently, questions have arisen about possible impacts of ocean pH change on the physiological processes that are dependent on stable intracellular chemistry (Tanhua et al., 2015). Tanhua gives the example that "marine organisms would have to use more energy to regulate chemistry within their cells, which would result in less energy available for other biological processes such as growth or reproduction." Godbold and Calosi (2013) argue that ocean acidification and climate affect marine organisms from tropical to high-latitude ecosystems (see also Walther et al., 2002; Kroeker et al., 2013a) but make an important point that many current studies of ocean acidification focus on short-term (days to a few months) issues, while neglecting the possibilities of longer-term, gradual species gradual acclimatization, for example, changes due to phenotypic plasticity, or adaptation, which is the "the selection of extant genetic variation that moves the average phenotype of a population towards the fitness peak" (Godbold & Calosi, 2013); see also Ghalambor et al. (2007: 394) who argue that "different types of phenotypic plasticity can uniquely contribute to adaptive evolution when populations are faced with new or altered environments." Godbold and Calosi (2013) also point out that despite the gaps in knowledge, it is known that ocean acidification does not occur in isolation but integrated with other drivers of environmental modification such as warming, eutrophication, hypoxia, eutrophication, and pollution (see also Halpern et al., 2008).

Additionally, within the area of biotic mediation of ecosystem functioning, it is argued that ocean acidification and warming may influence dynamics such as biogeochemical cycling of macronutrients and decomposition, which are intrinsic to the food web. There are also the possibilities of phenotypic buffering and adaptation, which refer to the capacity of organisms to adapt their physiological status (including growth, reproduction, cell repair) in reaction to environmental dynamics, which is termed phenotypic plasticity. This area, of adaptation and associated and potential costs, require more research (Godbold & Calosi, 2013; see also Gaston, 2003; Bibby et al., 2007; Ghalambor et al., 2007; Dashfield et al., 2008; Godbold & Solan, 2009; Godbold et al., 2009, 2011; Munday et al., 2009; Bulling et al., 2010; Laverock et al., 2010; Hale et al., 2011; Hicks et al., 2011; Sanz-Lázaro et al., 2011; Donohue et al., 2012; Johnson et al., 2012; Landes & Zimmer, 2012; Lohbeck et al., 2012, 2013; Sunday et al., 2012; Calosi et al., 2013; Kroeker et al., 2013b, c; Pespeni et al., 2013; Tatters et al., 2013).

Methane (CH_4) is also discussed in the context of "micromolar concentrations under zero oxygen conditions, and in the surface mixed layer under oxic conditions at (sub)nanomolar concentrations" (Boyle, 2007. Boyle (2007: 305) describes Reeburgh's (2007) study as indicating "that the ocean is a large chemical reactor that oxidizes CH_4." A more recent study on oceanic methane (Valentine, 2011) notes that "our knowledge of physical, chemical, geological and biological processes affecting methane in the ocean and in underlying sediments is expanding at a rapid pace" as more recent research indicates that methane is affected by biogeochemical processes which are not fully understood, and these interactions span the deep-subsurface, near subsurface, and ocean waters, and relate to the production, consumption, and transport of methane (Valentine, 2011: 147).

Other elements have also been studied in the context of ocean environments. Mercury is one of the most important elements for such chemical cycling, largely because of its negative impact on living organisms (Lamborg et al., 2002; Selin et al., 2008; Campbell & Jha, 2021). The main topics are the movement of mercury between the ocean and other sources such as the atmosphere and the lithosphere, and the role of anthropogenic influences on the quantity of mercury in circulation. For example, Fitzgerald et al. (2007) studied the cycling of mercury (Hg) in the environment, with evidence that much of the oceanic mercury is derived from the atmosphere, and in the ocean, it then accumulates in the biota. Boyle (2007: 305) notes that at the time of writing "although we have excellent analytical data on Hg and its chemical species, we lack information on how it is chemically transformed and transported within the water column. In other instances, trace elements can be used to understand global processes as well as their own cycling."

Lamborg et al. (2014: 65) describe mercury as a toxic, bioaccumulating trace metal, and the emissions of this element into the environment have increased significantly due to human activities such as mining and fossil fuel combustion. Some recent models have estimated that increases of mercury emissions into the oceanic mercury inventory by 36–1313 million moles since the 1500 and "deep North Atlantic waters and most intermediate waters are anomalously enriched in mercury relative to the deep waters of the South Atlantic, Southern and Pacific oceans, probably as a result of the incorporation of anthropogenic mercury" (Lamborg et al., 2014: 65). The findings of this study indicate that anthropogenically derived perturbations to the global mercury cycle have increased the amount of mercury in thermocline waters by about 150% increase and tripled the mercury content of surface waters compared to preanthropogenic levels (Lamborg et al., 2014: 65).

Streets et al. (2011: 10485) argue that understanding of comparatively little known past mercury emissions is important for the assessment of current trends, including in remote regions such as the cited Arctic and Himalayas. Their study developed a trend of anthropogenic emissions of mercury to the atmosphere from 1850 to 2008. This period is covered by relatively complete data sets, and there is also available data on estimates of anthropogenic emissions prior to 1850. The results indicated that global mercury emissions peaked in 1890 at 2600 Mg yr(-1), fell to 700–800 Mg yr(-1) in the interwar years, then experienced a constant rise after 1950 to the present levels of 2000 Mg yr(-1). Streets et al. (2011: 10485) conclude that "our estimate for total mercury emissions from human activities over all time is 350 Gg, of which 39% was emitted before 1850 and 61% after 1850. Using an eight-compartment global box model of mercury biogeochemical cycling, we show that these emission trends successfully reproduce present-day atmospheric enrichment in mercury."

Regarding the proportions of surface oceanic and atmospheric mercury, Mason et al. (1994: 3191) contend that "the atmosphere and surface ocean are in rapid equilibrium; the evasion of Hg° from the oceans is balanced by the total oceanic deposition from the atmosphere." Here, Hg° refers to a liquid or monoatomic vapor, and $Hg^{(II)}$ or Hg^{2} is the oxidized, mercuric ion. The authors acknowledge that the means whereby reactive Hg species are reduced to volatile Hg_0 in the oceans are not fully understood, but the contributory factor appears to be from biological factors. It is argued that the equilibrium of the surface oceans and the atmosphere, with the low level of Hg oceanic mercury sedimentation makes deposition on land the main sink for atmospheric Hg. Around 50% of the anthropogenically derived emissions may enter the global atmospheric cycle and the rest may be deposited locally, and the authors surmise anthropogenic emissions have tripled the concentrations of Hg in the atmosphere and ocean surfaces over the last 100 years. Mason et al. (1994: 3191) conclude that therefore two-thirds of the present Hg fluxes are directly or indirectly due to human actions. The elimination of the anthropogenic load in the ocean and atmosphere is estimated to take 15 to 20 years after the cancelation of all the anthropogenic factors for emissions.

Soerensen et al. (2010: 8574) in a study that promotes "an improved global model for air-sea exchange of mercury" taking the high concentrations over the North Atlantic as a case study argue that "globally, subsurface waters account for 40% of Hg inputs to the ocean mixed layer, and 60% is from atmospheric deposition." The study notes that the global oceans are a net sink for 3.8 MmolHg y^{-1}, but the North Atlantic is a net source to the atmosphere, which is potentially because of the enrichment of subsurface waters with mercury from historical human sources. The premise of this study is that "previous efforts to model Hg air-sea exchange and atmospheric transport have been unable to reproduce high atmospheric concentrations observed in the Northern Hemisphere marine boundary layer (MBL) during ocean cruises" (see also Liss & Merlivat, 1986; Nightingale et al., 2000; Rolfhus & Fitzgerald, 2001; Andersson et al., 2007; Kuss et al., 2009; Loux 2004; Sunderland et al., 2010).

A similar study by Sunderland and Mason (2007) develops an "empirically constrained multicompartment box model for mercury cycling in open ocean regions to investigate changes in concentrations resulting from anthropogenic perturbations of the global mercury cycle." The findings of the simulated analyses indicate that "anthropogenic enrichment in all surfaces (25%) and deep ocean waters (11%) is lower than global atmospheric enrichment (300–500%) and varies considerably among geographic regions, ranging from >60% in parts of the Atlantic and Mediterranean to <1% in the deep Pacific." Additionally, open ocean mercury concentrations do not quickly equilibrate with atmospheric deposition and usually may increase if the human-derived emissions stay at the present level (Sunderland & Mason, 2007). The authors conclude with an estimation that "the temporal lag between changes in atmospheric deposition and ocean mercury concentrations will vary from decades in most of the Atlantic up to centuries in parts of the Pacific" (Sunderland & Mason, 2007).

A study by Strode et al. (2010) was justified on the grounds that "the ocean plays a critical role in the biogeochemical cycling of mercury (Hg)," and according to Mason and Sheu (2002), the ocean reservoir contains 1440 Mmol of Hg, greater than the atmospheric reservoir of 25 Mmol, and the ocean emissions contribute about a third of the current Hg source to the atmosphere. The anthropogenic perturbation (relevant activities) to the global Hg cycle is argued to be significant, as historic records in lake sediments show a threefold increase in Hg deposition since preindustrial times (Strode et al., 2010, see also Swain et al., 1992; Lamborg et al., 2002). Strode et al. (2010) examined the vertical transport of mercury (Hg) in the oceanic environment using a simple box diffusion model to represent vertical water transport coupled with a particulate Hg flux. This particulate flux "assumed that the mercury content of marine particles is proportional to the mercury concentration of surface waters via a sorption equilibrium constant, Kd" (Strode et al., 2010).

The findings of the study by Strode et al. (2010) indicate that the modeled vertical profile of oceanic mercury had a subsurface maximum at about 500 m depth, due to the remineralization of mercury which was bound to sinking organic particles, and the surface (top 100 m) concentrations of mercury increased by 150%, compared with preindustrial times. It is argued that in the past 150 years, 280 Mmol of anthropogenic Hg have accumulated in the ocean, which represents a 18% increase in the total oceanic Hg content. Also, 36% of the anthropogenic Hg is found in the top 400 m of the ocean water, and 7% occurs below 1500 m. Over the industrial era, 14% of cumulative anthropogenic emissions accumulated in the ocean (Strode et al. (2010). The authors conclude that "half of the accumulation of anthropogenic Hg in the ocean is due to sinking on particulates" (Strode et al., 2010; see also Swain et al., 1992, Lamborg et al., 2002, Mason & Sheu, 2002).

Amos et al. (2013: 410) related oceanic mercury, especially the formation of toxic methylmercury levels, to global health concerns, presenting a "global biogeochemical model with fully coupled atmospheric, terrestrial, and oceanic Hg reservoirs to better understand the human influence on Hg cycling and timescales for responses." The model is based on an historical inventory of human-derived emissions from 2000 BC to the time of writing (Amos et al., 2013: 410). The justification of the study is that over 50% of Hg in the ocean today is anthropogenic, anthropogenic enrichment is much greater than previously recognized, aggressive emission reductions are needed to stabilize ocean Hg concentrations, and "Hg cycles naturally through geochemical reservoirs, but human activities such as mining and more recently fossil fuel combustion have been increasing the Hg flux from the deep mineral reservoir to the atmosphere for millennia" (Amos et al., 2013, 410).

The findings of the study indicate that human-induced perturbations that are released to surface reservoirs, that is the atmosphere, marine, or terrestrial environments, may accumulate and remain in the subsurface ocean for up to centuries (Amos et al., 2013, 410). Additionally, "the simulated present-day atmosphere is enriched by a factor of 2.6 relative to 1840 levels, consistent with sediment archives, and by a factor of 7.5 relative to natural levels (2000 BC)" and the historic human-induced mercury re-emitted from surface reservoirs accounts for over 50% of current atmospheric deposition, compared to 27% from primary human-induced emissions, and 13% from natural sources (Amos et al., 2013, 410). Further, it is argued that only 17% of the current mercury in the surface ocean is natural and 50% of the anthropogenic enrichment is from pre-1950 emissions. Asia is currently the dominant contributor to primary anthropogenic emissions but contributes only 17% of the surface ocean mercury reservoir, as compared to the 30% of North American and European origin (Amos et al., 2013: 410). The authors note that accumulation of legacy (old) anthropogenic mercury points to possible future deposition even with constant current anthropogenic emissions, therefore requiring "aggressive global Hg emission reductions" to just maintain present levels of oceanic Hg concentrations (Amos et al., 2013: 410).

Integrating biological and chemical oceanography into inclusive marine biogeochemistry is obviously a complex and herculean task. Gruber and Doney (2019) note that "the complexity of the ocean biogeochemical/ecological problem is daunting, as it involves a complex interplay among biology, physical variability of the oceanic environment, and the interconnected cycles of a large number of bioactive elements, particularly those of carbon, nitrogen, phosphorus, oxygen, silicon, and iron. Furthermore, the ocean is an open system that exchanges mass and many elements with the surrounding realms, such as the atmosphere, the land, and the sediments." Writing about the links between marine ecology and biogeochemistry, Gruber and Doney (2019) note that the cycles of carbon, oxygen, phosphorus, nitrogen, and silicon are vital and the main engine of the oceanic biogeochemical cycles is the biological production of organic matter in the illuminated upper ocean (euphotic zone), some of which may sink into the deeper ocean and be degraded into inorganic materials. Gruber and Doney (2019) also maintain that some phytoplankton, including diatoms and coccolithophorids, produce shells made out of amorphous silicon and solid calcium carbonate, and the silicon and $CaCO_3$ cycles are closely associated with the organic matter cycle.

Physical Oceanography

Physical oceanography has been defined as the branch of oceanography dealing with the properties of seawater (temperature, density, pressure, and so on), its movement (waves, currents, and tides), and the interactions between the ocean waters and the atmosphere (Britannica, 2019). Bolles (2022a) defines physical oceanography as "the study of the physical properties and dynamic processes of the oceans" and "physical oceanographers study the interaction of the ocean with the atmosphere, how the ocean stores and releases heat, the physical properties (or chemical content) of water throughout the ocean, and the formation and movement of currents and coastal dynamics." Supportive of this definition, physical oceanography research at the National Aeronautical and Space Administration (NASA) records "ocean surface topography, sea surface temperature, the speed and direction of the winds over the ocean, and soon, sea surface salinity (salt content)" (Bolles, 2022a). The Woods Hole Oceanographic Institution remarks that physical oceanographers "make new observations of the oceans and use physics, mathematics, computer models and statistics to better understand how the oceans work and make more accurate predictions of how they may change in the future" (Woods Hole Oceanographic Institution, 2022).

Important topics usually covered within physical oceanography include the volume, classification, and distribution of the Earth's oceans, and the temperature, salinity, density, circulation, Coriolis effects, ocean–atmosphere interfaces, climate formations, ocean currents, temperature fluxes, and sea level changes, including tides and tsunamis (Gill, 1982; Hamblin & Christiansen, 1998; Stewart, 2007; Samelson, 2011; Talley et al., 2011; Su et al., 2014). Talley et al. (2011) list the topics of physical oceanography: "physical properties of seawater, heat and salt budgets, instrumentation, data analysis methods, introductory dynamics, oceanography and climate variability of each of the oceans and of the global ocean, and brief introductions to the physical setting, waves, and coastal oceanography."

Chelton (1994: 150) attempts a simple definition: "Physical oceanography is the study of the physics of the ocean. As such, the discipline encompasses a very broad diversity of phenomena, ranging from the smallest space and time scales of order 1 second and 1 cm associated with vertical turbulent mixing to the largest space and time scales of order centuries and 10,000 km associated with global climate variations." As it is the physics of the ocean, physical oceanography may also be termed marine physics (see Bangor University, 2020). The Scripps Institution of Oceanography (2022) defines physical oceanography as "the field of study that deals with mechanisms of energy transfer through the sea and across its boundaries, and with the physical interactions of the sea with its surroundings, especially including the influence of the seas on the climate of the atmosphere." Examples of research activities within a research group include general circulations of the oceans, including relations between ocean currents and the dynamics of ocean basins, current fluctuations and transport issues, wave motion, thermodynamics, optical and acoustic properties, and surf and coastal currents, impacts and sediment transport (Scripps Institution of Oceanography, 2022).

Other definitions emphasize the development of the subdiscipline of physical oceanography. For example, Thomson and Emery (2014) note that "physical oceanography is an ever-evolving science in which the instruments, types of observations, and methods of analysis undergo continuous advancement and refinement." Drijfhout et al. (2013: 257) add that "the field of physical oceanography has been transformed over the past two decades by the emergence of the modern global ocean observing system and of global eddy-permitting ocean circulation models. These observations and models have revealed the complexity of the ocean circulation across the spectrum of spatial and temporal scales." Van Haren (2018) describes physical oceanography as "a relatively young scientific branch, in comparison with marine biology or other oceanographic disciplines, and with most physics-related disciplines. However, because of its importance for fisheries and trading, surface ocean physics on currents and waves has been investigated for over centuries."

Research on physical oceanography may examine the links between this subdiscipline and related physical earth attributes. Olson et al. (2022) examined the ocean composition in present-day climates, with the justification that these oceanic factors are less studied than the role of atmospheric composition on the current climates. The oceanic variables were low versus high ocean salinity in response to Earth's present-day and Archean climate systems, using the ROCKE-3D ocean–atmosphere general circulation model. The Resolving Orbital and Climate Keys of Earth and Extraterrestrial Environments with Dynamics (ROCKE-3D) is a three-dimensional General Circulation Model (GCM) developed at the NASA Goddard Institute for Space Studies for the modeling of atmospheres of the Earth's immediate solar system and exoplanetary terrestrial planets (Way et al., 2017). Olson et al. (2022) found that "saltier oceans yield warmer climates in large part due to changes in ocean dynamics. Increasing ocean salinity from 20 to 50 g/kg results in a 71% reduction in sea ice cover in our present-day Earth scenario. This same salinity change also halves the pCO_2 threshold at which Snowball glaciation occurs in our Archean scenarios." The authors hypothesize that higher levels of greenhouse gases such as CO_2 and CH_4, with a saltier ocean, may result in a warm Archean Earth with only seasonal ice at the poles, despite a 20% decline in energy from the Sun. The authors concluded their study with a strong focus on the Archean eon "when Earth received ~20% less energy from the Sun, but there is nonetheless clear evidence for liquid water, a productive marine biosphere, and biological impacts on oceanic and atmospheric chemistry."

Hence, physical oceanography has important links with biogeochemistry and marine biogeography. Bolles (2022b) describes the impact of the parameters of physical oceanography on the carbon cycle, arguing that "physical oceanography influences the carbon cycle through its modulation of the biology and also through processes that control carbonate chemistry (e.g., temperature, alkalinity/salinity) and carbon dioxide flux rates between the air-sea interface (e.g., surface wind speeds)." What is termed the "ocean solubility pump" removes atmospheric carbon dioxide as air is mixed with and dissolved in the upper ocean water, and as carbon dioxide solubility increases in cold water, at the higher, colder latitudes the colder carbon dioxide dense water sinks into the deeper ocean levels and forms the deep ocean circulation "conveyor belt," over hundreds of years (Bolles, 2022b). Oceanic movements may eventually bring this water to the ocean surface distinct from the sinking area, for example in the tropical lower latitudes, and because warm waters cannot retain similar levels of carbon dioxide as colder waters, the carbon dioxide may be transferred into the atmosphere (Bolles, 2022b).

Bolles (2022b) also notes that physical oceanography links to biological oceanography (and hence, also to marine biology and marine biogeography; see Tables above and the references: Lalli & Parsons, 1993; Miller & Wheeler, 2012; Wierucka et al., 2019; Collins et al., 2020; Machovsky-Capuska & Raubenheimer, 2020; Baker et al., 2021; Catlett et al., 2022; Flowers et al., 2022; Liu et al., 2022; Nagaiwa et al., 2022; Nishizawa & Joseph, 2022; Takahashi et al., 2022; Shen et al., 2022; Coxe et al., 2023; Harris & Gibson, 2023; Mancini et al., 2023; Nel et al., 2023; Tovar Verba et al., 2023). The links between physical oceanography and biological oceanography are "through controls on mixed layer depths/dynamics, nutrient fluxes, and temperature which influences phytoplankton growth rates and primary production" (Bolles, 2022b). Bolles (2022b) gives the example of shallow mixed oceanic layers, where phytoplankton are more exposed to light than in deep mixed layers, and consequently grow faster with more light input. Where mixed layers deepen, "more nutrients are entrained from below, and phytoplankton can grow more rapidly, that is, less nutrient limitation. Thus, phytoplankton growth is optimal in situations where neither light nor nutrients are limited" (Bolles, 2022b). The case study is given of the spring phytoplankton population blooms in the North Atlantic, where the wintertime deep mixing has increased the surface nutrients, increased springtime solar illumination, and provided more light for photosynthesis, and heating has shallowed the mixed layer. In March, the bloom starts at mid-latitudes and moves northward, reaching high latitudes in June when the sun elevations increase (Bolles, 2022b).

Holdsworth et al. (2021) also give an incisive example of modeling projections of ocean circulation (an important aspect of physical oceanography) and biogeochemistry to examine large-scale climate changes, where the modeling used "atmospheric climatologies with augmented winds to simulate historical (1986–2005) and future (2046–2065) periods separately." Holdsworth et al. (2021) note that high-frequency wind variability is vital for freshwater distribution in the studied region. The projections suggested several outcomes: there would be increases in sea surface temperatures, while surface salinity decreases, depending on the assumptions of moderate or high emissions; stratification increases, with some nutrient limitation at the surface layers; increased primary production and phytoplankton productivity; denser water surfaces increase in depth, mostly due to surface water heating and freshening (Holdsworth et al., 2021). The saturation state changes are mostly due to "anthropogenic CO_2 with minor contributions from solubility, remineralization and advection," and "extreme states of temperature, oxygen, and acidification are projected to become more frequent and more extreme" (Holdsworth et al., 2021).

Skákala et al. (2022) examined another attribute of physical oceanography interactions with ocean biogeochemistry (mutual influences), and the consequences for modeling shelf seas, and some biogeographical issues such as algal blooms. The study used modeling tools to estimate the impact of biogeochemical tracers on vertical light and heat attenuation on the North-West European Shelf (NWES) (Skákala et al., 2022). In this case, the physical attributes of the ocean are heat and light, and the biological attributes are the phytoplankton bloom, as the physical attributes "influence mixing in the upper ocean and feed back into marine biology by influencing the timing of the phytoplankton bloom" (Skákala et al., 2022). The study compared different light schemes that represented the impact of biogeochemistry on physics and found that physics is sensitive to the spectral resolution of radiances and the optically active constituents. The justification for this study was that "within the Earth system, physics and biology mutually interact in many non-trivial ways. In the marine environment biological processes are driven by physical transport, mixing, temperature, salinity, and the incoming light, whereas biology impacts physics through its role in the carbon cycle…, oceanic albedo" and …, "underwater light attenuation" (Skákala et al., 2022).

Additionally, another justification was that the impact of biogeochemical light attenuation on marine physics has been examined for large, globally dominant areas, such as the North Atlantic, tropical Pacific and wider areas, with results indicating some regionally-dependent impacts, but less on the shelf sea environment, despite the importance of the latter for oceanic production, and consequently for the "study the complex interaction between biogeochemistry and physics" (Skákala et al., 2022; see also Oschlies, 2004; Manizza et al., 2005; Borges et al., 2006; Lengaigne et al., 2007; Jahnke, 2010). In support of their modeling study, Skákala et al. (2022) also argue that "the most obvious source of biogeochemical feedback to physics in coupled physical–biogeochemical ocean models is the attenuation of underwater radiances by optically active biogeochemical tracers and the subsequent impact on heat fluxes, temperature and mixed layer depth" and that it is of importance "to better understand both the biogeochemical impact on ocean physics, and the sensitivity of the simulated physics to how precisely such an impact is incorporated into the physical model."

Physical oceanography links with marine biogeography were also examined in a study by Hátún et al. (2022), which looked at the links between oceanic gyres and marine ecosystem distribution, focusing on the impact of the North Atlantic subpolar gyre (SPG) on the subpolar North Atlantic (SPNA). Oceanic gyres are defined as large systems of rotating ocean currents: the Marine Ocean Service (2021) notes that "there are five major gyres, which are large systems of rotating ocean currents. The ocean churns up various types of currents. Together, these larger and more permanent currents make up the systems of currents known as gyres." Hátún et al. (2022) focus on the dynamics of the North Atlantic subpolar gyre (SPG) and its interaction with the Atlantic inflow poleward, and the returning cold subarctic water masses flowing equatorward, which is referred to as the Atlantic Meridional Overturning Circulation (AMOC), and examined the "linkages between the changing physical and biogeochemical environment and the organisms inhabiting this mid- to high latitude area."

Hátún et al. (2022) further report that "the concept of biogeographical zones and ecoregions with large areas sharing similar physical, biogeochemical and biological characteristics is presently making a come-back in marine science." This is based on the phenomenon that when two different water masses lie next to each other, frontal zones are created that delineate biogeographic boundaries, and in these areas the physical properties and biological assemblages may vary fast over short distances. These water mass boundaries determine the outlines of oceanic gyres, which then create frontal zones and biogeographical boundaries. In their interactive and multiple papers, Hátún et al. (2022) point out that "all trophic levels are represented, with *physical oceanography and biogeochemistry* here referred to as the 0-th level, *primary producers* (e.g., phytoplankton, 1st level), *secondary producers* (e.g., zooplankton, 2nd level) and *fish/predators* (3rd level)." They assess their authors contributions on physical drivers in the SPNA as giving a "bottom-up" perspective on ecosystem changes in this "biologically productive region" (a limitation to the neglect of human-induced "top-down" effects), but their approach being justified as "the projected climate changes are expected to primarily disturb bottom-up processes" (Hátún et al., 2022).

Marine biogeography, physical oceanographic and biogeochemical studies may emphasize faunal, floral and ecosystem approaches, with both the faunal and floral studies examining relevant ecosystems. A faunal study is that of Hátún (2009: 149), located in the north-eastern North Atlantic ecosystem, a region which is characterized by cold subarctic fauna within the subpolar gyre, warm Lusitanian fauna from the Bay of Biscay and, between these two contrasting systems, a cold-temperate boreal fauna in the mixing area to the west of the British Isles and south of Iceland (Hátún, 2009: 149; see also Ekman, 1953; Fraser, 1958). Hátún (2009: 149) note that in this region the faunistic boundaries are not very clear, due largely to the powerful annual and interannual temperature changes in the boreal region, which is demarcated by temperature: the cold flank is termed Arctic-boreal, and the warm flank is termed Lusitanian-boreal (Ekman, 1953). The Hátún (2009: 149) study argues that major changes in fauna from the English Channel in the south to the Barents Sea in the north-east and near Greenland in the north-west have occurred periodically in the late 1920s, 1960s, and 1990s, and these events were due to exchanges of subarctic and subtropical water masses in the north-eastern North Atlantic Ocean, linked to changes in the power and extent of the subpolar gyre (Hátún, 2009: 149).

Hátún (2009: 149) points out that these exchanges may contribute to changing impacts exerted by the subarctic or Lusitanian biomes on the "intermediate faunistic zone in the north-eastern Atlantic." The study examined this "strong and persistent bottom-up bio-physical link" using a numerical ocean general circulation model and data covering four connected, food chain trophic levels – from lowest to highest, phytoplankton (plant species), zooplankton (animal species), blue whiting fish (*Micromesistius poutassou* A. Risso, 1827), and pilot whales (*Delphinus globiceps* Cuvier, 1812). The study findings indicate the possibility of using plankton and other data for simulation modeling of the dynamics of the subpolar gyre, which "suggests a potential for predicting the distribution of the main faunistic zones in the north-eastern Atlantic a few years into the future, which might facilitate a more rational management of the commercially important fisheries in this region."

A floral study is that of Aumont et al. (2003: 1–15) titled "an ecosystem model of the global ocean including Fe, Si, P colimitations," which argues that large areas of the global ocean have lower than expected chlorophyll concentrations given the ambient phosphate and nitrate levels. In these areas, termed High Nutrient-Low Chlorophyll regions (HNLC), it is hypothesized that limitations of phytoplankton growth by other nutrients like silicate or iron exist, a position supported by experiments. Additionally, a paradox is described where until the recent past, there is contrary evidence of the commonly held idea that the magnitude of the biological pump was mostly controlled by the high levels of macronutrients nitrate and phosphate and by sunlight levels and that with sufficient sunlight, there is usually a positive correlation between macronutrients concentrations and phytoplankton biomass, at least in the open ocean ecosystems. Aumont et al. (2003) argue that in three major areas, the subarctic Pacific, the eastern and central equatorial Pacific, and the Southern Ocean, this common view seems questionable.

The paradox of these areas has been examined with three main hypotheses: one is the iron limitation hypothesis, the possibility that "phytoplankton growth may be limited by other nutrients than phosphate or nitrate, in particular by the micronutrient iron," with silicon abundance vital due to the requirement of this element for diatoms, with examples of the equatorial Pacific ocean and the Antarctic Circumpolar Current (see Jacques, 1983; Sommer, 1986; Dugdale & Wilkerson, 1998) (Aumont et al., 2003). A second hypothesis is that increased grazing pressure may decimate much of the phytoplankton biomass (see also 1991). The third hypothesis is that an "unfavorable light-mixing regime may prevent phytoplankton from blooming, especially in the Southern Ocean where the summer mixed layer depth is about 60–120 m" (Aumont et al., 2003; see also Nelson & Smith, 1991). Aumont et al. (2003) point out that "the consensus is now that iron or silicate is limiting the growth of large cells, mainly diatoms. The predominant small-sized phytoplankton which has low iron requirement is maintained at modest levels by efficient grazing by microzooplankton" (see also Chavez et al., 1991; Landry et al., 1995; Fitzwater et al., 1996; Verity et al., 1996).

A study by Aumont et al. (2003) used a nine-component ecosystem model in the Hamburg model of the oceanic carbon cycle (HAMOCC5), a model that includes phosphate, silicate, dissolved iron, phytoplankton size fractions (nanophytoplankton and diatoms), zooplankton size fractions (microzooplankton and mesozooplankton), detritus, and semilabile (longer term) dissolved organic matter. The model simulates the Southern Ocean and the equatorial Pacific, but "in the subarctic Pacific, silicate and phosphate surface concentrations are largely underestimated because of deficiencies in ocean dynamics" (Aumont et al., 2003). Aumont et al. (2003) argue that the low chlorophyll concentrations in HNLC areas are explainable by the traditional hypothesis of a simultaneous iron-grazing limitation: "Diatoms are limited by iron whereas nanophytoplankton is controlled by very efficient grazing by microzooplankton… The model predicts that the ocean carries with it about 75% of the phytoplankton demand for new iron, assuming a 1% solubility for atmospheric iron" and "a higher supply of iron to surface water leads to a higher export production but paradoxically to a lower primary productivity."

Another floral-based study is that of Nissen et al. (2018), which asserts that the biogeography of Southern Ocean phytoplankton exerts control over the local biogeochemistry and the export of macronutrients to lower latitudes and depths and that there is a competitive interaction between coccolithophores and diatoms, with the former commoner along the "Great Calcite Belt" (40–60°S), and the diatoms dominating the regions south of 60°S. This study had a complex modeling background, to analyze and assess the factors that control coccolithophore (an important phytoplankton group) distribution and the competition between coccolithophores and diatoms. The authors argue that "this modeling study is the first to comprehensively assess the importance of both bottom-up and top-down factors in controlling the relative importance of coccolithophores and diatoms in the SO over a complete annual cycle" (Nissen et al., 2018). The model was a regional high-resolution model (ROMS–BEC) for the Southern Ocean (24–78°S), with modifications to include representation of coccolithophores. The study also assessed the comparative importance of the bottom-up (temperature, nutrients, light) and the top-down (grazing by zooplankton) factors that control the Southern Ocean coccolithophore biogeography during the growing season. Additionally, the role of diatoms in limiting coccolithophore abundance and the importance of microzooplankton and macrozooplankton grazing for the relative importance of coccolithophores and diatoms in the study area was investigated.

The findings of the simulated study were that coccolithophores contribute 17% to annually integrated net primary productivity south of 30°S; hence, they are an important part of the Southern Ocean phytoplankton community. The highest coccolithophore biomass is north of 50°S in late austral summer, a time of high light levels, when combined with silicic acid limiting diatoms (Nissen et al., 2018). The authors conclude that top-down factors are an important control on the relative abundance of diatoms and coccolithophores in the Southern Ocean, and "consequently, when assessing potential future changes in Southern Ocean coccolithophore abundance, both abiotic (temperature, light, and nutrients) and biotic factors (interaction with diatoms and zooplankton) need to be considered" (Nissen et al., 2018). It is concluded that the study is vital, as "Coccolithophores impact biogeochemical cycles, especially organic matter cycling, carbon sequestration, and oceanic carbon uptake both via photosynthesis and calcification, leading to cascading effects on the global carbon cycle and hence climate" (Nissen et al., 2018).

Geological Oceanography or Marine Geology

Geological oceanography and marine geology are seen as synonymous by most authorities, but occasionally, a distinction emerges. For example, Bangor University (2020) hints that marine geologists and geological oceanographers are two groups, defining marine geology as "concerned with the processes that shape the ocean basins, determine the structure and composition of the Earth's crust and control features of the continental margins" and geological oceanography as "concerned with marine sediments" which include the understanding of the processes that govern the origin, erosion, transport, and deposition of sediments in the ocean. The University of Victoria (2022) describes an undergraduate course on geological oceanography as studying "geological processes that shape the ocean basins, oceanic plate creation and structure, hydrothermal circulation at mid-ocean ridges, bathymetry, ocean islands, different types of ocean margins, ocean sediments, the sedimentary record of past ocean circulation, coastal geology" and a similar level course on marine geology as "covering modern marine geological processes in a wide range of oceanic environments: mid-ocean ridges, mid-plate volcanoes and hot spots, coastlines, continental margins and abyssal plains."

The Journal of Marine Biology & Oceanography (2022) writes "geological oceanography or Marine geology involves the study of geophysical, sedimentological, geochemical, and paleontological investigations of the ocean floor and coastal margins" but defines marine geology as "the branch of geology which studies the geological structure and development of the part of Earth's crust that makes up the floors of the seas and oceans" and "marine geology uses the methods and findings of geomorphology, geophysics & geochemistry" (Journal of Marine Biology & Oceanography (2022). Marine geology may be inferred to have slight differences from geological oceanography, these being similar to the differences between marine biology and biological oceanography: in such double worded subjects, the second word is more important. Kwader (2021) makes a similar point; in comparing hydrogeology vs. geohydrology, "the second half of the word is the science used to understand the first half of the word" hence "a hydrogeologist uses the physical geological characteristics of the strata or rock to infer the hydrologic characteristics of the material" and "geohydrology (less commonly used) is the interpretation of the geology from the hydrologic characteristics of the strata or rock."

The breadth of the field of geological oceanography was pointed out in the aims and scope of Marine Geology, "the premier international journal on marine geological processes in the broadest sense" (Marine Geology, 2022). Topics include "papers on subjects as diverse as seafloor hydrothermal systems, beach dynamics, early diagenesis, microbiological studies in sediments, palaeoclimate studies and geophysical studies of the seabed… the influence of anthropogenic processes on coastal/marine geology and coastal/marine geoarchaeology… papers are concerned with the marine realm and that they deal with geology: with rocks and/or sediments now in the ocean (not on land), and physical and chemical processes affecting them." Kontakiotis et al. (2022) also remark on the breath of this subject, in the Journal of Marine Science and Engineering, noting that "research into modern oceans, seas, and their coastal zones, as well as marine ecosystems, provides valuable information for deciphering the geological dynamics," this breadth including analysis of the seafloor using stratigraphic and sedimentary models, interpretations of the past through analyses of ichnological (the branch of paleontology studying fossilized tracks, trails, burrows, borings, or other trace fossils) interpretations, marine deposits, paleoenvironments, fossils, frameworks for analyzing mineral and hydrocarbon deposits, and newer subjects such as geoheritage studies.

Their edited collection (see Kontakiotis et al., 2022, a special Issue comprising twenty-one research papers, in four themes) includes articles covering, geological processes in the modern marine environment (Theme 1: Hayat et al., 2022; Hussain & Al-Ramadan, 2022, Petropoulos et al., 2022, Yutsis et al., 2022); the evolution of marine environments and ecosystems during the Phanerozoic, 541 million years BP until present (Theme 2: Ali et al., 2021, Garcia et al., 2021, Johnson, 2021, Wang, He, et al., 2021a, Ruban, 2022a, b, Tolokonnikova & Ruban, 2022); Quaternary marine studies (Theme 3: Giamali et al., 2021, Wang et al., 2021b, Arce-Chamorro et al., 2022); the application of marine geological processes to petroleum geology (Theme 4: Khan et al., 2021, Janjuhah et al., 2021, Dai et al., 2022; Dubinya et al., 2022, Fazal et al., 2022, Li et al., 2022, Maravelis et al., 2022, Ruban, 2022a, b).

Geological oceanography has important links with oceanic and global biogeochemistry, partly through the fields of marine geochemistry and marine biology (Goldberg, 1954; Henderson, 2016). The main issues concern the links between the circling of chemical elements and compounds between the oceans geological base, in many cases sediments, the water of the ocean and in some cases the atmosphere and dry land including rivers and waters connected to the sea. This cycling may also affect living organisms. For example, Henderson (2016) mentions the quantities of the elements in the oceans, especially sodium (Na), chloride (Cl) (more than 90 of this dissolved mass, and together with magnesium (Mg), sulfur (S), and calcium (Ca), represent 99%). Henderson (2016) adds that "add only three more elements – K, Br, and C – and you have more than 999 parts in a 1000 of the dissolved mass in seawater." There are 82 other naturally occurring elements, which are found in ocean water at low concentrations, an example being iron, which comprises one-billionth of the dissolved mass in ocean waters, but is an essential nutrient for oceanic life, and may be the nutrient which limits marine with its elimination. Therefore, "without knowledge of the cycle of Fe, we cannot understand the controls on ecosystems in the ocean, nor their role in the global cycling of carbon and other elements" (Henderson, 2016).

An old study by Goldberg (1954: 249) mentions notes that "the ability of the hydrated oxides of manganese and iron to adsorb ions from solution (scavenging)" as important for some issues in marine geology, chemistry, and biology. The example is taken of the ferruginous sediments of the Pacific Ocean, where iron oxides may be associated with titanium, cobalt, and zirconium in amounts proportional to the iron content and copper and nickel are related to the manganese content (Goldberg, 1954: 249). In this study, marine sediments are classified by the geosphere in which the solid phases originate. The author notes that "the distribution of certain ionic species in sea water between the solid and aqueous phases is considered on the basis of scavenging and co-ordination compound theory" and "the concentration of minor elements by members of the marine biosphere is explained either by the direct uptake of the element or by the uptake of iron or manganese oxides with the accompanying scavenged element" (Goldberg, 1954).

Another important issue concerns the complex relations between solid earth (the ocean floor) and sea water, and sometimes atmospheric gases. For example, one concern is the formation of oceanic phosphorite, a compound including phosphorus, which is defined as "P-rich sedimentary rocks, usually composed of carbonate hydroxyl fluorapatite ($Ca_{10}(PO_4CO_3)_6F_{2-3}$) which occurs as nodules and crusts originally formed in oceanic environments" (Douglas & Beveridge, 1998; see also Southgate, 1986). Wang et al. (2014) write that phosphorite "occurs in lenses in sediments off the coasts of the Chinese mainland and the Korean Peninsula. The Yellow River delta extends along more than 100 km of the coast, and river sediment can be traced all the way out to the edge of the continental shelf. Sediment wedges defining ancient deltas are prime prospects for the discovery of offshore placer and phosphorite deposits" (Ludwig & Valencia, 1993). Additionally, "phosphorite is also formed in regions of oceanic upwelling where large volumes of phosphate-rich cold water rise from great depths to the surface. In the warmer surface waters, phosphate precipitates out of solution and then sinks to the seafloor forming deposits of phosphorite. There is a possibility that large reserves may be located off the east coast of Vietnam where a seasonal localized upwelling formed by the SCS's monsoon-driven current system generates the required oceanic conditions" (Ritterbush, 1978).

Another issue concerns carbon compound movements between the ocean floor and the oceanic water, which may be affected by oceanic geological episodes. For example, Nature (2011) reports that "carbon bound up in seafloor sediments is thought to have played a key role in past episodes of climate change. The breakup of the northeast Atlantic continents around 55 million years ago was associated with the injection of large amounts of molten magma into carbon-rich sediments on the seafloor. The heating of these sediments may have triggered the release of large quantities of methane and carbon dioxide into the overlying ocean and atmosphere." However, Nature (2011) notes that "the ultimate impact of the carbon emitted from these emerging and distinct seafloor sources, and its contribution to the global carbon budget, is uncertain… There is evidence that carbon trapped beneath the seabed continues to influence carbon dynamics, at least in the deep ocean."

Nature (2011) also mentions the role of ocean spreading based the parting of plates, due to tectonic movements, which are "hotbeds of magmatic activity, owing to the eruption of molten magma from below. Hydrothermal vents are common at these spreading sites. These vents emit large quantities of geothermally heated, mineral-rich fluids. The minerals and energy contained in these fluids support a huge diversity of microbes and invertebrates, such as giant tube worms and clams, in an otherwise barren sea floor." One result appears to be release of methane: for example, "in the geologically young Guaymas basin in the Gulf of California, magma has intruded into vast areas of the thick, organic sediments at the bottom of the basin. The heating of these sediments – that had been thought to sequester rather than emit carbon – seems to have resulted in the release of methane into the overlying ocean." Jahncke et al. (2005, 691) note that "methane hydrates constitute a large global carbon reservoir that is vulnerable to destabilization via ocean-floor disruption. Such destabilization has important implications for the global cycles of both carbon and methane, especially with respect to atmospheric chemistry and potential climate warming." The authors argue that in their study that there is evidence of "how major tectonic events led to a methane hydrate release in the early Aptian

(Early Cretaceous), as evidenced by a new high-resolution terrestrial δ13C (carbon stable isotope) record." And there is the possibility that "plate tectonics, a fundamental Earth process, may control methane hydrate reservoir disruption over long timescales via deformation of the continental margins."

Geomatics and Ocean Biogeochemistry

Geomatics techniques, especially based on remote sensing methods, may be used in the study of ocean biogeochemistry, by measuring the parameters that may be used to assess relevant geochemical and biological features (Campbell, 2018). LiDAR systems, described in Chap. 4, are emerging as important tools, currently and potentially may be used for measuring both water types and water organisms (Gray et al., 2015; Hostetler et al., 2018); Collister et al., 2020). Roddewig et al. (2017) note that "LiDAR systems have been developed for looking into seawater, primarily for profiling the seafloor surface and, sometimes, also for measuring the optical properties of the water column. LiDARs also have been developed for airborne or shipborne measurements of water clarity and attenuation, spatial and temporal variations in oceanic scattering layers and plankton distributions, internal waves, chlorophyll content, and fish… producing global maps of phytoplankton biomass and total particulate organic carbon" (see also Collins et al., 1984; Guenther et al., 1996; Lillycrop et al., 1996; Fry et al., 1997; Bukin et al., 1998; Churnside et al., 1998, 2001, 2012; Vasilkov et al., 2001; Allocca et al., 2002; Lee et al., 2002; Chekalyuk et al., 2003; Lin, 2003; LaRocque et al., 2004; Popescu et al., 2004; Tuell et al., 2005, 2010; Churnside & Donaghay, 2009; Kampel et al., 2009; Reineman et al., 2009; Schorstein et al., 2009; Ramnath et al., 2010; Behrenfeld et al., 2013; Rudolf & Walther, 2014).

Churnside and Thorne (2005: 5504) write that "airborne lidar has the potential to survey large areas quickly and at a low cost per kilometer along a survey line." Gray et al., 2015 note that "active source systems such as LiDAR can probe the subsurface structure of oceans." Such active systems may derive range information by "range-gated systems, which uses a pulsed laser and a range-gated receiver to temporally resolve the laser pulse" and "optically modulated, continuous-wave (OMCW) systems, in which a continuous wave laser is modulated in either frequency or intensity and range is determined by calculating the phase shift of the modulation envelope" (Gray et al. (2015). Such LiDAR techniques have been applied to the measurement of "profiles of attenuation and backscattering, chlorophyll fluorescence, thin phytoplankton layers, and fish stocks," with laser wavelength at 532 nm, which is suitable for research in coastal waters, but shorter wavelengths may be more suitable for the open ocean. Mindful of the differences between multiwavelength and single-wavelength LiDAR, Gray et al. (2015) pose the question "What benefits can a multiwavelength LiDAR provide over a single-wavelength one?" Moreover, their research explores the quantitative advantage of the optimally penetrating wavelength in different water types.

This study by Gray et al. (2015) gives an example of the use of multiwavelength oceanographic LiDAR for measurement of water types, "in water tanks with optical conditions simulating both sediment-laden and biologically rich water types." The authors conclude that their "measurements and models show that using a laser wavelength of 470–490 nm in the open ocean leads to an improvement factor of 1.50–1.75 compared to a 532 nm system. In more turbid areas using a laser wavelength of 560–580 nm leads to an improvement factor of 1.25" and "using multiple LiDAR wavelengths can help detect and characterize constituents in the water column" (Gray et al., 2015). The study findings indicated that multiple wavelengths improve research, although 532 nm systems are suitable in clearer coastal waters, they may be less optimal for other turbidity levels. Crucially, "the optimal wavelength follows a fairly well defined progression from blue to red as the turbidity increases: blue in the open ocean, green in relatively clear coastal waters, yellow green in bloom conditions, and yellow orange in waters with a high sediment load. In exceptionally turbid waters, generally rivers with high sediment load…" (Gray et al., 2015). The findings indicated that the maximum penetrating wavelength redshifted from 532 nm as turbidity increased, suggesting that LiDAR systems designed for open ocean waters should have a wavelength from 470 and 490 nm, and in turbid water conditions perhaps a wavelength between 560 and 580 nm would be more suitable, with a multiwavelength laser, with to change wavelength reflecting environmental changes (Gray et al., 2015).

In some studies, applications to water optical and chemical dynamics are undertaken. For example, Collister et al. (2020: 4650) present a study and point out that "the goal of this project was to explore the utility of LiDAR for describing the distribution of upper ocean optical and biogeochemical properties over a wide range of oceanographic conditions, including turbid, eutrophic coastal waters, optically clear oligotrophic waters, and calcite-rich coccolithophore bloom waters." They argue that "oceanographic LiDAR can provide remote estimates of the vertical distribution of suspended particles in natural waters, potentially revolutionizing our ability to characterize marine ecosystems and properly represent them in models of upper ocean biogeochemistry." Their study was based on "a ship-based system to measure the LiDAR attenuation coefficient (α) and linear depolarization ratio (δ) across a variety of optically and biogeochemically distinct water masses, including turbid coastal waters, clear oligotrophic waters, and calcite rich waters associated with a mesoscale coccolithophore bloom" (Collister et al., 2020: 4650).

The sea surface-inherent optical properties (IOPs, which refer to the "scattering and absorption properties of seawater and its dissolved and suspended materials"; see Zaneveld, 2013) were measured to determine the response of α and δ to variation in particle abundance and composition. The magnitude of α was consistent with the diffuse attenuation coefficient (Kd, which refers to "a measure of how light dissipates with depth in water" see National Geodetic Survey, 2022), and the α versus Kd relationship was not linear, and δ was "positively related to the scattering optical depth and the calcite fraction of backscattering" (Collister et al., 2020). Collister et al. (2020) argue that such LiDAR-based studies are useful, as "the substrates and energy sources that fuel upper ocean biogeochemistry exhibit strong spatial gradients that play an important role in the structure and function of marine ecosystems." And hence "oceanographic LiDAR has the potential to revolutionize our understanding of marine ecosystems and their role in elemental cycling and global climate processes by providing a means to measure the vertical distribution of upper ocean biomass at spatiotemporal resolutions that are unachievable using standard in situ sampling techniques" (Collister et al., 2020).

For measurements in water, LiDAR, which is based on the measurement of laser pulses and the detection of backscattering from particles in the water column, may measure the "biogeochemically relevant materials in the ocean," and this "makes LiDAR the only presently available candidate for making above-water measurements of both the horizontal and vertical distribution of upper ocean ecosystems" (Collister et al., 2020). Current contributions of LiDAR-based methods, some in their infancy, include high-resolution, vertical and horizontal measurements of phytoplankton biomass, a more accurate recording for models of upper ocean primary production through a more accurate representation of biomass distributions, better measurement of distributions of upper trophic level aquatic organisms, recording of the links between upper ocean ecosystems and atmospheric features like cloud and aerosol formation, and the prediction of global climate dynamics (Collister et al., 2020; see also Churnside & Thorne, 2005; Shaw et al., 2008; Churnside & Donaghay, 2009; Hill & Zimmerman, 2010; Churnside et al., 2011; Roddewig et al., 2017; Schulien et al., 2017; Behrenfeld et al. 2019a, b).

There are also studies of LiDAR applications to marine organism distribution patterns. For example, Churnside and Donaghay (2009: 778) present a study of LiDAR applications to plankton studies, applied to oceanic and coastal waters, with examples from "upwelling regions, areas with large fresh-water influx, and warm-core eddies." The findings indicated that the scattering detected by LiDAR was likely caused by plankton, and there was spatial variability in the plankton layers, which suggest the possibility that airborne lidar can be used to detect and map the spatial extent of thin scattering layers and to link these to larger scale physical processes and can also be used to track their development and to guide ship-based sampling. This technological assessment "will be crucial in future studies designed to test the hypothesis that thin plankton layers have the spatial extent and intensity to play a key role in controlling the recruitment of fish larvae, biogeochemical cycling, trophic transfer processes, plankton biodiversity, and harmful algal bloom dynamics" (Churnside & Donaghay, 2009: 778).

Hostetler et al. (2018: 121) present another study of LiDAR applications to research marine organism distributions and ecologies. They argue that although satellite passive ocean color technology has "provided an unbroken ~20-year record of global ocean plankton properties" there is evidence that "this measurement approach has inherent limitations in terms of spatial-temporal sampling and ability to resolve vertical structure within the water column" (Hostetler et al., 2018: 121). However, LiDAR-based techniques may be promoted as a solution, where measurements from a spaceborne lidar are combined with ocean color data. Hostetler et al. (2018: 121) further justify the use of satellite LiDAR, as airborne LIDARs have been used for many years for the study of ocean subsurface properties, but recent technological developments have demonstrated that satellite LiDAR can measure plankton properties. The "satellite LiDAR era in oceanography has arrived" although "oceanographic research with satellite lidars is in its infancy" (Hostetler et al., 2018). Examples include the Cloud-Aerosol Lidar with Orthogonal Polarization (CALIOP) instrument, located in space from 2006 on the Cloud-Aerosol Lidar and Infrared Pathfinder Satellite Observation (CALIPSO) platform (see also Hu, 2009; Winker et al., 2009; Lu et al., 2016). The advantage of LiDAR for such work is that "unlike the ocean color technique's reliance on the sun as a source of light (thus, 'passive' remote sensing), a LiDAR uses lasers to generate its own photons that are ultimately scattered back to the instrument's receiver (thus, 'active' remote sensing)" (Hostetler et al., 2018).

Larger marine animals may also be studied using LiDAR methods (Churnside & Thorne, 2005; Dawson et al., 2017). For example, Behrenfeld et al. (2019a: 257) argue that LiDAR measurements may be used for the study of marine mammals that feed on plankton, pointing out that "every night across the world's oceans, numerous marine animals arrive at the surface of the ocean to feed on plankton after an upward migration of hundreds of meters. Just before sunrise, this migration is reversed, and the animals return to their daytime residence in the dark mesopelagic zone (at a depth of 200–1000 m)." The authors refer this "daily excursion" as diel vertical migration (DVM) or diurnal vertical migration and may be an attempt to avoid predators in the sunlit ocean surface. Such DVMs are often recorded by acoustic systems mounted on surface vessels (for example, acoustic Doppler current profilers) and "these data show that night-time arrival and departure times are highly conserved across ocean regions and that daytime descent depths increase with water clarity, indicating that animals have faster swimming speeds in clearer waters" (Behrenfeld, Gaube, et al., 2019a: 257). These may not provide sufficient information, as "after decades of acoustic measurements, vast ocean areas remain unsampled and places for which data are available typically provide information for only a few months, resulting in an incomplete understanding of DVMs" and "addressing this issue is important, because DVMs have a crucial role in global ocean biogeochemistry" (Behrenfeld, Gaube, et al., 2019a: 257).

Additionally, "night-time feeding at the surface and daytime metabolism of this food at depth provide an efficient pathway for carbon and nutrient export" (Behrenfeld, Gaube, et al., 2019a: 257). A satellite-mounted LiDAR instrument was used to record the global distributions of optical signals from the DVM animals that reach the night ocean surface. The findings indicated the animals usually comprised a larger fraction of the plankton abundance in the clear subtropical gyres. This was consistent with the hypothesis that such animals avoided predators in such regions. The total DVM biomass was higher in more productive regions with higher food availability, and the "the 10-year satellite record reveals significant temporal trends in DVM biomass and correlated variations in DVM biomass and surface productivity" and the authors conclude that "these results provide a detailed view of DVM activities globally and a path for refining the quantification of their biogeochemical importance" (Behrenfeld, Gaube, et al., 2019a: 257).

Churnside et al. (2011: 621) present a similar LiDAR-based study, noting that "intense, ephemeral foraging events within localized hot spots represent important trophic transfers to top predators in marine ecosystems, though the spatial extent and temporal overlap of predators and prey are difficult to observe using traditional methods." The case study of the southeastern Bering Sea is a region that includes high marine productivity, and a hot spot, where three foraging events were recorded. The methodology used airborne LiDAR, as well as aerial surveys, visual counts, ship-based acoustics and net catches, and space-based synthetic aperture radar. The recorded foraging events (including buildups of concentrations of euphausiids, copepods, herring, other small pelagic fishes, humpback whales, Dall's porpoise, short-tailed shearwaters, northern fulmars, and other pelagic seabirds) lasted for 1 to 8 days, and "build-up and break down of the events occurred on 24 h time scales, and diameters ranged from 10 to 20 km" (Churnside et al., 2011: 621). The findings from the LiDAR and acoustic remote sensing data indicated that the prey densities inside the events were much higher than those outside the events. The authors conclude that "this implies a need for either very intensive traditional surveys covering large expanses or for adaptive surveys guided by remote sensing. To our knowledge, this is the first time that an Alaskan hot spot was monitored with the combination of airborne and satellite remote sensing" (Churnside et al., 2011: 621; see also Pauly et al., 1998; Sinclair & Stabeno, 2002; Hunt Jr & Stabeno, 2005, Jahncke et al., 2005, Ladd et al., 2005, Vlietstra et al., 2005, Batten et al., 2006, Palacios et al., 2006, Piatt et al., 2006, Zainuddin et al., 2006).

Conclusions

This chapter has examined the relations between the biogeochemical and oceanographic sciences, in terms of their relative contributions to these links and their roles in the explanation of global ecologies. Key complexes emerge from this brief but detailed summary. Oceanography has at least a third of the components that comprise global biogeochemistry and includes the cycling of major elements such as oxygen, nitrogen, hydrogen, and sulfur. The science of oceanography is a vast undertaking, with substantial transdisciplinary and interdisciplinary foci, such as biological oceanography and marine biology and marine biogeography, geological oceanography and marine geology, physical oceanography, and marine physics, and chemical oceanography and marine chemistry and biochemistry. In all these clusters, it is arguable that the branch termed "oceanography" tends toward broader, environmental linkages, while the subdiscipline with the prefix "marine" is possibly more specialized, more fully linked with the basic science and its application to the seawater and its underlying geology. High-level technology, including LiDAR within the field of remote sensing, is also developing to enable more sophisticated analyses of oceanographic data. The current status of the field of oceanography is therefore an important indicator and complement for the development of biogeochemistry, including all of the studies of biogeochemistry of the oceans and some of the atmospheric and terrestrial studies. Considering the important role of the oceans in global environmental change, this type of study contributes to the examination of future developments in the field of biogeochemistry.

References

Ali, S. K., Janjuhah, H. T., Shahzad, S. M., Kontakiotis, G., Saleem, M. H., Khan, U., Zarkogiannis, S. D., Makri, P., & Antonarakou, A. (2021). Depositional sedimentary facies, stratigraphic control, paleoecological constraints, and paleogeographic reconstruction of Late Permian Chhidru formation (Western Salt Range, Pakistan). *Journal of Marine Science and Engineering, 9*(12), 1372. https://doi.org/10.3390/jmse9121372

Allocca, D. M., London, M. A., Curran, T. P., Concannon, B. M., Contarino, V. M., Prentice, J., Mullen, L. J., & Kane, T. J. (2002). Ocean water clarity measurement using shipboard lidar systems. In R. J. Frouin & G. D. Gilbert (Eds.), *Proceedings Volume 4488: International symposium on optical science and technology* (pp. 106–114). https://doi.org/10.1117/12.452807

Amos, H. M., Jacob, D. J., Streets, D. G., & Sunderland, E. M. (2013). Legacy impacts of all-time anthropogenic emissions on the global mercury cycle. *Global Biogeochemical Cycles, 27*(2), 410–421. https://doi.org/10.1002/gbc.20040

Andersson, M. E., Gardfeldt, K., Wangberg, I., Sprovieri, F., Pirrone, N., & Lindqvist, O. (2007). Seasonal and daily variation of mercury evasion at coastal and offshore sites from the Mediterranean Sea. *Marine Chemistry, 104*(3-4), 214–226. https://doi.org/10.1016/j.marchem.2006.11.003

Antoine, D., André, J. M., & Morel, A. (1996). Oceanic primary production: 2. Estimation at global scale from satellite (coastal zone color scanner) chlorophyll. *Global biogeochemical cycles, 10*(1), 57–69. https://doi.org/10.1029/95GB02832

Arce-Chamorro, C., Vidal-Romaní, J. R., & Sanjurjo-Sánchez, J. (2022). New model of coastal evolution in the Ria de Vigo (NW Spain) from MIS2 to present day based on the Aeolian sedimentary record. *Journal of Marine Science and Engineering, 10*(10), 1350. https://doi.org/10.3390/jmse10101350

Arctic Biodiversity Data Service. (2016). *Longhurst Marine Provinces*. Retrieved from https://geo.abds.is/geonetwork/srv/api/records/e903ac58-ff9b-40aa-b2f6-1592a46b4f41#:~:text=At%20the%20first%20level%20of,in%20every%20major%20ocean%20basin

Aumont, O., Maier-Reimer, E., Blain, S., & Monfray, P. (2003). An ecosystem model of the global ocean including Fe, Si, P colimitations. *Global Biogeochemical Cycles, 17*(2), 1–15. https://doi.org/10.1029/2001GB001745

Baker, B. J., Appler, K. E., & Gong, X. (2021). New microbial biodiversity in marine sediments. *Annual Review of Marine Science, 13*, 161–175. https://doi.org/10.1146/annurev-marine-032020-014552

Bangor University. (2020). *What is ocean sciences?* Retrieved from https://www.bangor.ac.uk/oceansciences/about/what.php.en

Bates, N. R., Astor, Y. M., Church, M. J., Currie, K., Dore, J. E., González-Dávila, M., Lorenzoni, L., Muller-Karger, F., Olafsson, J., & Santana-Casiano, J. M. (2014). A time-series view of changing ocean chemistry due to ocean uptake of anthropogenic CO_2 and ocean acidification. *Oceanography, 27*(1), 126–141. https://doi.org/10.5670/oceanog.2014.16

Batten, S. D., Hyrenbach, K. D., Sydeman, W. J., Morgan, K. H., Henry, M. F., Yen, P. P., & Welch, D. W. (2006). Characterising meso-marine ecosystems of the North Pacific. *Deep sea research. Part II: Topical studies in oceanography, 53*(3–4), 270–290. https://doi.org/10.1016/j.dsr2.2006.01.004

Behrenfeld, M. J., & Falkowski, P. G. (1997). A Consumer's guide to phytoplankton primary productivity models. *Limnology and Oceanography, 42*, 1479–1491. https://doi.org/10.4319/lo.1997.42.7.1479

Behrenfeld, M. J., Randerson, J. T., McClain, C. R., Feldman, G. C., Los, S. O., Tucker, C. J., Falkowski, P. G., Field, C. B., Frouin, R., Esaias, W. E., Kolber, D. D., & Pollack, N. H. (2001). Biospheric primary production during an ENSO transition. *Science (New York, NY), 291*(5513), 2594–2597. https://doi.org/10.1126/science.1055071

Behrenfeld, M., Hu, Y., Hostetler, C., Dall'Olmo, G., Rodier, S., Hair, J., & Trepte, C. (2013). Space-based lidar measurements of global ocean carbon stocks. *Geophysical Research Letters, 40*(16), 4355–4360. https://doi.org/10.1002/grl.50816

Behrenfeld, M. J., Gaube, P., Della Penna, A., O'Malley, R. T., Burt, W. J., Hu, Y., Bontempi, P. S., Steinberg, D. K., Boss, E. S., Siegel, D. A., Hostetler, C. A., Tortell, P. D., & Doney, S. C. (2019a). Global satellite-observed daily vertical migrations of ocean animals. *Nature, 576*, 257–261. https://doi.org/10.1038/s41586-019-1796-9

Behrenfeld, M. J., Moore, R. H., Hostetler, C. A., Graff, J., Gaube, P., Russell, L. M., Chen, G., Doney, S. C., Giovannoni, S., Liu, H., et al. (2019b). The North Atlantic Aerosol and Marine Ecosystem Study (NAAMES): Science motive and mission overview. *Frontiers in Marine Science, 6*. https://doi.org/10.3389/fmars.2019.00122

Ben-David, M., & Flaherty, E. A. (2012). Stable isotopes in mammalian research: A beginner's guide. *Journal of Mammalogy, 93*(2), 312–328. https://doi.org/10.1644/11-MAMM-S-166.1

Bibby, R., Cleall-Harding, P., Rundle, S., Widdicombe, S., & Spicer, J. (2007). Ocean acidification disrupts induced defences in the intertidal gastropod Littorina littorea. *Biology Letters, 3*(6), 699–701. https://doi.org/10.1098/rsbl.2007.0457

Bolles, D. (2022a). *Physical ocean*. Retrieved from https://science.nasa.gov/earth-science/oceanography/physical-ocean

Bolles, D. (2022b). *Carbon cycle. NASA Science*. Retrieved from https://science.nasa.gov/earth-science/oceanography/ocean-earth-system/ocean-carbon-cycle#:~:text=of%20biological%20oceanography.-,Physical%20oceanography%20influences%20the%20carbon%20cycle%20through%20its%20modulation%20of,e.g.%2C%20surface%20wind%20speeds

Borges, A. V., Schiettecatte, L.-S., Abril, G., Delille, B., & Gazeau, F. (2006). Carbon dioxide in European coastal waters. *Estuarine, Coastal and Shelf Science, 70*(3), 375–387. https://doi.org/10.1016/j.ecss.2006.05.046

Boyle, E. A. (2007). Introduction: Chemical oceanography. *Chemical Reviews, 107*(2), 305–307. https://doi.org/10.1021/cr0502596

Briggs, J. C. (1974). *Marine zoogeography*. McGraw-Hill.

Britannica, T. Editors of Encyclopaedia. (2019). *Oceanography*. Encyclopedia Britannica. https://www.britannica.com/science/oceanography

Bukin, O. A., Major, A. Y., Pavlov, A. N., Shevtsov, B. M., & Kholodkevich, E. D. (1998). Measurement of the light scattering layers structure and detection of the dynamic processes in the upper ocean layer by shipborne lidar. *International Journal of Remote Sensing, 19*(4), 707–715. https://doi.org/10.1080/014311698215946

Bulling, M. T., Hicks, N., Murray, L., Paterson, D. M., Raffaelli, D., White, P. C., & Solan, M. (2010). Marine biodiversity-ecosystem functions under uncertain environmental futures. *Philosophical Transactions of the Royal Society of London. Series B, Biological Sciences, 365*(1549), 2107–2116. https://doi.org/10.1098/rstb.2010.0022

Burdige, D. J. (2007). Preservation of organic matter in marine sediments: Controls, mechanisms, and an imbalance in sediment organic carbon budgets? *Chemical Reviews, 107*(2), 467–485. https://doi.org/10.1021/cr050347q

Calosi, P., Rastrick, S. P., Graziano, M., Thomas, S. C., Baggini, C., Carter, H. A., Hall-Spencer, J. M., Milazzo, M., & Spicer, J. I. (2013). Distribution of sea urchins living near shallow water CO_2 vents is dependent upon species acid-base and ion-regulatory abilities. *Marine Pollution Bulletin, 73*(2), 470–484. https://doi.org/10.1016/j.marpolbul.2012.11.040

Campbell, M. (2018). *Geomatics and conservation biology*. Nova Science Publishers.

Campbell, M., & Jha, K. K. (2021). *Critical research techniques in animal and habitat ecology: International examples*. Nova Science Publishers.

Carroll University. (2022). *Oceanography*. Retrieved from https://www.carrollu.edu/academics/arts-sciences/marine-sciences/oceanography

Catlett, D., Siegel, D. A., Matson, P. G., Wear, E. K., Carlson, C. A., Lankiewicz, T. S., & Iglesias-Rodriguez, M. D. (2022). *Plumes and Blooms: Microbial eukaryote diversity and composition ver 1*. Environmental Data Initiative. https://doi.org/10.6073/pasta/bbe75db57c4ce7f7842e22923ae6badf

Chavez, F. P., Buck, K. R., Coale, K. H., Martin, J. H., DiTullio, G. R., Welschmeyer, N. A., Jacobson, A. C., & Barber, R. T. (1991). Growth rates, grazing, sinking, and iron limitation of equatorial Pacific phytoplankton. *Limnology and Oceanography, 36*(8), 1816–1833. https://doi.org/10.4319/lo.1991.36.8.1816

Chekalyuk, A., Hoge, F., Swift, R., & Yungel, J. (2003). New technological developments for ocean LIDAR biomonitoring. In *Proceedings of SPIE – The International Society for Optical Engineering. Ocean Remote Sensing and Imaging II* (p. 5155). https://doi.org/10.1117/12.507658

Chelton, D. B. (1994). Physical oceanography: A brief overview for statisticians. *Statistical Science, 9*(2), 150–166. https://doi.org/10.1214/ss/1177010480

Churnside, J. H., & Donaghay, P. L. (2009). Thin scattering layers observed by airborne lidar. *ICES Journal of Marine Science, 66*(4), 778–789. https://doi.org/10.1093/icesjms/fsp029

Churnside, J. H., & Thorne, R. E. (2005). Comparison of airborne lidar measurements with 420 kHz echo-sounder measurements of zooplankton. *Applied Optics, 44*(26), 5504–5511. https://doi.org/10.1364/AO.44.005504

Churnside, J. H., Tatarskii, V. V., & Wilson, J. J. (1998). Oceanographic lidar attenuation coefficients and signal fluctuations measured from a ship in the Southern California Bight. *Applied Optics, 37*(15), 3105–3112. https://doi.org/10.1364/ao.37.003105

Churnside, J., Wilson, J., & Tatarskii, V. (2001). Airborne lidar for fisheries applications. *Optical Engineering, 40*(3), 406–414. https://doi.org/10.1117/1.1348000

Churnside, J. H., Brown, E. D., Parker-Stetter, S., Horne, J. K., Hunt, G. L., Hillgruber, N., Sigler, M. F., & Vollenweider, J. J. (2011). Airborne remote sensing of a biological hot spot in the southeastern Bering Sea. *Remote Sensing, 3*(3), 621–637. https://doi.org/10.3390/rs3030621

Churnside, J. H., Marchbanks, R. D., Lee, J. H., Shaw, J. A., Weidemann, A., & Donaghay, P. L. (2012). Airborne lidar detection and characterization of internal waves in a shallow fjord. *Journal of Applied Remote Sensing, 6*(1), 063611. https://doi.org/10.1117/1.JRS.6.063611

Cochran, J. K. (2014). Chemical oceanography. *Reference Module in Earth Systems and Environmental Sciences*. https://doi.org/10.1016/B978-0-12-409548-9.09152-1

Collins, D., McDermid, I. S., Breckinridge, J., Bell, J., & Zanoni, R. (1984). Recent progress in the measurement of temperature and salinity by optical scattering. Ocean optics VII. *International Society for Optics and Photonics, 489*. https://doi.org/10.1117/12.943311

Collins, S., Boyd, P. W., & Doblin, M. A. (2020). Evolution, microbes, and changing ocean conditions. *Annual Review of Marine Science, 12*, 181–208. https://doi.org/10.1146/annurev-marine-010318-095311

Collister, B. L., Zimmerman, R. C., Hill, V. J., Sukenik, C. I., & Balch, W. M. (2020). Polarized lidar and ocean particles: Insights from a mesoscale coccolithophore bloom. *Applied Optics, 59*(15), 4650–4662. https://doi.org/10.1364/AO.389845

Coxe, N., Casas, S., Marshall, D., La Peyre, M., Kelly, M., & La Peyre, J. (2023). Differential hypoxia tolerance of eastern oysters from the northern Gulf of Mexico at elevated temperature. *Journal of Experimental Marine Biology and Ecology, 559*, 151840. https://doi.org/10.1016/j.jembe.2022.151840

Dai, H., Li, H., & Li, Y. (2022). Fragmentation characteristics of seafloor massive sulfides: A coupled fluid-particle flow simulation. *Journal of Marine Science and Engineering, 10*(9), 1306. https://doi.org/10.3390/jmse10091306

Dashfield, S. L., Somerfield, P. J., Widdicombe, S., Austen, M. C., & Nimmo, M. (2008). Impacts of ocean acidification and burrowing urchins on within-sediment pH profiles and subtidal nematode communities. *Journal of Experimental Marine Biology and Ecology, 365*(1), 46–52. https://doi.org/10.1016/j.jembe.2008.07.039

Dawson, S. M., Bowman, M. H., Leunissen, E., & Sirguey, P. (2017). Inexpensive aerial photogrammetry for studies of whales and large marine animals. *Frontiers of Marine Science, 4*, 366. https://doi.org/10.3389/fmars.2017.00366

Department of Marine Sciences. (2022). *Chemical oceanography*. Retrieved from https://www.marsci.uga.edu/research/content/chemical-oceanography#:~:text=Chemical%20oceanography%20encompasses%20the%20study,material%2C%20and%20the%20ocean%20sediments

DeVries, T., Primeau, F. W., & Deutsch, C. A. (2012). The sequestration efficiency of the biological pump. *Geophysical Research Letters, 39*(13), 1–5. https://doi.org/10.1029/2012GL051963

Dixon, R. W., O'Brien, S. R., & Hodge, J. B. (2021). The oceanography of Cuchlaine A.M. King. *Progress in Physical Geography: Earth and Environment, 45*(3), 460–463. https://doi.org/10.1177/0309133321100528

Donat, J. R., & Bruland, K. W. (1995). Trace elements in the oceans. In *Trace elements in natural waters*. CRC Press.

Doney, S. C. (2010). The growing human footprint on coastal and open-ocean biogeochemistry. *Science (New York, NY), 328*(5985), 1512–1516. https://doi.org/10.1126/science.1185198

Donohue, P., Calosi, P., Bates, A. H., Laverock, B., Rastrick, S., Mark, F. C., Strobel, A., & Widdicombe, S. (2012). Physiological and behavioural impacts of exposure to elevated pCO2 on an important ecosystem engineer, the burrowing shrimp Upogebia deltaura. *Aquatic Biology, 15*, 73–86. https://doi.org/10.3354/ab00408

Douglas, S., & Beveridge, T. J. (1998). Mineral formation by bacteria in natural microbial communities. *FEMS Microbiology Ecology, 26*(2), 79–88. https://doi.org/10.1111/j.1574-6941.1998.tb00494.x

Drijfhout, S., Marshall, D., & Dijkstra, H. (2013). Chapter 11 – Conceptual models of the wind-driven and thermohaline circulation. *International Geophysics., 103*, 257–282. https://doi.org/10.1016/B978-0-12-391851-2.00011-8

Dubinya, N., Bayuk, I., Hortov, A., Myatchin, K., Pirogova, A., & Shchuplov, P. (2022). Prediction of overpressure zones in marine sediments using rock-physics and other approaches. *Journal of Marine Science and Engineering, 10*(8), 1127. https://doi.org/10.3390/jmse10081127

Dugdale, R. C., & Wilkerson, F. P. (1998). Silicate regulation of new production in the equatorial Pacific upwelling. *Nature, 391*(6664), 270–273. https://doi.org/10.1038/34630

Ekman, S. (1953). *Zoogeography of the Sea*. Sidgwick and Jackson.

Ellis, E. (2007). What is marine biology? Defining a science in the United States in the mid 20th century. *History and Philosophy of the Life Sciences, 29*(4), 469–493.

Environmental Science.Org. (2022a). *What is an oceanographer?* https://www.environmental-science.org/career/oceanographer

Environmental Science.Org. (2022b). *What is a biological oceanographer?* https://www.environmentalscience.org/career/biological-oceanographer

Falkowski, P. G., Greene, R. M., & Geider, R. J. (1992). Physiological limitations on phytoplankton productivity in the ocean. *Oceanography, 5*(2), 84–91. https://doi.org/10.5670/oceanog.1992.14

Falkowski, P. G., Barber, R. T., & Smetacek, V. V. (1998). Biogeochemical controls and feedbacks on ocean primary production. *Science (New York, NY), 281*(5374), 200–207. https://doi.org/10.1126/science.281.5374.200

Fazal, A. G., Shah, F., Miraj, M. A. F., Janjuhah, H. T., Kontakiotis, G., & Jan, A. K. (2022). Geochemical analysis of cretaceous shales from the Hazara Basin, Pakistan: Provenance signatures and paleo-weathering conditions. *Journal of Marine Science and Engineering, 10*(6), 800. https://doi.org/10.3390/jmse10060800

Field, C. B., Behrenfeld, M. J., Randerson, J. T., & Falkowski, P. (1998). Primary production of the biosphere: Integrating terrestrial and oceanic components. *Science (New York, NY), 281*(5374), 237–240. https://doi.org/10.1126/science.281.5374.237

Fitzgerald, W. F., Lamborg, C. H., & Hammerschmidt, C. R. (2007). Marine biogeochemical cycling of mercury. *Chemical Reviews, 107*(2), 641–662. https://doi.org/10.1021/cr050353m

Fitzwater, S. E., Coale, K. H., Gordon, R. M., Johnson, K. S., & Ondrusek, M. E. (1996). Iron deficiency and phytoplankton growth in the equatorial Pacific. *Deep Sea Research Part II: Topical Studies in Oceanography, 43*(4–6), 995–1015. https://doi.org/10.1016/0967-0645(96)00033-1

Florida Institute of Technology. (2022). *What's the difference between oceanography and marine biology?* Retrieved from https://news.fit.edu/archive/difference-between-oceanography-and-marine-biology/#:~:text=So%2C%20what's%20the%20difference%20between,%2C%20physiology%2C%20and%20life%20history

Flowers, K. I., Babcock, E. A., Papastamatiou, Y. P., Bond, M. E., Lamb, N., Miranda, A., Nuñez, R., Valentin-Albanese, J., Clementi, G. M., Kelley, M. C., & Chapman, D. D. (2022). Varying reef shark abundance trends inside a marine reserve: Evidence of a Caribbean reef shark decline. *Marine Ecology Progress Series, 683*, 97–107. https://doi.org/10.3354/meps13954

Fraser, J. H. (1958). The drift of the planktonic stages of fish in the northeast Atlantic and its possible significance to the stocks of commercial fish. *International Commission for the Northwest Atlantic Fisheries Special Publication, 1*, 289–310.

Fry, E. S., Emery, Y., Quan, X., & Katz, J. W. (1997). Accuracy limitations on Brillouin lidar measurements of temperature and sound speed in the ocean. *Applied Optics, 36*(27), 6887–6894. https://doi.org/10.1364/ao.36.006887

Gantt, B., & Meskhidze, N. (2013). The physical and chemical characteristics of marine primary organic aerosol: A review. *Atmospheric Chemistry and Physics, 13*, 3979–3996. https://doi.org/10.5194/acp-13-3979-2013

Garcia, G. G., Garcia, A. J. V., Henriques, M. H. P., Marques, R. M., & Pena dos Reis, R. (2021). Taphofacies and petrofacies theoretical marine models applied to the Coquina of the Amaral Formation (Lusitanian Basin, Portugal). *Journal of Marine Science and Engineering, 9*(12), 1319. https://doi.org/10.3390/jmse9121319

Gaston, K. J. (2003). *The structure and dynamics of geographic ranges*. Oxford University Press.

Ghalambor, C. K., McKay, J. K., Carroll, S. P., & Reznick, D. N. (2007). Adaptive versus non-adaptive phenotypic plasticity and the potential for contemporary adaptation in new environments. *Functional Ecology, 21*(3), 394–407. https://doi.org/10.1111/j.1365-2435.2007.01283.x

Giamali, C., Kontakiotis, G., Antonarakou, A., & Koskeridou, E. (2021). Ecological constraints of plankton bio-indicators for water column stratification and productivity: A case study of the Holocene North Aegean Sedimentary Record. *Journal of Marine Science and Engineering, 9*(11), 1249. https://doi.org/10.3390/jmse9111249

Gill, A. E. (1982). *Atmosphere-ocean dynamics*. Academic.

Godbold, J. A., & Calosi, P. (2013). Ocean acidification and climate change: Advances in ecology and evolution. *Philosophical transactions of the Royal Society of London. Series B, Biological Sciences, 368*(1627), 20120448. https://doi.org/10.1098/rstb.2012.0448

Godbold, J. A., & Solan, M. (2009). Relative importance of biodiversity and the abiotic environment in mediating an ecosystem process. *Marine Ecology Progress Series, 396*, 273–282. https://doi.org/10.3354/meps08401

Godbold, J. A., Solan, M., & Killham, K. (2009). Consumer and resource diversity effects on marine macroalgal decomposition. *Oikos, 118*(1), 77–86. https://doi.org/10.1111/j.1600-0706.2008.17072.x

Godbold, J. A., Bulling, M. T., & Solan, M. (2011). Habitat structure mediates biodiversity effects on ecosystem properties. *Proceedings of the Royal Society B: Biological Sciences, 278*(1717), 2510–2518. https://doi.org/10.1098/rspb.2010.2414

Goldberg, E. D. (1954). Marine Geochemistry 1. Chemical scavengers of the Sea. *The Journal of Geology, 62*(3), 249–265. https://doi.org/10.1086/626161

Gray, D. J., Anderson, J., Nelson, J., & Edwards, J. (2015). Using a multiwavelength LiDAR for improved remote sensing of natural waters. *Applied Optics, 54*(31), F232–F242. https://doi.org/10.1364/AO.54.00F232

Gruber, N., & Doney, S. C. (2019). Modeling of ocean biogeochemistry and ecology. In J. K. Cochran (Ed.), *Encyclopedia of Ocean Sciences* (Vol. 5, 3rd ed., pp. 291–302). Elsevier. https://doi.org/10.1016/B978-0-12-409548-9.11409-5

Guenther, G. C., Thomas, R. W., & LaRocque, P. E. (1996). Design considerations for achieving high accuracy with the shoals bathymetric lidar system. In V. I. Feigels & Y. I. Kopilevich (Eds.), *CIS selected papers: Laser remote sensing of natural waters: From theory to practice* (pp. 54–71). https://doi.org/10.1117/12.258353

Hale, R., Calosi, P., McNeill, L., Mieszkowska, N., & Widdicombe, S. (2011). Predicted levels of future ocean acidification and temperature rise could alter community structure and biodiversity in marine benthic communities. *Oikos, 120*(5), 661–674. https://doi.org/10.1111/j.1600-0706.2010.19469.x

Halpern, B. S., Walbridge, S., Selkoe, K. A., Kappel, C. V., Micheli, F., D'Agrosa, C., Bruno, J. F., Casey, K. S., Ebert, C., Fox, H. E., Fujita, R., Heinemann, D., Lenihan, H. S., Madin, E. M., Perry, M. T., Selig, E. R., Spalding, M., Steneck, R., & Watson, R. (2008). A global map of human impact on marine ecosystems. *Science, 319*(5865), 948–952. https://doi.org/10.1126/science.1149345

Hamblin, W. K., & Christiansen, E. H. (1998). *Earth's dynamic systems* (8th ed.). Prentice-Hall.

Hansen, J., Sato, M., Ruedy, R., Lo, K., Lea, D. W., & Medina-Elizade, M. (2006). Global temperature change. *Proceedings of the National Academy of Sciences, 103*(39), 14288–14293. https://doi.org/10.1073/pnas.0606291103

Harris, L. G., & Gibson, J. L. (2023). Contrasting decadal recruitment patterns in the sea urchin Strongylocentrotus droebachiensis in the Gulf of Maine. *Journal of Experimental Marine Biology and Ecology, 558*, 151832. https://doi.org/10.1016/j.jembe.2022.151832

Hátún, H., Payne, M. R., Beaugrand, G., Reid, P. C., Sandø, A. B., & Drange, H. (2009). Large biogeographical shifts in the north-eastern Atlantic Ocean: From the subpolar gyre, via plankton, to blue whiting and pilot whales. *Progress in Oceanography, 80*(3-4), 149–162. https://doi.org/10.1016/j.pocean.2009.03.001

Hátún, H., Silva, T. S. G., Skagseth, Ø., & Grønkjær, P. (2022). Editorial: Physical drivers of biogeographical shifts in the Northeastern Atlantic – And adjacent shelves. *Frontiers in Marine Science, 9*, 1062204. https://doi.org/10.3389/fmars.2022.1062204

Hayat, S., Skampa, E., Gogou, A., Stavrakakis, S., Parinos, C., & Triantaphyllou, M. (2022). Seasonal variability in present-day coccolithophore fluxes in deep Eastern Mediterranean Sea: A multi-year study (2015–2017) of coccolithophore export in SE Ionian Sea at 4300 m depth. *Journal of Marine Science and Engineering, 10*(11), 1761. https://doi.org/10.3390/jmse10111761

Hedgpeth, J. W. (1957). Marine biogeography. In J. W. Hedgpeth (Ed.), *Treatise on marine ecology and palaeoecology* (pp. 359–382). Geological Society of America.

Henderson, G. M. (2016). Ocean trace element cycles. *Philosophical Transactions. Series A, Mathematical, Physical, and Engineering Sciences, 374*(2081), 20150300. https://doi.org/10.1098/rsta.2015.0300

Hicks, N., Bulling, M. T., Solan, M., Raffaelli, D., White, P. C., & Paterson, D. M. (2011). Impact of biodiversity-climate futures on primary production and metabolism in a model benthic estuarine system. *BMC Ecology, 11*(7). https://doi.org/10.1186/1472-6785-11-7

Hill, V. J., & Zimmerman, R. C. (2010). Estimates of primary production by remote sensing in the Arctic Ocean: Assessment of accuracy with passive and active sensors. *Deep Sea Research Part I: Oceanographic Research Papers, 57*(10), 1243–1254. https://doi.org/10.1016/j.dsr.2010.06.011

Holdsworth, A. M., Zhai, L., Lu, Y., & Christian, J. R. (2021). future changes in oceanography and biogeochemistry along the Canadian Pacific continental margin. *Frontiers in Marine Science, 8*, 602991. https://doi.org/10.3389/fmars.2021.602991

Hopkins, J. B., 3rd, & Ferguson, J. M. (2012). Estimating the diets of animals using stable isotopes and a comprehensive Bayesian mixing model. *PLoS One, 7*(1), e28478. https://doi.org/10.1371/journal.pone.0028478

Hostetler, C. A., Behrenfeld, M. J., Hu, Y., Hair, J. W., & Schulien, J. A. (2018). Spaceborne LiDAR in the study of marine systems. *Annual Review of Marine Science, 10*, 121–147. https://doi.org/10.1146/annurev-marine-121916-063335

Hoyle, B. D., & Lerner, K. L. (2023). *Oceanography, chemical. Water Encyclopedia.* http://www.waterencyclopedia.com/Oc-Po/Oceanography-Chemical.html#:~:text=Chemical%20oceanography%20can%20be%20further,found%20on%20the%20ocean%20floor

Hu, Y. (2009). Ocean, land, and meteorology studies using space-based LiDAR measurements. In *5th WSEAS international conference on remote sensing (REMOTE"09)* (No. LF99-9425) https://ntrs.nasa.gov/api/citations/20090037431/downloads/20090037431.pdf

Hunt, G. L., Jr., & Stabeno, P. J. (2005). Oceanography and ecology of the Aleutian Archipelago: Spatial and temporal variation. *Fisheries Oceanography, 14*, 292–306. https://doi.org/10.1111/j.1365-2419.2005.00378.x

Hussain, A., & Al-Ramadan, K. (2022). Organic matter burial in deep-sea fans: A depositional process-based perspective. *Journal of Marine Science and Engineering, 10*(5), 682. https://doi.org/10.3390/jmse10050682

Jacques, G. (1983). Some ecophysiological aspects of the Antarctic phytoplankton. *Polar Biology, 2*(1), 27–33. https://doi.org/10.1007/BF00258282

Jahncke, J., Coyle, K. O., & Hunt, G. L., Jr. (2005). Seabird distribution, abundance, and diets in the eastern and central Aleutian Islands. *Fisheries Oceanography, 14*, 160–177. https://doi.org/10.1111/j.1365-2419.2005.00372.x

Jahnke, R. A. (2010). Global synthesis. In K. K. Liu, L. Atkinson, R. Quiñones, & L. Talaue-McManus (Eds.), *Carbon and nutrient fluxes in continental margins. Global change – The IGBP series*. Springer. https://doi.org/10.1007/978-3-540-92735-8_1

Janjuhah, H. T., Kontakiotis, G., Wahid, A., Khan, D. M., Zarkogiannis, S. D., & Antonarakou, A. (2021). Integrated porosity classification and quantification scheme for enhanced carbonate reservoir quality: Implications from the Miocene Malaysian carbonates. *Journal of Marine Science and Engineering, 9*(12), 1410. https://doi.org/10.3390/jmse9121410

Johnson, M. E. (2021). Geological oceanography of the Pliocene warm period: A review with predictions on the future of global warming. *Journal of Marine Science and Engineering, 9*(11), 1210. https://doi.org/10.3390/jmse9111210

Johnson, V. R., Brownlee, C., Rickaby, R. E. M., Graziano, M., Milazzo, M., & Hall-Spencer, J. M. (2012). Responses of marine benthic microalgae to elevated CO_2. *Marine Biology, 160*, 1813–1824. https://doi.org/10.1007/s00227-011-1840-2

Journal of Experimental Marine Biology and Ecology. (2022). *Journal of Experimental Marine Biology and Ecology*. Elsevier B.V.

Journal of Marine Biology & Oceanography. (2022). *Oceanology*. Retrieved from https://www.scitechnol.com/marinebiology/oceanology.php

Journal of the Marine Biological Association of the United Kingdom. (2022). *Journal of the Marine Biological Association of the United Kingdom*. Cambridge University Press.

Kampel, M., Lorenzzetti, J. A., Bentz, C. M., Nunes, R. A., Paranhos, R., Rudorff, F. M., & Politano, A. T. (2009). Simultaneous measurements of chlorophyll concentration by lidar, fluorometry, above-water radiometry, and ocean color MODIS images in the Southwestern Atlantic. *Sensors, 9*(1), 528–541. https://doi.org/10.3390/s90100528

Khan, U., Janjuhah, H. T., Kontakiotis, G., Rehman, A., & Zarkogiannis, S. D. (2021). Natural processes and anthropogenic activity in the Indus River sedimentary environment in Pakistan: A critical review. *Journal of Marine Science and Engineering, 9*(10), 1109. https://doi.org/10.3390/jmse9101109

Khatiwala, S., Tanhua, T., Mikaloff Fletcher, S., Gerber, M., Doney, S. C., Graven, H. D., Gruber, N., McKinley, G. A., Murata, A., Ríos, A. F., & Sabine, C. L. (2013). Global ocean storage of anthropogenic carbon. *Biogeosciences, 10*, 2169–2191. https://doi.org/10.5194/bg-10-2169-2013

Kontakiotis, G., Antonarakou, A., & Ruban, D. A. (2022). Geological oceanography: Towards a conceptual framework. *Journal of Marine Science and Engineering, 10*(12), 2027. MDPI AG. https://doi.org/10.3390/jmse10122027

Kroeker, K. J., Kordas, R. L., Crim, R., Hendriks, I. E., Ramajo, L., Singh, G. S., Duarte, C. M., & Gattuso, J. P. (2013a). Impacts of ocean acidification on marine organisms: Quantifying sensitivities and interaction with warming. *Global Change Biology, 19*(6), 1884–1896. https://doi.org/10.1111/gcb.12179

Kroeker, K. J., Gambi, M. C., & Micheli, F. (2013b). Community dynamics and ecosystem simplification in a high-CO_2 Ocean. *Proceedings of the National Academy of Sciences, 110*(31), 12721–12726. https://doi.org/10.1073/pnas.1216464110

Kroeker, K. J., Micheli, F., & Gambi, M. C. (2013c). Ocean acidification causes ecosystem shifts via altered competitive interactions. *Nature Climate Change, 3*(2), 156–159. https://doi.org/10.1038/nclimate1680

Kuss, J., Holzmann, J., & Ludwig, R. (2009). An elemental mercury diffusion coefficient for natural waters determined by molecular dynamics simulation. *Environmental Science & Technology, 43*(9), 3183–3186. https://doi.org/10.1021/es8034889

Kwader, T. (2021). Is it hydrogeology or geohydrology? *The Driller*. Retrieved from https://www.thedriller.com/articles/85612-is-it-hydrogeology-or-geohydrology

Ladd, C., Jahncke, J., Hunt, G. L., Jr., Coyle, K. O., & Stabeno, P. J. (2005). Hydrographic features and seabird foraging in Aleutian Passes. *Fisheries Oceanography, 14*, 178–195. https://doi.org/10.1111/j.1365-2419.2005.00374.x

Lalli, C. M., & Parsons, T. R. (1993). *Biological oceanography: An introduction*. Pergamon Press Ltd.

Lamborg, C. H., Fitzgerald, W. F., O'Donnell, J., & Torgersen, T. (2002). A non-steady state compartmental model of global-scale mercury biogeochemistry with interhemispheric atmospheric gradients. *Geochimica et Cosmochimica Acta, 66*(7), 1105–1118. https://doi.org/10.1016/S0016-7037(01)00841-9

Lamborg, C. H., Hammerschmidt, C., Bowman, K., Swarr, G. J., Munson, K. M., Ohemus, D. C., Lam, P. J., Heimbürger, L., Rijkenberg, M. J. A., & Saito, M. A. (2014). A global ocean inventory of anthropogenic mercury based on water column measurements. *Nature, 512*, 65–68. https://doi.org/10.1038/nature13563

Landes, A., & Zimmer, M. (2012). Acidification and warming affect both a calcifying predator and prey, but not their interaction. *Marine Ecology Progress Series, 450*, 1–10. https://doi.org/10.3354/meps09666

Landry, M. R., Kirshtein, J., & Constantinou, J. (1995). A refined dilution technique for measuring the community grazing impact of microzooplankton, with experimental tests in the central equatorial Pacific. *Marine Ecology Progress Series. Oldendorf, 120*(1), 53–63. https://doi.org/10.3354/meps120053

LaRocque, P. E., Banic, J. R., & Cunningham, A. G. (2004). Design description and field testing of the SHOALS-1000T airborne bathymeter. In G. W. Kamerman (Ed.), *Laser radar technology and applications IX. Proceedings Volume 5412 Defense and security*. https://doi.org/10.1117/12.564924

Laverock, B., Smith, C. J., Tait, K., Osborn, A. M., Widdicombe, S., & Gilbert, J. A. (2010). Bioturbating shrimp alter the structure and diversity of bacterial communities in coastal marine sediments. *ISME, 4*(12), 1531–1544. https://doi.org/10.1038/ismej.2010.86

Lee, K. J., Youngsikpark Bunkin, A., Nunes, R., Pershin, S., & Voliak, K. (2002). Helicopter based lidar system for monitoring the upper ocean and terrain surface. *Applied Optics, 41*(3), 401–406. https://doi.org/10.1364/ao.41.000401

Lengaigne, M., Menkes, C., Aumont, O., Gorgues, T., Bopp, L., André, J., & Madec, G. (2007). Influence of the oceanic biology on the tropical Pacific climate in a coupled general circulation model. *Climate Dynamics, 28*, 503–516. https://doi.org/10.1007/s00382-006-0200-2

Li, C., Zhan, L., & Lu, H. (2022). Mechanisms for overpressure development in marine sediments. *Journal of Marine Science and Engineering, 10*(4), 490. https://doi.org/10.3390/jmse10040490

Libes, S. M. (2009). *An introduction to marine biogeochemistry* (2nd ed.). Wiley.

Lillycrop, W. J., Parson, L. E., & Irish, J. L. (1996). Development and operation of the shoals airborne lidar hydrographic survey system. In V. I. Feigels & Y. I. Kopilevich (Eds.), *CIS selected papers: Laser remote sensing of natural waters: From theory to practice*. https://doi.org/10.1117/12.258351

Lin, C. S. (2003). Ocean lidar instrument and measurements. In C. R. Bostater Jr. & R. Santoleri (Eds.), *Remote sensing of the ocean and sea ice 2002. Proceedings Volume 4880 international symposium on remote sensing*. https://doi.org/10.1117/12.464286

Liss, P. S., & Merlivat, L. (1986). Air-sea exchange rates: Introduction and synthesis. In P. Buat-Menard (Ed.), *The role of air-sea exchange in geochemical cycling* (pp. 113–127). D Reidel Publishing Company.

Liu, X., Han, X., & Han, Z. (2022). Effects of climate change on the potential habitat distribution of swimming crab Portunus trituberculatus under the species distribution model. *Journal of Oceanology and Limnology, 40*, 1556–1565. https://doi.org/10.1007/s00343-021-1082-1

Lohbeck, K. T., Riebesell, U., & Reusch, T. B. (2012). Adaptive evolution of a key phytoplankton species to ocean acidification. *Nature Geoscience, 5*(5), 346–351. https://doi.org/10.1038/ngeo1441

Lohbeck, K. T., Riebesell, U., Collins, S., & Reusch, T. B. (2013). Functional genetic divergence in high CO2 adapted Emiliania huxleyi populations. *Evolution; International Journal of Organic Evolution, 67*(7), 1892–1900. https://doi.org/10.1111/j.1558-5646.2012.01812.x

Longhurst, A. R. (2007a). Biogeographic partition of the ocean. In A. R. Longhurst (Ed.), *Ecological geography of the sea* (pp. 19–34). Academic. https://doi.org/10.1016/B978-012455521-1/50003-6

Longhurst, A. R. (2007b). Biomes: The primary partition. In A. R. Longhurst (Ed.), *Ecological geography of the sea* (pp. 89–102). Academic. https://doi.org/10.1016/B978-012455521-1/50007-3

Longhurst, A. R. (2007c). Provinces: The secondary compartments. In A. R. Longhurst (Ed.), *Ecological geography of the sea* (pp. 103–114). Academic. https://doi.org/10.1016/B978-012455521-1/50008-5

Longhurst, A. R. (2010). *Ecological geography of the sea* (2nd ed.). Elsevier.

Loux, N. T. (2004). A critical assessment of elemental mercury air/water exchange parameters. *Chemical Speciation and Bioavailability, 16*(4), 127–138. https://doi.org/10.3184/095422904782775018

Lu, X., Hu, Y., Pelon, J., Trepte, C., Liu, K., Rodier, S., Zeng, S., Lucker, P., Verhappen, R., Wilson, J., Audouy, C., Ferrier, C., Haouchine, S., Hunt, B., & Getzewich, B. (2016). Retrieval of ocean subsurface particulate backscattering coefficient from space-borne CALIOP lidar measurements. *Optics Express, 24*(25), 29001–29008. https://doi.org/10.1364/OE.24.029001

Ludwig, N. A., & Valencia, M. J. (1993). Oil and mineral resources of the East China Sea: Prospects in relation to maritime boundaries. *GeoJournal, 30*, 381–387. https://doi.org/10.1007/BF00807218

Machovsky-Capuska, G. E., & Raubenheimer, D. (2020). The nutritional ecology of marine apex predators. *Annual Review of Marine Science, 12*, 361–387. https://doi.org/10.1146/annurev-marine-010318-095411

MacKenzie, K. M., Palmer, M. R., Moore, A., Ibbotson, A. T., Beaumont, W. R., Poulter, D. J., & Trueman, C. N. (2011). Locations of marine animals revealed by carbon isotopes. *Scientific Reports, 1*, 21. https://doi.org/10.1038/srep00021

Mancini, P. L., Valim, E. E. M., de Barros Bauer, A., & Fischer, L. G. (2023). Intraspecific trophic variation in brown booby (Sula leucogaster) from the Southwestern Atlantic. *Marine Biology, 170*, 1. https://doi.org/10.1007/s00227-022-04134-

Manizza, M., Le Quere, C., Watson, A. J., & Buitenhuis, E. J. (2005). Bio-optical feedbacks among phytoplankton, upper ocean physics and sea-ice in a global model. *Geophysical Research Letters, 32*(5), L05603. https://doi.org/10.1029/2004G. L020778.

Maravelis, A. G., Kontakiotis, G., Bellas, S., Antonarakou, A., Botziolis, C., Janjuhah, H. T., Makri, P., Moissette, P., Cornée, J.-J., Pasadakis, N., Manoutsoglou, E., Zelilidis, A., & Karakitsios, V. (2022). Organic geochemical signatures of the Upper Miocene (Tortonian—Messinian) sedimentary succession onshore Crete Island, Greece: Implications for hydrocarbon prospectivity. *Journal of Marine Science and Engineering, 10*(9), 1323. https://doi.org/10.3390/jmse10091323

Marine Biology 2022. Marine Biology. International Journal on Life in Oceans and Coastal Waters. : .

Marine Ecology Progress Series (MEPS). (2022). *Marine ecology progress series*. Inter-Research Science Publisher.

Marine Geology. (2022). *Aims & scope*. https://www.sciencedirect.com/journal/marine-geology

Marine Ocean Service. (2021). *What is a gyre?* https://oceanservice.noaa.gov/facts/gyre.html#:~:text=A%20gyre%20is%20a%20large%20system%20of%20rotating%20ocean%20currents.&text=The%20ocean%20churns%20up%20various,and%20salinity%20drive%20ocean%20currents

Mason, R. P., & Sheu, G.-R. (2002). Role of the ocean in the global mercury cycle. *Global Biogeochemical Cycles, 16*(4), 1093. https://doi.org/10.1029/2001GB001440
Mason, R. P., Fitzgerald, W. F., & Morel, F. M. (1994). The biogeochemical cycling of elemental mercury: Anthropogenic influences. *Geochimica et cosmochimica acta, 58*(15), 3191–3198. https://doi.org/10.1016/0016-7037(94)90046-9
Massachusetts Institute of Technology. (2022). *Fields of study*. Retrieved from https://mit.whoi.edu/academics/fields/
McNichol, A. P., & Aluwihare, L. I. (2007). The power of radiocarbon in biogeochemical studies of the marine carbon cycle: Insights from studies of dissolved and particulate organic carbon (DOC and POC). *Chemical Reviews, 107*(2), 443–466. https://doi.org/10.1021/cr050374g
Meskhidze, N., & Nenes, A. (2006). Phytoplankton and cloudiness in the Southern Ocean. *Science (New York, NY), 314*(5804), 1419–1423. https://doi.org/10.1126/science.1131779
Miller, C. B., & Wheeler, P. A. (2012). *Biological oceanography*. Wiley.
Millero, F. J. (2007). The marine inorganic carbon cycle. *Chemical Reviews, 107*(2), 308–341. https://doi.org/10.1021/cr0503557
Mlakar, M. (2022). *Special issue biogeochemistry of trace elements in the marine environment*. Retrieved from https://www.mdpi.com/journal/jmse/special_issues/marine_trace_elements
Mopper, K., Stubbins, A., Ritchie, J. D., Bialk, H. M., & Hatcher, P. G. (2007). Advanced instrumental approaches for characterization of marine dissolved organic matter: Extraction techniques, mass spectrometry, and nuclear magnetic resonance spectroscopy. *Chemical Reviews, 107*(2), 419–442. https://doi.org/10.1021/cr050359b
Morse, J. W., Arvidson, R. S., & Lüttge, A. (2007). Calcium carbonate formation and dissolution. *Chemical Reviews, 107*(2), 342–381. https://doi.org/10.1021/cr050358j
Mote Marine Laboratory & Aquarium. (2022). *Marine biogeochemistry*. Retrieved from https://mote.org/research/program/biogeochemistry
Munday, P. L., Dixson, D. L., Donelson, J. M., Jones, G. P., Pratchett, M. S., Devitsina, G. V., & Døving, K. B. (2009). Ocean acidification impairs olfactory discrimination and homing ability of a marine fish. *Proceedings of the National Academy of Sciences of the United States of America, 106*(6), 1848–1852. https://doi.org/10.1073/pnas.0809996106
Nagaiwa, R., Tachibana, A., & Moteki, M. (2022). Occurrence patterns of larval mesopelagic fishes in the mouth of highly eutrophic Tokyo Bay, Central Japan. *Journal of Oceanography*. https://doi.org/10.1007/s10872-022-00668-z
National Geodetic Survey. (2022). *Diffuse attenuation coefficient* (Kd). Retrieved from https://www.ngs.noaa.gov/RSD/topobathy/Diffuse%20Attenuation%20Coefficient.pdf
National Geographic. (2022). *Oceanography*. Retrieved from https://www.nationalgeographic.org/encyclopedia/oceanography/
National Oceanic and Atmospheric Administration. (2021a). *How much water is in the ocean?* Retrieved from https://oceanservice.noaa.gov/facts/oceanwater.html#:~:text=The%20ocean%20covers%20more%20than,in%20glaciers%20and%20ice%20caps
National Oceanic and Atmospheric Administration. (2021b). *What is marine biogeography?* Retrieved from https://oceanservice.noaa.gov/facts/biogeography.html#:~:text=Marine%20biogeography%20is%20the%20study,biogeography%E2%80%94is%20a%20tricky%20process
Nature. (2011). *Deep-sea discoveries*. Retrieved from https://www.nature.com/articles/ngeo1059.pdf
Nel, A., McQuaid, C. D., Duna, O. O., Giménez, L., & Porri, F. (2023). Similar metabolic responses of co-occurring post-settlement mussels to temperature change despite distinct geographical distributions. *Marine Biology, 170*, 2. https://doi.org/10.1007/s00227-022-04147-3
Nelson, D. M., & Smith, W. O. (1991). Sverdrup revisited: Critical depths, maximum chlorophyll levels, and the control of Southern Ocean productivity by the irradiance-mixing regime. *Limnology and Oceanography, 36*(8), 1650–1661. https://doi.org/10.4319/lo.1991.36.8.1650
Neuer, S., Benway, H., Bates, N., Carlson, C., Church, M., DeGrandpre, M., Dunne, J., Letelier, R., Lomas, M., Lorenzoni, L., Muller-Karger, F., Perry, M., & Quay, P. (2017). Monitoring ocean change in the 21st century. *Eos, 98*. https://doi.org/10.1029/2017EO080045

Newsome, S. D., Clementz, M. T., & Koch, P. L. (2010). Using stable isotope biogeochemistry to study marine mammal ecology. *Marine Mammal Science, 26*(3), 509–572. https://doi.org/10.1111/j.1748-7692.2009.00354.x

Nightingale, P., Malin, G., Law, C., Watson, A. J., Liss, P., Liddicoat, M., Boutin, J., & Upstill-Goddard, R. (2000). In situ evaluation of air-sea gas exchange parameterizations using novel conservative and volatile tracers. *Global Biogeochemical Cycles, 14*(1), 373–387. https://doi.org/10.1029/1999GB900091

Nishizawa, H., & Joseph, J. (2022). Differences in the morphological body condition index of sea turtles between species and size classes. *Journal of the Marine Biological Association of the United Kingdom, 102*(7), 479–485. https://doi.org/10.1017/S0025315422000765

Nissen, C., Vogt, M., Münnich, M., Gruber, N., & Haumann, A. (2018). Factors controlling coccolithophore biogeography in the Southern Ocean. *Biogeosciences Discussions*, 1–37. https://doi.org/10.5194/bg-2018-157

Olson, S. L., Jansen, M. F., Abbot, D. S., Halevy, I., & Goldblatt, C. (2022). The effect of ocean salinity on climate and its implications for earth's habitability. *Geophysical Research Letters, 49*(10), e2021GL095748. https://doi.org/10.1029/2021GL095748

Oschlies, A. (2004). Feedbacks of biotically induced radiative heating on upper-ocean heat budget, circulation, and biological production in a coupled ecosystem-circulation model. *Journal of Geophysical Research, 109*(C12). https://doi.org/10.1029/2004JC002430

Palacios, D. M., Bograd, S. J., Foley, D. G., & Schwing, F. B. (2006). Oceanographic characteristics of biological hot spots in the North Pacific: A remote sensing perspective. *Deep Sea Research Part II: Topical Studies in Oceanography, 53*(3-4), 250–269. https://doi.org/10.1016/j.dsr2.2006.03.004

Pauly, D., Trites, A. W., Capuli, E., & Christensen, V. (1998). Diet composition and trophic levels of marine mammals. *ICES Journal of Marine Science, 55*(3), 467–481. https://doi.org/10.1006/jmsc.1997.0280

Pespeni, M. H., Barney, B. T., & Palumbi, S. R. (2013). Differences in the regulation of growth and biomineralization genes revealed through long-term common-garden acclimation and experimental genomics in the purple sea urchin. *Evolution; International Journal of Organic Evolution, 67*(7), 1901–1914. https://doi.org/10.1111/evo.12036

Petropoulos, A., Kapsimalis, V., Evelpidou, N., Karkani, A., & Giannikopoulou, K. (2022). Simulation of the nearshore sediment transport pattern and beach morphodynamics in the semi-enclosed Bay of Myrtos, Cephalonia Island, Ionian Sea. *Journal of Marine Science and Engineering, 10*(8), 1015. https://doi.org/10.3390/jmse10081015

Phillips, D. L., & Eldridge, P. M. (2006). Estimating the timing of diet shifts using stable isotopes. *Oecologia, 147*(2), 195–203. https://doi.org/10.1007/s00442-005-0292-0

Piatt, J. F., Wetzel, J., Bell, K., DeGange, A. R., Balogh, G. R., Drew, G. S., Geernaert, T., Ladd, C., & Byrd, G. V. (2006). Predictable hotspots and foraging habitat of the endangered short-tailed albatross (Phoebastria albatrus) in the North Pacific: Implications for conservation. *Deep Sea Research Part II: Topical Studies in Oceanography, 53*(3-4), 387–398. https://doi.org/10.1016/j.dsr2.2006.01.008

Popescu, A., Schorstein, K., & Walther, T. (2004). A novel approach to a Brillouin–LIDAR for remote sensing of the ocean temperature. *Applied Physics B, 79*, 955–961. https://doi.org/10.1007/s00340-004-1666-4

Ramnath, V., Feygels, V., Kopilevich, Y., Park, J. Y., & Tuell, G. (2010). Predicted bathymetric lidar performance of Coastal Zone Mapping and Imaging Lidar (CZMIL). In S. S. Shen & P. E. Lewis (Eds.), *Algorithms and technologies for multispectral, hyperspectral, and ultraspectral imagery XVI. Proceedings Volume 7695 SPIE Defense, Security, and Sensing*. https://doi.org/10.1117/12.851978

Reeburgh, W. S. (2007). Oceanic methane biogeochemistry. *Chemical Reviews, 107*(2), 486–513. https://doi.org/10.1021/cr050362v

Reineman, B. D., Lenain, L., Castel, D., & Melville, W. K. (2009). A portable airborne scanning lidar system for ocean and coastal applications. *Journal of Atmospheric and Oceanic Technology, 26*(12), 2626–2641. https://doi.org/10.1175/2009JTECHO703.1

Ritterbush, S. W. (1978). Marine resources and the potential for conflict in the South China Sea. In *Fletcher forum of world affairs 1983–2000: Vol. 2. Winter*. http://hdl.handle.net/10427/76180

Rivadeneira, M. M. (2017). Marine Biogeography. *Oxford Bibliographies*. https://doi.org/10.1093/OBO/9780199874002-0167

Roddewig, M. R., Pust, N. J., Churnside, J. H., & Shaw, J. A. (2017). Dual polarization airborne lidar for freshwater fisheries management and research. *Optical Engineering, 56*(3), 031221. https://doi.org/10.1117/1.OE.56.3.031221

Rolfhus, K., & Fitzgerald, W. F. (2001). The evasion and spatial/temporal distribution of mercury species in Long Island Sound. *CT-NY. Geochimica et Cosmochimica Acta, 65*(3), 407–418. https://doi.org/10.1016/S0016-7037(00)00519-6

Ruban, D. A. (2022a). The Tanais Bay of the Eastern Paratethys Sea at the Sarmatian–Maeotian Transition (Late Miocene): Widespread desiccations and local uplifts in the light of historical information. *Journal of Marine Science and Engineering, 10*(7), 915. https://doi.org/10.3390/jmse10070915

Ruban, D. A. (2022b). Islands in the Caucasian Sea in three Mesozoic time slices: Novel dimension of geoheritage and geotourism. *Journal of Marine Science and Engineering, 10*(9), 1300. https://doi.org/10.3390/jmse9101109

Rudolf, A., & Walther, T. (2014). Laboratory demonstration of a Brillouin lidar to remotely measure temperature profiles of the ocean. *Optical Engineering, 53*(5), 051407. https://doi.org/10.1117/1.OE.53.5.051407

Sabine, C. L., Feely, R. A., Gruber, N., Key, R. M., Lee, K., Bullister, J. L., Wanninkhof, R., Wong, C. S., Wallace, D. W., Tilbrook, B., Millero, F. J., Peng, T. H., Kozyr, A., Ono, T., & Rios, A. F. (2004). The oceanic sink for anthropogenic CO_2. *Science (New York, NY), 305*(5682), 367–371. https://doi.org/10.1126/science.1097403

Samelson, R. M. (2011). *The theory of large-scale ocean circulation*. Cambridge University Press. https://doi.org/10.1017/CBO9780511736605

Sanz-Lázaro, C., Valdemarsen, T., Marín, A., & Holmer, M. (2011). Effect of temperature on biogeochemistry of marine organic-enriched systems: Implications in a global warming scenario. *Ecological Applications, 21*(7), 2664–2677. https://doi.org/10.1890/10-2219.1

Schorstein, K., Fry, E. S., & Walther, T. (2009). Depth-resolved temperature measurements of water using the Brillouin lidar technique. *Applied Physics B, 97*, 931. https://doi.org/10.1007/s00340-009-3661-2

Schulien, J. A., Behrenfeld, M. J., Hair, J. W., Hostetler, C. A., & Twardowski, M. S. (2017). Vertically-resolved phytoplankton carbon and net primary production from a high spectral resolution lidar. *Optics Express, 25*(12), 13577–13587. https://doi.org/10.1364/OE.25.013577

Scott, M. R. (2019). *Ocean biogeochemistry. Water: Science and issues*. Retrieved from https://www.encyclopedia.com/science/news-wires-white-papers-and-books/ocean-biogeochemistry

Scripps Institution of Oceanography. (2022). *Physical oceanography (Po)*. Retrieved from https://scripps.ucsd.edu/doctoral/program-areas/physical-oceanography-po

Selin, N. E., Jacob, D. J., Yantosca, R. M., Strode, S., Jaegle, L., & Sunderland, E. M. (2008). Global 3-D land-ocean-atmosphere model for mercury: Present-day versus preindustrial cycles and anthropogenic enrichment factors for deposition. *Global Biogeochemical Cycles, 22*(2). https://doi.org/10.1029/2007GB003040

Shaw, J. A., Churnside, J. H., Wilson, J. J., Lerner, N. E., Tiensvold, R. R., Bigelow, P. E., & Koel, T. M. (2008). *Airborne lidar mapping of invasive lake trout in Yellowstone Lake*. https://www.montana.edu/jshaw/documents/ILRC24_YstoneLidar.pdf

Shen, L., Huang, T., Chen, Y., Chu, Z., & Xie, Z. (2022). Diverse transformations of sulfur in seabird-affected sediments revealed by microbial and stable isotope analyses. *Journal of Oceanology and Limnology*. https://doi.org/10.1007/s00343-021-1173-z

Shiklomanov, I. (1993). World freshwater resources. In P. H. Gleick (Ed.), *Water in crisis: A guide to the world's fresh water resources*. Oxford University Press.

Sinclair, E. H., & Stabeno, P. J. (2002). Mesopelagic nekton and associated physics of the southeastern Bering Sea. *Deep Sea Research Part II: Topical Studies in Oceanography, 49*(26), 6127–6145. https://doi.org/10.1016/S0967-0645(02)00337-5

Skákala, J., Bruggeman, J., Ford, D., Wakelin, S., Akpınar, A., Hull, T., Kaiser, J., Loveday, B. R., O'Dea, E., Williams, C. A. J., & Ciavatta, S. (2022). The impact of ocean biogeochemistry on physics and its consequences for modelling shelf seas. *Ocean Modelling, 172*, 101976. https://doi.org/10.1016/j.ocemod.2022.101976

Smith, K. J., Trueman, C. N., France, C. A., Sparks, J. P., Brownlow, A. C., Dähne, M., Davison, N. J., Guðmundsson, G. A., Khidas, K., Kitchener, A. C., Langeveld, B. W., Lesage, V. S., Meijer, H. J., Ososky, J. J., Sabin, R. C., Timmons, Z. L., Víkingsson, G. A., Wenzel, F. W., & Peterson, M. J. (2021). Stable isotope analysis of specimens of opportunity reveals ocean-scale site fidelity in an elusive whale species. *Frontiers in Conservation Science*. https://doi.org/10.3389/fcosc.2021.653766

Soerensen, A. L., Sunderland, E. M., Holmes, C. D., Jacob, D. J., Yantosca, R. M., Skov, H., Christensen, J. H., Strode, S. A., & Mason, R. P. (2010). An improved global model for air-sea exchange of mercury: High concentrations over the North Atlantic. *Environmental Science & Technology, 44*(22), 8574–8580. https://doi.org/10.1021/es102032g

Solomon, S., Qin, D., Manning, M., Averyt, K., & Marquis, M. (Eds.). (2007). *Climate change 2007-the physical science basis: Working Group I contribution to the fourth assessment report of the IPCC* (Vol. 4). Cambridge University Press.

Sommer, U. J. M. B. (1986). Nitrate-and silicate-competition among Antarctic phyto-plankton. *Marine Biology, 91*(3), 345–351. https://doi.org/10.1007/BF00428628

Southgate, P. N. (1986). Cambrian phoscrete profiles, coated grains, and microbial processes in phosphogenesis; Georgina Basin, Australia. *Journal of Sedimentary Research, 56*(3), 429–441. https://doi.org/10.1306/212F893B-2B24-11D7-8648000102C1865D

Stewart, R. H. (2007). *Introduction to physical oceanography*. Texas A&M University.

Streets, D. G., Devane, M. K., Lu, Z., Bond, T. C., Sunderland, E. M., & Jacob, D. J. (2011). All-time releases of mercury to the atmosphere from human activities. *Environmental Science & Technology, 45*(24), 10485–10491. https://doi.org/10.1021/es202765m

Strode, S., Jaeglé, L., & Emerson, S. (2010). Vertical transport of anthropogenic mercury in the ocean. *Global Biogeochemical Cycles, 24*(4). https://doi.org/10.1029/2009GB003728

Su, J., Li, T., & Zhang, R. (2014). The initiation and developing mechanisms of Central Pacific El Niños. *Journal of Climate, 27*(12), 4473–4485. https://doi.org/10.1175/JCLI-D-13-00640.1

Sun, C., Huang, D., Xu, Q., Gao, F., Li, X., & Wang, A. (2022). Diverse habitat preferences of two sea cucumber species and the seasonal change in a coral reef area. *Journal of Oceanography and Limnology, 40*, 1578–1591. https://doi.org/10.1007/s00343-021-1254-z

Sunday, J. M., Bates, A. E., & Dulvy, N. K. (2012). Thermal tolerance and the global redistribution of animals. *Nature Climate Change, 2*(9), 686–690. https://doi.org/10.1038/nclimate1539

Sunderland, E. M., & Mason, R. P. (2007). Human impacts on open ocean mercury concentrations. *Global Biogeochemical Cycles, 21*(4). https://doi.org/10.1029/2006GB002876

Sunderland, E. M., Dalziel, J., Heyes, A., Branfireun, B. A., Krabbenhoft, D. P., & Gobas, F. (2010). Response of a macrotidal estuary to changes in anthropogenic mercury loading between 1850 and 2000. *Environmental Science & Technology, 44*(5), 1698–1704. https://doi.org/10.1021/es9032524

Swain, E. B., Engstrom, D. R., Brigham, M. E., Henning, T. A., & Brezonik, P. L. (1992). Increasing rates of atmospheric mercury deposition in midcontinental North America. *Science, 257*(5071), 784–787. https://doi.org/10.1126/science.257.5071.784

Takahashi, K. D., Makabe, R., Takao, S., Kashiwase, H., & Moteki, M. (2022). Phytoplankton and ice-algal communities in the seasonal ice zone during January (Southern Ocean, Indian Sector). *Journal of Oceanography, 78*, 409–424. https://doi.org/10.1007/s10872-022-00649-2

Talley, L., Pickard, G. L., Emery, W., & Swift, J. H. (2011). *Descriptive physical oceanography: An introduction* (6th ed.). Academic.

Tanhua, T., Orr, J. C., Lorenzoni, L., & Hansson, L. (2015). Increasing ocean carbon and ocean acidification. *World Meteorological Bulletin, 64*(1), 48–51. Retrieved from https://public.wmo.int/en/resources/bulletin/monitoring-ocean-carbon-and-ocean-acidification-0

Tatters, A. O., Schnetzer, A., Fu, F., Lie, A. Y., Caron, D. A., & Hutchins, D. A. (2013). Short-versus long-term responses to changing CO2 in a coastal dinoflagellate bloom: Implications for interspecific competitive interactions and community structure. *Evolution; International Journal of Organic Evolution, 67*(7), 1879–1891. https://doi.org/10.1111/evo.12029

Texas A&M University. (2022). *What is oceanography?* Retrieved https://ocean.tamu.edu/about/what-is-oceanography/index.html

Thomson, R. E., & Emery, W. J. (2014). *Data analysis methods in physical oceanography* (3rd ed.). Elsevier B.V. https://doi.org/10.1016/C2010-0-66362-0

Thurman, H. V., & Burton, E. A. (2001). *Introductory oceanography* (9th ed.). Prentice Hall.

Tolokonnikova, Z. A., & Ruban, D. A. (2022). Bryozoan diversity dynamics at the Devonian–carboniferous transition: Evidence from Transcaucasia. *Journal of Marine Science and Engineering, 10*(7), 959. https://doi.org/10.3390/jmse10070959

Tovar Verba, J., Stow, A., Bein, B., Pennino, M. G., Lopes, P. F. M., Ferreira, B. P., Mortier, M., Lima, S. K. Q., & Pereira, R. J. (2023). Low population genetic structure is consistent with high habitat connectivity in a commercially important fish species (Lutjanus jocu). *Marine Biology, 170*, 5. https://doi.org/10.1007/s00227-022-04149-1

Tuell, G. H., Feygels, V., Kopilevich, Y., Weidemann, A. D., Cunningham, A. G., Mani, R., Podoba, V., Ramnath, V., Park, J. Y., & Aitken, J. (2005). Measurement of ocean water optical properties and seafloor reflectance with scanning hydrographic operational airborne lidar survey (SHOALS): II. Practical results and comparison with independent data. In R. J. Frouin, M. Babin, & S. Sathyendranath (Eds.), *Proceedings of the SPIE* (Vol. 5885, pp. 115–127). https://doi.org/10.1117/12.619215

Tuell, G. H., Barbor, K., & Wozencraft, J. (2010). Overview of the coastal zone mapping and imaging lidar (CZMIL): A new multi-sensor airborne mapping system for the U.S. Army Corps of Engineers. In S. S. Shen & P. E. Lewis (Eds.), *Algorithms and technologies for multispectral, hyperspectral, and ultraspectral imagery XVI. Proceedings Volume 7695 SPIE Defense, Security, and Sensing*. https://doi.org/10.1117/12.851905

United States Geological Survey. (2022). *How much water is there on earth?* Retrieved from https://www.usgs.gov/special-topics/water-science-school/science/how-much-water-there-earth?qt-science_center_objects=0#overview

University of British Columbia. (2022). *Oceanography at UBC's Vancouver Campus*. Retrieved from https://you.ubc.ca/ubc_programs/oceanography/#:~:text=Oceanography%20is%20the%20study%20of,global%20climate%2C%20and%20impacts%20communities

University of Southampton. (2022). *Marine biogeochemistry*. Retrieved from https://www.southampton.ac.uk/research/groups/marine-biogeochemistry

University of Victoria. (2022). *Undergraduate calendar – September 2022*. University of Victoria.

Vaillancourt, R. (2016). *Re: How do you differentiate biological oceanography from marine biology?* Retrieved from: https://www.researchgate.net/post/How-do-you-differentiate-Biological-Oceanography-from-Marine-Biology/584efcec93553b04f952ad26/citation/download

Valentine, D. L. (2011). Emerging topics in marine methane biogeochemistry. *Annual Review of Marine Science, 3*, 147–171. https://doi.org/10.1146/annurev-marine-120709-142734

Van Haren, H. (2018). Grand challenges in physical oceanography. *Frontiers in Marine Science, 5*, 404. https://doi.org/10.3389/fmars.2018.00404

Vander Zanden, M. J., Clayton, M. K., Moody, E. K., Solomon, C. T., & Weidel, B. C. (2015). Stable isotope turnover and half-life in animal tissues: A literature synthesis. *PLoS One, 10*(1), e0116182. https://doi.org/10.1371/journal.pone.0116182

Vasilkov, A. P., Goldin, Y. A., Gureev, B. A., Hoge, F. E., Swift, R. N., & Wright, C. W. (2001). Airborne polarized lidar detection of scattering layers in the ocean. *Applied Optics, 40*(24), 4353–4364. https://doi.org/10.1364/ao.40.004353

Verity, P. G., Stoecker, D. K., Sieracki, M. E., & Nelson, J. R. (1996). Microzooplankton grazing of primary production at 140 W in the equatorial Pacific. *Deep Sea Research Part II: Topical Studies in Oceanography, 43*(4-6), 1227–1255. https://doi.org/10.1016/0967-0645(96)00021-5

Vlietstra, L. S., Coyle, K. O., Kachel, N. B., & Hunt, G. L., Jr. (2005). Tidal front affects the size of prey used by a top marine predator, the short-tailed shearwater (Puffinus tenuirostris). *Fisheries Oceanography, 14*, 196–211. https://doi.org/10.1111/j.1365-2419.2005.00369.x

Walther, G. R., Post, E., Convey, P., Menzel, A., Parmesan, C., Beebee, T. J., Fromentin, J. M., Hoegh-Guldberg, O., & Bairlein, F. (2002). Ecological responses to recent climate change. *Nature, 416*(6879), 389–395. https://doi.org/10.1038/416389a

Wang, P., Li, Q., & Li, C.-F. (2014). *Geology of the China seas*. Elsevier B.V.

Wang, F., He, G., Deng, X., Yang, Y., & Ren, J. (2021a). Fish teeth Sr isotope stratigraphy and Nd isotope variations: New insights on REY enrichments in deep-sea sediments in the Pacific. *Journal of Marine Science and Engineering, 9*(12), 1379. https://doi.org/10.3390/jmse9121379

Wang, H., Yi, L., Deng, X., & He, G. (2021b). Geochemical and mineral properties of quaternary deep-sea sediments in the Central-Tropical Pacific and its response to the Mid-Pleistocene Transition. *Journal of Marine Science and Engineering, 9*(11), 1254. https://doi.org/10.3390/jmse9111254

Way, M. J., Aleinov, I., Amundsen, D. S., Chandler, M. A., Clune, T., Del Genio, A. D., Fujii, Y., Kelley, M., Kiang, N. Y., Sohl, L., & Tsigaridis, K. (2017). Resolving orbital and climate keys of Earth and extraterrestrial environments with dynamics 1.0: A general circulation model for simulating the climates of rocky planets. *The Astrophysical Journal Supplement Series, 231*(1), 12. https://doi.org/10.3847/1538-4365/aa7a06

Wierucka, K., Barthes, N., Harcourt, R., Schaal, B., Charrier, I., & Pitcher, B. J. (2019). Chemical fingerprints suggest direct familiarisation rather than phenotype matching during olfactory recognition in Australian sea lions (Neophoca cinerea). *Journal of Experimental Marine Biology and Ecology, 517*(7), 49–53. https://doi.org/10.1016/j.jembe.2019.06.001

Winker, D. M., Vaughan, M. A., Omar, A., Hu, Y., Powell, K. A., Liu, Z., Hunt, W. H., & Young, S. A. (2009). Overview of the CALIPSO mission and CALIOP data processing algorithms. *Journal of Atmospheric and Oceanic Technology, 26*(11), 2310–2323. https://doi.org/10.1175/2009JTECHA1281.1

Wood, E., Armstrong, F., & Richards, F. (1967). Determination of nitrate in sea water by cadmium-copper reduction to nitrite. *Journal of the Marine Biological Association of the United Kingdom, 47*(1), 23–31. https://doi.org/10.1017/S002531540003352X

Woods Hole Oceanographic Institution. (2022). *Physical oceanography*. Retrieved from https://www.whoi.edu/what-we-do/understand/departments-centers-labs/po/

Yutsis, V., Levchenko, O., & Putans, V. (2022). Contourite and turbidite features in the Middle Caspian sea and their connection to geohazards derived from high-resolution seismic data. *Journal of Marine Science and Engineering, 10*(7), 990. https://doi.org/10.3390/jmse10070990

Zainuddin, M., Kiyofuji, H., Saitoh, K., & Saitoh, S. I. (2006). Using multi-sensor satellite remote sensing and catch data to detect ocean hot spots for albacore (Thunnus alalunga) in the northwestern North Pacific. *Deep Sea Research Part II: Topical Studies in Oceanography, 53*(3-4), 419–431. https://doi.org/10.1016/j.dsr2.2006.01.007

Zaneveld, J. (2013). Fifty years of inherent optical properties. *Methods in Oceanography, 7*, 3–20. https://doi.org/10.1016/j.mio.2014.03.002

Chapter 6
Biogeochemistry and Conservation Biology

Michael O'Neal Campbell

Abstract Biogeochemistry, encompassing nearly all the factors for plant and animal ecological, chemical, and physical relations, has a close and even overlapping relation with the conservation sciences. In this relation, both the natural processes and human-constructed systems have severely affected the status of many plant and animal species, including their physiological and ecological dynamics. However, the information on these relations is largely scattered, focusing on the impacts of particular chemicals on particular species, rather than on combined chemical groups on ecosystems, and the possibilities for conservation science and policy. This chapter examines the role of the understanding of biogeochemistry in the development and effective conservation management and policy and how this may inform biogeochemical research. Current research findings indicate a shift in the relevance from global scale chemical flows to smaller scales at the regional, local, and even micro level scenarios. Case studies are taken of the impacts of lead, zinc, mercury, cadmium, arsenic, chromium, copper, and selenium, pesticides (including insecticides and rodenticides), veterinary compounds (such as nonsteroidal anti-inflammatory drugs or NSAIDs and polychlorinated biphenyls or PCBs), and industrial pollutants such as perfluoroalkyl substances on terrestrial, aquatic, and marine life, and the impacts of ameliorative policy actions. Conservation policies have been evolved to remedy these events, but in many cases, more research is required to remedy the impacts of the chemical changes. Significantly, chemical systems are increasingly studied in conjunction with conservation issues, and these actions have contributed to positive results for conservation efforts, and knowledge of conservation issues.

M. O. Campbell (✉)
Simon Fraser University, Burnaby, BC, Canada
e-mail: evaz@torontomu.ca

M. O. Campbell, *Biogeochemistry and the Environment*,
https://doi.org/10.1007/978-3-031-47017-2_6

Introduction

Conservation efforts use the findings of numerous pure and applied sciences, including fields related to or within biogeochemistry (Campbell, 2017, 2018). Currently, conservation planning, management, and policy are based on interdisciplinary (disciplines sharing topics and methodologies of two or more disciplines), cross-disciplinary (the techniques of one discipline adopted by one or more other disciplines), and multidisciplinary (the creation of shared knowledge between two or more disciplines) knowledge clusters, which may transcend the biological, chemical, and earth sciences (Bianchi, 2021; Campbell, 2020). Disciplines incorporating one or more of these characteristics include those mixing social and natural sciences (e.g., political and cultural ecology, agricultural and environmental economics, environmental sociology and psychology, etc., and planning and management studies including environmental and natural hazards), linking of two natural sciences (e.g., biochemistry, biophysics, geochemistry, geophysics, geoecology, bioclimatology, and biological oceanography), and the more technological sciences (e.g., geomatics and engineering applications). While these disciplines may have a broad focus on issues unrelated to conservation, their methods may be used in conservation-related research, applications, or management (Campbell, 2018).

Recently, a new, broad discipline named conservation biology has developed, based on a biological perspective for conservation issues (Soule & Wilcox, 1980; Soule, 1986a, b). For some, the addition of "biology" to its name is a little inappropriate, as conservation is an extremely broad topic and biology is a narrow well-defined discipline (Campbell, 2015). Some also justify the inclusion of the term "biology" on the grounds that the discipline examines biological conservation. These debates are ongoing, reflecting the modern origin of the discipline (Sodhi & Ehrlich, 2010). Current studies are based on three main components: the importance of conservation, the methodology for such conservation applications, and the importance of biology. Definitions of conservation biology indicate the recent development and breadth of this subject. Meine et al. (2006) argue that the discipline is too recent a development to be clearly defined. Some definitions emphasize the conservation objectives of the discipline: for example, Soulé and Wilcox (1980) define conservation biology as "a mission driven discipline comprising both pure and applied science"; and "is a new field, or at least a new rallying point for biologists wishing to pool their knowledge and techniques to solve problems."

Work (2015: 368) points out that "the field of conservation biology has grown substantially from its inception in 1985 through a proliferation of journals, research articles, societies, and institutions of higher education that offer conservation biology courses or programs." On the breadth theme, Meine (2010: 7) describes the "main thread" of conservation biology as "the description, explanation, appreciation, protection, and perpetuation of biological diversity." This author acknowledges that some of the conservation biology topics have previously been developed and even owned by other disciplines and knowledge systems, such as "wilderness protection, sustained yield, wildlife protection and management, the diversity-stability hypothesis, ecological restoration, sustainability, and ecosystem health" (Meine, 2010: 7).

The history of conservation is the basis of the development of conservation biology, including the interlinked developments of the strands of conservation, including the philosophy, policy, science, and practice aspects (Meine, 2004; Campbell, 2015). Much of the substance of current conservation biology developed during the nineteenth century within the disciplinary boundaries of biology and geography and pieces of legislation were reinforced by the attention of social science research within sociology, political ecology, and planning (Campbell, 2015). The format of conservation biology, in synthesizing the techniques of the biological and social sciences oriented to conservation objectives, slowly emerged from these roots (Campbell, 2015). One of the founding literature texts was that of George Perkins Marsh, written during the post–Civil War United States (an era of industrialization and competing social attention possible negative impacts), titled *Man and Nature*, in which the second chapter ("Transfer, Modification, and Extirpation of Vegetable and of Animal Species") was a founding attempt to analyze nature–society relations, with people defined as "a new geographic force" (Meine et al., 2006: 633; see also Hays, 1959; Kolko, 1963; Leonard, 2009).

As conservation topics developed, a division was discerned between two approaches with core anthropocentric philosophies; the utilitarian Resource Conservation Ethic, which was based on scientific natural resource management, and the Romantic-Transcendental Preservation Ethic, more focused on the spiritual and aesthetic benefits of nature–society relations (Hays, 1959; Callicott, 1990). With time, in the latter half of the twentieth century, a stronger natural environment focus emerged, resulting in the beginnings of rudimentary conservation biology (Campbell, 2015). Such developments were supported by new literature such as those of Ehrenfeld (1970), MacArthur and Wilson (1967), and the journals of Conservation Biology and Biological Conservation (Meine et al., 2006; Campbell, 2015).

New, related sciences, practices, and structures also emerged, such as island and historical biogeography, population biology, genetics, variants of ecology (e.g., ecosystem and landscape ecology, zoological parks), applied subjects such as forestry, wildlife, range and fisheries management, interdisciplinary subjects such as environmental ethics, economics and history, and technological subjects such as remote sensing and geographical information systems (Crisci et al., 2003; Meine, 2004; Posadasa et al., 2006). Events included the First International Conference on Conservation Biology in September 1978, published as Conservation Biology: An Evolutionary Ecological Perspective (Soulé & Wilcox, 1980); a Second International Conference on Conservation Biology convened at the University of Michigan in May 1985 (Soulé, 1986a, b); the Society for Conservation Biology (SCB) was approved at the end of the meeting (Soulé, 1986a, b); and the National Forum on BioDiversity, September 21–24, 1986 in Washington, DC, published as Biodiversity (Wilson & Peter, 1988). In June 1987, the first annual meeting of the Society for Conservation Biology was held in Bozeman, Montana (Meine, 2010).

Conservation biology had developed as a discipline by the latter quarter of the twentieth century, and this discipline "reflected essential qualities that set it apart from predecessor and affiliated fields" including its breadth with a "primary focus on the conservation of genetic, species, and ecosystem diversity (rather than those ecosystem components with obvious or direct economic value)" (Meine et al., 2006: 639). Conservation biology overrode the narrower subjects with strong disciplinary boundaries and replaced them with a broader and more inclusive science and methodology, also linked with the social sciences and humanities, including associated paradigms based on social power and ethics (Roebuck & Phifer, 1999; Campbell, 2015). Involved scholars and practitioners included sociologists, economists, political scientists, members of nongovernmental organizations, policymakers, social workers, managers, businesspeople, and (Barry & Oelschlaeger, 1996; Soule & Orians, 2001).

These developments ensured that conservation biology became increasingly practical, goal-oriented, and globally aware (Campbell, 2017, 2018). Ehrenfeld (1992: 1625) pointed out that "conservation biology is not defined by a discipline but by its goal – to halt or repair the undeniable, massive damage that is being done to ecosystems, species, and the relationships of humans to the environment." Additionally, "many specialists in a host of fields find it difficult, even hypocritical, to continue business as usual, blinders firmly in place, in a world that is falling apart" (Ehrenfeld, 1992: 1625). Meine et al. (2006) add that conservation biology has progressed from the initial focus on "genetics and demographics of small populations, population and habitat viability, landscape fragmentation, reserve design, and management of natural areas and endangered species" to include issues such as the permeability and connectivity of landscapes, species interactions, global warming, and geomatics techniques, these enabling new applications (see also Odenbaugh, 2003; Proulx et al., 2014).

Conservation biology principally focuses on larger animals and their habitats, as these larger species may require more extensive requirements that may conflict with human life spaces and needs (Campbell, 2015; 2017, 2018, 2020, 2022; Campbell & Jha, 2021; Jha & Campbell, 2021). Large, mostly charismatic wildlife, such as the great cats, large, armored reptiles, and massive herbivores (e.g., elephants, rhinoceroses, hippopotamuses, giraffes, and bovids), are slaughtered as food, for skins, and as dangers in human life spaces (Campbell & Jha, 2021; Daskin & Pringle, 2016; Young et al., 2014, 2015). Here, "large" wildlife, although subjective, refers to birds, mammals, and reptiles zrecognized as large (e.g., for birds, approximately raven-sized; for mammals, about cougar-sized; and for reptiles, approximately a medium-sized alligator, taking North American examples). Large wildlife, especially mammals, has been described as megafauna: Museum Victoria (2012) defines megafauna as "big animals, generally animals with a body mass of over 40 kilograms." Barnosky (2008: 11543) defines megafauna as those "weighing at least 44 kg (roughly the size of sheep to elephants)" and including people. Dirzo et al. (2014) offer a body weight greater than 15 kg and argue that extinction or reduction of such species has pronounced impacts on ecosystems.

Large wildlife has also been exemplified with large mammals such as White-tailed deer (*Odocoileus virginianus*, Zimmermann, 1780), moose (*Alces alces*, Linnaeus, 1758), black bear (*Ursus americanus*, Pallas, 1780), and wolves (*Canis Lycaon*, Schreber, 1775) (Carruthers & Gunson, 2015) or the tiger (*Panthera tigris*, Linnaeus, 1758) and leopard (*Panthera pardus*, Linnaeus, 1758) in India (Acharya et al., 2016; Madhusudan, 2003). Carnivorous mammals and birds are threatened due to fear and control issues (Campbell, 2015). Their large size and real and imagined threats toward people and companion and agricultural animals may contribute to negative public attitudes toward their conservation (Saenz & Carrillo, 2002;

Conforti & Azevedo, 2003; Kleiven et al., 2004; Hoogesteijn & Hoogesteijn, 2008; Treves & Bruskotter, 2014). The situation of such large wildlife is "a global extinction crisis," termed "global defaunation," which affects larger species more than smaller species, because the large species are preferred by hunters, and they have larger or more complex habitat requirements and slower population growth rates (Young et al., 2014: 7036). The important ecological role of large mammal species indicates their decline and/or extinction is "often associated with pronounced effects on other aspects of community composition and structure, ecosystem function and even evolutionary trajectories" (Young et al., 2014: 7036).

The rest of this chapter examines the biogeochemistry of conservation biology, centering on the impacts of chemicals on animals. As conservation issues largely relate to larger animals, as the principal targets of human actions, the main focus will be on the direct and indirect impacts on such animals, including poisoning animal bodies and habitats, encouragement of chemically negative environmental change, and the role of biogeochemical cycles in ecological change. These events and impacts are scaled, from impacts on individual animals and patch-sized plant communities, to continental-sized and intercontinental regions, linked by biomes, migration patterns, and mobile air masses. Examples are taken from biomes and various animal (bird, mammal, reptile, fish) species. The objective is to document the impact of chemicals on various animals and ecosystems and assess the work required for the solution of these issues. The main focus is on the deliberate and accidental poisoning of animals with chemical elements and compounds.

Deliberate Poisoning of Animals

Animals may be poisoned deliberately for their elimination (due to safety or competitive concerns for people, human life spaces, companion and/or agricultural animals), using chemical compounds of varying environmental impact and toxicity (Campbell, 2015). Ogada (2014: 1) writes that "poisons have long been used to kill wildlife throughout the world. An evolution has occurred from the use of plant- and animal-based toxins to synthetic pesticides to kill wildlife, a method that is silent, cheap, easy, and effective." Berny (2007, 93) writes that "the use of pesticides has dramatically increased over the last 50 years to enhance agricultural production" and "pesticides are widely used to control agricultural pests. Unfortunately, because of their biological activity, they may have detrimental effects on nontarget species." Such poisoning may affect single or related species, usually those determined as pests and/or dangerous carnivores. The chemicals used, usually complex compounds such as dichlorodiphenyltrichloroethane (DDT, $C_{14}H_9Cl_5$) or dichlorodiphenyldichloroethylene (DDE, $C_{14}H_8Cl_4$), may remain in the environment and affect unintended species (Campbell, 2022). Such poisoning is still an important ecological factor for animal species' presence. For example, Gil-Sánchez et al. (2021, 1201) write that "intentional poisoning is a global wildlife problem and an overlooked risk factor for public health. Managing poisoning requires unbiased and high-quality data through wildlife monitoring protocols, which are largely lacking" with the findings of the study indicating "several animal groups (e.g., domestic mammalian carnivores and vultures) tended to be over-represented in the poisoning national database, while others (e.g., corvids and small mammals) were underrepresented" and "the probability of a given species being overrepresented was higher for heaviest, aerial, and cryptic species" but the authors conclude that "monitoring poisoned fauna based on heterogeneous sources may produce important biases in detection rates; thus, such information should be used with caution by managers and policy-makers."

Several sources take continent-wide examples of the impact management of such environmental poisons (Campbell, 2015, 2022). For example, Grilo et al. (2021) write that "poisoning of wild and domestic animals is a problem in all the countries in the world that has been described by several authors from ancient times to nowadays" and "nowadays the intentional poisoning of domestic and wild animals is a crime in the European Union (EU), but as in the past the poison is still used in rural areas of a number of European countries to kill animals that were considered harmful for human activities." In this study, the Laboratory of Pharmacology and Toxicology of the Faculty of Veterinary Medicine (LFT-FMV) studied poison levels in 503 wildlife (red foxes, birds of prey, lynx, and wild boar), and domestic animal samples (dog, cat, sheep, cows, and horses), between January 2014 and October 2020. Pesticide residues were detected in 239 of these samples. The pesticides included, in order of quantity molluscicides, carbamates, rodenticides (e.g., the banned strychnine and Aldicarb), and organophosphates. This study allowed to realize that a great number of positive samples involved banned pesticides (i.e., Aldicarb and Strychnine, and the commercially available Methiocarb and Anticoagulant rodenticides). In some areas, domestic species were more affected, while in some areas wild animals were more affected (Grilo et al., 2021).

Raptors in Europe have also been directly targeted with pesticides and poisons, with poisons such as Carbofuran and even pesticides, placed to be consumed by scavengers. For example, Vandenbroucke et al. (2010: 266) write that in Belgium "in the last few years, corpses of birds of prey and other predators have frequently been suspected of malicious poisoning." Carbofuran is the most common poison, followed by aldicarb, carbaryl, strychnine, and chlorophacinone (Vandenbroucke et al. (2010: 266). The chemical formulae and impacts of these compounds on raptors are shown in Table 6.1. In Flanders (northern Belgium), 162 Common buzzards (*Buteo buteo*, Linnaeus, 1758) were examined (2003–2006), and 91 were determined to have been deliberately killed, by poisoning ($n = 79$) or shooting ($n = 12$). Other species poisoned included the Common kestrel (*Falco tinnunculus*, Linnaeus, 1758), Eurasian sparrowhawk, *Accipiter nisus,* Linnaeus, 1758), Red kite (*Milvus milvus*, Linnaeus, 1758), Eurasian magpie (*Pica pica*, Linnaeus, 1758), Carrion crow (*Corvus corone*, Linnaeus, 1758), and Common pheasant (*Phasianus colchicus*, Linnaeus, 1758) (Vandenbroucke et al. (2010, 266). Hernández and Margalida (2008) also describe the pesticide poisoning of the Cinereous vulture (*Aegypius monachus*, Linnaeus, 1766) in Spain from 1990 to 2006, which involved 241 incidents and 464 vultures, with Carbofuran, aldicarb, and strychnine found in 88% of

Table 6.1 Pesticides toxic to birds

Compound	Formula	Research assessment for impacts on raptors/birds
Carbofuran	$C_{12}H_{15}NO_3$	Carbofuran is highly toxic to birds.[1]
Aldicarb	$C_7H_{14}N_2O_2S$	Indeterminate impacts on birds. [2]
Carbaryl	$C_{12}H_{11}NO_2$	Slight toxicity on birds.[3]
Strychnine	$C_{21}H_{22}N_2O_2$	Highly toxic, but indeterminate results of poisoning.[4]
Chlorophacinone	$C_{23}H_{15}ClO_3$	Serious possibilities for physiology.[5]

Sources: 1. National Pesticide Information Center (1996); 2. Moore et al. (2010); 3. Walker et al. (1991); 4. Wobeser and Blakley (1987). Schmutz et al. (1989). 5. Rattner et al. (2015)

the recorded incidents. Of the recorded incidents, 98% represented deliberate poisoning, and 83% resulted in the bird's death (Hernández & Margalida, 2008).

Ogada (2014: 10) describes the deliberate poisoning of African eagles, "the most targeted," including the Martial, Crowned, Tawny, and Verreaux's eagles (Anderson, 2000; Brown, 1991; Odino & Ogada, 2008). Other victims were nontargeted eagles (e.g., the target of the poisoning may be mammal carnivores) and were Tawny and Bateleur eagles (Brown, 1991; Mijele, 2009; Ogada & Kibuthu, 2012; Watson, 1986). Meyburg et al. (2020) also describe the poisoning of Greater-spotted eagles. Ogada (2014: 1) contends that there has been a "steep increase has recently been observed in the intensity of wildlife poisonings, with corresponding population declines" and where despite the fact that "under national laws, it is illegal to hunt wildlife using poisons in 83% of African countries" it is also apparent that "pesticide regulations are inadequate, and enforcement of existing legislation is poor. Few countries have forensic field protocols, and most lack storage and testing facilities" (Ogada, 2014: 1).

Carbofuran ($C_{12}H_{15}NO_3$) is determined as the most widely used and abused pesticide in Africa, and poisoning techniques are listed as including baiting carcasses, soaking grains in pesticide solutions, mixing pesticides to create salt licks, and poisoning waterholes. These are to kill damage-causing animals; harvest fish and land animals for bushmeat, traditional medicine, and wildlife products; and to kill wildlife sentinels such as the vultures, which may cluster and hence attract the attention of observers to poachers' activities. The author concludes that as "populations of scavengers, particularly vultures, have been decimated by poisoning," there should be actions "banning pesticides, improving pesticide regulations and controlling distribution, better enforcement and stiffer penalties for offenders, increasing international support and awareness, and developing regional pesticide centers" (Ogada, 2014: 1).

Gulliford (2015) takes the example of the history of the United States of America, noting that for 2015, "this year, for example, marks the 100th anniversary of the founding of the Bureau of Biological Survey." Congress created this agency to trap, poison, and kill predators and "varmints" across the West. The result was an "ecological holocaust of strychnine-ridden carcasses and indiscriminate destruction up the food chain. We tried to kill coyotes; we brought death to eagles instead"

(Gulliford, 2015). The characteristics of the legislative instruments included the elimination of the predators of deer and elk; to encourage the populations of the latter; the "agency's massive onslaught of poisons and steel traps," the perception of predators like wolves as "beasts of waste and desolation" (a quotation of Theodore Roosevelt); a lack in understanding that "healthy predator-prey relationships helped maintain healthy ecosystems"; the use of facilities such as the Biological Survey's poison laboratory in Denver, which experimented with strychnine, arsenic, and cyanide; the employment of hunters, whose goal was to ensure that "predatory animals are destroyed by trapping, shooting, den hunting during the breeding season, and poisoning" (one survey listed predator killings as "849 wolves, 20,241 coyotes, 85 mountain lions, 3432 bobcats, 30 lynxes, and 41 bears"); and random poisoning using mostly strychnine in carcasses, which killed mammal predators and raptors and eagles; and also use of Compound 1080, toxic to mammal (banned in 1972, before the passage of the Endangered Species Act) (Gulliford, 2015). Gulliford (2015) concludes "how ironic that the same agency that sponsored decades of predator control – the Bureau of Biological Survey – evolved into the U.S. Fish and Wildlife Service. Congress gave the newly named agency a mandate to protect endangered species, including some of the very species the government had spent years killing off. Colorado Parks and Wildlife, for example, even brought back lynx, animals that had previously been poisoned and trapped."

Some argue that the situation has not markedly improved (Campbell, 2017). For example, Waage (2014), examining current scenarios, contends that "the U.S. Department of Agriculture's Wildlife Services program is in dire need of reform – particularly when it comes to its practice of killing predators that play key ecological roles across the American landscape. Wildlife Services can immediately take a big step in the right direction by ceasing the use of two deadly predator poisons, sodium cyanide and Compound 1080, both of which are unnecessary, inhumane, dangerous to people, and environmentally destructive." It is noted that the poisons are mostly used as requested by private livestock owners, and in 2012, more than 14,500 carnivores were killed using the two poisons. Waage (2014) argues that "especially disturbing is that Wildlife Services has never weighed the true costs of these practices, in terms of threats to both public safety and wildlife, against perceived benefits accrued to a few private agricultural interests."

Additionally, despite the efforts of government reviewers to end the use of sodium cyanide, and Compound 1080, among others, there has not been a serious, effective ban. Examples of efforts include a 1963 criticism by the Advisory Group on Wildlife and Game Management, under the United States Department of Interior (DOI), concerning the nonselective impacts of poisoning, and a recommendation for more selective and/or nonlethal effective methods (see also Cain et al., 1971; Leopold et al., 1964). Another effort was the 1972 recommendation by the Advisory Committee on Predator Control that poisons must not be used that may kill unintended species. This led to a ban of Compound 1080, strychnine, sodium cyanide, and thallium) on federal lands by Executive Order, which was rescinded in 1982. Also in 2004, the United States Department of Agriculture (USDA) Inspector

General criticized the Wildlife Services' storage, security, and management of hazardous chemicals, including sodium cyanide and Compound 1080.

Poisoning is an important factor in the decline of Hooded vultures across sub-Saharan Africa, especially in Cameroon, Uganda, Kenya, and Southern Africa (Mundy, 1997; Ogada et al., 2010; Ogada & Buij, 2011; Ssemmanda & Pomeroy, 2010; Thiollay, 2001; Virani et al., 2011; Zimmerman et al., 1996). Vultures are sometimes poisoned because poachers think vultures will reveal illegal hunting kills (Hancock, 2009a, b, 2010). Poisoning targeting animals such as lions (*Panthera leo*, Linnaeus, 1758), hyenas (*Crocuta crocuta*, Erxleben, 1777), and jackals (*Canis mesomelas*, Schreber, 1775) (Bridgeford, 2004a, b) may affect vultures (Anderson, 1999; Pain et al., 2003; Roche, 2006; Verdoorn et al., 2004). Where vultures were not poisoned, their populations did not decline (Mundy et al., 1992). Van Wyk et al. (2001a: 243) detected "the presence of quantifiable residues of 14 persistent chlorinated hydrocarbon pollutants," after collecting "whole blood, clotted blood, heart, kidney, liver and muscle samples" from White-backed, Cape and Lappet-faced vultures. Comparisons between wild adults and nestlings, and captive birds showed lower pesticide levels for captive birds (pesticides included gamma-BHC (lindane), alpha(cis)-chlordane, and alphaendosulfan) (Van Wyk, Bouwman, et al., 2001a: 243).

Unintentional Poisoning with Chemical Elements

This type of poisoning usually occurs when elements (including heavy metals), intended for nonpoisoning objectives, eventually enter the food chain, and may prove lethal for some unintended organisms, as the chemicals migrate up the food chain (Campbell, 2017). These elements may combine with other elements to create compounds, which may have the same impacts as the elements. Research findings indicate that animals may ingest and retain several elements in combination, so detecting the most dangerous element may be difficult. Therefore, some studies seek to discover the impacts of one element on the selected species, some others seek to discover all the chemical elements in the studied organisms (Campbell, 2015, 2022). Typically affected birds are listed in Table 6.2. Such compounds may accumulate toward the top of food chains, and knowledge of these effects currently dominates the discourses on environmental and ecological pollution (Campbell, 2022). Raptors and waterbirds at the top and lower levels of the food chain are especially vulnerable to the chemicals ingested by prey species due to the processes of concentration and accumulation, e.g., lead from bullets fired at prey species, which are then consumed by predators, or the pollution of prey species foods from industrial pollutants (Campbell, 2015, 2022).

Rattner (2009: 773) includes such events in the general field of wildlife toxicology, citing the "unintentional poisoning of birds from ingestion of spent lead shot and predator control agents, alkali poisoning of waterbirds, and die-offs from maritime oil spills." Wildlife toxicology is described as "driven by chemical use and misuse, ecological disasters, and pollution-related events affecting humans"

Table 6.2 Birds commonly affected by environmental chemicals

Orders	Families	English names
Birds of prey or raptors		
Accipitriformes	*Accipitridae*, Vieillot, 1816	Eagles, hawks, vultures
	Pandionidae, Sclater & Salvin, 1873	Ospreys
	Sagittariidae, Finsch & Hartlaub, 1870	Secretary birds
Falconiformes	*Falconidae*, Leach, 1820	Falcons, caracaras
Cathartiformes or Accipitriformes	*Cathartidae*, Lafresnaye, 1839	New World vultures
Strigiformes	*Strigidae*, Leach, 1820	Typical owls
	Tytonidae, Ridgway, 1914	Barn owls
Waterbirds		
Charadriiformes	*Laridae*, Rafinesque, 1815	Gulls, terns
	Charadriidae, Leach, 1820	Plovers, lapwings
Anseriformes	*Anatidae*, Leach, 1820	Ducks, geese, swans
Ciconiiformes	*Ciconiidae*, J. E. Gray, 1840	Storks
Pelecaniformes	*Pelecanidae*, Rafinesque, 1815	Pelicans
	Ardeidae, Leach, 1820	Herons, egrets, bitterns
Suliformes	*Sulidae*, Reichenbach, 1849	Gannets, boobies
	Phalacrocoracidae, Reichenbach, 1850	Cormorants, shags
Gaviiformes	*Gaviidae*, Coues, 1903	Loons
Coraciiformes	*Alcedinidae*, Rafinesque, 1815	Kingfishers
Gruiformes	*Rallidae*, Rafinesque, 1815	Rails, crakes, coots

Source: Campbell (2022). Classification of common raptors and waterbirds. The waterbirds have nearly 20 families, so the commoner families are presented.

(Rattner, 2009: 773). Historically, the main point was the development of an international infrastructure for conservation, manifested in stages by the mid-century investigation of the effects of pesticides on domestic and wild animals, public debates on pollutant hazards, the development of contaminant monitoring programs, population-level studies of DDT impacts on raptorial and fish-eating birds, assessment of the global nature of organochlorine pesticide contamination, and the development of ecotoxicology in 1969 and of the Society of Environmental Toxicology and Chemistry in 1979.

Huang and Mayer (2019) argue that "birds are extremely sensitive to changes in the environment, so much so that they have been used as sentinels for poisonous gas detection in coal mines and heavy metal, toxin, and/or chemical detection in bodies of water. They are sensitive to heavy metals (e.g., lead, zinc, cadmium, copper, mercury, iron) in their environment." Birds higher up in the food chain, such as eagles, smaller raptors, and waterbirds, share prey species in living and carcass form and are usually the main obligate (full-time) and facultative (part-time) scavengers in terrestrial marine, riverine, vegetated, and urban areas; hence, they may ingest chemicals from carcasses and other dead organic matter (Campbell, 2015). Fish-eating eagles such as the Bald eagle, White-tailed eagle, and Steller's Sea eagle, consume some of the same marine species (dead or alive) as ospreys, gulls, and waders. Szumiło-Pilarska et al. (2017) studied the diets of the White-tailed eagle, the Great cormorant (*Phalacrocorax carbo*, Linnaeus, 1758), Herring gull (*Larus argentatus* Pontoppidan, 1763), mallard (*Anas platyrhynchos*, Linnaeus, 1758), and common merganser (*Mergus merganser*, Linnaeus, 1758), on the Polish coast (see also Falandysz et al., 1988; Misztal-Szkudlińska et al., 2011; Szumiło et al., 2013; Szumiło-Pilarska et al., 2016; Kalisińska et al. 2013, 2014). In some cases, eagles such as the Bald eagle may prey on waterbirds and acquire any poisons from their bodies (Buehler, 2020; Middleton et al., 2018). Most raptors also take small mammals such as rodents, snakes, and small birds. The existence of chemical pollutants in the ecosystem (i.e., plants, water, and prey species such as small fish, small mammals such as rodents, and small birds such as passerines and pigeons) therefore begins a process to avian toxicity and may have consequences for any animals at the higher levels of the food chain (Szumiło-Pilarska et al., 2017; Baudrot et al., 2018: 33).

By the 1980s and 1990s, there was global attention to a series of events, including increased heavy metal pollution from mining and smelting, agrichemical technologies and selenium toxicosis, and environmental disasters including the Chernobyl disaster and the Exxon Valdez accident. Noticed chemical elements and compounds included heavy metals, such as lead, iron, and zinc, and chemicals manufactured for agricultural and industrial use, such as anti-inflammatory drugs (NSAIDs), polychlorinated biphenyls (PCBs), insecticides, rodenticides, and insecticides (Campbell, 2018). Rattner (2009: 773) cites "biomarker development, endocrine disruption, population modeling, and studies with amphibians and reptiles" as among the major issues during the 1990s. Additionally, the synthetic pesticides developed during the mid-twentieth century and the ecological impacts of these compounds have spawned powerful attention from numerous stakeholders, which has initiated conservation-oriented actions. Rattner (2009: 773) argues for current challenges, which include "the need to more thoroughly estimate and predict exposure and effects of chemical-related anthropogenic activities on wildlife and their supporting habitat." During the twenty-first century, there was interest in new and emerging compounds (pharmaceuticals, flame retardants, surfactants), and potential population-level effects of some compounds (Rattner, 2009).

Current research on wildlife toxicology is heavily concerned with the impacts of heavy metals on animals (Campbell, 2022). Lenntech (2022) defines heavy metals as "any metallic chemical element that has a relatively high density and is toxic or poisonous at low concentrations. Examples of heavy metals include mercury (Hg), cadmium (Cd), arsenic (As), chromium (Cr), thallium (Tl), and lead (Pb)." Rahman and Singh (2019) assess the relative impacts of five toxic heavy metals (THMs) (arsenic (As), cadmium (Cd), chromium (Cr)(VI), mercury (Hg), and lead (Pb) on the environment, as these metals "are non-threshold toxins and can exert toxic effects at very low concentrations. These heavy metals are known as most problematic heavy metals and as toxic heavy metals." It is further argued that "these THMs have severe deteriorating effects on various microorganisms, plants, and animals" (Rahman & Singh, 2019). Tchounwou et al. (2012: 133) point out that "heavy metals are naturally occurring elements that have a high atomic weight and a density at least five times greater than that of water." The wide use of these metals in industrial, agricultural, medical, technological, and domestic developments has led to their ubiquitous distribution in natural and built environments, with consequences for human and nonhuman life forms. Table 6.3 shows heavy metals.

Kitowski et al. (2016) argue that despite the prosecution of detailed surveys, some aspects of heavy metal accumulation and transmission in ecosystems are vague or unknown, as has also been noted by Helander et al. (2009), Stankovic et al. (2014), and Kitowski et al. (2017). Kitowski et al. (2016) argue for raptor models for the study of chemical pollution of bird physiology, as they are frequently apex

Table 6.3 The main chemical elements and their impacts

Element	Symbol	Threshold	Impact on raptors, waterbirds, and prey species (e.g., *Passerines*)
Lead	Pb	2–6 ppm dw[1,5]	"Gradual weight loss, polyuria/polydipsia, depression, weakness, anorexia, and/or rapid anemia"[3]
Zinc	Z	2 ppm[3]	"Passive regurgitation, lethargy, weakness, weight loss, anemia, cyanosis, polyuria/polydipsia, diarrhea, regurgitation, neurologic signs, feather picking, anorexia, and/or hemoglobinuria."[3]
Mercury	Hg	1+ ppm ww[2,5]	Reduced reproductive output, weak immune function, avoidance of high-energy behaviors, death with higher doses[4]
Cadmium	Cd	3 ppm dw[1]	"Affects the endocrine system, kidneys, reproduction, moulting, haemoglobin formation and growth"[5]
Arsenic	As	0.5 ppm ww[6]	Liver, kidney problems[6] "Muscular incoordination, debility, slowness, jerkiness, falling hyperactivity, fluffed feathers, drooped eyelid, huddled position, unkempt appearance, loss of righting reflex, immobility, seizures"[7]
Chromium	Cr	4 ppm dw[7]	Gastrointestinal and immune system, respiratory tract, male reproductive system damage, miscarriage, low birth weight, and development problems of the skeleton, reproductive system[8]
Copper	Cu	4–6 ppm[8]	Gastrointestinal tract damage, hyperplasia of the forestomach mucosa, liver and kidney damage, immune system damage, decreased embryo and fetal growth.[8]
Selenium	Se	3 ppm ww[11]	Emaciation, breast muscle atrophy, absence of fat, lung fluids, excess body gelatinous or serous fluid, swollen livers, flaccid heart, congested and enlarged spleens, head feather loss.[9]

The threshold represents the level at which the chemical begins to cause toxic reactions in the animals. Kitowski et al. (2016) suggest that the conversion from wet (ww) to dry weight (dw) 4.0. Sources: 1. Pérez-López et al. (2008), Scheuhammer (1987), Nighat et al. (2013); 2. Weir et al. (2018); 3. Huang and Mayer (2019); 4. Whitney and Cristol (2017); 5. Gruz et al. (2019); 6. National Research Council, 1977; 7. Eisler (1988); 8. Agency for Toxic Substances and Disease Registry, 2019, 2020a; 9. Albers et al. (1996), O'Toole & Raisbeck, (1997). Note that, for example 10 mg/kg = 10 μg/g = 10 ppm predators, sedentary, wide-ranging and are herbivore consumers, which ingest contaminants from plants and water (also cited are Martin et al., 2008; Castro et al., 2011; Rajamani & Subramanian, 2015; Golden et al., 2016). Waterbirds may play the same role as raptors in riverine and coastal areas (Campbell, 2015). Basically, heavy metal concentrations in avian bodies are strong indicators of heavy metal levels in their diet, and in the contaminated ecosystem (Kitowski et al., 2016, 146, see also Castro et al., 2011; Campbell 2015, 2022).

Raptors and waterbirds may also acquire toxins, not only from the fish but also from the sea. Heavy metals, such as mercury and cadmium may be acquired by this method (Campbell, 2015: 2022). Kalisinska et al. (2014: 858) wrote that during the 1950s and after, research indicated that among the various Hg species present in nature, MeHg is the main factor for neurological, neuromotorical, behavioral, and reproductive disorders in warm-blooded vertebrates (see also Clarkson & Magos, 2006; Rutkiewicz et al., 2011). Mercury and cadmium occur naturally in oceanic waters, with concentration levels increasing through both natural causes and human-induced contamination. Fish-eating seabirds tend to contain higher levels of mercury in their livers than other birds for this reason (Parslow et al., 1973). According to Kalisinska et al. (2014: 858), Mercury (Hg) contamination in piscivorous (fish-eating) birds, especially methylmercury (MeHg), has been globally documented regarding its bioaccumulation and biomagnification in food chains. Examples of such piscivorous species are the white-tailed eagle and osprey, which are exposed to chronic or acute mercury poisoning, especially from methylmercury (MeHg), in their prey diet (Kalisinska et al., 2014: 858; see also Holt et al., 1979).

Such heavy metals may accumulate in bird livers, kidneys, gastrointestinal (digestive) tracts, nervous systems, eggs, bones, and feathers, resulting in severe illness, incapacity, and death (the last implying lethal toxicity) (Frantz et al., 2012: 23; Kramar et al., 2019). Other effects include impacts on hormones, respiratory and reproduction systems, and migration (Haschek et al., 2013; Kitowski et al., 2016; Scheuhammer, 1987; Williams et al., 2014). Variables in lethal toxicities may concern the species, body organ affected, the avian ecosystem, and the toxicity of the chemical considered (Campbell, 2015; Kramar et al., 2019). Sublethal levels of heavy metal toxicity are also common; records for impacts on some passerine species (often raptor prey) include reduced plumage intensity (especially due to copper, zinc, nickel and lead ingestion) (Eeva et al., 1998), reduced genetic diversity (especially due to copper, zinc cadmium and lead) (Eeva et al., 2006), reduced fledging success (especially from mercury, lead and cadmium) (Evers et al., 2008; Nam & Lee, 2006) and reduced humoral immune responsiveness (especially from lead and cadmium (Snoeijs et al., 2004). Other impacts included "aberrant incubation behaviour, lethargy and wing area asymmetry" (Frantz et al., 2012: 23) mostly from mercury poisoning (Evers et al., 2008). Lead poisoning may also induce bone mineralization, "the process by which the organic bone matrix becomes filled with calcium phosphate nanocrystals" (Frantz et al., 2012; Gangoso et al., 2009; Kuhn, 2001). This is vital for bone formation, which is "accomplished by coordinated multicellular activity," basically "a cell-mediated process involving the specific, highly ordered deposition of a unique form of calcium phosphate confined to precise locations within the organic matrix... essential for bone to meet its multi-functional property requirements" (Kuhn, 2001: 50).

Within the field of wildlife toxicology, new concepts have emerged that emphasize a broader range of documented and possible effects (Campbell, 2017). For example, Burger (1995: 369) determines most studies by ecotoxicologists and ecologists have looked at the effects of pollutants on individuals and populations mostly considering one or only a few impacts. Recently, some studies have counterbalanced this approach, using "a holistic approach to ecological risk assessment," which "suggests that a rigorous paradigm should be applied to toxicants, from hazard identification to risk characterization" (Burger, 1995: 369). A conceptual risk assessment model is evolved, using a flexibly developed environmental health risk assessment paradigm to ascertain the lead risk to birds, based on available information. The model includes a hazard identification phase, a target identification phase (including the recording of receptors, endpoints, relations, spatial and temporal scales, and some indicators), which are specific to the particular hazard, species, population, or community under examination (Burger, 1995: 369). Lead was the focus of the case study because it has important, extreme effects on avian species; these impacts include declining populations of associated species, the migration of toxins up the food chain, reductions in clutch and egg size, and changing migration patterns. The research on lead poisoning is based on information on lead quantities in tissues, exposure times, and foraging and reproductive behavioral impacts. Differences between exposure in the wild and the laboratory are crucial, especially if the wild birds may have means of lead removal that may not be available or may be less apparent to laboratory birds (Burger, 1995: 369).

Some studies also examine the impacts of chemical elements, while others focus on chemical compounds composed of several elements (Campbell, 2022). Studies of one element include that of Kramar et al. (2019), which looked at only mercury contamination in Bald Eagles, using blood and feather samples from nestling Bald Eagles in the Virginian coastal plain, piedmont, and western regions to determine regional variations of Hg toxicity. The Hg levels were analyzed using a Milestone DMA-80. As an example of looking for several elements or compounds, Kitowski et al. (2017) examined chemical impacts on white-tailed sea eagles, using livers from 22 dead eagles in Eastern Poland. The eagles were classified by sex, age (adults over 2 years old, immature less than 2 years old), based on iris color, plumage, and gonad development (see also Forsman, 1999; Helander et al., 1989, 2009; Kitowski et al., 2017). After mineralization using laboratory techniques, inductively coupled plasma optical emission spectrometry (ICP-OES) with a Thermo Scientific iCAP Series 6500 analyzer with a charge injection device (CID) was used for the identification of elements and toxic chemicals.

Kitowski et al. (2017) judged the extracted chemicals through referral to published findings regarding the toxicity levels. For example, lead hepatic (liver) concentrations above 6 mg/kg^{-1} dw (6 ppm) have resulted in subclinical toxicity (below the level for visible symptoms) (Helander et al., 2009; Pain et al., 1995). Lead levels equal to or above 10–15 mg/kg^{-1} dw (dry weight) (10–15 ppm) usually indicate poisonous levels (Franson, 1996; Martin et al., 2008). These are supported by research linking hepatic lead concentrations in raptors to lead poisoning and subclinical effects (Franson, 1996; Kitowski et al., 2017; Martin et al., 2008; Pain et al., 1995). For cadmium, levels at ≥3 mg/kg^{-1} dw may result from increased environmental exposure (Battaglia et al., 2005; Burgat, 1990). For mercury, Shore et al. (2011) argued that hepatic mercury concentrations greater than 6.7 mg/kg^{-1} dw (6.7 ppm) may affect avian reproduction, and mercury values above 67.0 mg/kg^{-1} dw may kill nonmarine birds. For selenium, Kitowski et al. (2017) argue that "selenium at the level of 4–10 mg/kg^{-1} dw is considered the background concentration in avian liver tissue" and cite a study by Ohlendorf (1989), where "a hepatic Se concentration above 10 mg/kg^{-1} dw was taken as exceeding the background level." For zinc, Kitowski et al. (2017) point to the study by Honda et al. (1990) in which 200 mg/kg^{-1} dw was the background concentration in liver tissue (Kitowski et al., 2017).

Poisoning with a Range of Heavy Metals

Case studies illustrate some of the main points, one issue being the greater impacts of some toxic poisons on birds than mammals (Campbell, 2022). Guitart et al. (2010: 260) note in Europe, using "a collation of published and non-published wildlife poisoning data from Belgium, France, Greece, Italy and Spain over the last 10 years," that "birds, particularly waterfowl and raptors, were more commonly reported as victims of poisoning than wild mammals" and metals (particularly lead arising from sporting/hunting activities) and pesticides (mainly anticholinesterases and anticoagulants) are frequent causes of poisoning, and often have fatal consequences." In some cases, considering the broad range of bird species and types of pollutants, representative species may be selected, as indicators or representative species for overall assessment (Campbell, 2017, 2018). For example, Badry et al. (2020: 139198) argue for biomonitoring in raptors, with the objective being the discovery of which raptor species in the case study (Europe, where potential pollutant regulation is harmonized) are the best biomonitors when the pollutant quantification is based on the analysis of tissues. Considering "four priority pollutant groups: toxic metals (lead and mercury), anticoagulant rodenticides, pesticides and medicinal products" and raptor behavior variants (food web, foraging trait, diet, preferred habitat, and migratory behavior), the relatively common Common buzzard (*Buteo buteo*, Linnaeus, 1758) buteo) and/or Tawny owl (*Strix aluco*, Linnaeus, 1758) were found to be the most suitable indicator species. The authors therefore "recommend that the common buzzard and tawny owl should be the initial focus of any pan-European raptor monitoring," with the qualification that "other species may be better sentinels for certain pollutants, such as the golden eagle (Aquila chrysaetos) for lead, the northern goshawk (Accipiter gentilis) for mercury across areas including Northern Europe, and vultures (where they occur in Europe) are likely best suited for monitoring non-steroidal anti-inflammatory drugs (NSAIDs)" (Badry et al., 2020: 139198).

As noted above, several studies examine the impacts of multiple pollutants on birds near the top of the food chain (Campbell, 2015, 2022). For example, Kim and Oh (2012: 1091–1097) used a study of the occurrence of iron, manganese, copper, lead and cadmium in the livers, muscles, kidneys, and bones of Eurasian Eagle Owls (*Bubo bubo*, Linnaeus, 1758), Brown Hawk Owls (*Ninox scutulata*, Raffles, 1822), and Collared Scops Owls (*Otus lempiji*, actually appears to be *Otus semitorques*, Temminck & Schlegel, 1844; see Hong et al., 2013) from Korea. The findings of this study indicated that iron concentrations by tissue within species were similar, but among tissues, there were significant differences across all species. Manganese and copper concentrations in muscles, kidneys, tissues, and bones showed interspecies differences. Livers did not show such differences (Kim & Oh, 2012: 1091). The authors concluded that the lead and cadmium contamination in Eurasian Eagle Owls could be derived from a Korean source, and those of Brown Hawk Owls may be from Korean and wintering sites. Collared Scops Owl contamination may be from breeding and/or wintering sites (Kim & Oh, 2012: 1091).

There were interspecies differences in the level of toxic absorption of manganese and copper concentrations, mostly below the toxic levels. There were also interspecies variations for lead and cadmium concentrations in livers, bones, and tissues, but the levels were within expected background levels for wild birds (Kim & Oh, 2012: 1091). The authors however note that "for some Eurasian Eagle owls and Collared Scops Owls, lead concentrations were at an acute exposure level, whilst lead concentrations were at a chronic exposure level in Brown Hawk owls. Cadmium concentrations were at a chronic exposure level in all three owl species. Acute and chronic poisoning was significantly correlated between indicator tissues." In conclusion, "these results suggest that pollutant levels vary with regard to migratory patterns. Eurasian Eagle owls are residents and reflect Korean contamination. Brown Hawk owls are summer visitors and pollutant levels may reflect Korean and wintering site exposure, and Collared Scops owls are passage migrants and may reflect contaminant exposure at breeding and/or wintering sites" (Kim & Oh, 2012: 1096).

In another study by Grúz et al. (2018: 6378), the concentrations of several heavy metals (As, Cd, Cr, Cu, Hg, Pb, Zn) were examined in the feathers of ten avian species, with samples from Hungary: Long-eared owl (*Asio otus*, Linnaeus, 1758) Little owl (*Athene noctua*, Scopoli, 1769), Tawny owl (*Strix aluco*, Linnaeus, 1758), Barn owl (*Tyto alba*, Scopoli, 1769), Eurasian sparrowhawk (*Accipiter nisus*, Linnaeus, 1758), rook (*Corvus frugilegus*, Linnaeus, 1758), Hooded crow (*Corvus cornix*, Linnaeus, 1758), Carrion crow (*Corvus corone*, Linnaeus, 1758), Common buzzard (*Buteo buteo*, Linnaeus, 1758), and Barn swallow (*Hirundo rustica*, Linnaeus, 1758). The feathers were analyzed using inductively coupled plasma optical emission spectroscopy (ICP-OES). The findings were that Little owls had the highest As concentrations, barn swallows had the highest Cd, Cr, and Pb concentrations, Sparrow hawks had the highest Cu and Hg concentrations, and owls had the highest Zn concentrations. The Cd levels were higher in adult crows than in juveniles. The authors conclude that the studied area was safe for wildlife, as it was not contaminated by the mentioned heavy metals on that level, which can cause any adverse impact on birds (Grúz et al., 2018: 6378).

A similar, multiple heavy metal–based study was that of Kitowski et al. (2016), which examined concentrations of lead, cadmium, nickel, chromium, and mercury in the livers of six species of raptors, represented by 58 specimens captured in Eastern Poland. The six species, in order of size were the white-tailed eagle (*Haliaeetus albicilla*, Linnaeus, 1758) (2 specimens); common buzzard (*Buteo buteo*, Linnaeus, 1758) (31 specimens); Goshawk (*Accipiter gentilis* Linnaeus, 1758) (9 specimens); Marsh harrier (*Circus aeruginosus*, Linnaeus, 1758) (3 specimens), Common kestrel (*Falco tinnunculus*, Linnaeus, 1758) (3 specimens) and Sparrowhawk (*Accipiter nisus*, Linnaeus, 1758) (10 specimens). The findings indicated that higher mercury and chromium levels were associated with passerine predation and higher hepatic concentrations of lead and cadmium were associated with small mammal predation. The implications were that the passerine birds more likely ingested mercury and chromium (possibly from seeds or invertebrates) and the small mammals more likely ingested lead and cadmium (possibly from seeds, leaves, and small invertebrates). Sparrowhawk livers varied from Common buzzard livers in lead levels, but not in cadmium levels. Goshawks had lower levels of mercury than Sparrowhawks (possibly due to the use of mercury in antifungal products for seed dressing, the seeds consumed by small birds) although both raptors are bird hunters (Kitowski et al., 2016).

The main source of mercury was not water-sourced prey, as consumption of larger birds reduced mercury ingestion and accumulations (e.g., in goshawks), larger birds accumulated less mercury (e.g., the common buzzard and white-tailed eagle), cadmium and lead accumulations were linked to small- and medium-sized mammal consumption, and the study concluded that the species' prey preference affects the heavy metal contamination for raptor species (Kitowski et al., 2016: 5). Evidence indicated White-tailed eagles responded to declining fish availability by consuming game carrion, which increased their ingestion of lead. The study concluded that raptors may vary food preferences, altering their vulnerability to metal poisoning (Helander et al., 2009; Nadjafzadeh et al., 2013).

Another study that found more serious impacts from heavy metals is that of Beyer et al. (2005). In the Tri-State Mining District (Oklahoma, Kansas, Missouri), mining, milling, and smelting have contributed to contamination with Pb, Cd, and Zn. The authors point out that milled mine waste ("chat") has polluted the district, as flotation tailings and from smelters as aerial deposition or slag (Beyer et al., 2005: 108). The findings indicated that American robins (*Turdus migratorius*, Linnaeus, 1766), northern cardinals (*Cardinalis cardinalis*, Linnaeus, 1758), and waterfowl suffered from Pb tissue concentrations significantly higher than those of reference birds, and comparable to levels recorded at other, severely Pb contaminated sites. Some birds had tissue concentrations of Pb associated with serious health effects. The main impacts were: "impaired biological functions and external signs of poisoning"; severe Zn poisoning effects, including increased pancreatitis and increased zinc concentrations in the liver and kidney of waterfowl; increased Cd in the kidneys of songbirds, although "no proximal tubule cell necrosis associated with Cd poisoning was observed"; and "mean activities of the Pb-sensitive enzyme delta-aminolevulinic acid dehydratase (ALAD) were decreased by >50% in red blood cells in these birds ($p < 0.05$)" (Beyer et al., 2005: 108).

Another similar study is that of Snoeijs et al. (2004, 399), which examined the effects of heavy metal contamination on free-living great tits (*Parus major*, Linnaeus, 1758), based on a study area composed of four study sites along a pollution gradient near a metallurgic smelter. The study findings indicated that despite the homogeneity of the body condition or hematocrit values (volume percentage of red blood cells in blood) among great tits from the four study sites, the heavy metal exposure affected the individual birds' humoral immune responsiveness (also termed antibody-mediated immunity, which is based on antibody molecules that are secreted by plasma cells). The authors noted that the Great tits recorded farthest from the smelter complex were significantly higher in immune responsiveness than the birds from the areas closest to the metallurgic smelter. It is concluded that more research is required to determine any causal association between heavy metal contamination and immunosuppression (Snoeijs et al., 2004: 399).

Other studies have examined the impacts of heavy metals on marine ecologies and environments. For example, Ishii et al. (2013, S75) investigated the accumulation patterns of metals, including cadmium and mercury, in tissues among four species of seabirds" these being the Horned puffin (*Fratercula corniculate*, Naumann, JF, 1821), Thick-billed murre or Brünnich's guillemot (*Uria lomvia*, Linnaeus, 1758), the Short-tailed shearwater or Slender-billed shearwater (*Ardenna tenuirostris*, Temminck, 1836), the Northern fulmar or Arctic fulmar (*Fulmarus glacialis*, Linnaeus, 1761), and the Slaty-backed gull (*Larus schistisagus*, Stejneger, 1884). The authors argue that geographic variations of metal levels may contribute to variations in metal accumulation levels in the studied bird species; hence, the concentrations of metals in seabirds are assessed to consider to reflect their habitat. There are interspecies differences in the accumulation pattern of the studied specimens and the high accumulation of metals could have impacts on seabirds even with no visible symptoms (Ishii et al., 2013: S75).

One study examined the novel topic of chemical poisoning from indoor materials, on incarcerated birds. Guthrie et al. (2020: 371) examined heavy metal toxicosis in six adult Waldrapp ibis (*Geronticus eremita*, Linnaeus, 1758) due to paint chip ingestion when the birds were placed in an indoor enclosure, which was undertaken due to a "regional highly pathogenic avian influenza epornitic and government restrictions." The results included: normal conditions for two birds; "clinical disease signs between 45 and 64 days of confinement, including weakness, lethargy, inability to fly, low body condition, and bilateral wing droop in four birds"; the death of one bird (before chelation therapy attempts), with increased concentrations of lead and zinc in the liver; lead toxicosis in five birds, detected by plasma lead analysis, with about 40 days of hospitalization and treatment with sodium calcium edetate and fluid therapy, which solved the toxicosis; clinicopathological findings including "elevated plasma aspartate aminotransferase, creatine kinase, glutamate dehydrogenase, and bile acids values"; and erythrocyte morphological changes including hypochromasia, decreased size of polychromatophils with some cytoplasmic stippling, abnormal shapes, and D cells" (Guthrie et al., 2020: 371). There was also "particulate radio-opaque material in the ventriculus of all affected birds" detected by whole-body radiographic imaging" (Guthrie et al., 2020: 371). Additionally, "plasma biochemistry, complete blood counts, and blood lead values were performed repeatedly on all 5 birds and were used to guide therapy and monitor treatment response" (Guthrie et al., 2020: 371). The authors concluded that these study results point to the importance of early ameliorative management of the treatment of lead toxicosis in birds, with awareness of long-term clinical monitoring and hematology (Guthrie et al., 2020: 371).

Another study tackled the similarly novel topic of heavy metal poisoning of raptors engaged in falconry (Pappalardo et al., 2021). Here, falconry refers to the training of raptors (falcons, hawks, eagles, hawks, falcons), to hunt game animals; this is broadly synonymous with hawking (Campbell, 2022; Carroll, 1996). Pappalardo et al. (2021: 797) take an example from the United Arab Emirates (UAE), where falconry is a "long standing tradition" and the wellbeing of falconry raptors is important for the local culture, and plasma chemistry analysis is used in monitoring, assessing, and managing diseases in birds. The authors argue that the imbalances of Cu, Mg, and Zn have been linked to several diseases in avian species, and knowledge of the reference intervals (RIs) of these minerals is vital for the clinical management of the falcon species (Pappalardo et al., 2021: 797). Here the reference interval refers to the normal level of occurrence of the event: "The reference interval principle is usually to take the central 95% of a reference population" (Jones & Barker, 2008: S93). The findings indicated that using reference intervals based on Cu, Mg, and Zn in captive falcons according to the Quality Assurance and Laboratory Standards Committee of the American Society for Veterinary Clinical Pathology, and the Clinical Laboratory Standards Institute guidelines, and comparing these with blood chemistry analyses on 252 clinically healthy falcons from the Dubai Falcon Hospital in the UAE (about half being gyrfalcons, the rest other falcon species), here were no significant interspecies or intersex variations in RIs of Cu (1.5–7.0 μmol/L), Mg (0.49–0.78 mmol/L), or Zn (11.8–34.1 μmol/L).

Other studies have examined the impacts of single elements on animals in accidental poisoning, such elements commonly being lead, zinc, mercury, cadmium, chromium, copper, selenium, and arsenic. Some of these studies are reviewed below, comprising both applications to birds and mammals. A common thread running through these studies concerns the relationship between the toxicity of the elements, and their migration through the food chain, involving apex animals and prey species, and the plant and water environment. Due to the prevalence of such elements in the environment, ecological studies cannot ignore the role of these chemical elements and in some cases derived compounds in animal ecology. The list starts with lead, possibly one of the most dangerous and important elements in food chain poisoning (Campbell, 2015, 2017, 2018, 2022).

Poisoning from Lead

Lead constitutes one of the main sources of poisoning, especially for birds high in the food chain, such as predators and scavengers, with a principal lead source being lead bullets in the carcasses of hunted animals (Campbell, 2015, 2017, 2022). Lead toxicity is also among the most dangerous forms of toxicity for birds (Cruz-Martinez et al., 2012). Lead accumulates in the bones, flesh, and the nervous system, impairing brain and neurological functions, contributing to behavioral change and, in extreme cases, death. Lead in the digestive tract may be absorbed into the blood and alter the role of calcium in chemical processes, disrupting the functions of calcium homeostasis, signal transmissions, cholinergic nerve cells, and the cerebellum and creating anemia through blood cell impacts (Campbell, 2022; Haig et al., 2014; Redig & Arent, 2008). Visible behavioral impacts include flight, walking, foraging, digestion, and strength impairment (De Francisco et al., 2003). Wani et al. (2015: 55) noted that lead is very persistent in the environment and the levels are rising, due to the continuous use of compounds containing this element in most countries, which pose serious threats, as there are lead toxicity impacts on most bodily functions, including those of the digestive system, nervous system, respiratory system, and reproductive system (Wani et al. (2015: 62).

Numerous studies have documented the impact of lead on animals, especially birds (Campbell, 2015, 2022). Several parameters have been constructed for evaluating lead levels in animals. For example, Lumeij (1985: 133) conducted an incisive review of the clinicopathologic aspects of lead poisoning in birds, using "the diagnostic value of several clinicopathologic parameters for the diagnosis of plumbism (lead poisoning) in birds." One parameter was venous (the network of veins that deliver deoxygenated blood back to the heart) lead concentration. To test such lead levels in the veins, blood samples may be collected by "venipuncture from the jugular, basilic, or metatarsal vein into a heparinized syringe" (see Liebezeit & Sheaffer, 2014). Another parameter is the plasma delta-aminolevulinic acid dehydratase concentration (ALA-D), where the ALAD gene makes the delta-aminolevulinate dehydratase enzyme. The enzyme produces the heme molecule, which is important for body organs, but is found mostly in blood, the liver, and in the bone marrow (MedicinePlus, 2009). MedicinePlus (2009) documents heme is an essential component of several iron-containing proteins termed hemoproteins, including hemoglobin (the protein that carries oxygen in the blood). Lead has a negative effect in that "delta-aminolevulinic acid dehydratase (ALAD, EC 4.2.1.4) is a zinc metalloenzyme whose inhibition by lead is the first and most sensitive indicator of lead exposure and whose decreased activity has been clearly implicated in the pathogenesis of lead poisoning" (Wetmur, 2015). In an experimental study, Cantrell et al. (1977: 110) found that "enzyme activity fell rapidly to a steady state as blood lead levels rose above normal. Exogenous zinc was shown to activate the enzyme..." in the study of baboons. Scinicariello et al. (2007: 35) noted that lead deposition in the body occurs in blood, bone, and soft tissues, lead inhibits δ-aminolevulinic acid dehydratase (ALAD), and as lead has "the greatest effect on ALAD, measurement of ALAD activity can be used as a marker of effect of lead exposure."

Martínez-López et al. (2004: 270) in a study of lead levels found in the feathers of adults and blood of nestlings of Booted eagle (*Hieraaetus pennatus*, Gmelin, 1788), European buzzard (*Buteo buteo*, Linnaeus, 1758), and goshawk (*Accipiter gentilis*, Linnaeus, 1758), sampled during the 1999 and 2000 breeding seasons in the Province of Murcia (Southeastern Spain), found a significant relationship between blood delta-ALAD and lead (rho = −568), this exhibiting a negative correlation with blood lead levels even when the lead concentrations were below 5 microg/dl. The authors argued these results indicate the usefulness of this enzymatic biomarker for low lead exposure in the studied raptors, and there were positive relationships between the lead concentrations in the blood of the nestlings and the lead concentrations in the feathers of parental birds, (rho = 0.817), which suggested the advantage of using feathers as parameters for the for biomonitoring environmental exposure to lead (Martínez-López et al., 2004: 270).

A similar study by Espín et al. (2015: 185) examined the "the δALAD activity and δALAD ratio in the blood of four free-living bird species" (the Griffon vulture, Eagle owl, Slender-billed gull and Audouin's gull) and investigated the correlations between δALAD activity/ratio and lead concentrations in blood samples, with the findings indicating a decrease in δALAD activity in lead-exposed Griffon vultures and Eagle owls, and very low blood lead concentrations and no correlations were found for the gull species (Espín et al., 2015: 185). The findings also indicated that Eagle owls were more sensitive to δALAD enzymatic inhibition by lead than Griffon vultures. The authors concluded that "further studies are recommended to provide new data on Pb concentrations at which δALAD activity is affected in different wild bird species and to elucidate why different species tolerate Pb in different ways" (Espín et al., 2015: 185).

Another indicator of lead poisoning is aminolevulinic-acid (ALA) excretion in urine or renal kidney tissue (Lumeij, 1985: 133). Stone et al. (1977: 174) examined red blood cell δ-aminolevulinic acid dehydratase (RBC-ALAD) in Japanese quail (*Coturnix japonica*, Temminck & Schlegel, 1848). This study on quails was to assess the impact of lead on blood in birds, which had been already studied in in people, where the lowered enzyme activity and negative correlation to blood lead concentrations are common. The findings indicated that comparing lead-exposed and nonlead-exposed birds, for the lead-exposed birds, the renal, hepatic, duodenal, and tibial lead concentrations were significantly ($P < 0.001$) higher. This showed that in the studied species, the RBC-ALAD was an indicator of lead exposure (Stone et al., 1977: 174). Such urine-based studies are more common with mammals than birds (Korbel & Kösters, 1994). For example, Green et al. (1978) examined dogs (*Canis familiaris*, Linnaeus, 1758) after intravenous lead administration, which produced "an acute syndrome of lead intoxication characterized by depression, vomiting, anorexia and weight loss," and chelation therapy reduced lead levels, as urine delta-aminolevulinic acid was eventually higher in untreated lead exposed dogs than treated dogs, varying depending on the type of chelation treatment. The authors concluded that "urinary delta-aminolevulinic acid was therefore recommended as an additional laboratory parameter which improved assessment of lead exposure in dogs, particularly in determining adequacy of chelation therapy" (Green et al., 1978).

Another indicator of lead poisoning is free erythrocytic protoporphyrin (FEPP or FEP) and zinc protoporphyrin (ZPP or ZnP, formula $C_{34}H_{32}N_4O_4Zn$) concentrations in venous blood (Lumeij, 1985: 133). For example, Harada and Miura (1984: 365) studied free erythrocyte protoporphyrin and zinc protoporphyrin and noted that in a fluorometric study of FEP and zinc protoporphyrin (ZnP) in erythrocytes of lead-poisoned rabbits and lead workers, for the rabbits the coexistence of ZnP and FPP (free type protoporphyrin) was recorded in the severe stage of acute lead poisoning. With the progress of lead intoxication, the ratio of ZnP to FEP in erythrocytes declined and the FPP was greater than the ZnP when the severe stage of intoxication was reached. The Association for Clinical Biochemistry and Laboratory Medicine (2018) writes that concentrations of ZPP are increased in incidents of lead poisoning and iron deficiency. This process affects health, as heme is an essential part of hemoglobin, which is the protein in red blood cells (RBCs) that transports oxygen from the lung's tissues and cells. Heme formation involves several reactions that end with the iron insertion into a molecule termed protoporphyrin. When there is insufficient iron, protoporphyrin combines with zinc to form zinc protoporphyrin (ZPP), which is useless, as it cannot transport oxygen. Lead stops iron, not zinc from attaching to protoporphyrin; hence, zinc protoporphyrin increases with increased lead poisoning (Association for Clinical Biochemistry and Laboratory Medicine 2018).

Another indicator of lead poisoning is the percentage of fluorocytes, reticulocytes, and erythrocytes with basophilic stippling in peripheral blood (Lumeij, 1985: 133). Basophilic stippling is defined as "the presence in RBCs (red blood cells) of inclusions of aggregated damaged ribosomes… coarse basophilic stippling is suggestive of lead poisoning" (Rashidi & Green, 2018). Such basophilic stippling is "indicative of disturbed rather than increased erythropoiesis (note: Erythropoiesis is the production of mature, haemoglobin-rich, red cells that carry oxygen to the tissues of the body, Frewin, 2014). It occurs in many blood diseases: thalassaemia, megaloblastic anaemias, infections, liver disease, poisoning by lead and other heavy metals, unstable haemoglobins, and pyrimidine-5′-nucleotidase deficiency" (Bain, 2017). Erythrocytes are "biconcave enucleate red blood cells responsible for transport of O2/CO2 between the body's tissues and lungs; their oxygen-carrying capacity is due to their high hemoglobin content" (Ingley & Klinken, 2006). Reticulocytes are immature red blood cells that are going through development, contrasted with the mature erythrocytes (Riley et al., 2001). Rai et al. (2022) note that "an increase or decrease in reticulocyte count can be an indicator of erythropoiesis activity or failure, especially relative to anemias and bone marrow dysfunction." This refers to Erythropoietin (EPO), which is a hormone produced mostly by interstitial cells in the kidney, protecting red blood cells and stimulating stem cells in bone marrow to produce red blood cells (erythropoiesis) (Kasper, 2003).

The $CaNa_2$ EDTA lead mobilization test is another indicator of lead poisoning (Lumeij, 1985: 133). Wedeen (1985: 127) records that in earlier studies, lead poisoning was recorded when expected symptoms of acute intoxication were detected, and there was elevation of blood lead. The ethylenediamine tetracetatic acid (EDTA) test is presently the most reliable method for detecting excessive lead contamination. Puschner and Poppenga (2009) describe $CaNa_2EDTA$, and succimer (another chelator) as the best chelators for treating birds, with the goal being "to chelate for the minimum amount of time necessary to resolve the intoxication (based on a decline in blood lead to an undetectable concentration)." The authors caution that "neurologic signs may initially worsen in birds treated with $CaNa_2EDTA$.30 Theoretically, this could be due to $CaNa_2EDTA$-induced mobilization of lead from bone. Thus, birds with chronic lead exposure and potentially higher bone lead concentrations may be more likely to be affected than acutely intoxicated birds. However, this has not been shown experimentally."

These background studies of the measurable parameters for lead poisoning in birds and mammals illustrate some of the complexities of lead poison measurement, and the importance of this issue. For these reasons, numerous studies have evaluated the importance of lead poisoning and the ecological impacts and scientific solutions. Many lead-based studies focus on either Eurasian birds or North American raptors. An example of the former is the work of Pain et al. (2019: 935), who noted that wild bird poisoning following lead ingestion from ammunition is acknowledged and researched, with recent research on terrestrial birds such as including raptors and scavengers. Recent research has expanded the number of taxa and the geographical regions of the studies. Studies have also recorded the effects on lead from embedded ammunition fragments in injured birds, these being noted to result in sublethal and welfare impacts. Additionally, lead poisoning is may kill one million wildfowl annually in Europe and contribute to the deaths of another ≥ 3 million. Modeling and correlative studies examined possible population-level effects of lead poisoning in terrestrial birds (Pain et al., 2019: 935). Causes of lead poisoning are listed as the direct ingestion of spent lead gunshot and ammunition in food, and the effects may be compounded in bird species that have a muscular gizzard and eat grit to help grind their food, and in raptors and scavengers. An extensive literature links avian lead poisoning to ammunition sources. The authors argue that there is strong support for the thesis that ammunition-derived lead is the main contributor to higher lead concentrations in the tissues of wild birds. Compared with the management of relative success of lead poison management for people, the prevention of lead poisoning in wildlife has been less successful, despite the fact that lead gunshot has been banned and replaced with nontoxic ammunition types in some locations and uses (e.g., for wildfowl hunting in the USA from 1991/92 and shooting in Denmark from 1996) (Pain et al., 2019: 935).

Another Eurasian, lead-based study of predators was conducted by Bassi et al. (2021: 146130), which examines the hypothesis that lead ammunition in hunting is the key factor for contamination for avian scavengers. This study, like many others, studied both predators (e.g., eagles) and scavengers (e.g., vultures), largely because both groups may eat the flesh of the same animals (the vultures mostly when the animals are already dead) and eagles, unlike vultures which are obligate (permanent) scavengers, may be facultative (part-time or occasional), scavengers (Campbell, 2015, 2022). Bassi et al. (2021: 146130) argue that raptors with scavenging habits are especially vulnerable to lead ingestion in intensive hunting locations and are indicators of the poisoning risk from lead ammunition. Their study examined 595 tissue samples from 252 carcasses of 4 species (golden eagle, bearded vulture, griffon vulture, cinereous vulture) in areas of intensive hunting in south-central Europe, between 2005 and 2019. The findings indicated little interspecies variation in lead concentrations, lead concentrations were mostly evident in long and small bones, followed by the liver and kidneys, and least in the brain, with 44.0% of the specimens having lead concentration above background thresholds in at least one tissue, indicating clinical poisoning. Golden eagles and Griffon vultures had higher tissue lead concentrations and higher clinical and subclinical poisoning than Bearded and Cinereous vultures, which the authors attributed to variations in feeding habits (Bassi et al., 2021: 146130). The study findings were that for all the studied species there was a rapid increase in lead values with age, but age-related differences were significant for only the golden eagle. As the birds with X-ray were detected with lead fragments in their digestive tract and had higher median lead concentrations, it could be inferred that hunting ammunition was the main source of lead poisoning. These findings implied that such lead poisoning affects the ecology and populations of the studied species, especially with delayed sexual maturity and low reproduction rate. The authors suggest that there should be a transition toward lead-free bullets and gunshots across Europe (Bassi et al., 2021: 146130).

A concurring study is that of Helander et al. (2021: 148799), which acknowledged both lethal and sublethal avian lead exposures from ammunition as factors for the health of scavenging and raptorial birds. The study examined the background and elevated Pb levels in the liver of 116 Golden eagles and 200 white-tailed sea eagles from Sweden during 2003–2011, and previous data on eagle owls. The findings indicated that for both eagle species, there were seasonal variations, and lead levels did not exceed "a generally accepted threshold for subclinical effects during summer but strongly elevated levels from October" (Helander et al., 2021: 148799). Additionally, the fledged juveniles had lower lead levels than the other age classes, although during the hunting season during the autumn (fall), the levels of these juveniles reach those of the older birds. The Eagle owl, which is not a scavenger, there were no seasonal changes and no link with ammunition was detected, but lead levels were like those of juvenile eagles before the hunting season in October. For total figures, 15% of the White-tailed eagle and 7% of the Golden eagles were at the level of lethal poisoning, although in high exposure areas, up to 24% of the White-tailed Sea eagles had lethal lead levels, and for both species, the lethal levels were 7% in low exposure intensity areas. Partial bans over shallow waters (2002) did not reduce the levels of lethal poisoning, and lead levels increased for the White-tailed Sea eagle from 1981 to 2011.

Another multispecies study was conducted by Descalzo et al. (2021: 142260), based on 16 bird species found dead or sick in Spain between 2004 and 2020, the species including the Eurasian griffon vulture (*Gyps fulvus*, Hablitz, 1783), Bearded vulture (*Gypaetus barbatus*, Linnaeus, 1758), Golden eagle (*Aquila chrysaetos*, (Linnaeus, 1758)), Spanish imperial eagle (*Aquila adalberti*, C. L. Brehm, 1861), Red kite (*Milvus milvus*, Linnaeus, 1758), and Northern goshawk (*Accipiter gentilis*, Linnaeus, 1758). The methodology used measurements of lead in the livers and blood of the sampled birds, and "biomarkers of haem biosynthesis, phosphorus (P) and calcium (Ca) metabolism, oxidative stress and immune function in individuals" (Descalzo et al., 2021: 142260). The findings indicated that lead poisoning is "a significant cause of death and sublethal effects on haem biosynthesis, P metabolism and oxidative stress in birds of prey in Spain," with some sampled individuals having lead levels equivalent to severe clinical poisoning (Descalzo et al., 2021: 142260). The blood lead levels increased with age, and seasonally with regard to the hunting season. An important finding concerned the impacts on the Eurasian Griffon vulture, as higher blood lead levels were associated with lower δ-ALAD activity in blood and phosphate levels in plasma, and higher blood lipid peroxidation and plasma carotenoid levels, which agree with the results of other studies of avian lead poisoning (Descalzo et al., 2021: 142260).

A single species study is provided by Isomursu et al. (2018: 858), covering the impacts of lead and mercury poisoning in White-tailed eagles, using 123 carcasses collected during 2000–2014 in the Baltic Sea area. Other factors for eagle mortality were also recorded, as human-related factors accounted for 60% of the causes of death, including spent lead ammunition poisoning (31%, mostly from scavenging behavior) and accidents such as electric power line, wind turbine and traffic collisions (24%). Other listed problems for this species with a gradually increasing population included intentional persecution (poisoning, shooting), construction, and forestry (Bevanger et al., 2010; Helander et al., 2002, 2008; Helander & Stjernberg, 2002; Herrmann et al., 2011; Krone et al., 2006; Krone et al., 2009; Saurola et al., 2013; Tiainen et al., 2016). Isomursu et al. (2018: 858) note that "scavenging raptors like the White-tailed eagle are at risk when they feed on offal and carcasses with lead bullet fragments and shots" (see also Nadjafzadeh et al., 2013; Pain et al., 2009). It is further noted that "the transition to non-toxic ammunition has met resistance and recently there has been opposite development as Norway lifted the ban on lead ammunition in terrestrial hunting" (Isomursu et al. (2018: 858; see also Arnemo et al., 2016).

White-tailed and Steller's Sea eagles are threatened by lead poisoning across their Eurasian range (Fisher et al., 2006; see for example, Falandysz et al., 1988; Kim et al., 1999; Kurosawa, 2000; Iwata et al., 2000; Ueta & Masterov, 2000; Kenntner et al., 2001, 2005; Krone et al., 2004). Fisher et al. (2006, 426) noted that these two eagle species "much work on poisoning from bullets comes from Japan, where it is customary for hunters to leave whole and partial shot deer carcasses in the field. These then become a source of bullet fragments that are ingested by scavenging raptors during and after the hunting season, most notably Stellar's sea-eagles… and white-tailed eagles…" (see also Falandysz et al., 1988; Iwata et al., 2000; Kim et al., 1999). One record reported the death of 39 Stellar's and 15 White-tailed eagles in Hokkaido, from 1996 to 1998 (Kurosawa, 2000). Commenting on this study, Fisher et al. (2006, 426) noted that "if this rate of mortality is an accurate measure of the proportion of deaths due to lead poisoning, then it represents a three-fold increase in the known yearly eagle mortality." Lead poisoning of White-tailed eagles has also been reported in Germany and Austria (Kenntner et al., 2001, 2004). Kenntner et al. (2005: 1831), for example, reported that 24% of 277 White-tailed eagles recovered "dead or moribund" in Germany from 1979 to 2005 had liver lead levels representative of acute poisoning.

In another Eurasian study located in the Alps, the main issue was golden eagles (*Aquila chrysaetos*, Linnaeus, 1758) scavenging on either nonretrieved carcasses or offal, discarded by hunters, and the key question concerned the frequency of such occurrences in the Alpine region (Madry et al., 2015). The body burdens (body burden is the term for the concentration (or amount) of chemicals in the body at any given time, ScienceDirect, 2020) in golden eagles (36 specimens) were measured in relation to their prey and compared with those of nearby eagle owls (19 specimens) in relation to their prey (marmots *Marmota marmota*, Linnaeus, 1758 and *Capra ibex*, Linnaeus, 1758) for eagles, and birds, rodents and snow hares *Lepus timidus*, Linnaeus, 1758) for eagle owls). The findings indicated that soft tissue lead levels were higher in golden eagles than in eagle owls, bone lead levels were 10 higher in eagles than in owls, and both eagle and owl prey species had low lead concentrations. The findings of the study indicated that Alpine eagles acquire lead from spent ammunition in both carcasses and offal in sublethal and lethal quantities (Madry et al., 2015).

In Europe, Spanish Greater Spotted eagles have been recorded with lead poisoning, possibly due to their mostly avian diet (Pérez-García et al., 2020). Pérez-García et al. (2020, 224) in southeast Spain indicated an avian diet with 18 bird species accounting for about 73% of prey items, the rest mostly mammals. Common prey species included the Common moorhen (*Gallinula chloropus* Linnaeus, 1758), Common Teal (*Anas crecca*, Linnaeus, 1758), and rats (*Rattus* spp.). A significant proportion of regurgitated pellets (42.3%, of which 64% contained lead shot) and prey remains (40.0%, of which 25% contain lead shot) contained lead shot. The authors concluded that the high lead shot quantities in pellets and prey remains of wintering Greater Spotted Eagles shows a high lead poisoning risk. Factors such as feeding behavior, the large space–time overlap between the raptor presence and the waterbird hunting season and non-compliance with the ban on the use of lead ammunition may contribute to high lead exposure" (Pérez-García et al., 2020, 224).

Spanish imperial eagles (*Aquila adalberti*, C. L. Brehm, 1861) often share carcasses with Griffon vultures (*Gyps fulvus*, Hablitz, 1783) and may therefore be exposed to the same level of poisoning risk (Mateo-Tomás et al., 2016). Griffon and Cinereous vultures were exposed to poison risk, and "this exposure may have increased after the ban on abandoning carcasses of domestic ruminants in the field due to the bovine spongiform encephalitis (BSE) crisis, both because the vultures consume hunting bag residues more frequently and because malnutrition may lead to mobilisation of lead stores" and lead toxicosis led to severe hypochromic anemia, disorientation, ataxia, impaired landing, and posterior paresis (Rodriguez-Ramos et al., 2008: 235; see also Iñigo & Atienza, 2007).

In Africa, both Old World *Acciptrid* vultures and similarly ranged eagles have been exposed to lead poisoning (Cade, 2007; Dobrowolska & Melosik, 2008; Gangoso et al., 2009; García-Fernández et al., 2008). Perhaps "vultures are more vulnerable to ingesting lead from ammunition than birds of prey such as falcons, hawks and eagles that often do their own hunting" (Verdoorn, G. in Torchia, 2018).

But "Griffon vultures feed on carcasses often consumed by many other species, including threatened ones (e.g., Spanish imperial eagle Aquila adalberti)" (Mateo-Tomás et al., 2016). These species have mostly European ranges but often have links with Africa (Campbell, 2015, 2022). The scavenging habits of eagles and vultures may occasionally be shared: An example was of the nonlead poisoning of elephant carcasses in Botswana, where 537 vultures and two Tawny eagles died (BBC, 2019).

Krüger and Amar (2018) discuss Bearded vulture (*Gypaetus barbatus*, Linnaeus, 1758) poisoning in southern Africa, using blood lead levels from dead and living specimens. The findings supported the conclusion that "the SE levels of lead may be detrimental to the survival and fecundity of this small and declining population" of bearded vultures with recommendations being the banning of "hunting and culling with lead ammunition, which is the most likely source of this pollutant. Such actions may reduce the population's susceptibility to other threats, which may be compounded by high lead levels, and help ensure the success of planned reintroduction programs" (Krüger & Amar, 2018: 491). Donázar et al. (2002: 89) also recorded subclinical and clinical intoxication lead levels in Egyptian vultures (*Neophron percnopterus*, Linnaeus, 1758) in the Canary Islands, possibly from ingesting lead shot. About 16% of the surveyed birds had over 0.2 ppm of lead, approaching fatal doses. This may affect other eagle species in the Canaris, such as the rare or vagrant Booted, Bonelli's and Short-toed eagles (Lepage, 2020). Calls have requested the replacement of lead pellets with steel or molybdenum/tungsten alloy bullets (Donázar et al., 2002; Campbell, 2015).

Lead toxicity has also been identified as a neurotoxin for some North American raptors perhaps because of the high levels of industrial pollution and lead bullet use hunting (Jacobson et al., 1977; Kaiser et al., 1980; Pattee et al., 1981; Reichel et al., 1984; Frenzel & Anthony, 1989; Craig et al., 1990; Langelier et al., 1991; Elliott et al., 1992; Gill & Langelier, 1994; Wayland & Bollinger, 1999; Miller et al., 2000, 2001; Wayland et al., 2003; Campbell, 2015). Due to these occurrences, in 1986, a United States Federal Law banned the use of lead ammunition for waterfowl hunting, as lead-poisoned waterfowl was a factor in the decline of eagle populations. This law was later considered ineffective (Kramer & Redig, 1997) and poison fatalities continued, despite extensive hunter compliance. When poison sources were investigated, it was established that the main lead sources were deer carcasses and deer gut piles left by hunters.

Cruz-Martinez et al. (2012) investigated the hypothesis that deer carcasses with lead bullets were the main contributory factor for lead poisoning of eagles, covering the period 1996–2009. The four epidemiological parameters of the study were: (1) links between poisoned eagle numbers and the deer hunting season in the US states of Minnesota, Wisconsin, and Iowa; (2) any correlations between animal recovery locations within deer-hunting zones; (3) lead isotope ratio analysis of metal fragments in eagle gastrointestinal tracts and blood lead levels; and (4) comparisons of eagle kidney copper concentrations in lead-exposed and nonexposed birds (Cruz-Martinez et al., 2012). Extracted Bald eagle livers were examined to determine the lead levels in Bald eagles after the lead shot ban of white-tailed deer (*Odocoileus virginianus*, Zimmermann, 1780).

The findings of Cruz-Martinez et al.'s (2012) study indicated that 334 of the 1277 Bald eagles in rehabilitation from January 1996 through December 2009 had higher lead levels. Eagles with high lead levels were also checked for copper levels, as many of the high-velocity rifle bullets used for deer hunting have copper jackets. The lead levels and poisoning were spatially and temporarily correlated with the hunting seasons, deer hunting rifle zones, and adult eagle age, and the high levels of matching isotopic signatures in the lead samples (from ingestion of lead) and blood and fragment samples within the isotope ratio of ammunition (Cruz-Martinez et al., 2012; see also Church et al., 2006). The copper levels correlated with lead levels, with high kidney copper levels occurring in lead-exposed eagles (see the section on copper). The study concluded that poisoning occurred from ammunition in deer and shotgun pellets in smaller animals, such as pheasants, squirrels, and rabbits, and hence, lead "is an important, if not the primary, source of lead exposure for bald eagles" (Cruz-Martinez et al., 2012). Cruz-Martinez et al. (2012) also examined Bald eagles admitted to a veterinary clinic over 40 years and wrote that "90% of the bald eagles received each year (120–130) for all types of problems have elevated lead residues in their blood…20–25% of these eagles have sufficiently high levels to cause clinical lead poisoning. Most of these birds die or are euthanized. In the last 24 years, over 500 eagles received or admitted to our clinic have either died or had to be euthanized due to lead poisoning" (Cruz-Martinez et al., 2012).

A similar single species study by Manning et al. (2019: 289–299) on Bald eagles in North America examined the effects of lead toxicity from spent ammunition, focusing on gross and histologic lesions, which are "underreported" for this species, and "to describe the associated lesions and their prevalence and to compare the lesions with blood, liver, kidney, and/or bone lead concentrations." The findings (based on 93 sampled eagles) indicated that the gross lesions linked to lead toxicity commonest within the heart (55% of the eagles, consisting of "multifocal myocardial pallor and rounding of the apex"), followed by the brain (14% of the eagles, consisting of petechiae or hemorrhagic necrosis) (Manning et al., 2019: 289). Histologic lesions occurred within the heart (82% of the eagles), brain (63% of the eagles), and eyes (28% of 87 eagles). The authors list the serious impacts of lead toxicity on Bald eagles: fibrinoid necrosis of small- to medium-caliber arteries, impacting the heart, brain, and eyes; and gross and histologic lesions, including lesions in the heart. The authors conclude that "the presence of fibrinoid arterial necrosis and parenchymal degeneration, necrosis, and/or hemorrhage within the heart, brain, and/or eyes is suggestive of lead toxicity in bald eagles and warrants evaluation of liver or bone lead concentrations" (Manning et al., 2019: 289).

Clark and Scheuhammer (2003: 23) conducted a broad multispecies study of avian lead impacts, examining lead exposure in 184 dead raptors of 16 species from Canada, the commonest species being the 131 Red-tailed hawks (*Buteo jamaicensis*, Gmelin, 1788), Great horned owls (*Bubo virginianus*, Gmelin, 1788), and Golden eagles (*Aquila chrysaetos*, Linnaeus, 1758). The findings indicated that most of the examined birds had very low lead levels, and 3–4% of total mortality in the 3 common species listed was due to lead poisoning. One of the 9 Bald eagles located away from an aquatic environment was lead poisoned, as was one Turkey Vulture. The authors concluded that their study substantiates the findings of other published research, that indicate that upland-foraging species of birds of prey and scavengers that generally predate game birds and mammals, risk lead poisoning from lead projectile ingestion from ammunition used in upland hunting. It is remarked that using nonlead ammunition for hunting upland game would remove this threat, which is the "only serious source of high lead exposure and lead poisoning for upland-foraging raptors" (Clark & Scheuhammer 2003: 23).

In a Californian multispecies study, Kelly et al. (2014) examined 48 carcasses (21 golden eagles, 23 turkey vultures, and 4 common ravens), recovered alive (45) or dead (3, two golden eagles and one turkey vulture); of the 45 living birds, 14 died after recovery and 31 were killed based on their condition. Blood samples were collected, and the measurable parameters for individual eagles were accession date, extraction location, admission factor, age class, clinical presentation, treatments, and outcome. Mortality was determined using the location and severity of lesions, radiography screening (for trauma and metallic opacities in the gastrointestinal tract, and heavy metal analysis), and "microscopic examination of tissues, including the brain, spinal cord, peripheral nerves, bone, muscle, tongue, heart, trachea, lung, liver, kidney, gonads, spleen, adrenals, thyroid/parathyroid glands, oesophagus, ventriculus, proventriculus, pancreas, intestines and bursa of fabricius" (Kelly et al. (2014, 2). The presence of rodenticides, organophosphate pesticides, strychnine, zinc and other substances was also investigated. Mortality due to lead poisoning occurred in 14% of the eagles and 17% of the turkey vultures, with some birds recoding pathological changes, elevated hepatic lead concentrations, and anticoagulant rodenticide intoxication. Four turkey vultures (8% of the total sample) died of anticoagulant rodenticide intoxication (Kelly et al., 2014).

Canadian studies of Golden and Bald eagles have revealed similar results (Campbell, 2015, 2022). For example, Owen (2019) describing the Golden eagles treated for lead poisoning at the Orphaned Wildlife Rehabilitation Society in Delta, B.C., pointed out that the Canadian government banned lead for hunting waterfowl 20 years before their writing and introducing regulations requiring nontoxic shot to hunt most migratory game birds across Canada. Lead shot has also been banned for all hunting in 55 national wildlife areas, but in some cases, pellets and bullets containing lead may be used for the hunting of large game and upland game birds, including grouse and pheasant. Additionally, the "Canadian government estimates that 40 to 80 tons of lead are used every year for hunting activities in Canada" (Owen, 2019). A cited study in western Canada (1986–1998) indicated 10% of 546 eagles were lead-poisoned and golden eagles were more affected by lead poisoning than bald eagles (Owen, 2019).

Lead pollution from lead tackles used in angling (fishing) also affects North American eagles (Bedrosian et al., 2012). Grade et al. (2017, 155) argue that "ingested lead fishing tackle has been documented in 28 species of North American birds, as a risk factor for 75 species and as a leading cause of death for common loons" (*Gavia immer*, Brunnich, 1764) (see also Pokras & Chafel, 1992; Blus, 1994; United States Environmental Protection Agency, 1994; Scheuhammer & Norris,

1995; Anderson et al., 2000; Stone & Okoniewski, 2001; Scheuhammer et al., 2002; Franson et al., 2003; Sidor et al., 2003; Strom et al., 2009; Grade, 2011). Bald eagles prey on young loons and may take lead from them (Evers et al., 2010). Grade et al. (2017, 157) used the methodology of Sidor et al. (2003) and Franson et al. (2003) to assess the lead poisoning of loons using dead loon specimens, and measurements of lead tackle parts present in the gizzard and gastrointestinal tracts of the loons. A weakness of this methodology was cited, as "ingested tackle erodes in the gizzard because of grinding action and stomach acids; and, in the case of jigs, hooks or other attachments break off and often dissolve or are passed prior to death" (Grade et al., 2017: 157; see also Pokras et al., 2009). Lead tackle toxicosis was blamed for the loon deaths when five conditions were present. These were when "tissue, blood, or body fluids were tested for lead and exceeded thresholds at which clinical signs of lead poisoning have been observed" (Grade et al., 2017: 157; see also Sidor et al., 2003, Franson & Pain, 2011). Also, when the necropsy form showed the death as due to lead toxicosis, with lead tackle in the gastrointestinal tract, or clinical pathology identifiable as lead toxicosis. Thirdly, when a lead tackle item was found in the loon's gastrointestinal tract, and fourth, when a radiograph revealed a tackle part, e.g., a lead sinker (a weight used with a fishing hook to increase its sinking rate, rate of sink, anchoring ability or casting distance) or jig (a hooked, lead sinker, covered to attract fish, jerked in vertical motion) in the loon's gastrointestinal tract. Finally, when a report included evidence of lead toxicosis (Grade et al., 2017, 157). The records were from the gizzard, using LeadCheck SwabsTM (3M, St. Paul, MN, USA), which can detect lead in metals at levels >0.2% (2000 ppm; 3M 2011) (Grade et al., 2017).

North American *Cathartid* vultures are also affected by lead poisoning (Campbell, 2015, 2022). The common, smaller Turkey (*Cathartes aura*, Linnaeus, 1758) and Black vultures (*Coragyps atratus*, Bechstein, 1793) may serve as indicator species for lead poisoning, and rarer species such as the California condor (*Gymnogyps californianus*, Shaw, 1797) are very vulnerable and therefore serve as conservation priorities (Campbell, 2015, 2022). For example, lead poisoning in California condors and Golden eagles in shared ranges has been documented (Pattee et al., 1990) as did ravens in the Greater Yellowstone Area (Craighead et al., 2009). One instructive publication was by Bloom et al. (1989), entitled "Lead contamination of golden eagles (Aquila chrysaetos) within the range of the Californian condor." Cade (2007: 2125) pointed out that "the scientific evidence that California condors (*Gymnogyps californianus,* Shaw, 1797) are frequently sickened and killed by lead poisoning from spent ammunition supports the conclusion that current levels of lead exposure are too high to allow reintroduced condors to develop self-sustaining populations in the wild in Arizona and, by inference, in California." This lead-related mortality reduced California condor numbers during the 1980s, and even during the post-reintroduction period (Kelly & Johnson, 2011; Snyder & Snyder, 2000). Lead accumulation in condor blood of condors has been recorded during and outside hunting seasons, influencing bans on lead ammunition in the condor's range (Hall et al., 2007; Hunt et al., 2007; Sorensen et al., 2007).

The Turkey vulture is often affected by lead ammunition in animal carcasses, incurring symptoms of weakness and uncoordinated movements, and even death (Carpenter et al., 2003; Campbell, 2015). Californian Turkey vultures have recorded higher lead poisoning levels, mostly correlated with the deer hunting season and the intensity of wild pig hunting, despite the possibly violated ban on lead ammunition in Californian hunting (Kelly & Johnson, 2011). Canadian case studies record similar issues (Campbell, 2015, 2022). Martin et al. (2008) report lead levels in the bones, kidneys, and livers of 225 carcasses from 19 species of raptors in southern Ontario, among which turkey vultures had the "highest mean concentrations of lead in bone and kidney compared to other raptor species" (Martin et al., 2008: 96). A study by Clark and Scheuhammer (2003) in Canada reported 184 carcasses of 16 species (including Golden eagles, Turkey vultures, Red-tailed hawks, and Great horned owls) evidently poisoned from game carcass consumption.

Lead poisoning also affects mammals, but studies of poisons in mammals may be fewer than those concerning birds (Campbell, 2015). For example, Chiverton et al. (2022) write that "while the range of negative impacts of Pb exposure to humans and avian wildlife are relatively well documented, little is known about risks to wild mammals despite recent scientific interest and publications"; however, "lead poisoning in captive wild mammals has been reported following consumption of lead-shot meat," with the example of "with post-mortem procedures and radiographs identifying metallic particulate Pb within the gastrointestinal (GI) tract" (see Burco et al., 2012; Hivert et al., 2018; North et al., 2015). Burco et al. (2012: 216) in their study of a cougar (*Puma concolor*, Linnaeus, 1771) note that lead toxicity has been historically documented, with acknowledgment that lead is a significant health issue, especially for water birds and avian scavengers, but only a small number of studies have recorded toxic effects on higher mammalian carnivores.

North et al. (2015) contend that "there is surprisingly little literature detailing the risk to mammalian scavengers and captive carnivores fed hunted meat" despite the commoner studies on scavenging birds and waterfowl, some findings of which are common knowledge. North et al. (2015) present a study of the death of two captive cheetahs (*Acinonyx jubatus jubatus*, Schreber, 1775), which followed the onset of acute nervous symptoms, where "clinical signs included hyper-excitability, seizures, arched back, tail held abnormally high and hyper-salivation." In one example, Hivert et al. (2018) studied the Tasmanian Devil (*Sarcophilus harrisii*, Boitard, 1841), noting that in their study area, "captive and semi-wild Tasmanian devils are fed with possum and wallaby meat provided by local hunters, who use lead ammunition. Lead ingestion can cause acute toxicity, including ataxia, coma and death, or chronic subclinical deleterious effects including decreased fertility."

Another similar study concerned the poisoning of raccoons, was that of Diters and Nielsen (1978: 187), which described a wild raccoon (Procyon lotor), which had clinical signs, including histopathologic and ultrastructural lesions which were indicative of lead intoxication. This diagnosis was confirmed by chemical analyses of liver and kidney tissues, revealing 35 ppm of lead in wet tissues. A further survey of hepatic lead concentrations in 13 other raccoons was carried out, and the findings indicated that raccoons in Connecticut are exposed to lead, with sources including garbage cans, which may be a source of lead. The authors concluded that raccoons can withstand high levels of lead without developing the lesions of plumbism, but it may not be known how lead might affect the stressed raccoon. However, the study revealed a high standard deviation of mean liver lead concentrations obtained in the study, which may point to the fact that the raccoons were not uniformly exposed to lead contamination (Diters & Nielsen, 1978: 187). L

Lead also affects domestic animals, including cattle (*Bos taurus*, Linnaeus, 1758), dogs (*Canis familiaris*, Linnaeus, 1758) including strays, and cats (*Felis catus*, Linnaeus, 1758) (Campbell, 2015). For cattle, a medical report for the Queensland Government (2022a) observed that cattle are the commonest food-producing animals affected and a similar report (State of Queensland, 2022b) describes the symptoms of acute lead poisoning in such animals as including "colic, staggering gait, rolling eyes, slobbering, muscle spasms, blindness, uncoordinated attempts to climb obstacles, excessive response to external stimuli, head pressing, convulsions." Chronic (more serious, longer term) lead poisoning symptoms include death, the acute symptoms listed above, or "loss of appetite, wasting, anemia, constipation, recumbency, difficulty breathing" (State of Queensland, 2022b). Subacute lead poisoning symptoms include "dullness, loss of appetite, abdominal pain, diarrhoea" (State of Queensland, 2022b). Sources of lead in this context include "discarded sump oil and oil filters, lead batteries (the most common cause of cases), flaking lead-based paint or old paint tins, lead shot and fishing sinkers, linoleum, grease, putty, metallic lead" (State of Queensland, 2022b).

Domestic and other animals may be used as indicators for human problems with environmental lead. For example, Bischoff et al. (2010: 185) argue that as human and nonhuman animals often share the same environment, both may be exposed to toxicants, and consequently, "domestic animals can be used as sentinels for exposure of people to these agents." Their case study concerned an agricultural area where cattle deaths resulted from lead poisoning based on elevated postmortem tissue lead concentrations. The research findings indicated that the remaining cattle (in addition to a dog, a cat, and a pregnant woman on the farm) had higher than average blood lead concentrations with the range of blood lead concentrations being 8.42 in the cat, to 85.41 μg/dL in a calf, but there were no visible clinical lead poisoning symptoms. The findings further indicated that the possible lead exposure source was paint in the barn and home. The authors conclude that "household dogs and cats have been used as sentinels for lead poisoning in humans, but cattle may also act as a sentinel species for environmental lead contamination" (Bischoff et al., 2010: 185).

A related but urban-based study was that of Morgan (1994: 18), in which cases of lead poisoning were studied for small companion animals, at a lead toxicosis monitoring program at Angell Memorial Animal Hospital, Boston MA, from 1987 to 1992. The findings indicated that the number of cases of lead poisoning declined during the study period, with the affected animals comprising 53 dogs, 20 birds, 8 cats, 3 rabbits, and 1 iguana. The main ailments were gastrointestinal and neurologic signs, with blood lead concentrations ranging from 40 to 620 micrograms/dl. Most of the affected animals could be treated with chelation therapy. Paint was the commonest (32.4%) lead source. The majority (57%) of the studied animals were traceable to three 3 inner city neighborhoods (Morgan, 1994: 18).

Esposito et al. (2019: 3473) also examined the occurrence of lead (Pb) and cadmium (Cd) in the livers and kidneys of dead stray dogs and dead cats found from 2014 to 2017 in Naples, Italy, using atomic absorption spectrophotometry (AAS) to detect heavy metal concentrations. Lead was detected in livers and kidneys analyzed, with lead concentration variations based on age class, gender, and tissue type. For lead levels, "the highest concentrations were detected in the liver (3.45 mg/kg in dog and 5.93 in cat, respectively) followed by the renal tissue, with no significant difference depending on the animal gender" (Esposito et al., 2019: 3473). The authors conclude that the study "can be considered the first one in Italy regarding stray dogs and cats as bio-indicators of environmental contamination due to lead and cadmium, suggesting that pets could be sentinel animals to evaluate human exposure to these heavy metals" (Esposito et al., 2019: 3473).

Poisoning from Mercury

Mercury is an extremely dangerous chemical element to animals and people (Clarkson and Magos 2006; Campbell, 2015, 2022). Henny and Elliott (2015: 337) assess that mercury toxicity depends on whether the mercury is in organic or inorganic form. Inorganic mercury is only minimally absorbed, but organic mercury is almost totally absorbed by the animal's intestine. Inorganic mercury produces methylmercury (MeHg), which fish accumulate from water and their diet, and comprises nearly all the mercury in fish flesh. MeHg can have severe impacts on developing neural tissue in birds, and due to the high mercury content of some fish flesh, fish-eating birds are especially vulnerable. In both animals and people, mere inhalation of mercury vapor can result in severe symptoms and in some cases death (Brean, 2019). Mercury poisoning is important for animals near the top of the food chain, due to accumulation up the chain (Campbell, 2015; Henny & Elliott, 2015).

One early source of mercury in the environment was the use of alkyl mercury as a fungicide for agricultural seed dressing during the 1960s and 1970s (Berg et al., 1966; Jenson et al., 1972). High concentrations of mercury-based fungicides in grains were consumed by small birds, including passerines that were predated by raptors, from Sparrow hawks to eagles (Kitowski et al., 2016, see also Olendorff, 1976; Tjernberg, 1981; Watson, 2010). Alkyl mercury was also used during the 1940s but was banned as a seed dressing in Sweden in 1966 (Johnels et al., 1979). The main recent mercury sources include deposition from coal-fired power plants (especially in Arctic and northeastern North America, which poison fish and fish-eaters) and mining in the Amazon basin and elsewhere usually for mercury or gold and silver (Henny & Elliott, 2015, see also Nriagu et al., 1992).

Among both terrestrial and marine environment raptors, mercury ascends the food chain, with similar effects (Campbell, 2022). For example, peregrine falcons which hunt mostly birds may prey on small birds poisoned by mercury-infected grains, and the peregrine falcons may also be predated by Golden eagles (Craig & Enderson, 2004; Ferguson-Lees & Christie, 2001; Gainzaraian et al., 2000; Sergio et al., 2004; Sergio & Hiraldo, 2008). Barnes et al., 2019 tested 700 peregrines for mercury in Washington, Maryland, Southern Nevada, and the Gulf Coast of Texas, and discovered all the birds tested positive for mercury. Barnes et al. (2019) argued that Peregrines are one of the most contaminated species, because they feed on birds exposed to mercury. Peregrines in the Lake Mead, Colorado River (Arizona and Nevada), acquire mercury through common predation on the Eared grebe (*Podiceps nigricollis*, Brehm, 1831), which may be exposed to mercury in Utah's Great Salt Lake. Peregrines are potentially useful as biomonitors, due to their global distribution, long lives and breeding frequency, diverse diet, and the wide range of their nesting habitat, both variable aquatic and terrestrial habitat types (e.g., freshwater and marine waterbodies, wetlands, arctic tundra, temperate forests, and desert) (Barnes et al., 2019 see also White et al., 2013).

White-tailed eagles are also frequently poisoned with mercury (Falandysz et al., 2001; Kalisinska et al., 2014). Similar poisoning has been recorded for sparrowhawks (*Accipiter nisus*, Linnaeus, 1758) in Poland, buzzards (*Buteo*, Lacépède, 1799) in Spain and the Czech Republic and Spanish and German goshawks (*Accipiter gentilis*, Linnaeus, 1758) (Castro et al., 2011; Houserova et al., 2005; Kitowski et al., 2016). Waterbodies and waterborne prey are also common sources of mercury poisoning in raptors (Kitowski et al. (2016, see also Kalisinska et al., 2014). Kitowski et al. (2016) discerned a difference between goshawks and sparrowhawks in mercury poisoning: "the strong correlation between mercury and the consumption of small birds…implies that consumption of bigger avian prey by Goshawks allows them to avoid the accumulation of large amounts of mercury in the livers, in contrast to the Sparrowhawks specialized in catching small birds (Kitowski et al., 2016; see also Newton, 1986).

Another study by Weir et al. (2018) examined the impact of the rodenticide brodifacoum (formula $C_{31}H_{23}BrO_3$) and mercury on raptor species, including Red-tailed hawks (*Buteo jamaicensis*, Gmelin, 1788), Red-shouldered hawks (*Buteo lineatus*, Gmelin, 1788), Barred owls (*Strix varia*, Barton, 1799), Great horned owls (*Bubo virginianus*, Gmelin, 1788), and ospreys *Pandion haliaetus*, Linnaeus, 1758). Weir et al. (2018) examined the birds' livers using a direct mercury analyzer approach and liquid chromatography–mass spectrometry (LC-MS) (to detect five anticoagulant rodenticides) and found that mercury residues were high fish-eating birds (ospreys 15.09 mg/kg and great blue herons 6.93 mg/kg, dry weight) and "relatively high in red-shouldered hawks and one eastern screech owl tested" (Weir et al., 2018). Six samples (three ospreys, one each of great blue heron, red-shouldered hawk, eastern screech owl) had levels above the health threshold of 1 mg/kg (wet weight). Brodifacoum, the only rodenticide consistently detected, exceeded 75% in barred owls, great horned owls, and red-shouldered hawks. The study concluded that brodifacoum was a greater risk than mercury for most raptor species (with the possible exceptions of the osprey, and less the red-shouldered hawk, which ingested less mercury) (Weir et al., 2018).

Mercury is also a dangerous poison for mammals, usually from the same sources as those for predatory birds, that is from animals and plants lower in the food chain, or from the physical environment (Campbell, 2015). For example, a study by Kalisińska et al. (2009: 345) examined mercury concentrations in the kidney, liver, and pectoral muscle of wild carnivorous mammals from Poland, comprising 10 red foxes (*Vulpes vulpes*, Linnaeus, 1758), 3 raccoon dogs (*Nyctereutes procyonoides* Gray, 1834), 2 badgers (*Meles meles* Linnaeus, 1758), 3 pine martens (*Martes martes* Linnaeus, 1758), and 1 polecat (*Mustela putorius* Linnaeus, 1758). The findings indicated that foxes located in Mielin Island fed mostly on black cormorant nestlings as well as on fish and their remains, and "the highest mean of mercury concentrations was revealed in the Mielin red fox juveniles: 5.11, 4.52, and 1.56 mg/kg d.w. being recorded" in the kidneys, liver and pectoral muscle: notably "no significant differences in mercury concentrations in the respective tissues were found between the remaining immature and adult red foxes; their mercury concentrations were several times lower than those of the Mielin individuals" (Kalisińska et al., 2009: 345). Finally, for all the sampled animals except the Mielin foxes, the mercury concentrations in the kidneys, liver, and pectoral muscle were less than 1.3, 1.0, and 0.5 mg/kg d.w., respectively, with the highest records in badgers (mostly predators of soil invertebrates), followed by pine martens and then the red fox and raccoon dog (Kalisińska et al., 2009: 345).

Poisoning from Chromium

Chromium toxicity results from organism exposure to types of chromium, for such as hexavalent chromium (chromium (VI), which destroys red blood cells (hemotoxic or hemotoxic) (Merriam-Webster, 2020a), genotoxic ("the process by which an agent produces a deleterious effect on DNA and other cellular targets that control the integrity of genetic material," Lopez et al., 2012), and carcinogenic (producing or tending to produce cancer, Merriam-Webster, 2020b). Hexavalent chromium affects blood cells by contributing to oxidation reactions, which can produce hemolysis and kidney and liver failure (Xie et al., 2017). Speer and Wise (2018) noted that "hexavalent chromium [Cr(VI)] is a valuable industrial resource but ubiquitous environmental contaminant. Humans can be exposed to Cr(VI) occupationally and through environmental sources. It is known Cr(VI) causes lung cancer and other health issues including reproductive and immune dysfunction."

Falandysz et al. (2001) studied the impact of chromium concentrations of 0.087 mg/kg dw in White-tailed eagles from the South Baltic and recorded values below those of Kitowski et al. (2016). Manning et al. (2008) also found low chromium hepatic values for an Australian, White-bellied sea eagle (*Haliaeetus leucogaster*, Gmelin, 1788) of 0.04 mg/kg ww (about 0.16 mg/kg dw). Van Wyk et al. (2001b) estimated a higher average concentration of 19.57 mg/kg dw of hepatic Cr. for the White-backed vulture (*Gyps africanus*, Salvadori, 1865) in the Republic of South Africa. Smaller raptors, especially those predating birds, recorded low levels of chromium. Kitowski et al. (2016) noted that in several studies, marsh harriers, kestrels, buzzards, and sparrowhawks in southern Italy (predators of passerines) had low hepatic concentrations of chromium, always below 0.40 mg/kg dw (see Zaccaroni et al., 2008). Tawny owls (*Strix aluco*, Linnaeus, 1758), had average hepatic concentrations just over 0.30 mg/kg dw. Little owls (*Athene noctua*, Scopoli, 1769) in northern Italy had concentrations of 0.29 mg/kg ww (about 1.16 mg/kg dw.) (Zaccaroni et al., 2003). These two owl species predated birds, an important source of chromium, as did Sparrowhawks (Kitowski et al., 2016).

Mammals also suffer from chromium poisoning (Campbell, 2022). For example, a study by Outridge and Scheuhammer (1993: 31) examined chromium ingestion in birds and mammals, arguing that "chromium(VI) compounds are absorbed significantly more efficiently (2–10% of dose) from the GI tract than inorganic Cr(III) compounds (0.5–3%), due to the increased membrane permeability of the former." However, it is noted that "Cr concentration data for mammalian and avian wildlife species and their potential food organisms are scarce. Worldwide, fewer than 50 species of free-living mammals and birds have been surveyed with regard to tissue Cr concentrations" (Outridge & Scheuhammer, 1993: 31). Further, it is recorded that higher concentrations occur in bone tissue than other tissues in animals chronically exposed to Cr. Measuring concentrations only in the liver and/or kidneys has been a common practice, yet these organs failed to show evidence of extant Cr contamination in some cases. It is recommended that analysis of the bone, liver, and kidneys be a minimum requirement for future Cr biomonitoring studies (Outridge & Scheuhammer, 1993: 31). The authors conclude that dietary Cr concentrations > or = 10 micrograms/g DW in food are potentially harmful to the health and reproductive success of predators of wildlife consumers, but although some fish and aquatic invertebrates may be sources of chromium, the removal of these organisms from Cr contaminated environments may be negative for wild birds and mammals that eat these organisms (Outridge & Scheuhammer, 1993: 31).

Poisoning from Nickel

Nickel occurs widely (usually bound with oxygen or sulfur, through oxides or sulfides), mostly in soils, volcanic emissions, industrial products, recycling and disposal mining, power plant operation and/or incinerator emissions, and industrial production of rubber, plastic, nickel-cadmium battery, and electroplating industries (Mudd et al., 2010; Das et al., 2018). Nickel is an essential micronutrient for maintaining health in some plants, invertebrates, birds, and mammals, including humans (Stangl & Kirchgessner, 1996, 1997). Eisler (1998) stated that nickel carbonyl $(Ni(CO)_4$ is the only nickel compound known to cause severe acute impacts, such as pulmonary damage and death, while strong toxic effects of other nickel compounds are less common. Buxton et al. (2019) contend that "notable human health toxicity effects identified from human and/or animal studies include respiratory cancer, non-cancer toxicity effects following inhalation, dermatitis, and reproductive effects.

These effects have thresholds, with indirect genotoxic and epigenetic events underlying the threshold mode of action for nickel carcinogenicity."

Nickel, like other heavy metals, may climb the food chain, and affect mammals and birds on this chain, an important recognized ecological problem (Campbell, 2015). For example, DeForest et al. (2012: 107) examined the European Union's Existing Substances regulation (EEC 793/93), which "was developed to assess the ecological risks posed by chemical substances such as Ni…" and also looked at "a generic framework for assessing secondary poisoning risks and prescribes the following terrestrial food chain: soil → earthworm → worm-eating bird or mammal." The Nickel Institute (2018) studied nickel in marine, fresh water, and soil, using a framework of the ratio of dietary predicted exposure concentrations (PEC) and dietary predicted no-effect concentrations (PNEC) (European Chemicals Bureau, 2003). The argument was that "the PEC in this assessment was the estimated Ni concentration in the diet of the representative predatory birds or mammals in each of the food chains evaluated, while the PNEC was the Ni concentration in bird and mammal diets at which no adverse effects are expected" (Nickel Institute, 2018). The ratio of the PEC to PNEC created Risk Characterization Ratio, or RCR, where an RCR <1 indicates low potential risk and an RCR >1 indicates a stronger risk. The formula is: a PEC-to-PNEC ratio, or RCR:RCR = PEC × RAF/PNEC, where RAF is relative bioavailability.

The bioaccumulation of nickel from soil or water into organisms at the bottom of the food chain is low, and there is little evidence that Ni biomagnifies in aquatic or terrestrial food chains (i.e., Ni concentrations do not increase with increasing trophic level). This contrasts with other chemicals, such as methyl mercury, which tends to biomagnify across multiple trophic levels, thereby resulting in higher trophic level organisms being potentially more susceptible to mercury poisoning" (Nickel Institute, 2018). Possibly, nickel may have an opposite reaction, termed "biodilution" where the metal reduces as it moves up the food chain (see Lapointe & Couture, 2006). A novel research method used by the Nickel Institute (2018) is a tiered approach, including three tiers, which vary in assumptions on diet breadth (narrower diets may be dominated by likely poisoned food items like the common cockle *Cerastoderma edule*, Linnaeus, 1758) and hence risks of poisoning. Broader diets have less risk, as more harmless foods may be included. It may be argued that mollusk-eating birds in freshwater ecosystems are at risk from such a narrow diet, but broader diets may be less risky. Similarly, worm-eating birds in soil ecosystems may be at risk, which would vary with wider diets (Nickel Institute, 2018).

Some raptors are recorded to have strong nickel ingestion (Campbell, 2022). For example, Falandysz et al. (2001) studied White-tailed eagles in the Baltic Sea area and found the sampled birds had liver-nickel concentrations 56 times higher (13.0 mg/kg dw) than those of the study by Kitowski et al. (2016) in east Poland. Hence, Falandysz et al. (2001) speculated that the White-tailed eagles from east Poland may have fewer opportunities to forage over highly mercury- and nickel-contaminated areas than the raptors from the south Baltic Sea coast (Kitowski et al. (2016, 11). For smaller raptors, Kitowski et al. (2016) examined sparrowhawks and goshawks, which accumulated similar levels of nickel (Ni 0.3 mg/kg dw) due to their common diet of birds. Also cited are the low nickel levels in one peregrine falcon (0.107 mg/kg dw (Kalisińska et al., 2008) in northwest Poland.

For mammals, there are serious effects caused by inhalation of nickel compounds other than nickel carbonyl, which may occur with aerosols of both soluble and insoluble nickel compounds (Eisler, 1998; see also USEPA, 1980). Eisler (1998) emphasizes that nickel carbonyl is extremely lethal to animals and is "the most rapidly and completely absorbed nickel compound in mammals. Additionally, for mammals, nickel toxicity is due to the chemical form of nickel, the dose, and the type of exposure. It is maintained that nickel exposure by inhalation, injection, or cutaneous contact is more significant than oral nickel exposure. In the study by Eisler (1998), the toxic effects of nickel on human and nonhuman animals are recorded for "respiratory, cardiovascular, gastrointestinal, hematological, musculoskeletal, hepatic, renal, dermal, ocular, immunological, developmental, neurological, and reproductive systems," and "nickel-inhalable dust, nickel subsulfide, nickel oxide, and, especially, nickel carbonyl induce acute pneumonitis, central nervous system disorders, skin disorders such as dermatitis, and cancer of the lungs and nasal cavity" (see also Benson et al., 1995).

Eisler (1998) also reports that in mammals, nickel retention in the body is low, the half-time residence of soluble forms of nickel is only several days, and there is not much evidence of tissue accumulation, except for the lungs: "In mammals, clearance was fastest from serum, followed by kidney, muscle, stomach, and uterus; relatively slow clearance was evident in skin, brain, and especially lung" (see also Kasprzak, 1987). Other pointers were that some forms of nickel are carcinogenic to animals, such carcinogenicity varies with the chemical form of nickel, route of exposure, animal species (at inter- and intraspecies levels), dose, and exposure duration (see also USEPA, 1980). At the species level, mice recorded higher concentrations of nickel than rats and some other rodents, and in one study, beavers and minks recorded higher nickel concentrations in their liver than birds in the same localities (see also Chau & Kulikovsky-Cordeiro, 1995).

Another important point is that of the mammal's age: while "age is an important variable and nickel generally increases in various organs with increasing age of terrestrial and marine mammals," other factors are less relevant such as "trophic position in the food chain, sex, and reproductive state (Eisler, 1998; Outridge & Scheuhammer, 1993). Impacts on fish are listed as including surfacing, rapid mouth and opercular movements, convulsions and loss of equilibrium, decreased glycogen in muscle and liver, increased lactic acid and glucose in the blood, lowered hydrogen peroxide production in tissues, reduction in superoxide dismutase, and "contractions of vascular smooth muscle – signs similar to those associated with hypertension in mammals" (Eisler, 1998; see also Evans et al., 1990; Khangarot & Ray, 1990; Ghazaly, 1992; Bowser et al., 1994; Ellgaard et al., 1995).

Outridge and Scheuhammer (1993) listed six points of nickel toxicity to mammals: resistant mammal species can resist low nickel contamination; real environmental exposures are less toxic than laboratory experimental exposures; inhaled nickel is more toxic than ingested nickel, because it is absorbed faster from the lungs than from the gastrointestinal tract, with fatal results commoner from respiratory failure than from nervous system damage; closely related species such as rats and mice may differ in nickel sensitivity; threshold effects on lung function or morphology may occur at low airborne nickel concentrations; and juveniles are often more sensitive to nickel than adults. Eisler (1998) adds that "treatment of mammals suffering from nickel poisoning is usually through administration of various classes of chelating agents… In all cases, the agents accelerate urinary excretion of absorbed nickel before extensive tissue injury occurs" (see also USEPA, 1980).

Poisoning from Cadmium

Cullen and Maldonado (2013) describe cadmium as comparatively mobile and acutely toxic to most life forms and that "the relatively high volatility of Cd metal, its large ionic radius, and its chemical speciation in aquatic systems makes Cd particularly susceptible to mobilization by anthropogenic and natural processes." It is also argued that the cadmium biogeochemical cycle is significantly changed by human inputs, especially after the beginning of the industrial revolution and the marked increases in fossil fuel burning and nonferrous metal mining (Cullen & Maldonado, 2013). Wayland and Scheuhammer (2011: 645) add that cadmium accumulates in avian tissues, especially through their diets. Some in vitro studies on embryonic chick femurs have indicated that cadmium increases resorption of calcium and phosphorus, impedes bone matrix formation, and decreases the bone mineral content (Wayland & Scheuhammer, 2011: 645). It is further reported that cadmium levels in birds may vary according to diet, ecosystem use, age, and physiological status (Wayland & Scheuhammer, 2011: 645).

For birds, cadmium can cause "sublethal behavioral effects at lower concentrations than lead and mercury" (Eisler, 1985). Kitowski et al. (2016) point out that cadmium concentrations in birds are usually strongest in the kidneys, intermediate in muscle, and lowest in the livers (see also Battaglia et al., 2005; Binkowski et al., 2013). However, Kitowski et al. (2016) argue that the liver can be a measurable area for measuring Cd exposure, as it accumulates about 50% of the total cadmium bodily deposit, and the body burden is stable in the liver, as the liver is an organ resistant to cadmium toxicity (Battaglia et al., 2005; Burgat, 1990).

Cadmium poisoning is frequently higher in seabirds, especially pelagic species, than in most terrestrial, freshwater, and shore birds (Wayland & Scheuhammer (2011: 645). Marine environments are frequently polluted with cadmium (Pacyna et al., 2010; Seshadri et al., 2010). For potential Sea eagle prey species, Parslow et al. (1973) report that gannets, and some other seabirds that forage mainly in offshore waters, usually have lower levels of cadmium in their livers (mean 4.7 ppm dry weight, seven adults) than do the mainly pelagic feeders. Examples are Fulmar (*Fulmarus glacialis*, Linnaeus 1761), mean 37 ppm, 11 individuals), some littoral and sublittoral mollusk feeders such as the Eurasian oystercatcher (*Haematopus ostralegus*, Linnaeus, 1758, mean 29 ppm) and Common eider (*Somateria mollissima*, Linnaeus, 1758, mean 10 ppm, 8 individuals). Cadmium is obtained from the invertebrates upon which the oceanic birds feed. Rainbow (1997: 195) supports the position of marine invertebrates being possible concentrators of environmental metals: "Trace metals are accumulated by marine invertebrates to body concentrations higher, in many cases orders of magnitude higher, than the concentrations in an equivalent weight of the surrounding seawater" (Parslow et al., 1973).

For smaller raptors preying on small mammals, Kitowski et al. (2016) recorded that the high concentration of cadmium in buzzard livers can be connected to their feeding on voles and other animals, as there is generally a higher cadmium level for small mammal consumption. This may point to voles and other small mammals feeding on fields fertilized with chemicals including cadmium (Louekari et al., 2000). Similarly, Falandysz et al. (2001) reported lower cadmium levels in the livers of Baltic White-tailed eagles than the studied goshawks. Cadmium levels in buzzards recorded Northern Italy, (Battaglia et al., 2005), Portugal (Carneiro et al., 2014) and Spain (Castro et al., 2011) were mostly lower than in East Poland (Kitowski et al., 2016). Other low levels of liver cadmium concentrations were recorded in goshawks in Germany (Kenntner et al., 2003) and buzzards in Spain (Castro et al., 2011). Kitowski et al. (2016) note that the levels of cadmium for peregrines in Sweden (Ek et al., 2004) were similar to those of goshawks in their studies.

For mammals, there are numerous impacts of cadmium poisoning. Marettová et al. (2015) report that for mammals, the impacts of cadmium poison depend on the exposure dose, routes, ways, and duration. Cooke (2011: 627) also writes that cadmium absorption and retention is clearly dose and duration dependent in small mammals and "controlled experimental studies involving laboratory rats and mice, in particular, have provided relevant data concerning the effects of cadmium in wild small mammals… focused on the cadmium concentrations in tissues, especially in kidney and liver." Larger mammals have also been studied from field sampling. For example, Esposito et al. (2019: 3473) found cadmium levels in the livers and kidneys of dead, stray dogs and cats between 2014 and 2017 in the city of Naples and its surrounding areas. Heavy metal concentrations (mostly lead and cadmium) were determined by using atomic absorption spectrophotometry (AAS). The findings for cadmium were that varying concentrations of Cd (ranging from 0.005 to 6.13 mg/kg) existed in livers and kidneys, and variations in cadmium levels were correlated with animal age class, gender, and kind of tissue for both elements, similar to other reports in published literature. For example, cadmium levels in the kidney were significantly higher in females than those in males for both dogs and cats (Esposito et al., 2019: 3473).

Marettová et al. (2015: 1) describe the effects of cadmium on mammals, taking the example of the structure of the seminiferous tubules, Leydig cells, and blood vessels in the testis. The main effect was the destruction of the seminiferous tubules with serious necrotic areas (these being death of tissue) and damage to "all stages of developing germ cells by inducing their structural changes and the apoptotic cell death" (Marettová et al. (2015: 1). It is further remarked that "sertoli supporting cells are considered the most vulnerable cells" as the damage of these cells may result in "cytoplasmic rearrangement and disruption of inter-Sertoli tight junctions resulting in increased permeability of the blood-testis barrier, structural changes in the Leydig cells and decreased testosterone secretion…an increase of the amount of interstitial connective tissue occurs." Cadmium impacts on blood vessels include morphological and physiological changes in vascular endothelial cells and smooth muscle cells (Marettová et al., 2015: 1). Summarizing, the impacts were listed as the necrosis of the sensitive cells, lesions in surrounding tissue and free cells, atrophy of the seminiferous tubules, Leydig cell regeneration and interstitial revascularization (Marettová et al., 2015: 1). These impacts in mammals were like the effects on birds, where "spermatogenic cells underwent irreversible degeneration or atrophy of seminiferous tubules in the absence of significant vascular lesions" (Marettová et al., 2015: 1).

Poisoning from Copper

Acute and chronic copper poisoning affects many animal species, but there may be interspecies variability (Blakley (2021). Chronic poisoning is common, manifested by high morbidity in species exhibiting the hemolytic crisis and severe liver damage. However, records indicate that copper toxicity may be less toxic for birds than lead toxicity (Campbell, 2015). Copper is "an essential cofactor for several oxidative stress-related enzymes including catalase, superoxide dismutase, peroxidase, cytochrome c oxidases, ferroxidases, monoamine oxidase, and dopamine βmonooxygenase" (Tchounwou et al., 2012, 134, see also Harvey & McArdle, 2008; Stern, 2010; Agency for Toxic Substances and Disease Registry, 2020a). Manto (2014, 327) writes that "copper is a cofactor of proteins and enzymes (called cuproenzymes) involved in fundamental mechanisms, such as energy generation, oxygen transportation, hematopoiesis, cellular metabolism and signal transduction." Copper is therefore not primarily a toxic element, being a nutrient within metalloenzymes in "hemoglobin formation, carbohydrate metabolism, catecholamine biosynthesis, and cross-linking of collagen, elastin, and hair keratin" (Tchounwou et al., 2012: 134). Toxicities may be generated when copper transitions between the oxidized Cu(II) and a reduced state Cu(I), and perhaps human cellular damage and Wilson disease (Chang et al., 1996; Harvey & McArdle, 2008; Tchounwou et al., 2012; Stern, 2010).

For birds, copper has been investigated as a toxic element, partly because the use of copper-jacketed bullets in human hunting of potential prey animals and the scavenging by raptors of the carcasses of such animals has encouraged studies on possible copper toxicity for raptors (Cruz-Martinez et al., 2012). Copper-jacketed lead bullets have been compared with lead bullets for avian toxicity (Kallok, 2013). Kallok's (2013) study examined injured, sick, and dead Bald eagles delivered to the University of Minnesota's Raptor Center, which were investigated for lead (from lead shot in carcasses), after, in general, copper jackets on soft lead bullets were hypothesized to be insufficient to prevent lead fragmentation into flesh after impact, which would poison scavenging eagles. A Raptor Center study detected copper in

63 out of 89 eagle carcasses, and alternative possibilities included copper bullets, which did not fragment into flesh on impact, unlike 12-gauge lead shotgun slugs (which left about 28 fragments per shot) and muzzleloader bullets (which left 82–141 lead fragments) several centimeters from the entry point of the wound (Kallok, 2013).

For smaller raptors, Franson et al. (2012) examined copper toxicity in American kestrels (*Falco sparverius*, Linnaeus, 1758), using bird ingestion of orally administered copper. The kestrels were orally administered 5 mg Cu/g body mass copper pellets nine times during a 38-day period. Most birds retained the pellets for over an hour, but 69% regurgitated the pellets in the 12h monitoring period. The findings indicated that hepatic copper concentrations were higher in Cu-administered kestrels than in the controls, but there was no difference in treated and control bird's blood Cu concentrations. Additionally, hepatic copper and the protein metallothionein concentrations were significantly correlated, but metallothionein was higher in male birds that received copper than in controls and for female birds that received copper were similar to controls. Clinical signs were absent: there was no treatment impact on body mass and no impacts from copper on hemoglobin and methemoglobin in the blood or in the kidney, plasma biochemistries, or hematocrit (the percentage of the blood that consists of red blood cells) (Franson et al., 2012: 145). The authors concluded that ingested Cu pellets did not pose a serious threat to American kestrels (and presumably phylogenetically related species), although the retention time of pellets in the stomach was not long (Franson et al., 2012: 145). Furthermore, kestrels and similar species that regurgitate Cu fragments would be unlikely to be seriously affected by Cu ingestion, but there are possibilities that some Cu fragments might be retained, with negative consequences (Franson et al., 2012: 145).

For mammals, there are indications of more severe copper poisoning. For example, Blakley (2021) argues that acute and chronic copper poisoning (toxicosis) is a global problem for many animal species. Blakley (2021) observes an inherited sensitivity to copper toxicosis similar to Wilson disease in humans exists in some breeds of dogs, notably Bedlington Terriers. Sheep are also most frequently affected (Blakley, 2021). Rankins and Pugh (2012: 18) report that copper toxicity is a more serious problem in sheep than in goats, as in sheep, the "magnitude of difference between copper deficiency and copper toxicity is quite small" (Rankins & Pugh, 2012: 18). Factors for such copper toxicosis (based on physiological absorption and/or retention of copper) include lowered levels of dietary molybdenum or sulfate, the herbivory of plants such as subterranean clover (*Trifolium subterraneum*), which may cause a mineral imbalance and excessive copper retention/chronic copper toxicosis (phytogenous toxicosis), and the consumption of other plants such as *Heliotropium europaeum* or *Senecio* spp. leading to hepatogenous chronic copper toxicosis (Blakley, 2021). The plants, which have average concentrations of copper and low concentrations of molybdenum, may contain hepatotoxic alkaloids, which contribute to retention of excess amounts of copper in the liver (Blakley, 2021).

Blakley (2021) gives the explanation of poisoning in the case of sheep: the main chronic copper toxicosis occurs when excessive copper is ingested over long periods, but the impact is subclinical until the accumulated liver copper is released into the bloodstream. Early indications of Cu toxicity include increased liver enzyme concentrations. Increased serum copper concentrations lead to lipid peroxidation (oxidants attack lipids) and intravascular hemolysis (the latter refers to the death of red blood cells, with the release of cell contents into the plasma). Furthermore, "this hemolytic crisis may be precipitated by numerous factors, including transportation, handling, weather conditions, pregnancy, lactation, strenuous exercise, or deteriorating nutrition. In sheep, dogs, or coyotes, such impacts are associated with chronic copper poisoning (Blakley, 2021). For sheep and cattle, "acute poisoning may follow intakes of 20–100 mg/kg of copper in sheep and young calves and of 200–800 mg/kg in mature cattle." Chronic poisoning of sheep has been linked with daily intakes of 3.5 mg/kg of copper, due to high levels of copper in grazing pastures that may contain 15–20 ppm of copper and low molybdenum concentrations. Impacts included animal deaths, sometimes after a few days of ingestion, sometimes due to renal failure (Blakley, 2021). In dogs, "with liver diseases such as chronic active hepatitis, the primary clinical signs may resemble those of chronic copper poisoning, which can be attributed to liver damage and subsequent retention of excessive copper"; however, Blakley (2021) notes that "it is not clear whether chronic active hepatitis causes the accumulation of copper in the liver or is the result of such accumulation" (Blakley, 2021).

The Australian Livestock Export Corporation Limited (2022) concurs with these findings, noting that sheep and goats are more vulnerable to copper poisoning than cattle, and sheep are more vulnerable than goats. In some examples, British breed sheep and crosses were more vulnerable to copper accumulation than pure Merino breeds. The main symptom of toxic copper poisoning was the sudden death of the animal, when "copper accumulation reaches a critical level, liver cells degenerate and there is a sudden release of copper into the blood stream causing massive haemolysis (breakdown of red blood cells) and death" (Australian Livestock Export Corporation Limited, 2022). The factors for such copper poisoning: hepatotoxic plant and fungal toxins, diets low in copper antagonists, such as molybdenum and sulfate, grazing grounds with lupin stubbles supporting growth of the fungus *Diaporthe toxica*, and plants containing pyrrolizidine alkaloids such as heliotrope and Paterson's Curse, all these sources being compounded by stressors that release accumulated copper from the liver, including strenuous exercise or heat and cold stress (Australian Livestock Export Corporation Limited, 2022).

Poisoning from Selenium

Selenium, a nonmetal element, commonly occurs in sulfide ores (where it may replace sulfur) and is generally considered vital for animal physiological health in small amounts, as selenium is an essential trace element, needed for the function of metabolism (Heinz & Hoffman, 1998; Spallholz &Hoffman, 2002; Tchounwou et al., 2012). Giri et al. (2021) report that "Selenium (Se), even though it is an important nutrient that animals require for better growth of the skeletal and cardiac muscle, reproductive, immunological response, etc., has now become toxic due to poisoning, selenium injection, fodders grown on selenium-toxic land, and indiscriminate minerals supplementation feeding." Selenium may form oxides with other elements (oxides include selenium dioxide, SeO_2; selenium trioxide, SeO_3, the former dissolving in water or nitric acid to form selenous acid, H_2SeO_3). Selenium is moderately toxic in large amounts, and selenium toxicity in humans causes gastrointestinal and behavioral disorders, neurological damage, liver cirrhosis, pulmonary edemas, or even death (Agency for Toxic Substances and Disease Registry, 2020b). Selenium studies focus on bioaccumulation through the food chain (Campbell, 2022).

Lemly (1996) points out that for such heavy metals, a "toxic threat is expressed as hazard and is based on the potential for food-chain bioaccumulation and reproductive impairment in fish and aquatic birds, which are the most sensitive biological responses for estimating ecosystem-level impacts of selenium contamination" (Lemly, 1996, 19). Five degrees of selenium hazard are possible, dependent on quantity and exposure of the occurring selenium: absent, minimal, low, moderate, and high selenium hazard. The five ecosystem components; water, sediments, benthic macroinvertebrates, fish eggs, and aquatic bird eggs, have variable hazard scores. Environmental threats were assessed by combining the five environmental types with the five ranks of selenium threat. The least hazardous environment would have each of the five ecosystem components with a rating of one, and the highest hazard would have a rating of 5 for each of these components. Based on these totals, a score of 5 would be no hazard, score of 6–8 would be minimal hazard, 9–11 would be low hazard, 12–15 would moderate hazard and 16 a high hazard (Lemly, 1996: 19). Lemly (1996: 35) concluded that this method is based on intensive field and laboratory research data on selenium cycling and toxicity in aquatic ecosystem, to assess the toxic threat of selenium to fish and wildlife."

Lemly (1993: 83) points out that "it is now possible to formulate diagnostic selenium concentrations in four distinct ecosystem-level components; water, food-chain, predatory fish (consuming fish or invertebrate prey), and aquatic birds." Additionally, even with low selenium toxicity, "waterborne selenium concentrations of 2 μg/l or greater (parts per billion; total recoverable basis in 0.45 μ filtered samples) should be considered hazardous to the health and long-term survival of fish and wildlife populations due to the high potential for food-chain bioaccumulation, dietary toxicity, and reproductive effects" (Lemly, 1993: 83). At the lower levels of the food chain (mostly small animals such as zooplankton, benthic invertebrates, and small fishes accumulate selenium with little effect up to 30 μg/g (even to 370 μg/g), but these levels are dangerous higher up in the food chain, where the tolerance level is only 3 μg/g for some fish and other wildlife species.

The thresholds of selenium tolerance for freshwater and anadromous (fish that migrate up rivers from the sea to spawn) fish are: "whole body = 4 μg/g; skeletal muscle = 8 μg/g; liver = 12 μg/g; ovaries and eggs = 10 μg/g" ... thresholds for aquatic birds are: liver = 10 μg/g; eggs=3 μg/g" (Lemly, 1993: 83). For the water-birds that predate fish, selenium concentrations in gravid ovaries and eggs offer "the most precise way to evaluate potential reproductive impacts to adult fish and aquatic bird populations" (Lemly (1993: 83). Therefore, "resource managers and aquatic biologists should obtain measurements of selenium concentrations present in water, food-chain organisms, and fish and wildlife tissues in order to formulate a comprehensive and conclusive assessment of the overall selenium status and health of aquatic ecosystems" (Lemly (1993, 83).

Spallholz and Hoffman (2002) list three forms of selenium toxicity: CH_3Se, which creates free radicals that affect enzymes and proteins or create superoxide and oxidative stress; excess selenocysteine formation and resultant inhibited selenium methylation metabolism, leading to the concentration of hydrogen selenide with toxicity to animals; and excess selenium linked to sulfur-containing enzymes and structural proteins creates avian teratogenesis (possibly from L-selenomethionine or to a lesser extent L-selenocysteine, most likely to be ingested by waterbirds). A teratogen is a substance that can affect physiological development in young animals, affecting the fetus and embryo (teratology is the study of abnormalities of physiological development in the young) (Bastow & Holmes, 2016; Heinz et al., 2011; Rogers & Kavlock, 1996). With variable opinions on the toxicity of L-selenomethionine to embryos, Spallholz and Hoffman (2002, 27) argue that "there is no suggestion from the literature that the levels of L-selenomethionine that would be expected to accumulate in eggs in the absence of environmental concentration of selenium pose harm to the developing embryo." This hints at environmental concentration, rather than physiological accumulation as a factor for any toxic results. Nevertheless, "for several species of aquatic birds, levels of Se as selenomethionine in the egg above 3 ppm on a wet weight basis result in reduced hatchability and deformed embryos" (Spallholz & Hoffman, 2002: 27).

Beckon and Maurer (2007) examined selenium toxicity in Bald eagles in the San Francisco Estuary, arguing that bioaccumulation dynamics and possible dietary toxicity thresholds of 3 μg/g, pose greater threats to fish eaters such as the Bald eagle more than waterbirds like mallards, as Bald eagles could ingest selenium from predating ducks. However, Bowerman et al. (1994) studied selenium and mercury and detected selenium in the studied feathers, concluding that as "Se concentrations were not significantly different across regions or between adult and nestling feathers, and Se concentrations ranged from 0.8 to 3.2 mg/kg" and there were "no significant relationships between adult or nestling feather concentrations and either measure of reproduction, productivity or nesting success, was found for either Hg or Se," "neither Hg nor Se concentrations are currently affecting bald eagle reproduction in the Great Lakes region."

Selenium toxicity impacts include "mortality, impaired reproduction with teratogenesis, reduced growth, histopathological lesions and alterations in hepatic glutathione metabolism" (Hoffman, 2002: 11). Hoffman (2002: 11) examined "adverse biological effects of Se in aquatic birds to altered glutathione metabolism and oxidative stress" and "oxidative stress in different stages of the mallard (*Anas platyrhynchos*) life cycle"; consequences of increased Se include increases in plasma and hepatic GSH peroxidase" and "dose-dependent increases in the ratio of hepatic oxidized to reduced glutathione (GSSG:GSH) and ultimately hepatic lipid peroxidation measured as an increase in thiobarbituric acid reactive substances (TBARS)." The results for mallard ducks (*Anas platyrhynchos*, Linnaeus 1758) showed teratogenesis (4.6 ppm wet weight Se in eggs), reduced duckling growth (15 ppm Se in liver), lowered immune function (5 ppm Se in liver) and histopathological lesions (29 ppm Se in liver) (Hoffman, 2002: 11).

Another study examined the impacts of selenium on mallard ducks (O'Toole & Raisbeck, 1997) and argued that "selenosis is thought to be a significant problem among waterfowl populations in selenium-contaminated wetlands in the western United States" (O'Toole & Raisbeck, 1997: 330). This study investigated the possibility that morphological criteria for selenosis could supplement chemical analysis, by feeding mallard ducks, with rations containing <1, 10, 25, and 60 ppm selenium (seleno-L-methionine), while other mallards were fed twice daily with either non- or selenium-supplemented ration (120 μg/g) for one of the two daily feedings. Selenium in the blood increased from the baseline concentrations, which were <0.4 μg/ml, to averages of 4.5 μg/ml (10 ppm group), 8.9 μg/ml (25 ppm group), and 16.0 μg/ml (60 ppm group). Mallards in the 60-ppm-group became emaciated and were either killed or died within 22 to 50 days of selenium exposure and also suffered from mild to moderate hepatopathy with single-cell necrosis, karyomegaly of hepatocytes, hyperplastic bile duct epithelium, and/or Kupffer cell iron accumulation. Lower dosages caused "lesions (integumentary structures of hard keratin), which developed into gross lesions of bilaterally symmetrical alopecia of the birds' scalp and the dorsal cervical midline, and also damaged digital nails, and "necrosis of the tip of the beak (maxillary nail)" which increased with increased dosages, and there were histologic lesions (O'Toole & Raisbeck, 1997: 330). The liver, kidneys, and feathers had high selenium concentrations, these findings indicating that selenium may be dangerous to animals at high concentrations and is especially so due to accumulation effects (Lemly, 1993; O'Toole & Raisbeck, 1997; Tchounwou et al., 2012; Ashbaugh et al., 2018; Campbell, 2022).

The Snowy plover (*Charadrius nivosus*, Cassin, 1858) populations in the southwestern United States declined, partly due to selenium toxicity (Ashbaugh et al., 2018). Selenium frequency was higher in this species, other heavy metals being vanadium, arsenic, cadmium, and lead in the blood, feathers, and potential prey of this species (e.g., of the tiger beetles, *Cicindela circumpicta*, Fitch, 1841, *Cicindela togata*, LaFerté-Sénectère, 1841). Ashbaugh et al. (2018: 703) argued that in "samples greater than instrumentation detection limits, 98% of snowy plover blood and 22% of feather samples were greater than Se toxicity thresholds of 1 ppm ww for blood and 5 ppm dw for feathers (blood quantifiable range: 0.83–15.12 ppm; feathers quantifiable range: 1.90–27.47 ppm)." Variables considered included sex, individual body condition, local tiger beetle Se concentrations, state, and presence of body molt. The authors argue that "these results provide some of the first evidence of Se exposure risk for interior snowy plovers nesting in saline lake and alkali flat environments of the SGP" (Ashbaugh et al., 2018: 703).

Selenium impacts on other waterbird species have also been documented, especially related to glutathione metabolism: growth reduction, possible immune function decline, increased hatchling liver to body weight and hepatic Glutathione disulfide (GSSG) to Glutathione (GSH) ratios in hatchings, when eggs contained 9 ppm Se in American avocets (*Recurvirostra americana*, Gmelin, 1789); decrease in hepatic GSH (Black-necked stilts, *Himantopus mexicanus*, P.L.S.Müller, 1776); emaciation, increased feather losses, histopathological lesions, and increased hepatic GSH peroxidase, loss of hepatic protein-bound thiols and total thiols, and minor increases in GSH in American coots (*Fulica americana*, Gmelin, 1789). A thiol is an organosulfur compound where sulfur replaces oxygen in the chemical structure. Other findings document a positive correlation between hepatic Se concentration and GSH peroxidase activity and a negative correlation between hepatic Se and GSH concentration in diving ducks (*Aythyini* spp.), and high blood levels linked to increased plasma GSH peroxidase activity and decreased plasma GSSG reductase activity in Emperor geese (*Anser canagicus*, Sevastianov 1802) (Hoffman, 2002: 11).

Selenium also affects mammals (Mihajlović, 1992). Hall (2022) contends that although all animal species are vulnerable to selenium poisoning, poisoning is commoner for herbivores (e.g., horses and other livestock) that graze selenium-containing plants, which may accumulate selenium from high soil levels, usually from alkaline soils with low rainfall. Chronic (long-term) selenium poisoning, also termed alkali disease, may develop the herbivores eat forages and grains containing selenium for extended periods. Hall (2022) describes affected animals as "dull, very thin, and lack energy… The hair coat becomes rough, and the long hairs of the tail or mane break off at the same level… Abnormal growth and structure of hooves result in circular ridges and cracking of the hoof wall… Other effects include anemia, liver disease, fluid build-up within the abdomen, a weakened heart, and reduced reproductive performance. The animal's breath may have a garlicky odor" (Hall, 2022; see also Bano et al., 2022).

Fish are an important selenium storage at the lower levels of the food chain (Campbell, 2022). For fish, selenium abstraction starts with the eggs, which are infected through the diet of the female fish. The selenium (usually at concentrations above 10 μg/g) is metabolized by the growing fish and "biochemical functions may be disrupted, and teratogenic deformity and death may occur" (Lemly, 2002: 39). Lemly (2002) found selenium in 20 fish species in Belews Lake, North Carolina, due to a coal-fired power plant during the mid-1970s, and 19 species became extinct. Physical impacts on the fish included "telangiectasia (swelling) of gill lamellae, elevated lymphocytes, reduced hematocrit and hemoglobin (anemia), corneal cataracts, exopthalmus (popeye), pathological alterations in liver, kidney, heart, and ovary (e.g., vacuolization of parenchymal hepatocytes, intracapillary proliferative glomerulonephritis, severe pericarditis and myocarditis, necrotic and ruptured mature egg follicles), reproductive failure (reduced production of viable eggs due to ovarian pathology, and post-hatch mortality due to bioaccumulation of selenium in eggs) and teratogenic deformities of the spine, head, mouth, and fins" (Lemly, 2002: 39). The author concluded that bioaccumulation in aquatic food chains enables the otherwise harmless selenium concentrations to develop to toxic levels, and the selenium contained in the contaminated sediments can continue to be cycled into food webs for long periods (Lemly, 2002: 39).

Arribére et al. (2008: 42) give an example of selenium and mercury impacts on fish, with a case study of National Parks of the northern patagonian Andean range, in South America. The fish species were catfish (*Diplomystes viedmensis* or *Olivaichthys viedmensis*, MacDonagh, 1931), creole perch (*Percichthys trucha*, Valenciennes, 1833), brown trout (*Salmo trutta*, Linnaeus, 1758), rainbow trout (*Oncorhynchus mykiss*, Walbaum, 1792), and brook trout (*Salvelinus fontinalis*, Mitchill, 1814). In this study, selenium and mercury were measured in muscle and liver tissues by instrumental neutron activation analysis, and the findings indicated that the average concentrations in muscle of mercury for all species, ages, and lakes were from 0.4 to 1.0 μg g^{-1} dry weight (DW) and for selenium from 0.8 to 1.5 μg g−1 DW. For liver concentrations, selenium concentrations in the liver of brown and rainbow trout, which were up to 279 μg g^{-1} DW, were above expected levels for local lakes, reaching selenium toxicity levels, about 20 μg g^{-1} DW. Mercury in all species ranged from 0.4 to 0.9 μg g^{-1} DW (the Brown trout, which was the top predator in the lakes, showed the lowest average mercury levels, also in the muscle). The authors argued that these species (brown and rainbow trout) "show lower Hg contents in muscle, suggesting a possible detoxification of Hg by a Se-rich diet. Creole perch and velvet catfish livers have lower Se concentrations, with a narrower span of values (2.3–8.5 μg g^{-1} and 3.3–5.5 μg g^{-1} DW respectively)" (Arribére et al. (2008: 42).

Selenium also affects invertebrates, lower on the food chain than fish. Debruyn and Chapman (2007: 1766) argue for the importance of and neglect of invertebrates, considering most studies of selenium pollution in freshwater ecosystems are on birds and fish, and invertebrates usually considered only as dietary sources of Se to higher trophic levels. Their study of 29 macro-invertebrate species, and links between toxic effects and 156 aqueous, dietary, or internal Se concentrations, found that internal concentrations had about 30-fold variation, while concentrations in the diets varied about 100-fold and those among taxa varied over 1000-fold. There was less variation in sublethal effects (about 10-fold lower concentrations) than in lethal effects. Sublethal effects in the studied invertebrates were at 1–30 microg Se/g dry weight in invertebrate tissue. These levels covered some dietary toxicity thresholds for waterbirds and fish. Therefore, the authors argued that Se may contribute to toxic impacts on some invertebrate species, even at concentrations that would be considered nontoxic for the organisms that consume the invertebrates (Debruyn & Chapman (2007: 1766).

Poisoning from Zinc

Benson (2022) describes zinc as an "essential trace element used by the body in the synthesis of several hundred enzymes. It plays an important role in an effective immune system and contributes to a number of normal bodily functions, including musculoskeletal growth, vision, and normal wound healing... Zinc toxicosis has been documented in humans and in a wide range of large, small, exotic, and wild animals." Zinc toxicity is among the more serious heavy metal impacts (Campbell, 2022). Huang and Mayer (2019) note that for birds, the heavy metals that most frequently cause toxicity are lead and zinc; with less common causes being mercury and copper and zinc toxicity resulting from chronic and/or repeated exposure. It is further noted that the clinical signs for zinc toxicosis include "passive regurgitation, lethargy, weakness, weight loss, anemia, cyanosis, polyuria/polydipsia, diarrhea, regurgitation, neurologic signs, feather picking, anorexia, and/or hemoglobinuria" (Huang & Mayer, 2019).

Bailey (2011) argues that although zinc is an essential element, "acute zinc poisonings are rather common in dogs, cats and in pet birds exposed to zinc-containing coins and galvanized metal objects such as bolts and nuts, jewelry, toys, and cages" and that "zinc is readily absorbed via the GI tract and most intoxications occur through the oral route." Benson (2022) contends that zinc poisoning in pet birds can result in anorexia, ataxia, vomiting and regurgitation, recumbency, weight loss, and polyuria/polydipsia can occur. Liver and kidney injury, pancreatitis, and death are possible, and in laying hens, zinc toxicosis can cause decreased egg production, decreased appetite, and weight loss. Zinc poisoning can also result from zinc phosphide, which is defined as a rodenticide, which is "relatively toxic in comparison with elemental zinc or zinc oxide" (Hudson et al., 1984). Hudson et al. (1984) present a study of zinc phosphide poisoning (between 16 and 47 mg/kg BW in the ring-necked pheasant (*Phasianus colchicus*, Linnaeus, 1758), golden eagle (*Aquila chrysaetos*, Linnaeus, 1758), mallard duck (*Anas platyrhynchos*, Linnaeus, 1758), and horned lark (*Eremophila alpestris*, Linnaeus, 1758). It is noted that "signs of zinc phosphide poisoning include excessive drinking, regurgitation, muscular incoordination, appetite loss, sluggishness, rapid breathing, and eyelid droop."

Pappalardo et al. (2021: 797) report that zinc toxicity is a recognized problem for birds, and serum or plasma concentration greater than 30.6–45.9 mmol/L (200–300 mg/dL) have been earmarked as diagnostic for Zn toxicosis in Psittacinea (parrots), but for other species, for example falcons, further studies would be necessary, to assess the level of zinc concentrations that would cause acute or chronic Zn toxicity (see also Osofsky et al., 2001). Puschner et al. (1999: 552) also examine zinc concentrations in caged and aviary birds, arguing that, mistakenly, "interpretation of zinc levels is often based on normal zinc concentrations established for poultry. Also, it is likely that intergenus differences exist in normal zinc concentrations of pet birds." The findings indicated that differences occur with respect to genera. For example, cockatoos and *Eclectus* parrots have higher normal zinc concentrations in serum or plasma than other Psittacinea, and the analysis of all the studied examples suggests that toxic zinc concentrations in livers of Psittacinea can be potentially below the range considered toxic in domestic fowl (>200 ppm) (Puschner et al., 1999: 552).

For mammals, including domestic and companion animals, zinc toxicity is common (Benson, 2022). Benson (2022) argues that "in small animals, zinc toxicosis is usually a result of ingestion of objects that contain or are coated with zinc, most commonly US pennies. In large animals, the main causes are contaminated pastures and excess dietary supplementation." It is further noted that clinical signs for zinc in such animals vary at the interspecies level, and the source, and amount of zinc, with signs including "intravascular hemolysis, organ damage, and disseminated intravascular coagulation (abnormal blood clotting throughout the blood vessels)." Benson (2022) further notes that "exposure to zinc in dogs, cats, and other small animals typically stems from dietary indiscretion involving metals that either contain zinc or are coated with zinc. Galvanized metals are metals that are coated with zinc to prevent rusting. US pennies, specifically pennies minted since 1982, are a key source because they contain 97.5% zinc" with other sources including "automotive parts, galvanized nails/nuts/bolts, wire/hardware, polyhedral game dice, zippers, toys, aluminum galvanized cookware, galvanized metal cages, BB pellets, jewelry, and zinc-coated garland and tinsel decorations. The primary sources of zinc for large/production animals are contaminated pastures and excess supplementation of zinc in the diet."

Benson (2022) lists the main impacts of zinc poisoning on mammals, especially noted companion and farm mammals, which may receive zinc from natural or artificial sources. These include nonspecific hepatomegaly, splenomegaly, pancreatic nodules, red-brown kidney, and inflammation, edema, or ulceration of the gastrointestinal mucosa (for dogs and cats), gastrointestinal bleeding, kidney injury, pancreatitis, and similar ailments for dogs and cats (for ferrets, *Mustela furo*, Linnaeus, 1758), joint erosions, weakened cartilage, lameness, and limb deformities (for foals), reduced weight gain, lameness, decreased litter size, gastrointestinal irritation/inflammation (for swine, *Sus domesticus*, Erxleben, 1777), and weight loss, diarrhea, decreased appetite, decreased milk production, polyuria with secondary dehydration, and cellular degeneration of the liver, pancreas, and kidneys (for ruminants, *Ruminantia*, Scopoli, 1777) (Benson, 2022).

Poisoning from Arsenic

Arsenic is an important element in people, animals, and plants (Campbell, 2022; National Research Council, 1977). Eisler (1988) describes arsenic (As) as a comparatively y common element, occurring in air, water, soil, and all living tissues, ranking 20th in abundance in the earth's crust, 14th in seawater, and 12th in the human body. Kaur et al. (2011: 769) put arsenic into context, arguing that "contamination of the environment with heavy metals has increased drastically over the last few decades" and "the heavy metals that are toxic include mercury, cadmium, arsenic, and selenium. Of these heavy metals, arsenic is one of the most important global environmental pollutants and is a persistent bioaccumulative carcinogen." Arsenic is described as a "toxic metalloid that exists in two major inorganic forms: arsenate and arsenite. Arsenite disrupts enzymatic functions in cells, while arsenate behaves as a phosphate analog and interferes with phosphate uptake and utilization. Despite its toxicity, arsenic may be actively sequestered in plant and animal tissues (Kaur et al., 2011: 769).

Arsenic is an element of vital concern for various reasons (Campbell, 2022). One reason is the role of arsenic in the spread of diseases in people and possibly animals. Eisler (1988) points out that "arsenic is a teratogen and carcinogen that can traverse placental barriers and produce fetal death and malformations in many species of mammals" and "although it is carcinogenic in humans, evidence for arsenic induced carcinogenicity in other mammals is scarce" although arsenic in small quantities is useful, as arsenic deficiency may contribute to "poor growth, reduced survival, and inhibited reproduction, have been recorded in mammals" and "at comparatively low doses, arsenic stimulates growth and development in various species of plants and animals." It is also concluded that "arsenic is bioconcentrated by organisms but is not biomagnified in the food chain" (Eisler, 1988).

These are obviously serious issues and justify the study of arsenic toxicity at local and global levels. There are also serious impacts on people. For example, Dangleben et al. (2013) argue that human exposure to arsenic (As) is a global problem because of its links with some cancers and other ailments, and the fact that globally, millions of people are exposed to arsenic regularly. Crucially, the authors note that there is increasing evidence indicating that arsenic impacts the immune system, "but its specific effects on immune function are poorly understood…the data show that chronic exposure to As has the potential to impair vital immune responses which could lead to increased risk of infections and chronic diseases, including various cancers" (Dangleben et al., 2013). Another problem concerns arsenic in groundwater, as reported by Shankar et al. (2014) "arsenic contamination of groundwater in different parts of the world is an outcome of natural and/or anthropogenic sources, leading to adverse effects on human health and ecosystem. Millions of people from different countries are heavily dependent on groundwater containing elevated level of As for drinking purposes."

For animals, there are numerous bad effects (Campbell, 2022). Tasneem et al. (2020) caution that "among metalloids, Arsenic (As) is recognized as a major threat to the environment in many countries" and "higher levels of As in birds can lead to an inability to coordinate muscle movements as well as weakness in the wings and legs. This leads to difficulty standing, walking, and flying, as well as adverse effects on the nervous and immune system that can cause an impairment in reproduction" (see also Misztal-Szkudlińska et al., 2011; Waseem et al., 2014; Sánchez-Virosta et al., 2015). Eisler (1988, 17) describes the symptoms of arsenic poisoning on birds as "liver, kidney problems, muscular incoordination, debility, slowness, jerkiness, falling hyperactivity, fluffed feathers, drooped eyelid, huddled position, unkempt appearance, loss of righting reflex, immobility" and "seizures." The impacts on mammals include "intense abdominal pain, staggering gait, extreme weakness, trembling, salivation, vomiting, diarrhea, fast and feeble pulse, prostration, collapse, and death" and "necropsy shows a reddening of gastric mucosa and intestinal mucosa, a soft yellow liver, and red edematous lungs" (Eisler (1988, 18).

Sánchez-Virosta et al. (2015: 506) describe arsenic as a "metalloid of high concern because of its toxic effects for plants and animals" and present a study of As exposure and effects in passerines, based on a literature search. as a result of an extensive search of the literature available. The findings indicated that published research examines As levels in internal tissues (37.5% of consulted studies), followed by feathers and eggs (32.5% each), feces (27.5%), and blood (15%). Common species selected for studies of passerines were the great tit (*Parus major*, Linnaeus,

1758; 50%) and the European Pied flycatcher (*Ficedula hypoleuca* Pallas, 1764; 22.5%). Differences in As exposure at the species level were due to factors such as diet and migratory status. Most such studies on passerines were untaken in the United States (30%), Belgium (22.5%), and Finland (20%), with a serious lack of studies on the countries of the southern hemisphere (Sánchez-Virosta et al., 2015).

For birds, Yang, et al. (2018: 777) point out that "birds are at a higher level in the food chain; however, the potential bioaccumulation and biotransformation of arsenic (As) in birds in As mines has rarely been studied"; their study concerned four passerine species: Tree sparrow (*Passer montanus*, Linnaeus, 1758), Light-vented bulbul (*Pycnonotus sinensis*, Gmelin, JF, 1789), Chinese hwamei (*Garrulax canorus*, Linnaeus, 1758), and magpie (*Pica pica*, Linnaeus, 1758) collected from an abandoned As mine in China. As concentrations were highest in muscles and feathers (4.95 mg/kg and 51.65 mg/kg, respectively). Their research methodology detected six As species (three dominant species (>92%) were arsenite (As(III)), arsenate (As(V)), and dimethylarsinic acid (DMA), and three minor species (<6.17%) were monomethylarsonic acid (MMA), arsenobetaine (AsB), and arsenocholine (AsC). Other findings indicated that revealed inorganic As species in the muscles and feathers of the birds from the As mine (Yang et al., 2018: 777).

Another bird impact study by Zhong et al. (2021) examined the effect of arsenic trioxide (ATO) on the gut–liver axis and resultant hepatotoxicity in waterfowl, using the example of the influence of ATO on duck intestines and livers, and the role of the gut–liver axis in ATO-induced hepatotoxicity and intestinal toxicity. The findings indicated ATO exposure led to intestinal damage, liver inflammatory cell infiltration and vesicle steatosis, and reduced α-diversity in the intestinal microbiota in the ATO-exposed ducks, with a changed bacterial composition. Other impacts were "reduced the expression of intestinal barrier-related proteins (Claudin-1, MUC2, ZO-1, and Occludin), resulting in increased intestinal permeability and elevated lipopolysaccharide levels" and the upregulation (increase in density of cell surface receptors) of "pyroptosis-related index levels in the liver and jejunum, and increased pro-inflammatory cytokine production (IFN-γ, TNF-α, IL-18, and IL-1β)." The authors concluded that "these results manifested that ATO exposure can cause liver and jejunal (note: jejunal refers to the middle part of the small intestine) inflammation and pyroptosis (cell death associated with inflammation), and the indirect gut-liver axis pathway may play an essential role in the potential mechanism of ATO-induced hepatotoxicity" (Zhong et al., 2021).

Sánchez-Virosta et al. (2020) present a study based on arsenic (As) impacts on great tits (*Parus major*, Linnaeus, 1758), on the premise that although arsenic is "broadly distributed due to natural and anthropogenic sources, and it may cause adverse effects in birds... research on other elements (Pb, Hg and Cd) has been prioritized, resulting in scarce data on As exposure and related effects in wild birds." Oxidative stress is factored as a mechanism responsible for As toxicity is oxidative stress, and objective of the study was "to investigate if environmentally relevant As levels affected oxidative stress biomarkers in great tits" (Sánchez-Virosta, 2020). The responses of the artificially exposed birds to the experimental manipulations (high, low, and control groups) were compared with birds in an As/metal-exposed population breeding near a Cu-Ni smelter in Finland (Smelter group). The findings indicated that arsenic administration had no significant effect on most of the biomarkers measured, the dose and duration of the As exposure was sufficient to induce oxidative damage in red cells of great tit nestlings. The authors, however, note that despite this "nestlings dosed with 1 $\mu g\ g^{-1}\ d^{-1}$ of sodium arsenite showed nonsignificantly higher oxidative stress biomarkers than controls, suggesting that we were close to an effect level for the redox-defense system. Oxidative effects at equivalent As levels combined with other stressors cannot be dismissed" (Sánchez-Virosta, 2020).

Kunito et al. (2008: 31) present a study of marine species, arguing that "although there have been numerous studies on arsenic in low-trophic-level marine organisms, few studies exist on arsenic in marine mammals, seabirds, and sea turtles." The authors argue that most of the arsenic detected in the livers of marine mammals, seabirds, and sea turtles is AB (Arsenobetaine AB), but this form is mostly undetected in the dugong (*Dugong dugon*, Müller, 1776). A paradox is noted, where there are low levels of arsenic in marine mammals, some seabirds, and some sea turtles, but the black-footed albatross and hawksbill and loggerhead turtles have high concentrations, like the levels in marine animals at lower trophic levels. Therefore, the latter animals may absorb the arsenic with another mechanism for accumulating arsenic. Kunito et al. (2008: 31) argue that "osmoregulation in these animals may play a role in the high accumulation of AB. Highly toxic inorganic arsenic is found in some seabirds and sea turtles, and some evidence suggests it may act as an endocrine disruptor, requiring new and more detailed studies for confirmation."

Unintentional Poisoning from Chemical Compounds

Many chemical compounds (a substance that is composed of more than one element, with atoms from more than one element connected by chemical bonds) also affect animals. Common chemical compounds affecting birds are illustrated in Table 6.4. These are mostly artificial compounds, produced for specific purposes, such as killing pests (pesticides), rodents (rodenticides), and veterinary compounds for livestock protection. The compounds may contain chemicals with different impacts on different animal species, which in combination may be as toxic as the chemical elements described above. Some more dangerous chemical compounds have been legally banned (Campbell, 2015).

Poisoning from Pesticides

Pesticides are composed of several connected elements, an example being dichlorodiphenyltrichloroethane (DDT) (chemical formula $C_{14}H_9CI_5$), which was invented by Paul Hermann Müller (1948), and described as "high efficiency… as a contact

Table 6.4 The main chemical compounds and their impacts

Compound	Formula	Impacts on raptors and waterbirds
DDT	$C_{14}H_9Cl_5$	Toxic to prey animals, reproductive degradation, shell thinning, endocrine problems.[1] Affects egg laying, pregnancy, lactation
DDE	$C_{14}H_8Cl_4$	Toxic to prey animals, shell thinning, embryonic and chick risk.[2] poor thyroid function, fertility decrease, immune function decline.[2]
Dieldrin/ Aldrin	$C_{12}H_8Cl_6O$	Fatal to raptors, including juveniles and breeding adults, levels correlated to levels in prey animals.[3]
PCB	$C_{12}H10_{-x}Cl_x$	Liver, immune, reproductive anemia; skin damage (chloracne); liver, stomach, thyroid injuries (hepatocarcinoma thymocyte apoptosis[4]
Diclofenac	$C_{14}H_{11}C_{12}NO_2$	Renal and visceral gout (crystallization of uric acid in tissues), enteritis[5]
PAH	$C_{12}H_{12}$	Toxic to embryos, birds, death, developmental abnormalities. Reduced eggs, hatching, more brood losses. Trophic level accumulation unlikely.[6]

Sources: 1. Colborn (1991), NPIC (1999); 2. Newton (1993); 3. Sibly et al. (2000); 4. Fernie et al. (2001), Crinnion (2011); 5. Campbell (2015); 6. Fingas (2017), Albers (2006). PAH refers to pulmonary hypertension, resulting from e.g., $C_{12}H_{12}$ 1,6-Dimethylnaphthalene

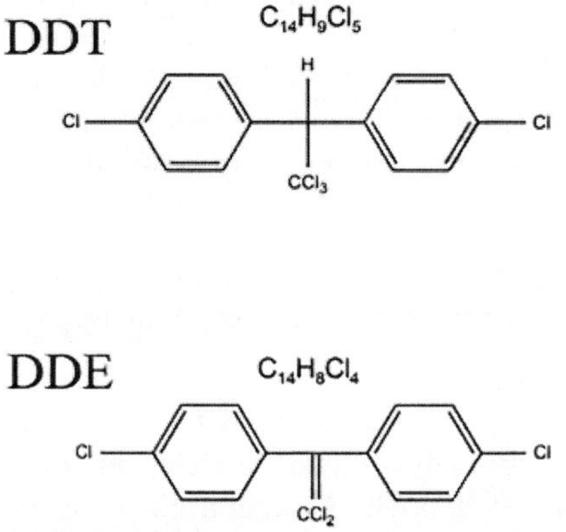

Fig. 6.1 Change from DDT to DDE

poison against several arthropods" (Nobel Foundation, 1948). DDT was an important pesticide for insect control in the United States until 1972, when it was banned by the United States Environmental Protection Agency (EPA). The National Pesticide Information Center (NPIC) (2011) noted that "bald eagles, ospreys and peregrine falcons were brought to the brink of extinction because their food sources (fish and other birds) were contaminated with DDE, the breakdown product of the insecticide DDT." DDE (dichlorodiphenyl dichloroethylene) is formed when hydrogen chloride (dehydrohalogenation) is lost from DDT (chemical formula $C_{14}H_8CI_{4)}$ (Walker et al., 1954). This change from DDT to DDE is shown in Fig. 6.1.

The National Pesticide Information Center (NPIC) (1999) describes DDT as "widely used… because it is effective, relatively inexpensive to manufacture, and lasts a long time in the environment." Additionally, DDT was banned because of its long persistence in the environment, its accumulation in fatty tissues, and its negative health effects on wildlife. Resistance occurs in some insects (e.g., the house fly *Musca domestica*, Linnaeus, 1758) which may develop the ability to quickly metabolize the DDT. Dieldrin, introduced as an alternative to DDT, is an organochloride developed in 1948 by J. Hyman & Co, Shell Chemical, Denver, then AMVAC, Los Angles as an insecticide, related to aldrin (with a similar formula, $C_{12}H_8Cl_6$) which oxidizes after insect ingestion to form dieldrin (Jorgenson, 2001). Dieldrin is strongly persistent environmentally and physiologically, and because it biomagnifies as it ascends the food chain, it is dangerous for predators. The National Center for Biotechnology Information (2022a) defines dieldrin as an "organochlorine compound resulting from the epoxidation of the double bond of aldrin. It is the active metabolite of the proinsecticde aldrin." This compound has a role as a xenobiotic and a carcinogenic agent and is an organochlorine insecticide, an organochlorine compound, and an epoxide. It is derived from aldrin, and both aldrin and dieldrin were used as pesticides for crops like corn and cotton. The EPA banned all uses of aldrin and dieldrin in 1974, except to control termites, and in 1987, the EPA banned all uses, due to concerns about damage to the environment and potentially to human health.

DDT impacts on animals included: nervous system and nerve impulse damage, hyperexcitability, tremors, incoordination, and convulsions; short-term liver lesions and longer-term liver changes, risk of live tumors; long persistence in the environment, the soil half-life being 2–15 years and in an aquatic environment about 150 years; and toxicity to birds, fish, and amphibians, affecting the heart, brain and other organs. There were strong impacts at the bottom of the food chain, especially for marine and aquatic animals preyed upon by waterbirds, raptors, and mammals (Campbell, 2015). DDT and DDE are more environmentally persistent than some other pollutants (Campbell, 2015). For example, García-Fernández et al. (2008: 432) concluded that "excepting DDT and DDE, most organochlorine-based pesticides in raptors from Spain have diminished over the last 2 decades. The concentrations of DDE found in the eggs of various species could in part explain problems in the reproductive success of raptors in Spain." Jürgens et al. (2016: 333) found high, sometimes dangerous, levels of DDT in fish in English ecosystems. Literature findings were compiled by Jürgens et al. (2016) on the threshold levels of DDT for prey animals and associated raptors and waterbirds. These are: "based on eggshell thinning in ducks" 0.014 ppm (reference, Canadian Council of Ministers of the Environment, 1999); "dietary effects concentration" in ospreys, 0.09–0.19 ppm (reference, Hinck et al. (2009); "dietary effects concentration" in bald eagles, 0.13 ppm (reference, Hinck et al. (2009); and "eggshell thinning in pelicans," 0.02 ppm (reference, Lazorchak et al., 2003).

DDT/DDE applications had impacts beyond the United States during the times of peak use (Campbell, 2022). For example, in Britain, impacts on raptors and waterbirds included those noted by Parslow et al. (1973) who concluded that "measurable residues of polychlorinated biphenyls (PCB) and DDE (the chief breakdown product of DDT found in birds) are present in virtually all British seabirds."

Some birds contain residues of other organochlorine materials, including dieldrin and hexachlorobenzene (HCB), generally in smaller quantities. PCB and HCB are usually produced from industrial sources, the others mainly from agricultural insecticides. As larger fish (examples of such fish are pollack *Pollachius pollachius* (Linnaeus, 1758), saithe *Pollachius virens* (Linnaeus, 1758), and haddock (*Melanogrammus aeglefinus*, Linnaeus, 1758) may carry more pollutants, the larger fish-eating birds like gannets that eat such fish are especially vulnerable (Parslow et al., 1973). The toxicity of these chemicals and subsequent animal mortality is partly dependent on the level of chemical compound ingested and the area of deposition, i.e., the subcutaneous fat or the liver, as with higher levels in the fat, stress may mobilize pollutants when fat is utilized, and the result may affect the liver and endocrine system (Parslow et al., 1973).

Other chemical compounds that affect animals include cyanide, strychnine (used to kill predators of livestock, such as coyotes), and rodenticides such as compound 1080 (Snyder & Snyder, 1991; Campbell, 2015). Salkowski and Penney (1994: 455) note that cyanide is "a deadly poison which animals and humans encounter from a number of sources." For cyanide, a common compound is sodium cyanide (NaCN), which is produced by treating hydrogen cyanide with sodium hydroxide (i.e., ($HCN + NaOH \rightarrow NaCN + H_2O$) (Rubo et al., 2006). This compound is used in the process of gold extraction, mostly from low-grade ore. More than 70% of cyanide consumption globally is used for this purpose. The process depends on the affinity of gold for cyanide, as the cyanide induces gold metal to oxidize and dissolve in the presence of oxygen and water, producing the sodium dicyanoaurate (C2AuN2Na) (National Center for Biotechnology Information, 2022c; Rubo et al., 2006).

In one study, Wiemeyer et al. (1986: 538) researched on the sodium cyanide "sensitivities" of six avian species, namely the Black vulture (*Coragyps atratus*, Bechstein, 1793), American kestrel (*Falco sparverius*, Linnaeus, 1758), Japanese quail (*Coturnix japonica*, Temminck & Schlegel, 1848), Domestic chicken (*Gallus gallus domesticus*, Linnaeus, 1758), Eastern screech-owl (*Otus asio*, Linnaeus, 1758, or *Megascops asio*, Linnaeus, 1758), and European starling (*Sturnus vulgaris*, Linnaeus, 1758). The methodology included the dosing with NsCN of three species, Domestic chickens, Black vultures, and Turkey vultures (*Cathartes aura*, Linnaeus, 1758) to assess the cyanide residues in the survivors and dead birds. The findings indicated that three meat-eating species (a scavenger, the Black vulture; and two predators the American kestrel, and Eastern screech-owl) had greater sensitivity to NaCN than three mostly herbivorous species (Japanese quail, Domestic chicken, and European starling) and that although there were higher concentrations of cyanide in the blood of the dead birds, the concentrations in dead birds overlapped with those in survivors, and blood was "superior to liver as the tissue of choice for detecting cyanide exposure" (Wiemeyer et al., 1986: 538).

Regarding mammals, Cope (2022) lists four main types of cyanide poisoning. One is classical acute cyanide poisoning, where "CN- binds to, and inhibits, the ferric (Fe3+) heme moiety form of mitochondrial cytochrome c oxidase" and the result is the blockage of the fourth step in the mitochondrial electron transport chain, which involves the reduction of O_2 to H_2O), and the overall result is the "arrest of aerobic metabolism, systemic hypoxia, and death from histotoxic anoxia," which may especially affect tissues such as the heart and brain, which depend on aerobic metabolism (Cope, 2022). There is also chronic cyanide poisoning-related hypothyroidism, which "is due to disruption of iodide uptake by the follicular thyroid cell sodium-iodide symporter by thiocyanate, a metabolite in the detoxification of cyanide" (Cope, 2022). There is also chronic cyanide and cyanide metabolite, which can include "diseases such as sorghum cystitis ataxia syndrome in horses, as well as various cystitis ataxia syndromes in cattle, sheep, and goats" (Cope, 2022). There is also chronic cyanogenic glycoside exposure (e.g., from *Sorghum* spp.)-associated musculoskeletal teratogenesis (birth defects or malformations in an embryo or fetus) (ankyloses, an inflammatory disease that causes vertebrae to fuse, or arthrogryposes, multiple joint stiffness) and fetal loss (Cope, 2022).

Pederson and Stokka (2020) also note that "a number of common plants may accumulate large quantities of cyanogenic compounds" and "cyanide poisoning most often is associated with sorghums and sudangrass... These cyanogenic compounds are in epidermal cells (outer tissue) of the plant, while the enzymes that enable cyanide production are in the mesophyll cells (leaf tissue)" and because of these plant-based poisons "once plants containing cyanide have been consumed, the toxin rapidly enters the blood stream and is transported throughout the body of the animal. Cyanide inhibits oxygen utilization by the cells in the animal's body. In essence, the animal suffocates. Ruminant animals (cattle and sheep) are more susceptible to cyanide poisoning than nonruminant animals because the ruminal microorganisms have enzymes that will release cyanide in the animal's digestive tract" (Pederson & Stokka, 2020).

Poisoning from Diclofenac Nonsteroidal Anti-Inflammatory Drug (NSAID)

Altman et al. (2015: 859) define diclofenac (formula $C_{14}H_{11}Cl_2NO_2$) as a "nonsteroidal anti-inflammatory drug (NSAID) of the phenylacetic acid class with anti-inflammatory, analgesic, and antipyretic properties... Similar to other NSAIDs, diclofenac is associated with serious dose-dependent gastrointestinal, cardiovascular, and renal adverse effects." When diclofenac was used for livestock, it became apparent that vultures were extremely vulnerable to this drug, when they fed on carcasses of diclofenac-treated livestock at carcass dumps, which was the traditional method of livestock disposal in South Asia (Green et al., 2004; Gross, 2006; Campbell, 2015). Vultures died one to two days after ingestion, after kidney failure and severe renal dysfunction, increased blood uric acid levels, renal and visceral gout (crystallization of uric acid in the tissues), and enteritis (inflammation of the intestine, especially the small intestine, and increased urates in the internal organs. These contributed to anorexia and emaciation and a chalky-white coating over the visceral organs, especially the liver and heart and in the body tissues (Oaks et al., 2001; Gilbert et al., 2002; Pain et al., 2002; Campbell, 2015).

The resultant Diclofenac Epidemic, which was published as an example of an avian and ecological catastrophe and human interference with the natural environment, was perhaps the most important poisoning episode of the twentieth century, an event termed the "global vulture crises" (Woodford et al., 2008; Campbell, 2015). This event, mostly affecting Asian vultures, resulted in the near extinction of six of the 23 recognized vulture species (especially five Indian species), listed as such on the International Union for the Conservation of Nature (IUCN) Red List of Threatened Species. The most affected species included the White-rumped vulture (*Gyps bengalensis*, Gmelin, 1788), Slender-billed vulture (*Gyps tenuirostris*, Hodgson, 1844), Indian vulture (*Gyps indicus*, Scopoli, 1786), and Red-headed vulture (*Sarcogyps calvus*, Scopoli, 1786) (Prakash, 1999; Prakash & Rahmani, 1999; Virani et al., 2001; Prakash et al., 2003; Clements et al., 2012; Gilbert et al., 2002, 2004). Between 1992 and 2007, the White-rumped vulture population declined to one-thousandth of the previous population, the Red-headed vulture's population to below 10% and that of the Slender-billed vulture to about 1000 birds (Cuthbert et al., 2006; Prakash et al., 2007). The Egyptian vulture (*Neophron percnopterus*) and red-headed vulture (*Sarcogyps calvus*) were classified as Endangered and Critically Endangered respectively (Cuthbert et al., 2006; Campbell, 2015).

Diclofenac was banned by legislation from 2006 to 2010 in India, Nepal, Pakistan, and Bangladesh, which allowed a slow vulture population recovery in India and later in Pakistan and Nepal. Meloxicam was promoted as a safe alternative to diclofenac, as studies concluded that "meloxicam is of low toxicity to Gyps vultures and that its use in place of diclofenac would reduce vulture mortality substantially in the Indian subcontinent. Meloxicam is already available for veterinary use in India" (Swan et al., 2006: 395). For these actions, European and African vultures were largely ignored, as were the Cathartid New World vultures in the Americas, as *Cathartid* vultures were not affected by diclofenac ingestion at levels over 100 times over the dosage that killed Asian vultures, with no records of toxicity, visceral gout, renal necrosis, or elevate plasma uric acid (Rattner et al., 2008; Campbell, 2015).

Some recent studies have indicated diclofenac impacts on eagles. For example, Sharma et al. (2014: 282) report that the Steppe eagle is susceptible. Although there is little evidence of "toxicity of diclofenac to the Accipitridae other than vultures… In this study, gross and microscopic lesions, and diclofenac tissue levels in Steppe Eagles *Aquila nipalensis* found at a cattle carcass dump in Rajasthan, India, show evidence of the toxicity of diclofenac for this species." These findings may indicate the possibility that diclofenac is toxic to nonvulture accipitrid raptors and a potential threat to a wider range of scavenging birds in South Asia. Mackenzie (2014) also recorded two dead Steppe eagles found with kidney failure, this incident indicating that if the Steppe eagle is vulnerable to diclofenac, then perhaps the other Aquila eagles may also be threatened (Mackenzie, 2014).

Poisoning from Polychlorinated Biphenyls (PCBs) and Perfluoroalkyl Substances (PFASs)

Polychlorinated biphenyls (PCBs, molecular formula $C_{12}H_5Cl_5$) are also important environmental pollutants, which may seriously impact on animal health. The National Center for Biotechnology Information (2022b) describes polychlorinated biphenyls as comprised of mixtures of up to 209 individual chlorinated compounds (known as congeners), with no known natural sources. PCBs are used as coolants and lubricants in transformers, capacitors, and some other types of electrical equipment, as they do not burn easily and are good insulators. PCB manufacture was stopped in the U. S. in 1977 with evidence they build up in the environment and can cause harmful health effects on people and animals (National Center for Biotechnology Information, 2022b).

Perfluoroalkyl substances, usually termed "perfluoroalkyl and polyfluoroalkyl substances" (PFASs), are also important environmental impactors (Campbell, 2022). Christensen et al. (2019: 147) define perfluoroalkyl substances as a type of contaminant used in industrial applications and consumer products, of which some are regulated or sometimes voluntarily limited because of concerns with environmental persistence and problematic health impacts. These compounds were described by the Environmental Protection Agency, 2022 as frequently utilized, environmentally persistent chemicals, which have been used especially since 1950, as PFASs and surfactants and polymers made with the aid of PFASs, in many industrial and commercial applications. It is further argued that "polymer applications include textile stain and soil repellents and grease-proof, food-contact paper," "as a consequence of the widespread use of PFASs and their resulting emissions, a broad range of these substances have been detected in the environment, wildlife, and humans" (Environmental Protection Agency (2022). A study by Giesy and Kannan (2001) first created awareness of "the global extent of such contamination was first demonstrated for perfluorooctane sulfonic acid, $C_8F_{17}SO_3H$ (PFOS) in wildlife."

For birds, PCBs have emerged as an important threat (Campbell, 2022). For example, an old study by Prestt et al. (1970) found that ingestion levels for PCBs by bird-feeding raptors in the liver were 70 ppm, similar to that of DDE. For freshwater fish-feeding waterbirds in the same region, the figure was 900 ppm. The authors argued that "most of the 559 specimens came from predatory birds obtained between April 1966 and August 1968." PCBs were recorded in terrestrial species from many regions of Britain, in all the individual and bulked samples of seabird eggs examined from one west coast location and two east coast colonies and in the majority of the freshwater species recorded in the midlands, east and south of England. These were like the levels of DDE; hence, possibly both chemical compounds had similar occurrences in ecosystems in the 1960s–1970s period. Testing used gas-liquid chromatography on livers and kidneys (Prestt et al., 1970). Another study by Parslow et al. (1973) concluded PCBs occurred at higher levels in gannets and other British seabirds than DDE, and gannets had higher mean liver levels (e.g., 200 ppm) than other seabirds (e.g., murres or guillemots *Uria aalge*, Pontoppidan,

1763, 116 ppm). In some dead gannets, PCB (also DDE) concentrations in the brain were a fraction of those in the liver. Residues of PCB and DDE also had a significant impact on eggshell thickness (Parslow et al., 1973).

As waterbirds species are frequently taken by White-tailed eagles after the eagles' reintroduction to Britain, there is a real threat to eagles as well, on both sides of the Atlantic (Love, 1983). In North America, high PCB levels have been reported Bald eagles (Campbell, 2022). Revkin (1997) reported that a young, dead Bald eagle was located on the upper Hudson River with high concentrations of PCB's, and this "finding, although limited to one eagle, is significant, the scientists said, because similar levels of PCB's in eagles or eagle eggs from polluted areas of the Great Lakes have been linked to reproductive problems and deformities in the birds." Revkin reported that the eagle's body contained 71 parts per million of PCB's, which was higher than the average recorded in three deformed fledgling eagles recorded from the Great Lakes area." It is argued that PCB concentrations accumulate mostly in the bird's brain and fat and originate in fish and aquatic insects and move up the food chain to the Bald eagles. As the recorded Bald eagle was young, the incident indicated a high accumulation PCBs have mostly been banned for their biological effects, but "the Hudson River is one of the worst PCB pollution sites in the world, with more than a million pounds of the chemicals released from two General Electric factories during a 30-year period ended in 1977" (Revkin, 1997).

In study in south China, Zhang et al. (2011: 556) examined the impacts of polychlorinated biphenyls and organochlorinated pesticides on mostly waterbird species (*Rallidae*, *Ardeidae*) and doves (Columbidae). This approach examined birds as indicators of the levels of organic pollutants, arguing that "there is exceedingly little data on organochlorinated pesticide (OCPs) and polychlorinated biphenyls (PCBs) residues in birds inhabiting in China." The findings described concentrations of PCBs, dichlorodiphenyltrichloroethane (DDTs), and hexachlorocyclohexanes (HCHs) in the muscle tissues, liver, and kidney of birds recorded near an e-waste recycling site. The findings indicated PCBs had higher concentrations than other compounds, with possible negative impacts on the target birds. Livers and kidneys had higher concentrations than muscle, and contaminant levels varied according to bird dietary composition and habitat, and PCB and DDT levels increased up the trophic levels. HCHs (hexachlorocyclohexane, formally benzene hexachloride, BHC) were mostly level at the interspecies level (Zhang et al., 2011: 556).

For PFASs, one study by Barghi et al. (2018: 220) contributed to the perceived gap in knowledge "on the accumulation and potential risks of persistent chlorinated and fluorinated contaminants in birds." Perfluoroalkyl substances (PFASs), organohalogen contaminants (OHCs), including PCBs, and organochlorine pesticides (OCPs), were measured, using the liver tissues of 69 specimens of 10 species of Korean birds. Jayaraj et al. (2016: 90) defined organochlorine (OCPs) pesticides as "synthetic pesticides" belonging "to the group of chlorinated hydrocarbon derivatives, which have vast application in the chemical industry and in agriculture…known for their high toxicity, slow degradation and bioaccumulation." The findings were that PFASs had the highest concentrations (Barghi et al., 2018). The concentrations of OHCs were higher in predatory than nonpredatory birds; hence, diet (on smaller, infected animals) and trophic position (the top of the food chain, allowing accumulation) were important issues. Further findings indicated concentrations of PCBs and DDTs were below toxic threshold levels, "PFOS concentrations in most of the predatory birds exceeded threshold values for adverse health effects…, implying potential health risks to Korean birds" (Barghi et al., 2018: 220). PFOs (perfluorooctane sulfonate) are a form of PFAS (National Institute of Environmental Health Sciences, 2020).

Another study by Luzardo et al. (2014) examined pollutants and bird specimens to measure contaminant levels and evaluate the studied species as regional bioindicators. These included 57 anthropogenic pollutants (23 organochlorine pesticides OCs), 18 polychlorinated biphenyls (PCBs), and 16 polycyclic aromatic hydrocarbons (PAHs)), using liver samples from 102 specimens of six raptor species from Wildlife Recovery Centers of the Canary Islands from 2009 and 2012. Polycyclic aromatic hydrocarbons (PAHs) are chemicals that occur naturally in coal, crude oil, and gasoline, also when these substances and others such as wood, tobacco, and garbage are burned, forming small air particles (Environmental Protection Agency, 2013). Examples are petrogenic alkyl polycyclic aromatic hydrocarbons, which are those derived from petroleum and pyrogenic polycyclic aromatic hydrocarbons, which are derived from combustion and anthropogenic sources (Paruk et al., 2014). The findings of the study by study by Luzardo et al. (2014) indicated that the Eurasian sparrowhawk (*Accipiter nisus*, Linnaeus, 1758), Barbary falcon (*Falco peregrinus pelegrinoides*, Temminck, 1829) and kestrel (*Falco tinnunculus*, Linnaeus, 1758) were more contaminated than the other species, mostly by high levels of DDT and dieldrin. Contamination by PCBs was comparatively low, perhaps below threshold levels of toxicity, as was the presence of polycyclic aromatic hydrocarbons (PAHs). The authors argue that their study "represents the first report of contamination by PAHs in all these species and is also the first report of PCB levels in barbary falcons" (Luzardo et al., 2014: 146).

Conclusions

This review of environmental toxicity in birds and mammal species illustrates the impacts and dangers of environmental chemistry and toxicology for animals, from the bottom to the top of food chains, the possibilities for human-deposited toxic chemical elements and compounds, and the migration of such chemicals up the food chain. This topic is very important for the biological end of biogeochemistry. Toxic elements and compounds have wide source networks, mostly enhanced by human action. Both birds and mammals are affected by such substances, but there are some interspecies variations in impacts. There must be more research on these topics, from a spatial, case study approach. Some consequences were severe enough to require the attempted restriction or abolition of chemical compounds, such as lead, diclofenac, and DDT/DDE. Other elements may require similar restrictions, but there may be variable likelihoods in the possible success of such restrictions, due to the range, toxicity, and sources of such elements. Some elements, which are otherwise useful at low concentrations, may be studied in more detail. The determination of toxic levels requires more research at inter- and intraspecies levels, acknowledging that sublethal and lethal definitions may vary by the organs in which they accumulate, such as embryos, livers, kidney and the gastrointestinal tracts, the toxicity element or compound and the likelihood of migration up the food chain. The replacement of toxic elements and compounds with safer duplicates is already current, especially for the most dangerous elements and compounds such as lead and diclofenac, the impacts of which border on disasters. Future scenarios may encourage the development of alternatives before serious consequences arise, based on more proactive models, and more sensitive environmental impact assessments of industrial and agricultural developments. This chapter used case studies of both birds and mammals for illustrative points. Future research may focus on a wider range of species including prey and parallel species and research tools.

References

Acharya, K. P., Paudel, P. K., Neupane, P. R., & Köhl, M. (2016). Human-wildlife conflicts in Nepal: patterns of human fatalities and injuries caused by large mammals. *PLOS One, 11*(9), e0161717. https://doi.org/10.1371/journal.pone.0161717

Agency for Toxic Substances and Disease Registry (ATSDR). (2019). *Toxicological profile for chromium.* Retrieved from https://www.atsdr.cdc.gov/ToxProfiles/tp7.pdf

Agency for Toxic Substances and Disease Registry (ATSDR). (2020a). *Toxicological profile for copper.* Retrieved from https://www.atsdr.cdc.gov/toxprofiles/tp132.pdf

Agency for Toxic Substances and Disease Registry (ATSDR). (2020b). *Selenium.* Retrieved from https://www.atsdr.cdc.gov/toxprofiles/tp92-c3.pdf

Albers, P. H. (2006). Birds and polycyclic aromatic hydrocarbons. *Avian and Poultry Biology Reviews, 17*, 125–140. https://doi.org/10.3184/147020606783438740

Albers, P. H., Green, D. E., & Sanderson, C. J. (1996). Diagnostic criteria for selenium toxicosis in aquatic birds: Dietary exposure, tissue concentrations, and macroscopic effects. *Journal of Wildlife Diseases, 32*(3), 468–485. https://doi.org/10.7589/0090-3558-32.3.468

Altman, R., Bosch, B., Brune, K., Patrignani, P., & Young, C. (2015). Advances in NSAID development: Evolution of diclofenac products using pharmaceutical technology. *Drugs, 75*(8), 859–877. https://doi.org/10.1007/s40265-015-0392-z

Anderson, M. D. (1999). Africa's hooded vulture: A dichotomy of lifestyle. *Vulture News, 41*, 3–5.

Anderson, M. D. (2000). Raptor conservation in the Northern Cape Province, South Africa. *Ostrich, 71*(1–2), 25–32. https://doi.org/10.1080/00306525.2000.9639859

Anderson, W. L., Havera, S. P., & Zercher, B. W. (2000). Ingestion of lead and nontoxic shotgun pellets by ducks in the Mississippi Flyway. *Journal of Wildlife Management, 64*(3), 848–857. https://doi.org/10.2307/3802755

Arnemo, J., Andersen, O., Stokke, S., Thomas, V. G., Krone, O., Pain, D. J., & Mateo, R. (2016). Health and environmental risks from lead-based ammunition: Science versus socio-politics. *EcoHealth, 13*, 618–622. https://doi.org/10.1007/s10393-016-1177-x

Arribére, M. A., Ribeiro Guevara, S., Bubach, D. F., Arcagni, M., & Vigliano, P. H. (2008). Selenium and mercury in native and introduced fish species of Patagonian Lakes, Argentina. *Biological Trace Element Research, 122*, 42–63. https://doi.org/10.1007/s12011-007-8059-6

Ashbaugh, H. M., Conway, W. C., Haukos, D. A., Collins, D. P., Comer, C. E., & French, A. D. (2018). Evidence for exposure to selenium by breeding interior snowy plovers (Charadrius nivosus) in saline systems of the Southern Great Plains. *Ecotoxicology, 27*(6), 703–718. https://doi.org/10.1007/s10646-018-1952-2

Australian Livestock Export Corporation Limited. (2022). *Copper poisoning*. Retrieved from http://www.veterinaryhandbook.com.au/Diseases.aspx?diseasenameid=58

Badry, A., Krone, O., Jaspers, V., Mateo, R., García-Fernández, A., Leivits, M., & Shore, R. F. (2020). Towards harmonisation of chemical monitoring using avian apex predators: Identification of key species for pan-European biomonitoring. *The Science of the Total Environment, 731*, 139198. https://doi.org/10.1016/j.scitotenv.2020.139198

Bailey, M. (2011). *Heavy metal intoxication in companion animals*. https://www.dvm360.com/view/heavy-metal-intoxication-companion-animal-proceedings

Bain, B. J. (2017). In B. J. Bain, I. Bates, & M. A. Laffan (Eds.), *Blood cell morphology in health and disease* (12th ed., pp. 61–92). Dacie and Lewis Practical Haematology. https://doi.org/10.1016/B978-0-7020-6696-2.00005-9

Bano, I., Skalickova, S., Arbab, S., Urbankova, L., & Horkyet, P. (2022). Toxicological effects of nanoselenium in animals. *Journal of Animal Science and Biotechnology, 13*, 72. https://doi.org/10.1186/s40104-022-00722-2

Barghi, M., Jin, X., Lee, S., Jeong, Y., Yu, J. P., Paek, W. K., & Moon, H. B. (2018). Accumulation and exposure assessment of persistent chlorinated and fluorinated contaminants in Korean birds. *Science of the Total Environment, 645*, 220–228. https://doi.org/10.1016/j.scitotenv.2018.07.040

Barnes, J. G., Doney, G. E., Yates, M. A., Seegar, W. S., & Gerstenberger, S. L. (2019). A broad-scale assessment of mercury contamination in peregrine falcons across the northern latitudes of North America. *Journal of Raptor Research, 53*(1), 1–13. https://doi.org/10.3356/JRR-18-0003

Barnosky, A. D. (2008). Megafauna biomass tradeoff as a driver of Quaternary and future extinctions. *PNAS, 105*(1), 11543–11548. https://doi.org/10.1073/pnas.0801918105

Barry, D., & Oelschlaeger, M. (1996). A science for survival: values and conservation biology. *Conservation Biology, 10*, 905–911.

Bassi, E., Facoetti, R., Ferloni, M., Pastorino, A., Bianchi, A., Fedrizzi, G., Bertoletti, I., & Andreotti, A. (2021). Lead contamination in tissues of large avian scavengers in south-central Europe. *The Science of the Total Environment, 778*, 146130. https://doi.org/10.1016/j.scitotenv.2021.146130

Bastow, B. D., & Holmes, J. L. (2016). *Teratology and drug use during pregnancy*. Medscape. WebMD. https://emedicine.medscape.com/article/260725-overview

Battaglia, A., Ghidini, S., Campanini, G., & Spaggiari, R. (2005). Heavy metal contamination in little owl (Athene noctua) and common buzzard (Buteo buteo) from northern Italy. *Ecotoxicology and Environmental Safety, 60*(1), 61–66. https://doi.org/10.1016/j.ecoenv.2003.12.019

Baudrot, V., Fritsch, C., Perasso, A., Banerjee, M., & Raoul, F. (2018). Effects of contaminants and trophic cascade regulation on food chain stability: application to cadmium soil pollution on small mammals – Raptor systems. *Ecological Modelling, 382*, 33–42. https://doi.org/10.1016/j.ecolmodel.2018.05.002

BBC. (2019). *Critically endangered vultures poisoned en masse in Botswana*. Retrieved from https://www.bbc.com/news/world-africa-48717018

Beckon, W. N., & Maurer, T. C. (2007). *Species at risk from selenium exposure in the San Francisco estuary*. U.S. Fish and Wildlife Service.

Bedrosian, B., Craighead, D., & Crandall, R. (2012). Lead exposure in Bald eagles from Big Game Hunting, the continental implications and successful mitigation efforts. *PLoS One, 7*(12), e51978. https://doi.org/10.1371/journal.pone.0051978

Benson, K. (2022). *Zinc toxicosis in animals*. Retrieved from https://www.merckvet manual.com/toxicology/zinc-toxicosis/zinc-toxicosis-in-animals

Benson, J. M., Cheng, Y. S., Eidson, A. F., Hahn, F. H., Henderson, R. H., & Pickrell, J. A. (1995). Pulmonary toxicity of nickel subsulfide in F344/N rats exposed for 1–22 days. *Toxicology, 103*, 9–22.

Berg, W., Johnels, A., Sjöstrand, B., & Westermark, T. (1966). Mercury content in feathers of Swedish birds from the past 100 years. *Oikos, 17*(1), 71–83. https://doi.org/10.2307/3564782

Berny, P. (2007). Pesticides and the intoxication of wild animals. *Journal of veterinary pharmacology and therapeutics, 30*(2), 93–100. https://doi.org/10.1111/j.1365-2885.2007.00836.x
Bevanger, K., Berntsen, F., Clausen, S., Dahl, E. L., Flagstad, O., Follestad, A., Halley, D., Hanssen, F., Johnsen, L., Kvaløy, P., Lund-Hoel, P., May, R., Nygård, T., Christian Pedersen, H., Reitan, O., Røskaft, E., Steinheim, Y., Stokke, B., & Vang, R. (2010). *Pre- and post-construction studies of conflicts between birds and wind turbines in coastal Norway (Bird-Wind)*. Report on findings 2007–2010. NINA Report 620. https://docs.wind-watch.org/norway-nina-report-01jan08.pdf
Beyer, W. N., Dalgarn, J., Dudding, S., French, J. B., Mateo, R., Miesner, J., Sileo, L., & Spann, J. (2005). Zinc and lead poisoning in wild birds in the tri-state mining district (Oklahoma, Kansas, and Missouri). *Archives of Environmental Contamination and Toxicology, 48*(1), 108–117. https://doi.org/10.1007/s00244-004-0010-7
Bianchi, T. S. (2021). The evolution of biogeochemistry: Revisited. *Biogeochemistry, 154*, 141–181. https://doi.org/10.1007/s10533-020-00708-0
Binkowski, Ł. J., Stawarz, R. M., & Zakrzewski, M. (2013). Concentrations of cadmium, copper, and zinc in tissues of mallard and coot from southern Poland. *Journal of Environmental Science and Health. Part B, Pesticides, Food Contaminants, and Agricultural Wastes, 48*(5), 410–415. https://doi.org/10.1080/03601234.2013.742725
Bischoff, K., Priest, H., & Mount-Long, A. (2010). Animals as sentinels for human lead exposure: A case report. *Journal of Medical Toxicology: official journal of the American College of Medical Toxicology, 6*(2), 185–189. https://doi.org/10.1007/s13181-010-0014-9
Blakley, B. R. (2021). *Copper poisoning in animals*. Merck Manual. https://www.merckvetmanual.com/toxicology/copper-poisoning/copper-poisoning-in-animals#:~:text=The%20sudden%20onset%20of%20clinical,membranes%2C %20hemoglobinuria%2C%20and%20jaundice
Bloom, P. H., Scott, J. M., Pattee, O. H., & Smith, M. R. (1989). In Meyburg & R. D. Chancellor (Eds.), *Lead contamination of Golden eagles (Aquila chrysaetos) within the range of the Californian condor (Gymnogyps californianus)*. Raptors in the Modern World, 481–482. Proceedings of the 3rd World Conference on Birds of Prey and Owls.
Blus, L. J. (1994). A review of lead poisoning in swans. *Comparative Biochemistry and Physiology Part C, 108*(3), 259–267. https://doi.org/10.1016/0742-8413(94)00021-2
Bowerman, W. W., Evans, E. D., Giesy, J. P., & Postupalsky, S. (1994). Using feathers to assess risk of mercury and selenium to Bald eagle reproduction in the Great Lakes region. *Archives of Environmental Contamination and Toxicology, 27*(3), 294–298. https://doi.org/10.1007/BF00213162
Bowser, D. H., Frenkel, K., & Zelikoff, J. T. (1994). Effects of in vitro nickel exposure on the macrophagemediated immune functions of rainbow trout (Oncorhynchus mykiss). *Bulletin of Environmental Contamination and Toxicology, 52*, 367–373.
Brean, H. (2019). *Peregrine falcons act as mercury bellwether, Nevada study shows*. Las Vegas Review-Journal. Retrieved from https://www.reviewjournal.com/local/local-nevada/peregrine-falcons-act-as-mercury-bellwethernevada-study-shows-1640798/
Bridgeford, P. (2004a). Lappet-faced vulture Torgos tracheliotus. In A. Monadjem, M. D. Anderson, S. E. Piper, & A. F. Boshoff (Eds.), *The vultures of Southern Africa – Quo vadis* (pp. 28–33). Birds of Prey Working Group.
Bridgeford, P. (2004b). Status of vultures in Namibia. In A. Monadjem, M. D. Anderson, S. E. Piper, & A. F. Boshoff (Eds.), *The vultures of Southern Africa – Quo vadis* (pp. 144–147). Birds of Prey Working Group.
Brown, C. J. (1991). Declining Martial Polemaetus bellicosus and Tawny Aquila rapax eagle populations and causes of mortality on farmlands in central Namibia. *Biological Conservation, 56*(1), 49–62. https://doi.org/10.1016/0006-3207(91)90088-Q
Buehler, D. A. (2020). Bald eagle (Haliaeetus leucocephalus), version 1.0. In A. F. Poole & F. B. Gill (Eds.), *Birds of the world*. Cornell Laboratory of Ornithology. https://doi.org/10.2173/bow.baleag.01
Burco, J., Myers, A. M., Schuler, K., & Gillin, C. (2012). Acute lead toxicosis via ingestion of spent ammunition in a free-ranging cougar (Puma concolor). *Journal of Wildlife Diseases, 48*(1), 216–219. https://doi.org/10.7589/0090-3558-48.1.216

Burgat, V. (1990). Un micropollutant: Le cadmium. *Bulletin Mensuel de l'Office National de la Chasse, 146*, 40–42.

Burger, J. (1995). A risk assessment for lead in birds. *Journal of Toxicology and Environmental Health, 45*(4), 369–396. https://doi.org/10.1080/15287399509532003

Buxton, S., Garman, E., Heim, K. E., Lyons-Darden, T., Schlekat, C. E., Taylor, M. D., & Oller, A. R. (2019). Concise review of nickel human health toxicology and ecotoxicology. *Inorganics, 7*(7), 89. https://doi.org/10.3390/inorganics7070089

Cade, T. J. (2007). Exposure of California Condors to lead from spent ammunition. *Journal of Wildlife Management, 71*(7), 2125–2133. https://doi.org/10.2193/2007-084

Cain, S. A., Kadlec, J. A., Allen, D. L., Cooley, R. A., Hornocker, M. H., & Wagner, F. H. (1971). *Predator control -1971. Report to the President's Council on Environmental Quality and the U.S. Department of the Interior by the Advisory Committee of Predator Control.* University of Michigan Press.

Campbell, M. (2015). *Vultures: Their evolution, ecology and conservation.* Taylor and Francis (CRC Press).

Campbell, M. (2017). *Biological conservation for the 21st century: A conservation biology of large wildlife.* Nova Science Publishers.

Campbell, M. (2018). *Geomatics and conservation biology.* Nova Science Publishers.

Campbell, M. (2020). *The physical geography of El Salvador: A geophysical and ecological approach.* Nova Science Publishers.

Campbell, M. (2022). The Great Eagles: Their Evolution, Ecology and Conservation. New York, NY: Nova Science Publishers.

Campbell, M., & Jha, K. K. (2021). *Critical research techniques in animal and habitat ecology: international examples.* Nova Science Publishers.

Canadian Council of Ministers of the Environment. (1999). *Canadian tissue Residue Guidelines for the Protection of Wildlife Consumers of Aquatic Biota! DDT.* Excerpt from: Canadian environmental quality guidelines, Winnipeg. http://www.ccme.ca/publications/ceqg_rcqe.html

Cantrell, A. C., Kilroe-Smith, T. A., Simões, M. M., & Border, E. A. (1977). The effect of zinc and pH on the behaviour of delta-aminolevulinic acid dehydratase activity in baboons exposed to lead. *British Journal of Industrial Medicine, 34*(2), 110–113. https://doi.org/10.1136/oem.34.2.110

Carneiro, M., Colaço, B., Brandão, R., Ferreira, C., Santos, N., Soeiro, V., Colaço, A., Pires, M. J., Oliveira, P. A., & Lavín, S. (2014). Biomonitoring of heavy metals (Cd, Hg, and Pb) and metalloid (As) with the Portuguese common buzzard (Buteo buteo). *Environmental monitoring and assessment, 186*(11), 7011–7021. https://doi.org/10.1007/s10661-014-3906-3

Carpenter, J. W., Pattee, O. H., Fritts, S. H., Rattner, B. A., Wiemeyer, S. N., Royle, J. A., & Smith, M. N. M. R. (2003). Experimental lead poisoning in Turkey vultures (Cathartes aura). *Journal of Wildlife Diseases, 39*(1), 96–104. https://doi.org/10.7589/0090-3558-39.1.96

Carroll, S. E. (1996). *Ancient & Medieval Falconry: Origins & functions in medieval England.* The Crown & Helm. http://www.r3.org/life/articles/falconry.html

Carruthers, B., & Gunson, K. E. (2015). *Development of a provincial wildlife mitigation strategy for large and small animals on highways in Ontario.* Presentation and Paper Submitted to the 2015 International Conference on Ecology and Transportation.

Castro, I., Aboal, J. R., Fernandez, J. A., & Carballeira, A. (2011). Use of raptors for biomonitoring of heavy metals: Gender, age, and tissue selection. *Bulletin of Environmental Contamination and Toxicology, 86*(3), 347–351. https://doi.org/10.1007/s00128-011-0220-4

Chang, L. W., Magos, L., & Suzuki, T. (Eds.). (1996). *Toxicology of metals.* CRC Press.

Chau, Y. K., & Kulikovsky-Cordeiro, O. T. R. (1995). Occurrence of nickel in the Canadian environment. *Environmental Reviews, 3*, 95–120.

Chiverton, L., Cromie, R., & Kock, R. (2022). European mammal exposure to lead from ammunition and fishing weight sources. *Heliyon, 8*(8), e10014. https://doi.org/10.1016/j.heliyon.2022.e10014

Christensen, K. Y., Raymond, M., & Meiman, J. (2019). Perfluoroalkyl substances and metabolic syndrome. *International Journal of Hygiene and Environmental Health, 222*(1), 147–153. https://doi.org/10.1016/j.ijheh.2018.08.014

Church, M. E., Gwiazda, R., Risebrough, R. W., Sorenson, K., Chamberlain, C. P., Farry, S., Heinrich, W., Rideout, B. A., & Smith, D. R. (2006). Ammunition is the principal source of lead accumulated by California condors re-introduced to the wild. *Environmental Science and Technology, 40*(19), 6143–6150. https://doi.org/10.1021/es060765s

Clark, A., & Scheuhammer, A. (2003). Lead poisoning in upland-foraging birds of prey in Canada. *Ecotoxicology, 12*, 23–30. https://doi.org/10.1023/A:1022576510445

Clarkson, T. W., & Magos, L. (2006). The toxicology of mercury and its chemical compounds. *Critical Reviews in Toxicology, 36*(8), 609–662. https://doi.org/10.1080/10408440600845619

Clements, T., Gilbert, M., Rainey, H. J., Cuthbert, R., Eames, J. C., Bunnat, P., Teak, S., Chansocheat, S., & Setha, T. (2012). *Vultures in Cambodia: Population, threats and conservation* (pp. 1–18). Bird Conservation International.

Colborn, T. (1991). Epidemiology of Great Lakes Bald eagles. *Journal of Toxicology and Environmental Health, 33*(4), 395–454.

Conforti, V. A., & Azevedo, F. C. C. (2003). Local perceptions of jaguars (Panthera onça) and pumas (Puma Concolor) in the Iguaçu National Park área, south Brazil. *Biological Conservation, 111*, 215–221. https://doi.org/10.1016/S0006-3207(02)00277-X

Cooke, J. (2011). Cadmium in small mammals. In W. N. Beyer & J. P. Meador (Eds.), *Environmental contaminants in biota interpreting tissue concentrations* (2nd ed., pp. 627–642). Taylor and Francis. https://doi.org/10.1201/b10598

Cope, R.B. (2022). *Cyanide poisoning in animals*. Retrieved from https://www.merckvet manual.com/toxicology/cyanide-poisoning/cyanide-poisoning-in-animals

Craig, R. C., & Enderson, J. H. (2004). *Peregrine falcon biology and management in Colorado, 1973–2001*. Technical Publication 43. Colorado Division of Wildlife.

Craig, T. H., Connelly, J. W., Craig, E. H., & Parker, T. L. (1990). Lead concentrations in Golden and Bald eagles. *Wilson Bulletin, 102*, 130–133.

Craighead, D., Bedrosian, B., Watson, R. T., Fuller, M., Pokras, M., & Hunt, W. (2009). *A relationship between blood lead levels of common ravens and the hunting season in the southern Yellowstone ecosystem. Ingestion of lead from spent ammunition: Implications for wildlife and humans*. Peregrine Fund.

Crinnion, W. J. (2011). Polychlorinated biphenyls: persistent pollutants with immunological, neurological, and endocrinological consequences. *Alternative Medicine Review, 16*(1), 5–13.

Cruz-Martinez, L., Redig, P. T., & Deen, J. (2012). Lead from spent ammunition: A source of exposure and poisoning in Bald eagles. *Human–Wildlife Interactions, 6*(1), 94–104.

Cullen, J. T., & Maldonado, M. T. (2013). Biogeochemistry of cadmium and its release to the environment. In A. Sigel, H. Sigel, & R. Sigel (Eds.), *Cadmium: From toxicity to essentiality. Metal ions in life sciences* (Vol. 11). Springer. https://doi.org/10.1007/978-94-007-5179-8_2

Cuthbert, R., Green, D. R., Ranade, S., SaravAn, S., Pain, D. J., Prakash, V., & Cunningham, A. A. (2006). Rapid population declines of Egyptian vulture (Neophron percnopterus) and Red-headed vulture (Sarcogyps calvus) in India. *Animal Conservation, 9*(3), 349–354. https://doi.org/10.1111/j.1469-1795.2006.00041.x

Dangleben, N. L., Skibola, C. F., & Smith, M. T. (2013). Arsenic immunotoxicity: A review. *Environmental Health, 12*(1), 73. https://doi.org/10.1186/1476-069X-12-73

Das, K. K., Reddy, R. C., Bagoji, I. B., Das, S., Bagali, S., Mullur, L., Khodnapur, J. P., & Biradar, M. S. (2018). Primary concept of nickel toxicity c – An overview. *Journal of Basic and Clinical Physiology and Pharmacology, 30*(2), 141–152. https://doi.org/10.1515/jbcpp-2017-0171

Daskin, J. H., & Pringle, R. M. (2016). Does primary productivity modulate the indirect effects of large herbivores? A global meta-analysis. *Journal of Animal Ecology, 85*(4), 857–868. https://doi.org/10.1111/1365-2656.12522

De Francisco, N., Troya, J. D. E., & Agüera, E. I. (2003). Lead and lead toxicity in domestic and free-living birds. *Avian Pathology, 32*(1), 3–13. https://doi.org/10.1080/0307945021000070660

DeBruyn, A. M. H., & Chapman, P. M. (2007). Selenium toxicity to invertebrates: Will proposed thresholds for toxicity to fish and birds also protect their prey? *Environmental Science & Technology, 41*(5), 1766–1770. https://doi.org/10.1021/es062253j

DeForest, D. K., Schlekat, C. E., Brix, K. V., & Fairbrother, A. (2012). Secondary poisoning risk assessment of terrestrial birds and mammals exposed to nickel. *Integrated Environmental Assessment and Management, 8*(1), 107–119. https://doi.org/10.1002/ieam.248

Descalzo, E., Camarero, P. R., Sánchez-Barbudo, I. S., Martinez-Haro, M., Ortiz-Santaliestra, M. E., Moreno-Opo, R., & Mateo, R. (2021). Integrating active and passive monitoring to assess sublethal effects and mortality from lead poisoning in birds of prey. *The Science of the Total Environment, 750*, 142260. https://doi.org/10.1016/j.scitotenv.2020.142260

Dirzo, R., Young, H. S., Galetti, M., Ceballos, G., Isaac, N. J. B., & Collen, B. (2014). Defaunation in the anthropocene. *Science, 345*, 401–406.

Diters, R. W., & Nielsen, S. W. (1978). Lead poisoning of raccoons in Connecticut. *Journal of Wildlife Diseases, 14*(2), 187–192. https://doi.org/10.7589/0090-3558-14.2.187

Dobrowolska, A., & Melosik, M. (2008). Bullet-derived lead in tissues of the wild boar (Sus scrofa) and red deer (Cervus elaphus). *European Journal of Wildlife Research, 54*(2), 231–235. https://doi.org/10.1007/s10344-007-0134-y

Donázar, J. A., Palacios, C. J., Gangoso, L., Ceballos, O., González, M. J., & Hiraldo, F. (2002). Conservation status and limiting factors in the endangered population of Egyptian vulture (Neophron percnopterus) in the Canary Islands. *Biological Conservation, 107*(1), 89–97. https://doi.org/10.1016/S0006-3207(02)00049-6

Eeva, T., Lehikoinen, E., & Ronkarönkä, M. (1998). Air pollution fades the plumage of the Great Tit. *Functional Ecology, 12*(4), 607–612. https://doi.org/10.1046/j.1365-2435.1998.00221.x

Eeva, T., Belskii, E., & Kuranov, B. (2006). Environmental pollution affects genetic diversity in wild bird populations. *Mutation Research/Genetic Toxicology and Environmental Mutagenesis, 608*(1), 8–15. https://doi.org/10.1016/j.mrgentox.2006.04.021

Ehrenfeld, D. W. (1970). *Biological conservation.* Holt, Rinehart and Winston.

Ehrenfeld, D. (1992). Conservation biology: its origins and definition. *Science (New York, N.Y.), 255*(5052), 1625–1626. https://doi.org/10.1126/science.255.5052.1625-a

Eisler, R. (1985). *Cadmium hazards to fish, wildlife and invertebrates: A synoptic review* (Vol. 85, p. 2). Biological Report, 1.2. United States Fish and Wildlife Service.

Eisler, R. (1988). *Arsenic hazards to fish, wildlife, and invertebrates: A synoptic review.* Biological Report 85 (1.12). Contaminant Hazard Reviews January 1988 Report No. 12. United States Fish and Wildlife Service, Patuxent Wildlife Research Center.

Eisler, R. (1998). *Nickel hazards to fish, wildlife, and invertebrates.* A Synoptic Review. Retrieved from https://www.pwrc.usgs.gov/eisler/CHR_34_Nickel.pdf

Ek, K. H., Morrison, G. M., Lindberg, P., & Rauch, S. (2004). Comparative tissue distribution of metals in birds in Sweden using ICP-MS and laser ablation ICP-MS. *Bulletin of Environmental Contamination and Toxicology, 47*(2), 259–269. https://doi.org/10.1007/s00244-004-3138-6

Ellgaard, E. G., Ashley, S. E., Langford, A. E., & Harlin, D. C. (1995). Kinetic analysis of the swimming behavior of the goldfish, Carassius auratus, exposed to nickel; hypoactivity induced by sublethal concentrations. *Bulletin of Environmental Contamination and Toxicology, 55*, 929–936.

Elliott, J. E., Langelier, K. M., Scheuhammer, A. M., Sinclair, P. H., & Whitehead, P. E. (1992). *Incidence of lead poisoning in Bald eagles and lead shot in waterfowl gizzards from British Columbia, 1988–91.* CWS Prog. Note 200, Canadian Wildlife Service.

Environmental Protection Agency. (2013). *Polycyclic Aromatic Hydrocarbons (PAHs).* Retrieved from https://www.epa.gov/sites/production/files/2014-03/documents/pahs_factsheet_cdc_2013.pdf

Environmental Protection Agency. (2022). *PFAS explained.* Retrieved from https://www.epa.gov/pfas/pfas-explained

Espín, S., Martínez-López, E., Jiménez, P., María-Mojica, P., & García-Fernández, A. J. (2015). Delta-aminolevulinic acid dehydratase (δALAD) activity in four free-living bird species exposed to different levels of lead under natural conditions. *Environmental Research, 137*, 185–198. https://doi.org/10.1016/j.envres.2014.12.017

Esposito, M., De Roma, A., Maglio, P., Sansone, D., Picazio, G., Bianco, R., De Martinis, C., Rosato, G., Baldi, L., & Gallo, P. (2019). Heavy metals in organs of stray dogs and cats from the city of Naples and its surroundings (Southern Italy). *Environmental science and pollution research international, 26*(4), 3473–3478. https://doi.org/10.1007/s11356-018-3838-5

European Chemicals Bureau. (2003). *Technical guidance document in support of Commission*. Directive 93/67/EEC on risk assessment for new notified substances, Commission Regulation (EC) 1488/94 on risk assessment for existing substances and Directive 98/8/EC of the European Parliament and of the Council concerning the placing of biocidal products on the market. European Commission, Joint Research Centre, European Chemicals Bureau.

Evans, D. H., Weingarten, K. E., & Walton, J. S. (1990). The effect of atropine on cadmium- and nickel-induced constriction of vascular smooth muscle of the dogfish shark ventral aorta. *Toxicology, 62*, 89–94.

Evers, D. C., Savoy, L. J., DeSorbo, C. R., Yates, D. E., Hanson, W., Taylor, K. M., Siegel, L. S., Cooley, J. H., Jr., Bank, M. S., Major, A., Munney, K., Mower, B. F., Vogel, H. S., Schoch, N., Pokras, M., Goodale, M. W., & Fair, J. (2008). Adverse effects from environmental mercury loads on breeding common loons. *Ecotoxicology, 17*(2), 69–81. https://doi.org/10.1007/s10646-007-0168-7

Evers, D. C., Paruk, J. D., McIntyre, J. W., & Barr, J. F. (2010). Common Loon (Gaviaimmer). In A. F. Poole (Ed.), *The Birds of North America*. Ithaca.

Falandysz, J., Jakuczun, B., & Mizera, T. (1988). Metals and organochlorines in four female White-tailed eagles. *Marine Pollution Bulletin, 19*(10), 521–526. https://doi.org/10.1016/0025-326X(88)90542-5

Falandysz, J., Ichihashi, H., Szymczak, K., Yamasaki, S., & Mizera, T. (2001). Metallic elements and metal poisoning among White-tailed Sea eagles from the Baltic Sea. *Marine Pollution Bulletin, 42*, 1190–1193. https://doi.org/10.1016/s0025-326x(01)00217-x

Ferguson-Lees, J., & Christie, D. A. (2001). *Raptors of the world*. Christopher Helm Publishers.

Fernie, K. J., Smits, J. E., Bortolotti, G. R., & Bird, D. M. (2001). Reproduction success of American Kestrels exposed to dietary polychlorinated biphenyls. *Environmental Toxicology and Chemistry, 20*(4), 776–781. https://doi.org/10.1002/etc.5620200412

Fingas, M. (2017). The fate of PAHs resulting from in-situ oil burns. *International Oil Spill Conference Proceedings, 1*, 1041–1056. https://doi.org/10.7901/2169-3358-2017.1.1041

Fisher, I. J., Pain, D. J., & Thomas, V. G. (2006). A review of lead poisoning from ammunition sources in terrestrial birds. *Biological Conservation, 131*(3), 421–432. https://doi.org/10.1016/j.biocon.2006.02.018

Forsman, D. (1999). *The raptors of Europe and the Middle East*. A handbook of field identification. Poyser.

Franson, J. C. (1996). Interpretation of tissue lead residues in birds other than waterfowl. In W. N. Beyer et al. (Eds.), *Environmental contaminants in wildlife: Interpreting tissue concentrations* (pp. 265–279). CRC Press, Lewis Publishers.

Franson, J. C., & Pain, D. J. (2011). Lead in birds. In W. N. Beyer & J. P. Meador (Eds.), *Environmental contaminants in Biota: Interpreting tissue concentrations* (2nd ed., pp. 563–593). CRC Press.

Franson, J. C., Hansen, S. P., Creekmore, T. E., Brand, C. J., Evers, D. C., Duerr, A. E., & DeStefano, S. (2003). Lead fishing weights and other fishing tackle in selected waterbirds. *Waterbirds, 26*(3), 345–352. https://doi.org/10.1675/1524-4695(2003)026[0345:LFWAOF]2.0.CO;2

Franson, J. C., Lahner, L. L., Meteyer, C. U., & Rattner, B. A. (2012). Copper pellets simulating oral exposure to copper ammunition: Absence of toxicity in American kestrels (Falco sparverius). *Archives of Environmental Contamination and Toxicology, 62*(1), 145–153. https://doi.org/10.1007/s00244-011-9671-1

Frantz, A., Pottier, M. A., Karimi, B., Corbel, H., Aubry, E., Haussy, C., Gasparini, J., & Castrec-Rouelle, M. (2012). Contrasting levels of heavy metals in the feathers of urban pigeons from close habitats suggest limited movements at a restricted scale. *Environmental Pollution, 168*, 23–28. https://doi.org/10.1016/j.envpol.2012.04.003

Frenzel, R. W., & Anthony, R. G. (1989). Relationship of diets and environmental contaminants in wintering Bald eagles. *Journal of Wildlife Management, 53*(3), 792–802. https://doi.org/10.2307/3809214

Frewin, R. (2014). *Chapter 27 – Biochemical aspects of anaemia* (3rd ed., pp. 515–532). Clinical Biochemistry: Metabolic and Clinical Aspects. https://doi.org/10.1016/B978-0-7020-5140-1.00027-4

Gainzaraian, J. A., Arambarri, R., & Rodríguez, A. F. (2000). Breeding density, habitat selection and reproductive rates of the Peregrine Falcon Falco peregrinus in Álava (northern Spain). *Bird Study, 47*, 225–231. https://doi.org/10.1080/00063650009461177

Gangoso, L., Alvarez-Lloret, P., Rodríguez-Navarro, A. A., Mateo, R., Hiraldo, F., & Donázar, J. A. (2009). Long-term effects of lead poisoning on bone mineralization in vultures exposed to ammunition sources. *Environmental Pollution, 157*(2), 569–574. https://doi.org/10.1016/j.envpol.2008.09.015

García-Fernández, A. J., Calvo, J. F., Martínez-López, E., María-Mojica, P., & Martínez, J. E. (2008). Raptor ecotoxicology in Spain: A review on persistent environmental contaminants. *AMBIO: A Journal of the Human Environment, 37*(6), 432–439. https://doi.org/10.1579/0044-7447(2008)37432

Ghazaly, K. S. (1992). Sublethal effects of nickel on carbohydrate metabolism, blood, and mineral contents of Tilapia nilotica. *Water, Air, and Soil Pollution, 64*, 525–532.

Giesy, J. P., & Kannan, K. (2001). Global distribution of perfluorooctane sulfonate in wildlife. *Environmental Science & Technology, 35*(7), 1339–1342. https://doi.org/10.1021/es001834k

Gilbert, M., Virani, M. Z., Watson, R. T., Oaks, J. L., Benson, P. C., Khan, A. A., et al. (2002). Breeding and mortality of Oriental White-backed vulture Gyps bengalensis in Punjab Province, Pakistan. *Bird Conservation International, 12*(4), 311–326. https://doi.org/10.1017/S0959270902002198

Gilbert, M., Oaks, J. L., Virani, M. Z., Watson, R. T., Ahmed, S., Chaudhry, J., et al. (2004). In R. C. Chancellor & Meyburg (Eds.), *The status and decline of vultures in the provinces of Punjab and Sind, Pakistan: A 2003 update* (pp. 221–234). Raptors worldwide. Proceedings of the 6th World Conference on Birds of Prey and Owls. WWGBP and MME/Birdlife Hungary.

Gill, C. E., & Langelier, K. M. (1994). British Columbia. Acute lead poisoning in a Bald eagle secondary to bullet ingestion. *Canadian Veterinary [Journal] = la Revue Veterinaire Canadienne, 35*(5), 303–304.

Gil-Sánchez, J. M., Aguilera-Alcalá, N., Moleón, M., Sebastián-González, E., Margalida, A., Morales-Reyes, Z., Durá-Alemañ, C. J., Oliva-Vidal, P., Pérez-García, J. M., & Sánchez-Zapata, J. A. (2021). biases in the detection of intentionally poisoned animals: Public health and conservation implications from a field experiment. *International journal of environmental research and public health, 18*(3), 1201. https://doi.org/10.3390/ijerph18031201

Giri, A., Ranjan, P., & Bharti, V. K. (2021). In P. Devi, P. Singh, A. Malakar, & D. Snow (Eds.), *Selenium toxicity in domestic animals*. Selenium Contamination in Water. https://doi.org/10.1002/9781119693567.ch4

Golden, N. H., Warner, S. E., & Coffey, M. J. (2016). A review and assessment of spent lead ammunition and its exposure and effects to scavenging birds in the United States. *Reviews of Environmental Contamination and Toxicology, 237*, 123–191. https://doi.org/10.1007/978-3-319-23573-8_6

Grade, T. J. (2011). *Effects of lead fishing tackle on common loons (Gavia immer) in New Hampshire. 1989–2010 [Thesis]*. University of Wisconsin.

Grade, T. J., Pokras, M. A., Laflamme, E. M., & Vogel, H. S. (2017). Population-level effects of lead fishing tackle on common loons. *Journal of Wildlife Management, 82*(1), 155–164. https://doi.org/10.1002/jwmg.21348

Green, R.A., Selby, L.A. & Zumwalt, R.W. (1978). *Experimental lead intoxication in dogs: A comparison of blood lead and urinary delta-aminolevulinic acid following intoxication and chelation therapy*. Retrieved from https://worldwidescience.org/topicpages/u/urine+delta-aminolevulinic+acid.html

Green, R. E., Newton, I., Shultz, S., Cunningham, A. A., Gilbert, M., Pain, D. J., & Prakash, V. (2004). Diclofenac poisoning as a cause of vulture population declines across the Indian subcontinent. *Journal of Applied Ecology, 41*(5), 793–800. https://doi.org/10.1111/j.0021-8901.2004.00954.x

Grilo, A., Moreira, A., Carrapiço, B., Belas, A., & São Braz, B. (2021). Epidemiological study of pesticide poisoning in domestic animals and wildlife in Portugal: 2014–2020. *Frontiers in Veterinary Science, 7*, 616293. https://doi.org/10.3389/fvets.2020.616293

Gross, L. (2006). Switching drugs for livestock may help save critically endangered Asian vultures. *PLoS Biology, 4*(3), e61. https://doi.org/10.1371/journal.pbio.0040061

Grúz, A., Déri, J., Szemerédy, G., Szabó, K., Kormos, É., Bartha, A., Lehel, J., & Budai, P. (2018). Monitoring of heavy metal burden in wild birds at eastern/north-eastern part of Hungary. *Environmental science and pollution research international, 25*(7), 6378–6386. https://doi.org/10.1007/s11356-017-1004-0

Grúz, A., Mackle, O., Bartha, A., Szabó, R., Déri, J., Budai, P., & Lehel, J. (2019). Biomonitoring of toxic metals in feathers of predatory birds from eastern regions of Hungary. *Environmental Science and Pollution Research International, 26*(25), 26324–26331. https://doi.org/10.1007/s11356-019-05723-9

Guitart, R., Sachana, M., Caloni, F., Croubels, S., Vandenbroucke, V., & Berny, P. (2010). Animal poisoning in Europe. Part 3: Wildlife. *Veterinary Journal, 183*(3), 260–265. https://doi.org/10.1016/j.tvjl.2009.03.033

Gulliford, A. (2015). *Looking back on a century of poisoning predators*. High Country News. Retrieved from https://www.hcn.org/articles/looking-back-on-a-century-of-poisoning-predators

Guthrie, A. L., Jayson, S. L., Strike, T. B., Sparrow, S. J., Flach, E. J., & Szladovits, B. (2020). Diagnosis and treatment of heavy metal toxicosis in six Waldrapp Ibis (Geronticus eremita). *Journal of Avian Medicine and Surgery, 34*(4), 371–380. https://doi.org/10.1647/1082-6742-34.4.371

Haig, S. M., D'Elia, J., Eagles-Smith, C., Fair, J. M., Gervais, J., Herring, G., … Schulz, J.H. 2014. The persistent problem of lead poisoning in birds from ammunition and fishing tackle. The Condor 116(3): 408–428. doi: https://doi.org/10.1650/CONDOR-14-36.1

Hall, J.O. (2022). *Selenium poisoning*. Retrieved from https://www.msdvetmanual.com/special-pet-topics/poisoning/selenium-poisoning#:~:text=Chronic%20(longterm)%20selenium%20poisoning%2C,the%20hair%20and%20the%20hooves

Hall, M., Grantham, J., Posey, R., & Mee, A. (2007). Lead exposure among reintroduced California condors in southern California. *California Condors in the 21st Century. Series in Ornithology, 2*, 163–184. Cambridge, MA: Nuttall Ornithological Club and American Ornithologists' Union.

Hancock, P. (2009a). *Botswana – Major poisoning incidents* (pp. 10–11). African Raptors.

Hancock, P. (2009b). Poisons devastate vultures throughout Africa. *African Raptors Newsletter, 2* (November).

Hancock, P. (2010). Vulture poisoning at Khutse. *Birds and People, 27*, 7. Retrieved from http://hdl.handle.net/1854/LU-1039067

Harada, K., & Miura, H. (1984). Free erythrocyte protoporphyrin (FEP) and zinc protoporphyrin (ZnP) as biological parameters for lead poisoning. *International Archives of Occupational and Environmental Health, 53*(4), 365–377. https://doi.org/10.1007/BF00380676

Harvey, L. J., & McArdle, H. J. (2008). Biomarkers of copper status: A brief update. *British Journal of Nutrition, 99*(Suppl. 3), S10–S13. https://doi.org/10.1017/S0007114508006806

Haschek, W. M., Rousseaux, C. G., & Wallig, M. A. (2013). *Haschek and Rousseaux's handbook of toxicologic pathology* (3rd ed.). Academic Press [Imprint]. Elsevier Science & Technology Books.

Hays, S. (1959). *Conservation and the Gospel of efficiency: The progressive conservation, 1890–1929*. Harvard University Press.

Heinz, G. H., & Hoffman, D. J. (1998). Methylmercury chloride and selenomethionine interactions on health and reproduction in mallards. *Environmental Toxicology and Chemistry, 17*(2), 139–145. https://doi.org/10.1002/etc.5620170202

Heinz, G. H., Hoffman, D. J., Klimstra, J. D., Stebbins, K. R., Kondrad, S. L., & Erwin, C. A. (2011). Teratogenic effects of injected methylmercury on avian embryos. *Environmental Toxicology and Chemistry, 30*(7), 1593–1598. https://doi.org/10.1002/etc.530

Helander, B. & Stjernberg T. (2002). *Action plan for the conservation of white-tailed sea eagle (Haliaeetus albicilla) Strasbourg: Birdlife International*. Retrieved from http://www.planetofbirds.com/ns/sap/White-tailed%20Eagle.pdf

Helander, B., Ekman, B., Hagerroth, J.-E., Hagerroth, P.-A., & Israelsson, J. (1989). Age-specific field characteristics of the White-tailed Sea eagle, Haliaeetus albicilla. *Var Vagelvarld, 48*, 319–334.

Helander, B., Olsson, A., Bignert, A., Asplund, L., & Litzén, K. (2002). The role of DDE, PCB, coplanar PCB and eggshell parameters for reproduction in the white-tailed sea eagle (Haliaeetus albicilla) in Sweden. *Ambio, 31*(5), 386–403. https://doi.org/10.1579/0044-7447-31.5.386

Helander, B., Bignert, A., & Asplund, L. (2008). Using raptors as environmental sentinels: monitoring the white-tailed sea eagle Haliaeetus albicilla in Sweden. *Ambio, 37*(6), 425–431. https://doi.org/10.1579/0044-7447(2008)37[425:uraesm]2.0.co;2

Helander, B., Axelsson, J., Borg, H., Holm, K., & Bignert, A. (2009). Ingestion of lead from ammunition and lead concentrations in White-tailed Sea eagles (Haliaeetus albicilla) in Sweden. *Science of the Total Environment, 407*(21), 5555–5563. https://doi.org/10.1016/j.scitotenv.2009.07.027

Helander, B., Krone, O., Räikkönen, J., Sundbom, M., Ågren, E., & Bignert, A. (2021). Major lead exposure from hunting ammunition in eagles from Sweden. *The Science of the total environment, 795*, 148799. https://doi.org/10.1016/j.scitotenv.2021.148799

Henny, C. J., & Elliott, J. E. (2015). Toxicology. In D. M. Bird & K. L. Bildstein (Eds.), *Raptor research and management techniques manual* (2nd ed., pp. 329–350). Hancock House Publishing, Ltd..

Hernández, M., & Margalida, A. (2008). Pesticide abuse in Europe: Effects on the Cinereous vulture (Aegypius monachus) population in Spain. *Ecotoxicology, 17*(4), 264–272. https://doi.org/10.1007/s10646-008-0193-1

Herrmann, C., Krone, O., Stjernberg, T., & Helander, B. (2011). *Population development of Baltic bird species: White-tailed Sea eagle (Haliaeetus albicilla)*. HELCOM Baltic Sea Environment Fact Sheet. https://helcom.fi/wp-content/uploads/2020/06/BSEFS-White_tailed-Eagle.pdf

Hinck, J. E., Norstrom, R. J., Orazio, C. E., Schmitt, C. J., & Tillitt, D. E. (2009). Persistence of organochlorine chemical residues in fish from the Tombigbee River (Alabama, USA), Continuing risk to wildlife from a former DDT manufacturing facility. *Environmental Pollution, 157*(2), 582–591. https://doi.org/10.1016/j.envpol.2008.08.021

Hivert, L. G., Clarke, J. R., Peck, S. J., Lawrence, C., Brown, W. E., Huxtable, S. J., Schaap, D., Pemberton, D., & Grueber, C. E. (2018). High blood lead concentrations in captive Tasmanian devils (Sarcophilus harrisii): A threat to the conservation of the species? *Australian Veterinary Journal, 96*(11), 442–449. https://doi.org/10.1111/avj.12753

Hoffman, D. J. (2002). Role of selenium toxicity and oxidative stress in aquatic birds. *Aquatic Toxicology, 57*(1–2), 11–26. https://doi.org/10.1016/S0166-445X(01)00263-6

Holt, G., Frøslie, A., & Norheim, G. (1979). Acta Veterinaria Scandinavica. *Mercury. France: DDE, and PCB in the avian fauna in Norway, 70*, 1–28. 1965–1976.

Honda, K., Marcovecchio, J. E., Kan, S., Tatsukawa, R., & Ogi, H. (1990). Metal concentrations in pelagic seabirds from the North Pacific Ocean. *Archives of Environmental Contamination and Toxicology, 19*(5), 704–711. https://doi.org/10.1007/BF01183988

Hong, Y. J., Kim, Y. J., Murata, K., Lee, H., & Min, M.-S. (2013). Determining the specific status of Korean Collared Scops Owls. *Animal Systematics, Evolution and Diversity, 29*(2), 136–143. https://doi.org/10.5635/ASED.2013.29.2.136

Hoogesteijn, R., & Hoogesteijn, A. (2008). Conflicts between cattle ranching and large predators in Venezuela: could use of water buffalo facilitate felid conservation? *Oryx, 42*(1), 132–138. https://doi.org/10.1017/S0030605308001105

Houserova, P., Hedbavny, J., Matejicek, D., Kracmar, S., Sitko, J., & Kuban, V. (2005). Determination of total mercury in muscle, intestines, liver and kidney tissues of cormorant (Phalacrocorax carbo), great crested grebe (Podiceps cristatus) and Eurasian buzzard (Buteo buteo). *Veterinarni Medicina, 50*, 61–68. https://doi.org/10.17221/5597-VETMED

Huang, J. & Mayer, J. (2019). *Lead and zinc toxicity in birds*. Retrieved from https://todaysveterinarypractice.com/lead-and-zinc-toxicity-in-birds/

Hudson, R. H., Tucker, R. K., & Haegele, M. A. (1984). *Handbook of toxicity of pesticides to wildlife* (2nd ed.). U.S. Fish and Wildlife Service Resource Publication 153.

Hunt, W. G., Parish, C. N., Farry, S. C., Lord, T. G., & Sieg, R. (2007). Movements of introduced California Condors in Arizona in relation to lead exposure. In A. Mee, L. S. Hall, & J. Grantham (Eds.), *California Condors in the 21st Century* (pp. 79–96). Nuttall Ornithological Club, American Ornithologists' Union, McLean.

Ingley, E., & Klinken, S. P. (2006). Erythrocytes. In G. J. Laurent & S. D. Shapiro (Eds.), *Encyclopedia of respiratory medicine* (pp. 142–146). Academic Press.

Iñigo, A., & Atienza, J. C. (2007). *Efectos del Reglamento 1774/2002 y las decisions adoptadas por la Comisión Europea en, 2003 y 2005 sobre las aves necrófagas en la peninsula Ibérica y sus posibles soluciones*. Official report to the European Commission. SEO/Birdlife 15-06-2007.

Ishii, C., Ikenaka, Y., Nakayama, S. M., Suzuki, Y., Watanuki, Y., Watanabe, Y., Fukuwaka, M. A., Yohannes, Y. B., Kawai, Y. K., & Ishizuka, M. (2013). Heavy metal pollution in Japanese seabirds. *The Japanese Journal of Veterinary Research, 61*(Suppl), S75–S76.

Isomursu, M., Koivusaari, J., Stjernberg, T., Hirvelä-Koski, V., & Venäläinen, E. R. (2018). Lead poisoning and other human-related factors cause significant mortality in white-tailed eagles. *Ambio, 47*(8), 858–868. https://doi.org/10.1007/s13280-018-1052-9

Iwata, H., Watanabe, M., Kim, E., Gotoh, R., Yasunaga, G., Tanabe, S., Masuda, Y., & Fujita, S. (2000). Contamination by chlorinated hydrocarbons and lead in Steller's Sea eagle and White-tailed Sea eagle from Hokkaido, Japan. In M. Ueta & M. J. Mcgrady (Eds.), *First Symposium on Steller's and White-tailed Sea eagles in East Asia* (pp. 91–106). Wild Bird Society of Japan.

Jacobson, E., Carpenter, J. W., & Novilla, M. (1977). Suspected lead toxicosis in a Bald eagle. *Journal of the American Veterinary Medical Association, 171*(9), 952–954.

Jayaraj, R., Megha, P., & Sreedev, P. (2016). Organochlorine pesticides, their toxic effects on living organisms and their fate in the environment. *Interdisciplinary Toxicology, 9*(3–4), 90–100. https://doi.org/10.1515/intox-2016-0012

Jenson, S., Johnels, A. G., Olsson, M., & Westermark, T. (1972). The avifauna of Sweden as indicators of environmental contamination with mercury and chlorinated hydrocarbons. *Proceedings of the International Ornithological Congress, 15*, 455–465.

Jha, K. K., & Campbell, M. (Eds.). (2021). *Critical research techniques in animal and habitat ecology examples from India*. Nova Science Publishers.

Johnels, A., Tyler, G., & Westermark, T. (1979). A history of mercury levels in Swedish fauna. *Ambio, 8*(4), 160–168.

Jones, G., & Barker, A. (2008). Reference intervals. The Clinical Biochemist. *Reviews 29 Suppl, 1*(Suppl 1), S93–S97.

Jorgenson, J. L. (2001). Aldrin and dieldrin: A review of research on their production, environmental deposition and fate, bioaccumulation, toxicology, and epidemiology in the United States. *Environmental Health Perspectives, 109*(Suppl. 1), 113–139. https://doi.org/10.1289/ehp.01109s1113

Jürgens, M. D., Crosse, J., Hamilton, P. B., Johnson, A. C., & Jones, K. C. (2016). The long shadow of our chemical past – High DDT concentrations in fish near a former agrochemicals factory in England. *Chemosphere, 162*, 333–344. https://doi.org/10.1016/j.chemosphere.2016.07.078

Kaiser, T. E., Reichel, W. L., Locke, L. N., Cromartie, E., Krynitsky, A. J., Lamont, T. G., Mulhern, B. M., Prouty, R. M., Stafford, C. J., & Swineford, D. M. (1980). Organochlorine pesticide, PCB, and PBB residues and necropsy data for Bald eagles from 29 states – 1975–77. *Pesticides Monitoring Journal, 4*, 145–149.

Kalisińska, E., Lisowski, P., Czernomysy-Furowicz, D., & Kavetska, K. (2008). Serratospiculiasis, mycosis, and haemosiderosis in wild peregrine falcon from Poland. *A case report. Bulletin of the Veterinary Institute in Pulawy, 52*, 75–79.

Kalisińska, E., Lisowski, P., Salicki, W., Kucharska, T., & Kavetska, K. (2009). Mercury in wild terrestrial carnivorous mammals from north-western Poland and unusual fish diet of red fox. *Acta Theriologica, 54*, 345–356. https://doi.org/10.4098/j.at.0001-7051.032.2008

Kalisińska, E., Gorecki, J., Lanocha, N., Okonska, A., Melgarejo, J. B., Budis, H., Rzad, I., & Golas, J. (2014). Total and methylmercury in soft tissues of White-tailed eagle (Haliaeetus albicilla) and osprey (Pandion haliaetus) collected in Poland. *Ambio – A Journal of the Human Environment, 43*(7), 858–870. https://doi.org/10.1007/s13280-014-0533-8

Kallok, M.A. (2013). *The case for copper: Research uncovers problems with lead bullets for deer hunting*. Minnesota Conservation Volunteer. Retrieved from https://www.dnr.state.mn.us/mcv-magazine/issues/2013/sep-oct/copper.html

Kasper, C. (2003). CHAPTER 7 – Erythropoietin. In A. W. Thomson & M. T. Lotze (Eds.), *The Cytokine Handbook (Fourth Edition) I* (pp. 149–166). Academic Press.

Kasprzak, K. S. (1987). Nickel. *Advances in modern environmental toxicology, 11*, 145–183.

Kaur, S., Kamli, M. R., & Ali, A. (2011). Role of arsenic and its resistance in nature. *Canadian Journal of Microbiology, 57*(10), 769–774. https://doi.org/10.1139/w11-062

Kelly, T. R., & Johnson, C. K. (2011). Lead exposure in free-flying Turkey vultures is associated with big game hunting in California. *PLoS One, 6*(4), e15350. https://doi.org/10.1371/journal.pone.0015350

Kelly, T. R., Poppenga, R. H., Woods, L. A., Hernandez, Y. Z., Boyce, W. M., Samaniego, F. J., Torres, S. G., & Johnson, C. K. (2014). Causes of mortality and unintentional poisoning in predatory and scavenging birds in California. *Veterinary Record Open, 1*(1), e000028. https://doi.org/10.1136/vropen-2014-000028

Kenntner, N., Tataruch, F., & Krone, O. (2001). Heavy metals in soft tissue of White-tailed eagles found dead or moribund in Germany and Austria from 1993 to 2000. *Environmental Toxicology and Chemistry, 20*(8), 1831–1837. https://doi.org/10.1002/etc.5620200829

Kenntner, N., Krone, O., Altenkamp, R., & Tataruch, F. (2003). Environmental contaminants in liver and kidney of free-ranging northern Goshawks (Accipiter gentilis) from three regions of Germany. *Archives of Environmental Contamination and Toxicology, 45*, 128–135. https://doi.org/10.1007/s00244-002-2100-8

Kenntner, N., Crettenand, Y., Fünfstück, H., Janovsky, M., & Tataruch, F. (2004). Lead poisoning and heavy metal exposure of Golden eagles (Aquila chrysaetos) from the European Alps. *Journal of Ornithology, 148*(2), 173–177. https://doi.org/10.1007/s10336-006-0115-zdoi

Kenntner, N., Tataruch, F., & Krone, O. (2005). Risk assessment of environmental contaminants in White-tailed Sea eagles (Haliaeetus albicilla) from Germany. In K. Pohlmeyer (Ed.), *Extended Abstracts of the XXVIIth Congress of the International Union of Game Biologists, Hanover* (pp. 125–127). DSV Verlag.

Khangarot, B. S., & Ray, P. K. (1990). Acute toxicity and toxic interaction of chromium and nickel to common guppy Poecilia reticulata (Peters). *Bulletin of Environmental Contamination and Toxicology, 44*, 832–839.

Kim, J., & Oh, J. M. (2012). Biological monitoring of heavy metal contaminations using owls. *Journal of Environmental Monitoring, 14*(3), 1091–1097. https://doi.org/10.1039/c2em10929h

Kim, E.-Y., Goto, R., Iwata, H., Masuda, Y., Tanabe, H., & Fujita, S. (1999). Preliminary survey of lead poisoning of Stellar's sea eagle (Haliaeetus pelagicus) and White-tailed eagle (Haliaeetus albicilla) in Hokkaido, Japan. *Environmental Toxicology and Chemistry, 18*(3), 448–451. https://doi.org/10.1002/etc.5620180312

Kitowski, I., Sujak, A., Wiącek, D., Strobel, W., Komosa, A., & Stobiński, M. (2016). Heavy metals in livers of raptors from Eastern Poland – The importance of diet composition. *Belgian Journal of Zoology, 146*(1), 3–13. https://doi.org/10.26496/bjz.2016.34

Kitowski, I., Jakubas, D., Wiącek, D., & Sujak, A. (2017). Concentrations of lead and other elements in the liver of the white-tailed eagle (Haliaeetus albicilla), a European flagship species, wintering in Eastern Poland. *Ambio, 46*(8), 825–841. https://doi.org/10.1007/s13280-017-0929-3

Kleiven, J., Bjerke, T., & Kaltenborn, B. P. (2004). Factors influencing the social acceptability of large carnivore behaviours. *Biodiversity and Conservation, 13*, 1647–1658. https://doi.org/10.1023/B:BIOC.0000029328.81255.38

Kolko, G. (1963). *The triumph of conservatism: A re-interpretation of American history, 1900–1916*. Free Press.

Korbel, R., & Kösters, J. (1994). Die Aktivitätsbestimmung der Delta-Aminolävulinsäure-Dehydratase – ein Verfahren zur Diagnose von Bleiintoxikationen bei Vögeln [Determination of the activity of delta-aminolevulinic acid dehydratase – a method for the diagnosis of lead poisoning in birds]. *Tierarztliche Praxis, 22*(4), 342–349.

Kramar, D. E., Carstensen, B., Prisley, S., & Campbell, J. (2019). Mercury concentrations in blood and feathers of nestling Bald eagles in coastal and inland Virginia. *Avian Research, 10*(1), 3. https://doi.org/10.1186/s40657-019-0142-0

Kramer, J., & Redig, P. T. (1997). Sixteen years of lead poisoning in eagles, 1980–1995: An epizootiologic view. *Journal of Raptor Research, 31*(4), 327–332.

Krone, O., Wille, F., Kenntner, N., Boertmann, D., & Tataruch, F. (2004). Mortality factors, environmental contaminants, and parasites of White-tailed Sea eagles from Greenland. *Avian Diseases, 48*(2), 417–424. https://doi.org/10.1637/7095

Krone, O., Stjernberg, T., Kenntner, N., Tataruch, F., Koivusaari, J., & Nuuja, I. (2006). Mortality factors, helminth burden, and contaminant residues in white-tailed sea eagles (Haliaeetus albicilla) from Finland. *Ambio, 35*(9), 8–104. https://doi.org/10.1579/0044-7447(2006)35[98:MFHBAC]2.0.CO;2

Krone, O., Kenntner, N., & Tataruch, F. (2009). Gefährdungsursachen des Seeadlers. *Denisia, 27*, 139–146. https://www.zobodat.at/pdf/DENISIA_0027_0139-0146.pdf

Krüger, S. C. & Amar, A. (2018). Lead Exposure in the Critically Endangered Bearded Vulture (Gypaetus barbatus) Population in Southern Africa. *Journal of Raptor Research, 52*(4), 491–499. https://doi.org/10.3356/JRR-17-86.1

Kuhn, L. T. (2001). In J. W. Martin (Ed.), *Bone mineralization* (2nd ed., pp. 47–54). Encyclopedia of Materials: Science and Technology. https://epdf.pub/queue/concise-encyclopedia-of-the-structure-of-materials.html

Kunito, T., Kubota, R., Fujihara, J., Agusa, T., & Tanabe, S. (2008). Arsenic in marine mammals, seabirds, and sea turtles. *Reviews of Environmental Contamination and Toxicology, 195*, 31–69. https://doi.org/10.1007/978-0-387-77030-7_2

Kurosawa, N. (2000). Lead poisoning in Steller's Sea eagles and White-tailed Sea eagles. In M. Ueta & M. J. McGrady (Eds.), *First Symposium on Stellar's and White-tailed Sea eagles in East Asia* (pp. 107–109). Wild Bird Society of Japan.

Langelier, K. M., Andress, C. E., Grey, T. K., Wooldridge, C., Lewis, R. J., & Marchetti, R. (1991). Lead poisoning in Bald eagles in British Columbia. *Canadian Veterinary Journal = la Revue Veterinaire Canadienne, 32*(2), 108–109.

Lapointe, D., & Couture, P. (2006). *Importance of the route of exposure in accumulation and subcellular partitioning of nickel in fathead minnows (Pimephales promelas).* Presentation P484 at the 27th Annual Meeting of the Society for Environmental Toxicology and Chemistry. November.

Lazorchak, J. M., McCormick, F. H., Henry, T. R., & Herlihy, A. T. (2003). Contamination of fish in streams of the mid-Atlantic region: An approach to regional indicator selection and wildlife assessment. *Environmental Toxicology and Chemistry, 22*, 545–553. https://doi.org/10.1002/etc.5620220312

Lemly, A. D. (1993). Guidelines for evaluating selenium data from aquatic monitoring and assessment studies. *Environmental Monitoring and Assessment, 28*(1), 83–100. https://doi.org/10.1007/BF00547213

Lemly, A. D. (1996). Assessing the toxic threat of selenium to fish and aquatic birds. *Environmental Monitoring and Assessment, 43*(1), 19–35. https://doi.org/10.1007/BF00399568

Lemly, A. D. (2002). Symptoms and implications of selenium toxicity in fish: the Belews Lake case example. *Aquatic toxicology (Amsterdam, Netherlands), 57*(1-2), 39–49. https://doi.org/10.1016/s0166-445x(01)00264-8

Lenntech. (2022). *Heavy metals*. Retrieved from https://www.lenntech.com/processes/heavy/heavy-metals/heavy-metals.htm#ixzz7ihZw0RUH.

Leonard, T. C. (2009). American economic reform in the progressive era: Its foundational beliefs and their relation to eugenics. *History of Political Economy, 41*(1), 109–141. https://doi.org/10.1215/00182702-2008-040

Leopold, A. S., Cain, S. A., Cottam, C. M., Gabrielson, I. N., & Kimball, T. L. (1964). *Predator and rodent control in the United States*. Wildlife Management Institute.

Lepage, D. (2020). *Avibase – Bird checklists of the world: Canary Islands*. Retrieved from https://avibase.bsc-eoc.org/checklist.jsp?region=esic&list=howard moore

Liebezeit, J. & Sheaffer, D. (2014). *Blood lead levels in avian scavengers: A comparison of urban and rural regions in the Pacific Northwest*. Retrieved from https://audubonportland.org/wp-content/uploads/2021/02/ASoP_Lead_preliminary_report_web_4.28.14.pdf

López, S. L., Aiassa, D., Benítez-Leite, S., Lajmanovich, R., Mañas, F., Poletta, G., Sánchez, N., Simoniello, M. F., & Carrasco, A. E. (2012). Chapter Two. Pesticides used in South American GMO-based agriculture: A review of their effects on humans and animal models. *Advances in Molecular Toxicology, 6*, 41–75. https://doi.org/10.1016/B978-0-444-59389-4.00002-1

Louekari, K., Mäkelä-Kurtto, R., Pasanen, J., Virtanen, V., Sippola, J., & Malm, J. (2000). *Cadmium in Fertilizers: Risks to human health and the environment*. Ministry of Agriculture and Forestry in Finland.

Love, J. A. (1983). *The return of the sea eagle*. Cambridge University Press.

Lumeij, J. T. (1985). Clinicopathologic aspects of lead poisoning in birds: A review. *The Veterinary Quarterly, 7*(2), 133–138. https://doi.org/10.1080/01652176.1985.9693969

Luzardo, O. P., Ruiz-Suárez, N., Henríquez-Hernández, L. A., Valerón, P. F., Camacho, M., Zumbado, M., & Boada, L. D. (2014). Assessment of the exposure to organochlorine pesticides, PCBs, and PAHs in six species of predatory birds of the Canary Islands, Spain. *Science of the Total Environment, 15*(472), 146–153. https://doi.org/10.1016/j.scitotenv.2013.11.021

MacArthur, R. H., & Wilson, E. O. (1967). *The theory of island biogeography*. Princeton University Press.

Mackenzie, D. (2014). *Europe's eagles under threat from vulture-killing drug*. Retrieved from https://www.newscientist.com/article/dn25650-europes-eagles-under-threat-from-vulture-killing-drug/

Madhusudan, M. D. (2003). Living amidst large wildlife: livestock and crop depredation by large mammals in the interior villages of Bhadra tiger reserve, South India. *Environmental Management, 31*, 466–475. https://doi.org/10.1007/s00267-002-2790-8

Madry, M. M., Kraemer, T., Kupper, J., Naegeli, H., Jenny, H., Jenni, L., & Jenny, D. (2015). Excessive lead burden among Golden eagles in the Swiss Alps. *Environmental Research Letters. PubMed, 10*(3). https://doi.org/10.1088/1748-9326/10/3/034003

Manning, T., Ross, G. A., & Symons, R. (2008). Environmental contaminants in White-bellied Sea eagles (Haliaeetus leucogaster) found in Sydney, Australia. *Australian Journal Ecotoxicology, 14*, 21–30.

Manning, L. K., Wünschmann, A., Armién, A. G., Willette, M., MacAulay, K., Bender, J. B., Buchweitz, J. P., & Redig, P. (2019). Lead Intoxication in Free-Ranging Bald Eagles (Haliaeetus leucocephalus). *Veterinary Pathology, 56*(2), 289–299. https://doi.org/10.1177/0300985818813099

Manto, M. (2014). Abnormal copper homeostasis: Mechanisms and roles in neurodegeneration. *Toxics, 2*(2), 327–345. https://doi.org/10.3390/toxics2020327

Marettová, E., Maretta, M., & Legáth, J. (2015). Toxic effects of cadmium on testis of birds and mammals: A review. *Animal Reproduction Science, 155*, 1–10. https://doi.org/10.1016/j.anireprosci.2015.01.007

Martin, P. A., Campbell, D., Hughes, K., & McDaniel, T. (2008). Lead in the tissues of terrestrial raptors in southern Ontario, Canada, 1995–2001. *Science of the Total Environment, 391*(1), 96–103. https://doi.org/10.1016/j.scitotenv.2007.11.012

Martínez-López, E., Martínez, J. E., María-Mojica, P., Peñalver, J., Pulido, M., Calvo, J. F., & García-Fernández, A. J. (2004). Lead in feathers and delta-aminolevulinic acid dehydratase activity in three raptor species from an unpolluted Mediterranean forest (Southeastern Spain). *Archives of Environmental Contamination and Toxicology, 47*(2), 270–275. https://doi.org/10.1007/s00244-004-3027-z

Mateo-Tomás, P., Olea, P., Jiménez-Moreno, M., Camarero, P. R., Sánchez-Barbudo, I. S., Martín-Doimeadios, R. C. R., & Mateo, R. (2016). *Mapping the spatio-temporal risk of lead exposure*

in apex species for more effective mitigation. Proceedings of the Royal Society of London Series B. https://doi.org/10.1098/rspb.2016.0662
MedicinePlus. (2009). *ALAD gene aminolevulinate dehydratase*. Retrieved from https://medlineplus.gov/genetics/gene/alad/
Meine, C. (2004). *Correction lines: Essays on land, leopold, and conservation*. Island Press.
Meine, C. (2010). Conservation Biology: Past and Present. In N. S. Sodhi & P. S. Ehrlich (Eds.), *Conservation biology for all* (pp. 7–26). Oxford University Press.
Meine, C., Soule, M., & Noss, R. F. (2006). A mission-driven discipline: The growth of conservation biology. *Conservation Biology, 20*(3), 631–651. https://doi.org/10.1111/j.1523-1739.2006.00449.x
Merriam Webster. (2020a). *Medical dictionary*. Hemotoxic. Retrieved from https://www.merriam-webster.com/medical/hemotoxic
Merriam Webster. (2020b). *Medical dictionary*. Carcinogenic. Retrieved from https://www.merriam-webster.com/dictionary/carcinogenic
Meyburg, B.-U., Kirwan, G. M., & Garcia, E. F. J. (2020). Greater Spotted eagle (Clanga clanga), version 1.0. In D. Hoyo et al. (Eds.), *Birds of the world*. Ithaca, NY. https://doi.org/10.2173/bow.grseag1.01
Middleton, H. A., Butler, R. W., & Davidson, P. (2018). Waterbirds alter their distribution and behavior in the presence of Bald eagles (Haliaeetus leucocephalus), Northwestern Naturalist, *99*(1), 21–30. https://doi.org/10.1898/NWN16-21.1
Mihajlović M. (1992). Toksicnost selena kod domaćih zivotinja [Selenium toxicity in domestic animals]. Glas. Srpska akademija nauka i umetnosti. *Odeljenje Medicinskih Nauka, 42*, 131–144.
Mijele, D. (2009). *Incidences of poisoning of vultures and lions in the Masai Maramara National Reserve*. Kenya Wildlife Service, Masai Mara Veterinary Report.
Miller, M. J. R., Wayland, M. E., Dzus, E. H., & Bortolotti, G. R. (2000). Availability and ingestion of lead shotshell pellets by migrant Bald eagles in Saskatchewan. *Journal of Raptor Research, 34*, 167–174. Retrieved from https://sora.unm.edu/sites/default/files/journals/jrr/v034n03/p00167-p00174.pdf
Miller, M. J. R., Wayland, M. E., & Bortolotti, G. R. (2001). Exposure of migrant Bald eagles to lead in prairie Canada. *Environmental Pollution, 112*(2), 153–162. https://doi.org/10.1016/S0269-7491(00)00122-6
Misztal-Szkudlińska, M., Szefer, P., Konieczka, P., & Namieśnik, J. (2011). Bio-magnification of mercury in trophic relation of Great Cormorant (Phalacrocorax carbo) and fish in the Vistula Lagoon, Poland. *Environmental Monitoring and Assessment, 176*(1–4), 439–449. https://doi.org/10.1007/s10661-010-1595-0
Moore, D. R., Teed, R. S., Rodney, S. I., Thompson, R. P., & Fischer, D. L. (2010). Refined avian risk assessment for aldicarb in the United States. *Integrated Environmental Assessment and Management, 6*(1), 83–101. https://doi.org/10.1897/IEAM_2009-022.1
Morgan, R. V. (1994). Lead poisoning in small companion animals: An update (1987–1992). *Veterinary and human toxicology, 36*(1), 18–22.
Mudd, G. M. (2010). Global trends and environmental issues in nickel mining: Sulfides versus laterites. *Ore Geology Reviews, 38*(1–2), 9–26. https://doi.org/10.1016/j.oregeorev.2010.05.003
Mundy, P. J. (1997). Hooded vulture Necrosyrtes monachus. In J. A. Harrison et al. (Eds.), *The atlas of southern African birds, 1: Non-passerines* (pp. 156–157). Birdlife South Africa.
Mundy, P., Butchart, D., Ledger, J., & Piper, S. (1992). *The vultures of Africa*. Academic Press.
Museum Victoria. (2012). *What does megafauna mean?* https://museumvictoria.com.au/about/mv-blog/apr-2012/what-does-megafauna-mean/
Nadjafzadeh, M., Hofer, H., & Krone, O. (2013). The link between feeding ecology and lead poisoning in white-tailed eagles. *Journal of Wildlife Management, 77*, 48–57. https://doi.org/10.1002/jwmg.440
Nam, D. H., & Lee, D. P. (2006). Reproductive effects of heavy metal accumulation on breeding feral pigeons (Columba livia). *Science of the Total Environment, 366*(2–3), 682–687. https://doi.org/10.1016/j.scitotenv.2006.02.004

National Center for Biotechnology Information. (2022a). *PubChem Compound Summary for CID 969491, Dieldrin*. Retrieved from https://pubchem.ncbi.nlm.nih.gov/compound/Dieldrin

National Center for Biotechnology Information. (2022b). *PubChem Compound Summary for CID 36231, 3,3',4,4',5,5'-Hexachlorobiphenyl*. Retrieved from https://pubchem.ncbi.nlm.nih.gov/compound/36231

National Center for Biotechnology Information. (2022c). *PubChem Compound Summary for CID 5460536, sodium dicyanidoaurate(I)*. Retrieved from https://pubchem.ncbi.nlm.nih.gov/compound/sodium-dicyanidoaurate_I

National Institute of Environmental Health Sciences. (2020). *Perfluoroalkyl and Polyfluoroalkyl Substances (PFAS)*. Retrieved from https://www.niehs.nih.gov/health/topics/agents/pfc/index.cfm

National Pesticide Information Center. (1996). *Carbofuran*. Retrieved from http://extoxnet.orst.edu/pips/carbofur.htm

National Research Council. (1977). *(US) Committee on Medical and Biological Effects of Environmental Pollutants*. Retrieved from https://www.ncbi.nlm.nih.gov/books/NBK231025/

Newton, I. (1986). *The Sparrow Hawk*. T and A D Poyser Ltd..

Newton, I. (1993). Causes of breeding failure in wild raptors: A review. In P. Y. Redig (Ed.), *Raptor, biomedicine* (pp. 62–70). University of Minnesota Press.

Nickel Institute. (2018). *Environmental Fact Sheet 6: Secondary poisoning risk assessment of birds and mammals exposed to nickel in their diets*. Retrieved from https://nickelinstitute.org/en/science/environmental-science-fact-sheets/fact-sheet-6-secondary-poisoning-of-birds-mammals/

Nighat, S., Iqbal, S., Nadeem, M. S., Mahood, T., & Shah, S. I. (2013). Estimation of heavy metal residues from the feathers of Falconidae, Accipitridae, and Strigidae in Punjab, Pakistan. *Turkish Journal of Zoology, 37*, 488–500. https://doi.org/10.3906/zoo-1112-1

Nobel Foundation. (1948). *Dichloro-Diphenyl-Trichloroethane and Newer Insecticides*. Retrieved from https://www.nobelprize.org/prizes/medicine/1948/muller/lecture/

North, M. A., Lane, E. P., Marnewick, K., Caldwell, P., Carlisle, G., & Hoffman, L. C. (2015). Suspected lead poisoning in two captive cheetahs (Acinonyx jubatus jubatus) in South Africa, in 2008 and 2013. *Journal of the South African Veterinary Association, 86*(1), E1–E5. https://doi.org/10.4102/jsava.v86i1.1286

NPIC. (1999). *DDT drug discovery*. Retrieved from http://npic.orst.edu/factsheets/ddtgen.pdf

NPIC. (2011). *Ecotoxicology*. Retrieved from http://npic.orst.edu/factsheets/ecotox.html

Nriagu, J. O., Pfeiffer, W. C., Malm, O., Magalhaes De Souza, C. M., & Mierle, G. (1992). Mercury pollution in Brazil. *Nature, 356*(6368), 389. https://doi.org/10.1038/356389a0

O'Toole, D., & Raisbeck, M. F. (1997). Experimentally induced selenosis of adult mallard ducks: Clinical signs, lesions, and toxicology. *Veterinary Pathology, 34*(4), 330–340. https://doi.org/10.1177/030098589703400409

Oaks, L., Rideout, B. A., Gilbert, M., Watson, R., Virani, M., & Ahmed Khan, A. (2001). Summary of diagnostic investigation into vulture mortality: Punjab Province, Pakistan, 2000–2001. Reports from the workshop on Indian Gyps vultures. In T. Katzner & J. Parry-Jones (Eds.), *Proceedings of the 4th Eurasian Congress on Raptors* (pp. 12–13). Estación Biológica Donaña, Raptor Research Foundation.

Odenbaugh, J. (2003). Values, advocacy, and conservation biology. *Environmental Values, 12*(1), 55–69. https://doi.org/10.3197/096327103129341225

Odino, M., & Ogada, D. L. (2008). *Furadan use in Kenya and its impacts on birds and other wildlife: A survey of the regulatory agency, distributors, and end-users of this highly toxic pesticide*. Report to the Bird Committee of Nature.

Ogada, D. L. (2014). The power of poison: pesticide poisoning of Africa's wildlife. *Annals of the New York Academy of Sciences, 1322*, 1–20. https://doi.org/10.1111/nyas.12405

Ogada, D. L., & Buij, R. (2011). Large declines of the Hooded vulture Necrosyrtes monachus across its African range. *Ostrich, 82*(2), 101–113. https://doi.org/10.2989/00306525.2011.603464

Ogada, D. L., & Kibuthu, P. M. (2012). Breeding ecology of Mackinder's eagle owls Bubo capensis mackinderi in farmlands of central Kenya. *Journal of Raptor Research, 46*(4), 327–335. https://doi.org/10.3356/JRR-11-90.1

Ogada, D.L., Thomsett, S., Virani, M.Z., Kendall, C. & Odino, M. 2010. Raptor road counts in Kenya: With emphasis on vultures. Report to the Royal Society for the Protection of Birds.

Ohlendorf, H. M. (1989). Bioaccumulation and effects of selenium in wildlife. In L. W. Jacobs (Ed.), *Selenium in Agriculture and the Environment* (pp. 133–177). Soil Science Society of America.

Olendorff, R. R. (1976). The food habits of North American Golden eagles. *American Midland Naturalist, 95*(1), 231–236. https://doi.org/10.2307/2424254

Osofsky, A., Jowett, P. L. H., Hosgood, G., & Tully, T. N. (2001). Determination of Normal Blood Concentrations of Lead, Zinc, Copper, and Iron in Hispaniolan Amazon Parrots (Amazona ventralis). *Journal of Avian Medicine and Surgery, 15*(1), 31–36. https://doi.org/10.1647/1082-6742(2001)015[0031:DONBCO]2.0.CO;2

Outridge, P. M., & Scheuhammer, A. M. (1993). Bioaccumulation and toxicology of chromium: implications for wildlife. *Reviews of Environmental Contamination and Toxicology, 130*, 31–77. https://doi.org/10.1007/978-1-4613-9763-2_2

Owen, B. (2019). *Switch ammo to stop lead poisoning in birds, B.C. wildlife experts tell hunters*. Canadian Press. Retrieved from https://www.cbc.ca/news/canada/british-columbia/switch-ammo-lead-poisoning-1.5361718

Pacyna, E. G., Pacyna, J. M., Sundseth, K., Munthe, J., Kindbom, K., Wilson, S., & Maxson, P. (2010). Global emission of mercury to the atmosphere from anthropogenic sources in 2005 and projections to 2020. *Atmospheric Environment, 44*(20), 2487–2499. https://doi.org/10.1016/j.atmosenv.2009.06.009

Pain, D. J., Sears, J., & Newton, I. (1995). Lead concentrations in birds of prey in Britain. *Environmental Pollution, 87*(2), 173–180. https://doi.org/10.1016/0269-7491(94)P2604-8

Pain, D. J., Prakash, V., Cunningham, A. A., & Ghalsasi, G. R. (2002). Vulture declines in India: patterns, causes, and spread. In T. Katzner & J. Parry-Jones (Eds.), *Conservation of Gyps vulture in Asia* (pp. 4–7). Third North American Ornithological Congress, September 24–28, 2002.

Pain, D. J., Cunningham, A. A., Donald, P. F., Duckworth, J. W., Houston, D. C., Katzner, T., Parry-Jones, J., Poole, C., Prakash, V., Round, P., & Timmins, R. (2003). Causes and effects of temporospatial declines of Gyps vultures in Asia. *Conservation Biology, 17*, 661–671.

Pain, D. J., Fisher, I. J., & Thomas, V. G. (2009). A global update of lead poisoning in terrestrial birds from ammunition sources. In R. T. Watson, M. Fuller, M. Pokras, & W. G. Hunt (Eds.), *Ingestion of Lead from Spent Ammunition: Implications for Wildlife and Humans* (pp. 28–301). The Peregrine Fund. https://doi.org/10.4080/ilsa.2009.0108

Pain, D. J., Mateo, R., & Green, R. E. (2019). Effects of lead from ammunition on birds and other wildlife: A review and update. *Ambio, 48*(9), 935–953. https://doi.org/10.1007/s13280-019-01159-0

Pappalardo, L., Silvanose, C. D., Beaufrère, H., Binoy, A., & Azmanis, P. (2021). Reference intervals for Cu, Mg, and Zn in captive gyrfalcons and other falcon species in the United Arab Emirates. *Journal of Veterinary Diagnostic Investigation: Official Publication of the American Association of Veterinary Laboratory Diagnosticians, Inc., 33*(4), 797–800. https://doi.org/10.1177/10406387211015655

Parslow, J. L. F., Jefferies, D. J., & Hanson, H. M. (1973). Gannet mortality incidents in 1972. *Marine Pollution Bulletin, 4*(3), 41–43. https://doi.org/10.1016/0025-326X(73)90303-2

Paruk, J. D., Long, D., Perkins, C., East, A., Sigel, B. J., & Evers, D. C. (2014). Polycyclic aromatic hydrocarbons detected in common loons (Gaviaimmer) wintering off coastal Louisiana. *Waterbirds, 37*, 85–93. https://doi.org/10.1675/063.037.sp111

Pattee, O. H., Wiemeyer, S. N., Mulhern, B. M., Sileo, L., & Carpenter, J. W. (1981). Experimental lead-shot poisoning in Bald eagles. *Journal of Wildlife Management, 45*(3), 806–810. https://doi.org/10.2307/3808728

Pattee, O. H., Bloom, P. H., Scott, J. M., & Smith, M. R. (1990). Lead hazards within the range of the California Condor. *Condor, 92*(4), 931–937. https://doi.org/10.2307/1368729

Pederson, L. & Stokka, G. (2020). *Cyanide poisoning*. North Dakota State University. Retrieved from https://www.ag.ndsu.edu/publications/livestock/cyanide-poisoning#:~:text=Cyanide%20inhibits%20oxygen%20utilization%20by,in%20the %20animal's%20digestive%20tract

Pérez-García, J. M., Marco-Tresserras, J., & Orihuela-Torres, A. (2020). Winter diet and lead poisoning risk of Greater Spotted eagles Clanga clanga in southeast Spain. *Bird Study, 67*(2), 224–231. https://doi.org/10.1080/00063657.2020.1810206

Pérez-López, M., De Mendoza, M. H., López-Beceiro, A., & Rodriguez, F. S. (2008). Heavy metal (Cd, Pb, Zn) and metalloid (As) content in raptor species from Galicia (NW Spain). *Ecotoxicology and Environmental Safety, 70*(1), 154–162. https://doi.org/10.1016/j.ecoenv.2007.04.016

Pokras, M. A., & Chafel, R. (1992). Lead toxicosis from ingested fishing sinkers in adult Common loons (Gaviaimmer) in New England. *Journal of Zoo and Wildlife Medicine, 23*, 92–97. Retrieved from http://www.jstor.org/stable/20460274

Pokras, M., Kneeland, M., Ludi, A., Golden, E., Major, A., Miconi, R., & Poppenga, R. H. (2009). Lead objects ingested by Common loons in New England. *Northeastern Naturalist, 16*(2), 177–182. https://doi.org/10.1656/045.016.0202

Posadasa, P., Crisci, J. V., & Katinas, L. (2006). Historical biogeography: A review of its basic concepts and critical issues. *Journal of Arid Environments, 66*, 389–403. https://doi.org/10.1016/j.jaridenv.2006.01.004

Prakash, V. (1999). Status of vultures in Keoladeo National Park, Bharatpur, Rajasthan with special reference to population crash in Gyps species. *Journal of the Bombay Natural History Society, 96*, 365–378.

Prakash, V., & Rahmani, A. R. (1999). Notes about the decline of Indian vultures, with particular reference to Keoladeo National Park. *Vulture News, 41*, 6–13.

Prakash, V., Green, R., Pain, D., Ranade, S., Saravanan, S., Prakash, N., Venkitachalam, R., Cuthbert, R., Rahmani, A., & Cunningham, A. (2007). Recent changes in populations of resident Gyps vultures in India. *Journal of the Bombay Natural History Society, 104*, 129–135.

Prakasha, V., Pain, D. J., Cunningham, A. A., Donald, P. F., Prakash, N., Verma, A., Gargia, R., Sivakumar, S., & Rahmania, A. R. (2003). Catastrophic collapse of Indian White-backed Gyps bengalensis and Long-billed Gyps indicus vulture populations. *Biological Conservation, 109*(3), 381–390. https://doi.org/10.1016/S0006-3207(02)00164-7

Prestt, I., Jefferies, D. J., & Moore, N. W. (1970). Polychlorinated biphenyls in wild birds in Britain and their avian toxicity. *Environmental Pollution, 1*(1), 3–26. https://doi.org/10.1016/0013-9327(70)90003-0

Proulx, R., Massicotte, P., & Pepino, M. (2014). Googling trends in conservation biology. *Conservation Biology, 28*, 44–51. https://doi.org/10.1111/cobi.12131

Puschner, B. & Poppenga, R. (2009). *Lead and zinc intoxication in companion birds*. Compendium. Retrieved from https://www.vetfolio.com/learn/article/lead-and-zinc-intoxication-in-companion-birds.

Puschner, B., St Leger, J., & Galey, F. D. (1999). Normal and toxic zinc concentrations in serum/plasma and liver of Psittacines with respect to genus differences. *Journal of veterinary diagnostic investigation: official publication of the American Association of Veterinary Laboratory Diagnosticians, Inc, 11*(6), 522–527. https://doi.org/10.1177/104063879901100606

Rahman, Z., & Singh, V. P. (2019). The relative impact of toxic heavy metals (THMs) (arsenic (As), cadmium (Cd), chromium (Cr)(VI), mercury (Hg), and lead (Pb)) on the total environment: An overview. *Environmental Monitoring and Assessment, 191*(7), 419. https://doi.org/10.1007/s10661-019-7528-7

Rai, D., Wilson, A. M., & Moosavi, L. (2022). *Histology, reticulocytes*. StatPearls Publishing. https://www.ncbi.nlm.nih.gov/books/NBK542172/

Rainbow, P. S. (1997). Trace metal accumulation in marine invertebrates: Marine biology or marine chemistry? *Journal of the Marine Biological Association of the United Kingdom, 77*(1), 195–210. https://doi.org/10.1017/S0025315400033877

Rajamani, J., & Subramanian, M. (2015). Toxicity assessment on the levels of select metals in the critically endangered Indian White-backed Vulture, Gyps bengalensis, in India. *Bulletin of Environmental Contamination and Toxicology, 94*(6), 722–726. https://doi.org/10.1007/s00128-015-1548-y

Rankins, D. L., & Pugh, D. G. (2012). Feeding and Nutrition. In D. G. Pugh & A. N. Baird (Eds.), *Sheep and goat medicine* (2nd ed., pp. 18–49). Amsterdam. https://doi.org/10.1016/C2009-0-60474-8

Rashidi, H. H., & Green, R. (2018). Red blood cell morphology and indices with clinical chemistry. In N. Rifai (Ed.), *Tietz textbook of clinical chemistry and molecular diagnostics, Sixth Edition* (Vol. 72, pp. 1731–1733.e12). Elsevier, Inc.

Rattner, B. A. (2009). History of wildlife toxicology. *Ecotoxicology (London, England), 18*(7), 773–783. https://doi.org/10.1007/s10646-009-0354-x

Rattner, B. A., Whitehead, M. A., Gasper, G., Meteyer, C. U., Link, W. A., Taggart, M. A., Meharg, A. A., Pattee, O. H., & Pain, D. J. (2008). Apparent tolerance of Turkey vultures (Cathartes aura) to the non-steroidal anti-inflammatory drug diclofenac. *Environmental Toxicology and Chemistry, 27*(11), 2341–2345. https://doi.org/10.1897/08-123.1

Rattner, B. A., Horak, K. E., Lazarus, R. S., Schultz, S. L., Knowles, S., Abbo, B. G., & Volker, S. F. (2015). Toxicity reference values for chlorophacinone and their application for assessing anticoagulant rodenticide risk to raptors. *Ecotoxicology, 24*(4), 720–734. https://doi.org/10.1007/s10646-015-1418-8

Redig, P. T., & Arent, L. R. (2008). Raptor toxicology. Veterinary clinics of North America. *Exotic Animal Practice, 11*(2), 261–282. https://doi.org/10.1016/j.cvex.2007.12.004

Reichel, W. L., Schmeling, S. K., Cromartie, E., Kaiser, T. E., Krynitsky, A. J., Lamont, T. G., & Swineford, D. M. (1984). Pesticide, PCB and lead residues and necropsy data for Bald eagles from 32 states – 1978–1981. *Environmental Monitoring and Assessment, 4*(4), 395–403. https://doi.org/10.1007/BF00394177

Revkin, A.C. (1997). *High PCB level is found in a hudson bald eagle*. Retrieved from https://www.nytimes.com/1997/09/17/nyregion/highpcb-level-is-found-in-a-hudson-bald-eagle.html

Reyes-Velarde, A. (2018). Young mountain lion found lounging in Pleasanton is released to the wild. Los Angeles Times. (latimes.com)

Riley, R. S., Ben-Ezra, J. M., Goel, R., & Tidwell, A. (2001). Reticulocytes and reticulocyte enumeration. *Journal of Clinical Laboratory Analysis, 15*(5), 267–294. https://doi.org/10.1002/jcla.1039

Roche, C. (2006). Breeding records and nest site preference of Hooded vultures in the Greater Kruger National Park. *Ostrich, 77*(1–2), 99–101. https://doi.org/10.2989/00306520609485515

Rodriguez-Ramos Fernandez, J., Gutiérrez, V., Höfle, U., Mateo, R., Monsalve, L., Crespo, E., & Blanco, J. (2008). *Lead in griffon and cinereous vultures in central Spain: correlations between clinical signs and blood lead levels*. Proceedings: Ingestion of Spent Lead Ammunition: Implications for Wildlife and Humans.

Roebuck, P., & Phifer, P. (1999). The persistence of positivism in conservation biology. *Conservation Biology, 13*(2), 444–446.

Rogers, J. M., & Kavlock, R. J. (1996). Developmental toxicology. In C. D. Klaassen (Ed.), *Casarett & Doull's toxicology* (5th ed., pp. 301–331). McGraw-Hill.

Rubo, A., Kellens, R., Reddy, J., Steier, N., & Hasenpusch, W. (2006). *Alkali metal cyanides. Ullmann's Encyclopedia of Industrial Chemistry*. Wiley-VCH.

Rutkiewicz, J., Nam, D. H., Cooley, T., Neumann, K., Padilla, I. B., Route, W., Strom, S., & Basu, N. (2011). Mercury exposure and neurochemical impacts in Bald eagles across several Great Lakes states. *Ecotoxicology, 20*(7), 1669–1676. https://doi.org/10.1007/s10646-011-0730-1

Sáenz, J. C., & Carrillo, E. (2002). Jaguares depredadores de ganado en Costa Rica: ¿un 982 problema sin solución? In R. Medellin, C. Equihua, C. Chetkiewicz, P. Crawshaw, A. Rabinowitz, K. F. Redford, J. Robinson, E. Sanderson, & A. Taber (Eds.), *El Jaguar en El Nuevo Milenio* (pp. 127–138). Fondo de cultura económica, Universidad Nacional Autónoma de México, Wildlife Conservation Society.

Salkowski, A. A., & Penney, D. G. (1994). Cyanide poisoning in animals and humans: A review. *Veterinary and Human Toxicology, 36*(5), 455–466.

Sánchez-Virosta, P., Espín, S., García-Fernández, A. J., & Eeva, T. (2015). A review on exposure and effects of arsenic in passerine birds. *The Science of the Total Environment, 512-513*, 506–525. https://doi.org/10.1016/j.scitotenv.2015.01.069

Sánchez-Virosta, P., Espín, S., Ruiz, S., Panda, B., Ilmonen, P., Schultz, S. L., Karouna-Renier, N., García-Fernández, A. J., & Eeva, T. (2020). Arsenic-related oxidative stress in experimentally dosed wild Great tit nestlings. *Environmental pollution (Barking, Essex: 1987), 259*, 113813. https://doi.org/10.1016/j.envpol.2019.113813

Saurola, P., Valkama, J., & Velmala, W. (2013). *The Finnish bird ringing Atlas*. Finnish Museum of Natural History and Ministry of Environment.

Scheuhammer, A. M. (1987). The chronic toxicity of aluminium, cadmium, mercury, and lead in birds: A review. *Environmental Pollution, 46*(4), 263–295. https://doi.org/10.1016/0269-7491(87)90173-4

Scheuhammer, A. M., & Norris, S. L. (1995). *A review of the environmental impacts of lead shotshell ammunition and lead fishing weights in Canada*. Canadian Wildlife Service Occasional Paper 88.

Scheuhammer, A. M., Money, S. L., & Kirk, D. (2002). *Lead fishing sinkers and jigs in Canada: Review of their use patterns and toxic impacts on wildlife*. Canadian Wildlife Service Occasional Paper 108.

Schmutz, J. K., Rose, K. A., & Johnson, R. G. (1989). Hazards to raptors from strychnine poisoned ground squirrels. *Journal of Raptor Research, 23*(4), 147–151. Retrieved from https://sora.unm.edu/sites/default/files/journals/jrr/v023n04/p00147-p00151.pdf

Scinicariello, F., Murray, H. E., Moffett, D. B., Abadin, H. G., Sexton, M. J., & Fowler, B. A. (2007). Lead and delta-aminolevulinic acid dehydratase polymorphism: where does it lead? A meta-analysis. *Environmental health perspectives, 115*(1), 35–41. https://doi.org/10.1289/ehp.9448

Sergio, F., & Hiraldo, F. (2008). Intraguild predation in raptor assemblages: A review. *Ibis, 150*, 132–145. https://doi.org/10.1111/j.1474-919X.2008.00786.x

Sergio, F., Marchesi, L., & Pedrini, P. (2004). Integrating individual habitat choices and regional distribution of a biodiversity indicator and top predator. *Journal of Biogeography, 31*(4), 619–628. https://doi.org/10.1046/j.1365-2699.2003.01002.x

Seshadri, B., Bolan, N. S., Naidu, R., & Brodie, K. J. (2010). The role of coal combustion products in managing the bioavailability of nutrients and heavy metals in soils. *Journal of Soil Science and Plant Nutrition, 10*(3), 378–398. https://doi.org/10.4067/S0718

Shankar, S., Shanker, U., & Shikha. (2014). Arsenic contamination of groundwater: a review of sources, prevalence, health risks, and strategies for mitigation. *The Scientific World Journal, 304524*. https://doi.org/10.1155/2014/304524

Sharma, A. K., Saini, M., Singh, S. D., Prakash, V., Das, A., Dasan, R. B., Pandey, S., Bohara, D., Galligan, T. H., Green, R. E., Knopp, D., & Cuthbert, R. J. (2014). Diclofenac is toxic to the Steppe eagle Aquila nipalensis: Widening the diversity of raptors threatened by NSAID misuse in South Asia. *Bird Conservation International, 24*, 282–286. https://doi.org/10.1017/S0959270913000609

Shore, R. F., Pereira, M. G., Walker, L. A., & Thompson, D. R. (2011). Mercury in nonmarine birds and mammals. In W. N. Beyer & J. P. Meador (Eds.), *Environmental contaminants in biota* (pp. 609–642). CRC Press.

Sibly, R. M., Newton, I., & Walker, C. H. (2000). Effects of dieldrin on population growth rates of sparrowhawks 1963–1986. *Journal of Applied Ecology, 37*(3), 540–546. https://doi.org/10.1046/j.1365-2664.2000.00516.x

Sidor, I. F., Pokras, M. A., Major, A. R., Poppenga, R. H., Taylor, K. M., & Miconi, R. M. (2003). Mortality of common loons in New England, 1987 to 2000. *Journal of Wildlife Diseases, 39*(2), 306–315. https://doi.org/10.7589/0090-3558-39.2.306

Snoeijs, T., Dauwe, T., Pinxten, R., Vandesande, F., & Eens, M. (2004). Heavy metal exposure affects the humoral immune response in a free-living small songbird, the great tit (Parus

major). *Archives of Environmental Contamination and Toxicology, 46*(3), 399–404. https://doi.org/10.1007/s00244-003-2195-6
Snyder, N. F. R., & Snyder, H. A. (1991). *Birds of prey: Natural history of North American Raptors*. Voyageur Press.
Snyder, N., & Snyder, H. (2000). *The California Condor: A saga of natural history and conservation*. Academic Press.
Sodhi, N. S., & Ehrlich, P. S. (2010). *Conservation biology for all*. Oxford University Press.
Sorenson, K., Burnett, J., Mee, A., & Hall, L. (2007). *Lead concentrations in the blood of Big Sur California Condors. California Condors in the 21st Century. Series in Ornithology, No. 2* (pp. 185–195). Nuttall Ornithological Club and American Ornithologists' Union.
Soulé. (1986a). What is conservation biology? *BioScience, 35*(11), 727–734. https://doi.org/10.2307/1310054
Soulé, M. E. (1986b). *Conservation biology: The science of scarcity and diversity*. Sinauer Associates.
Soulé, M. E., & Orians, G. H. (Eds.). (2001). *Conservation biology: Research priorities for the next decade*. Island Press.
Soulé, M. E., & Wilcox, B. A. (Eds.). (1980). *Conservation biology, an evolutionary ecological perspective*. Sinauer Associates.
Spallholz, J. E., & Hoffman, D. J. (2002). Selenium toxicity: Cause and effects in aquatic birds. *Aquatic Toxicology, 57*(1–2), 27–37. https://doi.org/10.1016/s0166-445x(01)00268-5
Speer, R., & Wise, J. P. (2018). *Current status on chromium research and its implications for health and risk assessment*. In: Reference Module in Chemistry, Molecular Sciences and Chemical Engineering. https://doi.org/10.1016/B978-0-12-409547-2.14283-0
Ssemmanda, R., & Pomeroy, D. (2010). Scavenging birds of Kampala: 1973–2009. *Scopus, 30*, 26–31.
Stangl, G. I., & Kirchgessner, M. (1996). Nickel deficiency alters liver lipid metabolism in rats. *Journal of Nutrition, 126*, 2466–2473.
Stangl, G. I., & Kirchgessner, M. (1997). Effect of nickel deficiency on fatty acid composition of total lipids and individual phospholipids in brain and erythrocytes of rats. *Nutrition Research, 17*, 137–147.
Stankovic, S., Kalaba, P., & Stankovic, A. R. (2014). Biota as toxic metal indicators. *Environmental Chemistry Letters, 12*(1), 63–84. https://doi.org/10.1007/s10311-013-0430-6
State of Queensland. (2022a). *Lead contamination in food-producing animals*. Retrieved from https://www.business.qld.gov.au/industries/farms-fishing-forestry/agriculture/land-management/health-pests-weeds-diseases/livestock/lead-contamination
State of Queensland. (2022b). *Causes and signs of lead poisoning and contamination in production animals*. Retrieved from https://www.business.qld.gov.au/industries/farms-fishing-forestry/agriculture/land-management/health-pests-weeds-diseases/livestock/lead-contamination/causes-signs
Stern, B. R. (2010). Essentiality and toxicity in copper health risk assessment: Overview, update and regulatory considerations. Journal of Toxicology and Environmental Health. *Part A, 73*(2), 114–127. https://doi.org/10.1080/15287390903337100
Stone, W. B., & Okoniewski, J. C. (2001). Necropsy findings and environmental contaminants in common loons from New York. *Journal of Wildlife Diseases, 37*, 178–184. https://doi.org/10.7589/0090-3558-37.1.178
Stone, C. L., Fox, M. R. S., Jones, A. L., & Mahaffey, K. R. (1977). δ-aminolevulinic acid dehydratase – A sensitive indicator of lead exposure in Japanese Quail. *Poultry Science, 56*(1), 174–181. https://doi.org/10.3382/ps.0560174
Strom, S. M., Langenberg, J. A., Businga, N. K., & Batten, J. K. (2009). Lead exposure in Wisconsin birds. In R. T. Watson, M. Fuller, M. Pokras, & W. G. Hunt (Eds.), *Ingestion of lead from spent ammunition: Implications for wildlife and humans* (pp. 1–8). The Peregrine Fund.
Swan, G., Naidoo, V., Cuthbert, R., Green, R. E., Pain, D. J., Swarup, D., Prakash, V., Taggart, M., Bekker, L., Das, D., Diekmann, J., Diekmann, M., Killian, E., Meharg, A., Chandra Patra,

R., Saini, M., & Wolter, K. (2006). Removing the threat of diclofenac to critically endangered Asian vultures. *PLoS Biology, 4*(3), 395–402. https://doi.org/10.1371/journal.pbio.0040066

Szumiło, E., Szubska, M., Meissner, W., Bełdowska, M., & Falkowska, L. (2013). Mercury in immature and adults Herring gulls (Larus argentatus) wintering on the Gulf of Gdańsk area. *Oceanological and Hydrobiological Studies, 42*(3), 260–267. https://doi.org/10.2478/s13545-013-0082-y

Szumiło-Pilarska, E., Grajewska, A., Falkowska, L., Hajdrych, J., Meissner, W., Frączek, T., Bełdowska, M., & Bzoma, S. (2016). Species differences in total mercury concentration in gulls from the Gulf of Gdansk (Southern Baltic). *Journal of Trace Elements in Medicine and Biology: Organ of the Society for Minerals and Trace Elements, 33*, 100–109. https://doi.org/10.1016/j.jtemb.2015.09.005

Szumiło-Pilarska, E., Falkowska, L., Grajewska, A., & Meissner, W. (2017). Mercury in feathers and blood of gulls from the Southern Baltic Coast, Poland. *Water, Air and Soil Pollution, 228*(4), 138. https://doi.org/10.1007/s11270-017-3308-6

Tasneem, F., Abbasi, N. A., Iqbal Chaudhry, M. J., Mashiatullah, A., Ahmad, S. R., Qadir, A., & Malik, R. N. (2020). Dietary proxies (δ15N, δ13C) as signature of metals and arsenic exposure in birds from aquatic and terrestrial food chains. *Environmental Research, 183*, 109191. https://doi.org/10.1016/j.envres.2020.109191

Tchounwou, P. B., Yedjou, C. G., Patlolla, A. K., & Sutton, D. J. (2012). Heavy metals toxicity and the environment. In A. Luch (Ed.), *Molecular, clinical and environmental toxicology. Experientia supplementum* (Vol. 101, pp. 133–164). Springer. https://doi.org/10.1007/978-3-7643-8340-4_6

Thiollay, J. M. (2001). Long-term changes of raptor populations in northern Cameroon. *Journal of Raptor Research, 35*, 173–186.

Tiainen, J., Mikkola-Roos, M., Below, A., Jukarainen, A., Lehikoinen, A., Lehtiniemi, T., Pessa, J., Rajasärkkä, A., Rintala, J., Sirkiä, P., & Valkama, J. (2016). *Suomen lintujen uhanalaisuus 2015 – the 2015 red list of Finnish bird species*. Ympäristöministeriö & Suomen ympäristökeskus.

Tjernberg, M. (1981). Diet of the Golden eagle Aquila chrysaetos during the breeding season in Sweden. Holarctic Ecology. *Nordic Society Oikos, 4*(1), 12–19. https://doi.org/10.1111/j.1600-0587.1981.tb00975.x

Torchia, C. (2018). *A new concern for African vultures: Bullet lead in carrion*. Retrieved from https://apnews.com/article/e3d078ac03b3491c98012ec2b0743b46

Treves, A., & Bruskotter, J. (2014). Tolerance for predatory wildlife. *Science, 344*, 476. https://doi.org/10.1126/science.1252690

Ueta, M., & Masterov, V. B. (2000). Estimation by a computer simulation of population trend of Steller's Sea Eagles. In M. Ueta & M. J. McGrady (Eds.), *First symposium on Steller's and White-tailed Sea Eagles in East Asia* (pp. 111–116). Wild Bird Society of Japan.

United States Environmental Protection Agency (USEPA). (1980). *Ambient water quality criteria for nickel*. EPA Report 440/5-80-060.

Van Wyk, E., Bouwman, H., van der Bank, H., Verdoorn, G. H., & Hofmann, D. (2001a). Persistent organochlorine pesticides detected in blood and tissue samples of vultures from different localities in South Africa. Comparative Biochemistry and Physiology. *Part C: Toxicology & Pharmacology, 129*(3), 243–264. https://doi.org/10.1016/S1532-0456(01)90201-7

Van Wyk, E., Van der Bank, F. H. &, Verdoorn, G. H. & Hofmann, D. (2001b). Selected mineral and heavy metal concentrations in blood and tissues of vultures in different regions of South Africa. South African Journal of Animal Science, 31(2), 57–64. doi: https://doi.org/10.4314/sajas.v31i2.3831.

Vandenbroucke, V., Van Pelt, H., De Backer, P., & Croubels, S. (2010). Animal poisonings in Belgium: A review of the past decade. *Vlaams Diergeneeskundig Tijdschrift, 79*, 259–268.

Verdoorn, G. H., van Zijl, N., Snow, T. V., Komen, L., & Marais, E. W. (2004). Vulture poisoning in southern Africa. In A. Monadjem, M. D. Anderson, S. E. Piper, & A. F. Boshoff (Eds.), *The Vultures of Southern Africa – Quo Vadis?* (pp. 195–201). Birds of Prey Working Group.

Virani, M., Gilbert, M., Watson, R., Oaks, L., Benson, P., Kham, A. A., & Baral, H.-S. (2001). Asian vulture crisis project: Field results from Pakistan and Nepal for the 2000–2001 field season. Reports from the workshop on Indian Gyps vultures. In T. Katzner & J. Parry-Jones (Eds.), *Proceedings of the 4th Eurasian Congress on Raptors* (pp. 461–469). Estación Biológica Donaña, Raptor Research Foundation.

Virani, M. M. Z., Kendall, C., Njoroge, P., & Thomsett, S. (2011). Major declines in the abundance of vultures and other scavenging raptors in and around the Masai Mara ecosystem, Kenya. *Biological Conservation, 144*(2), 746–752. https://doi.org/10.1016/j.biocon.2010.10.024

Waage, M. (2014). Stop predator poisons from killing wildlife and harming ecosystems: There's no place for these deadly poisons on American lands. In *Natural Resources Defense Council*. Retrieved from https://www.nrdc.org/sites/default/files/predator-poisons-FS.pdf

Walker, K. C., Goette, M. B., & Batchelor, G. S. (1954). Pesticide residues in foods, dichlorodiphenyltrichloroethane and dichlorodiphenyldichloroethylene content in prepared foods. *Journal of Agricultural and Food Chemistry, 2*(20), 1034–1037. https://doi.org/10.1021/jf60040a006

Walker, C. H., Brealey, C. J., Mackness, M. M. I., & Johnston, G. (1991). Toxicity of pesticides to birds: The enzymic factor. *Biochemical Society Transactions, 19*(3), 741–745. https://doi.org/10.1042/bst0190741

Wani, A. L., Ara, A., & Usmani, J. A. (2015). Lead toxicity: A review. *Interdisciplinary toxicology, 8*(2), 55–64. https://doi.org/10.1515/intox-2015-0009

Waseem, A., Arshad, J., Iqbal, F., Sajjad, A., Mehmood, Z., & Murtaza, G. (2014). Pollution status of Pakistan: a retrospective review on heavy metal contamination of water, soil, and vegetables. *BioMed Research International*, 813206. https://doi.org/10.1155/2014/813206

Watson, R. T. (1986). *The ecology, biology, and population dynamics of the Bateleur eagle Terathopius ecaudatus*. PhD Thesis. University of Witwatersrand, Johannesburg.

Watson, J. (2010). *The Golden Eagle*. A&C Black.

Wayland, M., & Bollinger, T. (1999). Lead exposure and poisoning in Bald eagles and Golden eagles in the Canadian prairie provinces. *Environmental Pollution, 104*(3), 341–350. https://doi.org/10.1016/S0269-7491(98)00201-2

Wayland, M., & Scheuhammer, A. M. (2011). Cadmium in birds. In W. N. Beyer & J. P. Meador (Eds.), *Environmental contaminants in biota interpreting tissue concentrations* (2nd ed., pp. 645–666). Taylor and Francis.

Wayland, M., Wilson, L. K., Elliott, J. R., Miller, J. E., Bollinger, T., McAdie, M., Langelier, K., Keating, J., & Froese, J. M. W. (2003). Mortality, morbidity, and lead poisoning of eagles in western Canada 1986–1998. *Journal of Raptor Research, 37*, 8–18. Retrieved from https://sora.unm.edu/sites/default/files/journals/jrr/v037n01/p00008-p00018.pdf

Wedeen, R. P. (1985). Use of the CaNa2EDTA Pb-mobilization test to detect occult lead nephropathy. *Uremia Investigation, 9*(2), 127–130. https://doi.org/10.3109/08860228509088200

Weir, S. M., Thomas, J. F., & Blauch, D. N. (2018). Investigating spatial patterns of mercury and rodenticide residues in raptors collected near the Charlotte, NC, USA, metropolitan area. *Environmental Science and Pollution Research, 33*, 33153–33161. https://doi.org/10.1007/s11356-018-3229-y

Wetmur, J. G. (2015). *Lead toxicity and the ala-dehydratase polymorphism*. https://www.novatitanaquatics.com/grant/NIH/R01-ES005046-06

White, C. M., Cade, T. J., & Enderson, J. H. (2013). *Peregrine Falcons of the World*. Lynx Edicions.

Whitney, M. C., & Cristol, D. A. (2017). Impacts of sublethal mercury exposure on birds: A detailed review. *Reviews of Environmental Contamination and Toxicology, 244*, 113–163. https://doi.org/10.1007/398_2017_4

Wiemeyer, S. N., Hill, E. F., Carpenter, J. W., & Krynitsky, A. J. (1986). Acute oral toxicity of sodium cyanide in birds. *Journal of wildlife diseases, 22*(4), 538–546. https://doi.org/10.7589/0090-3558-22.4.538

Williams, P. L., James, R. C., & Roberts, S. M. (2014). *Principles of toxicology: Environmental and industrial applications*. John Wiley & Sons.

Wilson, E. O., & Peter, F. M. (Eds.). (1988). *Biodiversity*. National Academy Press.

Wobeser, G., & Blakley, B. R. (1987). Strychnine poisoning of aquatic birds. *Journal of Wildlife Diseases, 23*(2), 341–343. https://doi.org/10.7589/0090-3558-23.2.341

Woodford, M. H., Bowden, C. G. R., & Shah, N. (2008). *Diclofenac in Asia and Africa –Repeating the same mistake? Harmonisation and improvement of registration and quality control of veterinary medicinal products in Africa*. OIE World Organisation for Animal Health.

Xie, Y., Holmgren, S., Andrews, D. M. K. & Wolfe, M. S. (2017). Evaluating the impact of the U.S. National Toxicology Program: A case study on Hexavalent Chromium. *Environmental Health Perspectives, 125*(2), 181–188. https://doi.org/10.1289/ehp21

Yang, F., Xie, S., Liu, J., Wei, C., Zhang, H., Chen, T., & Zhang, J. (2018). Arsenic concentrations and speciation in wild birds from an abandoned realgar mine in China. *Chemosphere, 193*, 777–784. https://doi.org/10.1016/j.chemosphere.2017.11.098

Young, H. S., Dirzo, R., Helgen, K. M., McCauley, D. J., Billeter, S. A., Kosoy, M. Y., Osikowicz, L. M., Salkeld, D. J., Young, T. P., & Dittmar, K. (2014). Declines in large wildlife increase landscape-level prevalence of rodent-borne disease in Africa. *Proceedings of the National Academy of Sciences of the United States of America, 111*(19), 7036–7041. https://doi.org/10.1073/pnas.1404958111

Young, H. S., McCauley, D. J., Dirzo, R., Goheen, J. R., Agwanda, B., Brook, C., Otárola-Castillo, E., Ferguson, A. W., Kinyua, S. N., McDonough, M. M., Palmer, T. M., Pringle, R. M., Young, T. P., & Helgen, K. M. (2015). Context-dependent effects of large-wildlife declines on small-mammal communities in central Kenya. *Ecological Applications, 25*, 348–360. https://doi.org/10.1890/14-0995.1

Zaccaroni, A., Amorena, M., Naso, B., Castellani, G., Lucisano, A., & Stracciari, G. L. (2003). Cadmium, chromium and lead contamination of Athene noctua, the Little owl, of Bologna and Parma, Italy. *Chemosphere, 52*(7), 1251–1258. https://doi.org/10.1016/s0045-6535(03)00363-1

Zaccaroni, A., Andreani, G., Ferrante, M. C., Carpene, E., Isani, G., & Lusiano, A. (2008). Metal concentrations in the liver and kidney of raptor species from the Calabria region, Italy. *Acta Veterinaria (Beograd), 58*(4), 315–324. https://doi.org/10.2298/AVB0804315Z

Zhang, X. L., Luo, X. J., Liu, J., Luo, Y., Chen, S. J., & Mai, B. X. (2011). Polychlorinated biphenyls and organochlorinated pesticides in birds from a contaminated region in South China: Association with trophic level, tissue distribution and risk assessment. *Environmental Science and Pollution Research International, 18*(4), 556–565. https://doi.org/10.1007/s11356-010-0397-9

Zhong, G., Wan, F., Lan, J., Jiang, X., Wu, S., Pan, J., Tang, Z., & Hu, L. (2021). Arsenic exposure induces intestinal barrier damage and consequent activation of gut-liver axis leading to inflammation and pyroptosis of liver in ducks. *The Science of the Total Environment, 788*, 147780. https://doi.org/10.1016/j.scitotenv.2021.147780

Zimmerman, D. A., Turner, D. A., & Pearson, D. J. (1996). *Birds of Kenya and Northern Tanzania*. Christopher Helm Publishers.

Chapter 7
Urban Biogeochemistry and Development: The Biogeochemical Impacts of Linear Infrastructure

Adil Khan and Indushree Maharana

Abstract This chapter describes the role of urban biogeochemistry as a crucial development within the larger science of biogeochemistry, with increasing relevance due to the increasingly important role of urban systems and their support within the global environment. It takes the novel perspective of examining the relations between urban biogeochemistry and the related biological, hydrological, chemical, and geological sciences, especially urban biogeography and urban geochemistry. With urban biogeochemistry comprising a vital, developing subdiscipline of biogeochemistry, and human demographics, urbanization, and infrastructural developments increasingly dominating biological systems, chemical cycling, and reservoirs, new scientific developments are necessary. Linear infrastructure developments, including roads and railways, are cited as the main urban structures, as these are fundamental to the development of residential cities and link such urbanized areas across less developed landscapes. Such developments are changing landscapes through compaction, creation of impermeable and nonporous surfaces, alteration of biogeographical patterns, and geochemical cycling, especially in fragile or intensively used landcover. Issues examined in this chapter include landcover denudation, topographical change and soils and/or water flows, flooding, artificial creation of steeper slopes, pipeline, and other pollution, increased weathering and erosion, landslide denudation, and vulnerability, and the consequences for biogeochemistry, biogeography, and geochemistry, and some requirements from urban infrastructural planning. Key examples are the impacts of urban infrastructural development in fragile, mountain environments, such as that of Joshimath (Garhwal Himalayas), Uttarakhand, India, an area where urban biogeochemical impacts are pronounced, along with earthquakes, climate effects, and biogeographical changes. The findings of this chapter indicate that construction and

The original version of the chapter "Urban Biogeochemistry and Development: The Biogeochemical Impacts of Linear Infrastructure" has been revised. A correction to this chapter can be found at https://doi.org/10.1007/978-3-031-47017-2_9

A. Khan · I. Maharana (✉)
Project Associates, Wildlife Institute of India, Dehradun, India

M. O. Campbell, *Biogeochemistry and the Environment*,
https://doi.org/10.1007/978-3-031-47017-2_7

infrastructure planning must consider environmental issues, including biogeographical and geochemical impacts on the biogeochemical system, for sustainable ecological outcomes. This contributes to the investigation of the biogeochemical bases of linear development projects and urban biogeochemistry in general.

Introduction

Urban biogeochemistry, basically defined as the study of biogeochemistry and its processes applied to an urban environment, is an increasingly important part of biogeochemistry, as urban environments are greatly expanding in the lithosphere, replacing natural vegetated and even desert landscapes, and increasingly participating in the chemical cycling between the lithosphere and the hydrosphere, atmosphere, biosphere and even the cryosphere (mainly through pollution, temperature change, physical-chemical waste, and biosphere extinctions) (Campbell, 2017; De Magalhães et al., 2017; Stewart & Kennedy, 2017; Chen et al., 2018a). Lin et al. (2014: 244) note that "urban ecosystems are unique in the sense that human activities are the major drivers of biogeochemical processes. Along with the demographic movement into cities, nutrients flow towards the urban zone (nutrient urbanization), causing the degradation of environmental quality and ecosystem health." Highlighting the complexity of urban biogeochemical systems, Douglas and Philip (2014) hold that "detailed enquiries into both the flows and stocks of individual nutrients and the interactions between the addition or removal, or accumulation, of any particular nutrient and the mobility or release to the atmosphere of other nutrients, are needed to establish the true nature of urban biogeochemistry."

Urban biogeochemistry may involve all the biologically related, chemical cycling and processes that affect urban and proximate areas, and the impacts in the areas exploited for urban sustainability, and the dumping grounds for shedding industrial and socially derived resource waste (Decker et al., 2002; Gu et al., 2012; Zhi et al., 2018). Considering the vast areas of the lithosphere (including the biosphere) and the hydrosphere and atmosphere that receive such waste, and the role of the biosphere as a carbon sink for mostly urban-derived carbon, the ambit of urban biogeochemistry is vast and is positioned in a central role in modern conservation science and management (Molina & Molina, 2004; Tarsitano, 2006; Kowarik, 2011; Manning, 2013; Kennedy et al., 2015). In terms of its geochemical background, urban biogeochemistry may be based on the main tenets of urban geochemistry, which is defined as "a unique discipline that is distinguished from general geochemistry by the complex infrastructure and intense human activities associated with concentrated population centers" and the aims of which have been defined as to "elucidate and quantify the sources, transport, transformations, and fate of chemicals in the urban environment, recognize the spatial and temporal (including legacies) variability in these processes, and integrate urban studies into global perspectives on climate change, biogeochemical cycles, and human and ecosystem health" (Chambers et al., 2016: 1).

Urban biogeochemistry has also been defined as related to urban ecology, and the study of the dynamics of urban contexts: "Urban ecology is an emerging discipline that seeks to understand these interactions, and one of the grand challenges for urban ecologists is to develop models that encompass the myriad influences of people on biogeochemistry" (Kaye et al., 2006: 192). Kaye et al. (2006: 192) add that urban biogeochemistry is a comparatively new discipline and is justified, because "existing models, developed primarily in unmanaged and agricultural ecosystems, work poorly in urban ecosystems because they do not include human biogeochemical controls such as impervious surface proliferation, engineered aqueous flow paths, landscaping choices, and human demographic trends." Additionally, although most people in many countries live in urban contexts, the urban landscapes "are a frontier for ecology, and there is little previous research available to guide urban ecologists. With little data or theory of its own, it is tempting for researchers of this young discipline to borrow models designed for agricultural or unmanaged ecosystems and to apply them to the urban setting" (Kaye et al., 2006: 192).

Urban areas are increasingly important, not only because of their expanding extent but also because of their impact on global ecosystems. Kaushal et al. (2014:1) argue that "urban biogeochemical cycles have exerted a major influence on the elemental composition of the Earth's surface from local to global scales." Grimm et al. (2008: 756) point out that urban ecology integrates the natural and social sciences to study urban areas, which are "radically altered local environments and their regional and global effects" and that urban areas are "hot spots that drive environmental change at multiple scales. Material demands for production and human consumption alter land use and cover, biodiversity, and hydro systems locally to regionally, and urban waste discharge affects local to global biogeochemical cycles and climate" (Grimm et al., 2008: 756). Molina and Molina (2004) point out that megacities (defined as cities with a human population over ten million) contribute significantly to atmospheric pollution. Although historically the main air pollutants were sulfur compounds, generated by burning coal, current evidence is that "photochemical smog – induced primarily from traffic, but also from industrial activities, power generation, and solvents – has become the main source of concern for air quality, while sulfur is still a major problem in many cities of the developing world" (Molina & Molina, 2004).

A major concern of urban biogeochemistry is the comparatively rapid changes that underpin urban systems, which may develop much faster than natural systems (Campbell, 2017). For example, Kaushal et al. (2014:1) propose the concept of "urban evolution" to represent such changes and argue that "the built environment often changes quickly in response to human activities, thus contributing to an urban evolution that affects structure, function, and ecosystem services of human settlements over time. Depending upon management, these changes can result in rapid losses of ecosystem functions/services or progress towards restoration." Changes and ongoing dynamics include over the long term, urban land development and nitrogen inputs, stormwater management and watershed drainage, human-accelerated weathering in urbanized watersheds, and salinization of freshwater in urbanizing areas. Related knowledge developments for urban infrastructural and ecological change include awareness of the issues of urban watersheds, the changes in urban hydrologic cycles based on infrastructure and human water consumption, the utility of a spatially and temporarily broader urban watershed study approach, salinization in relation to long-term watershed urbanization, human-accelerated weathering of construction materials that may create "urban karst," changes in the urban watershed carbon cycles, and long-term, distinct urban biogeochemical signatures of cities.

Urban biogeochemistry and related ecology are also concerned with energy and material flows through the cities and between the cities and their environments. Decker et al. (2002: 374) note that currently, the world's 25 largest cities include 4% of the global human population but have intensive impacts at multiple scales. The knowledge of the biogeochemistry of megacities includes climatic, demographic, and economic data, including food, water flows, fuels, and construction materials. The authors hold that water dwarfs the other inputs but is not retained in the urban system. Construction materials and food are dominant materials in the urban infrastructure and the waste stream. The fuels, which change into chemical wastes, have the most significant global impacts. Additionally, the urban area consumes from the geologic, hydrologic, atmospheric, and ecological systems, through processes at local, regional, and global levels. The authors discuss the concepts of urban metabolism and urban succession, which are "organizing concepts for data collection, analysis, and synthesis on urban systems" and conclude "that megacities are not the final stage of urban evolution; rather, the climax of urban development will occur at a global scale when human society is at steady state with resource supply rates" (Decker et al., 2002: 374).

In another example, based on energy and material flows through and between cities and their environments, Kennedy et al. (2015: 5985) point out that the management of global environment problems requires a deeper understanding of such flows in the context of urban areas. It is argued that "accessing, sharing, and managing energy and material resources is particularly critical for megacities, which face enormous social stresses because of their sheer size and complexity" (Kennedy et al., 2015: 5985). Their study examined the energy and material flows through the 27 so-called megacities in the world with 2010 populations above ten million people, and their findings were based on the urban resource flows, including electricity, heating and industrial fuel use, transportation energy use, water consumption, waste generation, and steel production in relation to heating, urban design, and economics and demographics. These findings identified the more resource-efficient cities, and the relationship between per capita energy use, urbanized extent per capita, and building floor area per capita was found to be correlated. Energy use (electricity and transportation) also increased with population, gross domestic product (GDP) and ground transportation fuel use in megacities grew at approximately half the rate of GDP growth (Kennedy et al., 2015: 5985).

Another study by Stewart and Kennedy (2017: 1159) examines urban "anthropogenic heating from building energy use, vehicle fuel consumption, and human metabolism" as an important term in the urban energy budget equation. The study argues that heating from human and domestic animal metabolism is frequently excluded from urban energy budgets as relatively unimportant, and the study "quantified all terms of the anthropogenic heat budget at metropolitan scale for the world's 26 largest cities, using a top-down statistical approach" (Stewart & Kennedy, 2017: 1159). The findings of the study, interestingly, indicated that the human metabolic heat releases in mid-latitude temperate cities accounted for less (4–8%) of annual anthropogenic heating than that of high-density tropical cities (10–45%). The animal heat release was less than 1% of the anthropogenic heating in all studied cities. The findings indicated that "heat flux density from human and animal metabolism combined is highest in Mumbai-the world's most densely populated megacity-at 6.5 W m-2, surpassing heat production by electricity use in buildings (5.8 W m-2) and fuel combustion in vehicles (3.9 W m-2)" (Stewart & Kennedy, 2017: 1159). The authors concluded that their findings, and those of related global climate models, "suggest that in the world's largest and most crowded cities, heat emissions from human metabolism alone can force measurable change in mean annual temperature at regional scale" (Stewart & Kennedy, 2017: 1159).

There are also issues linked to the construction industry, which is an important factor for urban biogeochemical processes and developments, mostly through waste production and management, and its links with the wider waste production system (Grimm et al., 2008). For example, De Magalhães et al. (2017: 265) argue that construction is a major waste producer, which increases with further urban development. There is insufficient knowledge of the methods for reducing urban waste and construction waste generation in urban infrastructure projects. The authors point out that effective management for the reduction of urban waste production must include the identification of building methods and the nature of building materials, the identification of cause-effect relationships between the designs and the generation of construction waste, and a structured checklist for control and evaluation for urban infrastructure designs, aiming at the reduction of urban waste. Douglas and Philip (2014) also look at wider anthropogenically generated compounds, on the premise that "the better urban biogeochemistry is understood, the more people, householders, communities, businesses, and local and national governments can do to improve the quality of urban life, avoid disease, and enhance urban ecosystem services and sustainability." Compounds include factory and vehicle emissions, waste collection and disposal, green space irrigation, sewage disposal, and drainage (Douglas & Philip, 2014).

The urban biological environment is another important part of the urban biogeochemical system, which puts the "bio" into urban biogeochemistry. The main topics concern the biological systems (urban green spaces such as gardens, treed avenues and parks, and any associated urban wildlife) that developed in urban areas, and their role in photosynthesis, oxygen production and temperature cooling, and the circulation of chemicals and wildlife conservation (Campbell, 2017). Kowarik (2011: 1974) argues that "with increasing urbanization, the importance of cities for biodiversity conservation grows... Cities can be richer in plant species, including in native species, than rural areas." The possibilities for urban ecosystems include the existence of various species of plants and animals, including "alien species can lead to both homogenization and differentiation among urban regions" and "self-sustaining populations of rare and endangered native species" the latter being unable to "replace the complete functionality of (semi-)natural remnants" (Kowarik, 2011: 1974). The author concludes that although conservation studies may focus on these relict habitats and native species in urban areas, there is a possibility for a "paradigm shift towards considering the whole range of urban ecosystems" and the consideration of "associated ecosystem services, social benefits, and possible contribution to biodiversity conservation" (Kowarik, 2011: 1974).

Tarsitano (2006: 799) describes the problems for biological systems in an urban setting, including air, ground, and water pollution, which create imbalances and the spread of harmful pathogens, which may affect not only people but also wild and domestic animals. The author argues that urban sustainability indicators tend to overemphasize the issues concerning the human population, to the neglect of the animal population. It is argued that although "cities need good waste, water, and air management, effective traffic planning, and good zoning of businesses, crafts, and services" it must be emphasized that "over and above these activities, cities also need for planning to take into account the existence of pets (dogs, cats, etc.) and non-pet animals (insects, birds, mice, etc.)" (Tarsitano, 2006: 799). The argument is for integrated-participatory control settings for urban animals, the development of integrated management for urban animal ecology, to better assess urban biodiversity and ecosystem functioning, and the negative impacts of human activities on urban biodiversity. The focus should be on "assessing and forecasting changes in urban biodiversity, structure, function, and dynamics of urban ecosystems, with relationships among society, economy, biodiversity and habitats" (Tarsitano, 2006: 799).

Urban and proximate soils are also important in the field of urban biogeochemistry, sometimes at the interface of biological, construction, and pollution systems (Campbell, 2017). For example, Lorenz and Lal (2009) argue that with the strong increase in human populations, there may be currently little-known impacts on biogeochemical carbon and nitrogen cycles. The biogeochemical cycles in urban ecosystems may be changed directly, including alteration of the biological, chemical, and physical soil properties and processes, or indirectly, through for example, the introduction of exotic plant and animal species and atmospheric pollutant deposition, or may be affected by the urban heat island and pollution island effects. Urban soils are potential reservoirs of soil organic carbon and impact atmospheric $CO_{(2)}$ concentrations, but the quantity of soil organic carbon is affected by urban landuse and soil parent material. The authors argue that "depending on the soil replaced or disturbed, urban soils may have higher or lower SOC pools, but very little is known"; and "in view of the increase in atmospheric $CO_{(2)}$ and reactive N concentrations as a result of urbanization, urban landuse planning must also include strategies to sequester C in soil, and also enhance the N sink in urban soils and vegetation" (Lorenz & Lal, 2009). It is concluded that enhanced knowledge on this topic "will strengthen soil ecological functions such as retention of nutrients, hazardous compounds and water, and also improve urban ecosystem services by promoting soil fertility" (Lorenz & Lal, 2009).

Pouyat et al. (2002: S107) provide a supporting study, which writes that "the transformation of landscapes from non-urban to urban landuse has the potential to greatly modify soil carbon (C) pools and fluxes" but for urban ecosystems, "very little data exists to assess whether urbanization leads to an increase or decrease in soil C pools." The findings of this study of urban soils and soils from an urban–rural gradient indicated that along the urban–rural landuse gradient, soil surface organic matter varied significantly, soils from urban forest stands had significantly higher organic C densities than those from suburban and rural areas, and in a multicity sample set, urban soils, loamy fill had the highest soil organic carbon, with the lowest in clean fill and old dredge materials, and soil organic C in residential areas was consistent for the studied cities. Additionally, low-density residential and institutional land uses had higher organic C densities than commercial landuse areas. The authors argue that these findings indicate "as adjacent land-use becomes more urbanized, forest soil C pools can be affected even in stands not directly disturbed by urban land development" and "that physical disturbances and inputs of various materials by humans can greatly alter the amount of C stored in these soils" (Pouyat et al., 2002: S107). The variations in soil carbon in terms of urban landuse variation may, however, vary regionally, as evidenced in another multicity study (Pouyat et al., 2006: 1566), where the authors concluded that the findings suggested that "urban soils have the potential to sequester large amounts of SOC, especially in residential areas where management inputs and the lack of annual soil disturbances create conditions for net increases in SOC" and "suggests the importance of regional variations of land-use and land-cover distributions, especially wetlands, in estimating urban SOC pools."

Urban Biogeochemistry and Other Sciences

Urban biogeochemistry, like its parent discipline biogeochemistry, is linked to several related disciplines, which provide supports that may enhance knowledge of the urban areas under study. From biological, chemical, and geological perspectives, the most important disciplines are geochemistry and biogeography. Geomatics may emerge as an important research tool for landscape evaluation and civil engineering, and for coastal or riverine cities, there may be links to hydrology, including limnology and oceanography where urban pollution and urban-derived chemical cycling intrude into coastal and marine environments. From the social science perspective, there may be links with urban and industrial geography. The importance of these disciplines may vary according to the local environment and the human activities in the area, as well as the ameliorative actions taken to repair or avoid the negative impacts on the ecological setting. Urban biogeography and urban geochemistry are examined below as the most important sciences required for the assessment of urban biogeochemistry dynamics and impacts.

Urban Biogeography

Urban biogeography refers to the science of biogeography applied to the urban context, including those ecosystems that play a part in urban biogeochemistry (Campbell, 2020). Biogeography may be defined as the study of "all aspects of the adaptations of an organism to its environment, considering systematically the origins, migrations, and associations of living things. Hence, it aims for a synthesis of data from non-biological and biological disciplines: geology, physical geography, geomorphology, climatology, and meteorology, on the one hand; biology, taxonomy, genetics, and physiology on the other" (Dansereau, 2015). Weeks (2010: 34) defines an urban area as "a place-based characteristic that incorporates elements of population density, social and economic organization, and the transformation of the natural environment into a built environment." The subdiscipline of urban biogeography therefore applies the definition of biogeography into that of urban areas, which is vital for the understanding of urban biogeochemistry because such biological systems are important agents for chemical cycling, such as carbon and nitrogen in urban contexts (Schwab, 2009; Churkina, 2012; Gong et al., 2023; Leng et al., 2023). Urban biological systems, commonly called green spaces, have become important urban landscape features, variants including large greenbelts and greenhearts, and smaller treed avenues, parks, and gardens, especially with pronounced urban expansion and recreation and conservation needs from stakeholders (Park & Lee, 2000; Kuhn, 2003; Morancho, 2003; Savard et al., 2000; Chiari & Seeland, 2004; White et al., 2005; Konijnendijk, 2017).

Pataki et al. (2011: 27) give the example of the role of urban green spaces in offsetting greenhouse gas (GHG) emissions, removing air and water pollutants, cooling local climates, and improving public health. The role of such green spaces in biogeochemical dynamics is however debatable, as "in some cases the environmental benefits of this infrastructure have been well documented, but they are often unclear, unquantified, and/or outweighed by potential costs" (Pataki et al., 2011: 27). The authors note that "quantifying biogeochemical processes in urban green infrastructure can improve our understanding of urban ecosystem services and disservices (negative or unintended consequences) resulting from designed urban green spaces" and suggest a framework that integrates the knowledge of the biogeochemical processes into the design and development of infrastructure, and give examples for greenhouse gas emission, reduction of stormwater runoff and improve air quality (Pataki et al., 2011: 27).

Reisinger et al. (2016) cite the example of nitrogen cycling in urban contexts, arguing that "nitrogen (N) pollution of freshwater, estuarine, and marine ecosystems is widespread and has numerous environmental and economic impacts" and that some of this nitrogen originates from natural and engineered ecosystems in urban watersheds which can change the downstream export of nitrogen compounds. The study argued that current research on urban nitrogen cycling frequently focuses on "specific ecosystems or on watershed-scale mass balances" but that "comparisons of specific N transformations across ecosystems are required to contextualize rates from individual studies" (Reisinger et al., 2016). Therefore, their study examined urban nitrogen cycling in terrestrial, aquatic, and engineered ecosystems, which were then compared with native reference ecosystems. The findings indicated that the net nitrogen mineralization and nitrification rates were more enhanced in the urban forest and riparian zones compared to the reference ecosystems. The findings also indicated that denitrification rates varied in the urban ecosystems, but there were no significant differences between the urban and reference rates. For urban streams, it was found that ammonium uptake was faster than nitrate uptake, and reduced stormwater runoff with potentially lowered nitrogen concentration "suggests that green infrastructure may reduce downstream N export" (Reisinger et al., 2016). The authors concluded that "despite multiple environmental stressors in urban environments, ecosystems within urban watersheds can process and transform N at rates similar to or higher than reference ecosystems" (Reisinger et al., 2016).

Another study examined carbon and nitrogen controls along riparian zones on an urban-to-rural gradient (Waters et al., 2014: 955), noting that this is important, because "denitrification is an anaerobic microbial process that transforms nitrate (NO) to nitrogen (N) gas, preventing the movement of NO into coastal waters where it can lead to eutrophication". Additionally, "urbanization can reduce the potential for denitrification in riparian zones and streams by altering the environmental conditions that foster denitrification" which include low oxygen and available carbon. The case study was of factors limiting denitrification potentials in forest and herbaceous vegetation along riparian and stream pool debris landcover, along urban to rural watersheds in a temperate ecosystem (Baltimore, Maryland). In June and November, the parameters measured were denitrification with carbon and nitric oxide additions, denitrification without such additions, the microbial biomass carbon and nitrogen content, the potential net mineralization and nitrification, the microbial respiration, and inorganic pools. The main findings were that denitrification potentials were highest in the herbaceous riparian soils and lowest in pool sediments, and the highest forested riparian soil denitrification potential was in the exurban watershed, in there was such in other by watershed type. Generally, most variables were higher in the June sample than in November. Carbon was also found to be stronger than nitrogen as a driver of denitrification potential. For the carbon and nitric oxide additions, the denitrifications were similar in the unamended and nitrogen treatments but were lower in the carbon amended and carbon plus nitrogen amended treatments. The authors concluded that "efforts to enhance denitrification in urban watersheds need to focus on the differential controls of denitrification across habitats, urban land use types, and seasons" (Waters et al., 2014: 955).

In a study on carbon, (Zhang et al., 2022) postulated that human activities can reduce or improve the storage of soil organic carbon, arguing that "urban expansion results in an initial loss of soil carbon, but long-term SOC changes during urban development are poorly understood." Their study was based on soil organic carbon variations in urban and suburban areas in several cities in China. The findings indicated soil organic carbon increased in highly urbanized areas and decreased in suburban areas, with the carbon changes being positively correlated with vegetation coverage and productive change, and these vegetation changes resulted from human activity changes and climate variation. The urban topsoils with increased vegetation were found to act as carbon sinks. The authors concluded that this knowledge would assist in the understanding of the role of the soil organic carbon content in cities, within the global carbon cycle.

Urban vegetation also protects the soil from erosion and therefore preserves the soils which are major pools of chemical elements and compounds (Morrow et al.,

2017). Ferreira et al. (2021: 481) cite the vulnerability of urban soils, which is acute because "urban areas comprise a variety of impervious and pervious surfaces leading to complex spatiotemporal rainfall-runoff processes." Such paved surfaces reduce water infiltration and increase rainfall-runoff, increasing flood risk, dependent on other factors such as "soil properties, landscape structure, and urban patterns" (Ferreira et al., 2021: 481). Long-term runoff impacts include soil erosion and degradation, movement of pollutants into water bodies that affect water quality, and in some cases strains on artificial drainage systems. The authors concluded that "with increasing urbanization and negative impacts of climate change, the status of aquatic ecosystems is expected to deteriorate, and thus, a paradigm shift on water management, focusing on nature-based solutions, is receiving increasing interest to achieve more sustainable development" (Ferreira et al., 2021: 481).

Urban Geochemistry

Urban geochemistry or urban environmental geochemistry, as the basis of urban biogeochemistry, is a subject termed by Thornton (1991) to describe the subject matter between geochemistry and urban pollution (Thornton, 2012; Chambers et al., 2016). Wong et al. (2006) define urban environmental geochemistry as "a field of scientific study that uses the chemistry of the solid earth, its aqueous and gaseous components, and life forms to examine the physical, chemical, and biological conditions of an urbanized environment" (see also Siegel, 2002). Thornton (1991) wrote that urban geochemistry "is concerned with the complex interactions and relationships between chemical elements and their compounds in the urban environment, the influence of past and present human and industrial activities on these, and the impacts or effects of geochemical parameters in urban areas on plant, animal and human health."

Urban geochemistry or urban environmental geochemistry provides the geological and chemical background of biogeochemical cycling in urban areas, the latter emerging when biological organisms interact with geochemical cycling (Wong et al., 2006; Campbell, 2020). Wong et al. (2006: 1) argue for the importance of urban geochemistry, pointing out that "the development of urban environmental geochemistry has yielded a significant volume of scientific information about geochemical phenomena found uniquely in the urban environment, such as the distribution, dispersion, and geochemical characteristics of some toxic and potentially toxic trace metals." Urban geochemistry may also be relevant to limnology and oceanography, the disciplines studying inland water and oceans, respectively, if urban-originated chemicals flow into waterbodies or enter the atmosphere and fall into water bodies, as an aspect of global geochemical cycling (Holland & Turiekian, 2014).

Long and Berry Lyons (2021: 235) discuss how urban geochemistry is at the "interface of environmental science and low-temperature geochemistry" and focuses on the "influence of urbanization on Earth's systems and subsequent impact on human and ecosystem health and the sources, transport, transformations, and fate of chemical constituents in urban environments." The authors explain that the urban geochemical environment can be more difficult than nonurban environments to study, "because of the complicated interactions of the various types of media, sources, transport, pathways, and inorganic and organic chemicals" (Long & Berry Lyons, 2021: 235). Four examples of current urban geochemical research are cited: (1) lead in the environment from smelters, leaded gasoline, and other sources; (2) vehicular pollutants in runoff, including petroleum and combustion products, and metals from vehicular parts; (3) salts, including from halite (NaCl) from road de-icer; and (4) new pollutants such as nanoplastics, pesticides, and pharmaceuticals. The combination of these geochemical developments and increased human urban population creates problems, the solution of which may require an interdisciplinary effort with collaboration between "environmental geochemists, ecologists, medical/public health professionals, engineers, politicians, and city planners" (Long & Berry Lyons, 2021: 235).

Gao et al. (2021) give a study of the types of trace metals that occur in urban areas, noting that "due to the rapid urbanization process, the consumption of trace and rare earth elements has dramatically increased" and examined the "urban geochemistry, source, and anthropogenic responding factor for 15 trace elements (Cd, Pb, Co, Sn, Cu, Ni, V, As, Mo, Sb, Al, Li, Fe, Zn, and Sr) and rare earth elements in surface water in a case study of Suzhou city." The findings indicated human action influences Cd, Co, Sn, Ni, As, Li, Fe, and Sr presence, and Pb, Cd, Gd, Li, Sr, Co, Fe, Ni, and Sn are influenced by urban sewage and industrial processes, while Zn, Cu, and Al are influenced by geologic sources, and V, Mo, As, and Sb) are influenced by agricultural and urban activities. The authors concluded that human activities are major determinants of the presence of trace elements (Gao et al., 2021).

Lead is a common element in urban geochemical systems, which is important due to the risk to biological systems (Campbell, 2017). For example, Wade (2021: 7981) examined the effect of urban soil pedogenesis on soil lead contamination from paint and gasoline in Durham, North Carolina. Soil samples were taken from "street sides, residential properties, and residual upland and floodplains" with strong variations in lead levels (from 6 to 8825 mg/kg, half between 50 and 200 mg/kg) (Wade, 2021: 7981). The findings indicated the highest lead concentrations were within one meter of residential foundations built before 1978, and the lead concentrations were negatively correlated with house age. The street side lead concentrations were higher than the lead in natural geology (which was <30 mg/kg), were correlated with the traffic flow, and were also lower than the Durham streetside soil samples from the 1970s. The latter result was attributed to urban pedogenesis, defined as the combination of natural and human processes that contribute to temporal change in soil constituents. The findings also indicated increased soil erosion redistributed old lead-containing sediments, with evidence of sedimentation rates under 4 mm annually. These redistribution processes on soils with high lead levels reduced the soil lead levels and were possibly more significant on street sides than soils near foundations. The authors conclude that "the development of an urban-pedogenesis framework can help guide public health approaches to Pb exposure by incorporating pedogenic processes that reduce and dissipate soil Pb contamination" (Wade, 2021: 7981).

These urban environmental changes are crucial for the changing urban biogeochemistry and management of many contexts, as biogeography and geochemistry form the backbone of urban biogeochemistry, and such changes are usually the result of urban infrastructural development. The rest of this chapter will examine the case study or urban linear infrastructural development from a biogeochemical perspective, supported by relevant literature a case study. The justification for the selection of a case study of linear infrastructure development lies in the major impact these developments have on the environment. For example, Wu and Li (2022) comment that "there is mounting evidence showing that linear infrastructures can increase habitat loss and fragmentation, landslide susceptibility, wildlife-vehicle and train collision, poaching, and illegible harvesting" and that "and intensity of the negative impacts of linear infrastructures are determined by various characteristics such as adjacent landscape, topography, hydrography, vegetation type, habitat quality" (see also Van der Ree et al., 2015; Kirschbaum & Stanley, 2018; Claireau et al., 2019). These issues cover the points raised above, concerning urban biogeography, urban geochemistry, and biogeochemical cycling.

Urban Biogeochemistry and Linear Infrastructure Development

Urban linear infrastructure development is among the most important issues to be studied in urban biogeochemistry, as transport and related networks strongly affect urbanized and nearby landcover, including fragile, rural contexts such as mountainous areas (Kaye et al., 2006; Raiter et al., 2018). Urban linear infrastructure development refers to the building, maintenance and landscaping of roads, railway tracks, pipelines, transmission lines, water canals, bridges, underground tunnels, communication lines, ropeways, etc., which may all be classified as linear infrastructure, and all of which may modify landcover to the extent of biogeochemical, biogeographical, ecological, and hydrological impacts (Raiter et al., 2018; CMS, 2020). As "bio-geo-chemistry" is the study of the interaction of biological, geological, chemical, and physical processes that shape natural environments over time and space (BIOGRIP, 2019), the main issue concerns how these developments may affect the vegetation and other biogeography, and the geochemical structures, namely the water flow possibilities, and soils that influence the biogeochemical cycling. Linear infrastructure development occurs through the expansion of urban and suburban areas, resulting in the increased size of the transportation network, and expanded energy distribution, mining, and logging activities, etc., to meet the needs of the growing human population (Laurence et al., 2009; Duniway & Herrick, 2011). Currently, few studies exist on the effects of linear infrastructure on the environment. Raiter et al. (2018: 446) note that "the extent of roads and other forms of linear infrastructure is burgeoning worldwide, but their impacts are inadequately understood and thus poorly mitigated," even though there are "many potential impacts, including alterations to the hydrological functions and soil processes upon which ecosystems depend." Studied topics include the effects of roads and railway lines on local ecosystems, including wildlife habitat fragmentation, direct wildlife mortality, noise pollution, and disruption of natural soil and vegetation systems, and on soil hydrology (Duniway & Herrick, 2011; Raiter et al., 2018).

Though the effects may not seem as obvious, studies have shown that even railways, which are thought to be a comparatively cleaner transportation option than roads, contribute a significant amount of polycyclic aromatic hydrocarbons (PAHs) and heavy metals into the environment (Wiłkomirski et al., 2011; Mętrak et al., 2015). Few studies have assessed the impacts of pipelines and even fewer on transmission lines. Linear infrastructure alters landscape ecological settings, including natural soil patterns and hydrology (Duniway & Herrick, 2011). In addition to the direct impacts, there are many indirect effects that may lead to cascade effects, further exacerbating the prevalent conditions of a given region (Duniway & Herrick, 2011; Raiter et al., 2018). Exotic species invasions are common in disturbed environments, contributing to changed ecological dynamics and processes for a given region (Laurence et al., 2009; Lorenz & Lal, 2009). Figures 7.1, 7.2, 7.3, 7.4, 7.5, 7.6, 7.7, 7.8, 7.9, 7.10, and 7.11 show urban developments.

The negative effects of linear infrastructure development may be categorized as follows:

- *Direct effects*, which are visible in the immediate vicinity of the structure, that is, physical area of disturbance. These effects have been studied in detail in previous reviews (Duniway & Herrick, 2011). The soil surface is scraped and soil horizons are disturbed, the topography is altered, and there are changes in hydrology, which affect the soil chemistry cycling, and biological systems (Laurence et al., 2009; Duniway & Herrick, 2011) and soil ecosystems (Lorenz & Lal, 2009; Raiter et al., 2018). Other impacts are physical barriers to wildlife movement, noise pollution and visual disturbances on animals, and habitat fragmentation by

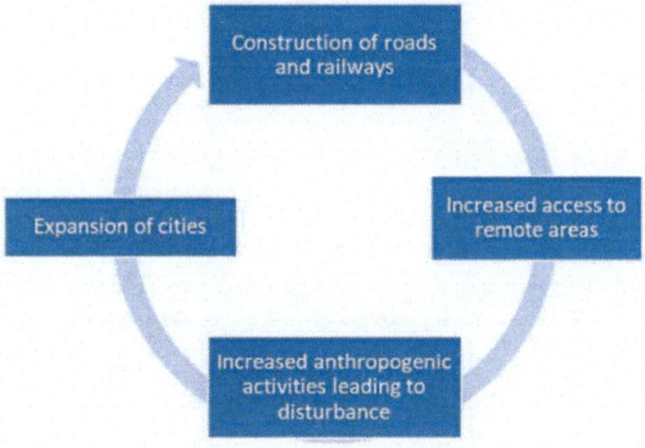

Fig. 7.1 The cycle of linear infrastructure expansion and its effects (Laurence et al., 2009)

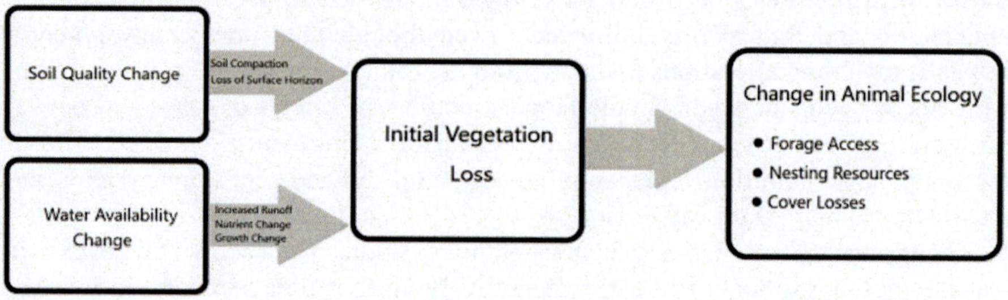

Fig. 7.2 An example of indirect effects of linear infrastructure (Duniway & Herrick, 2011)

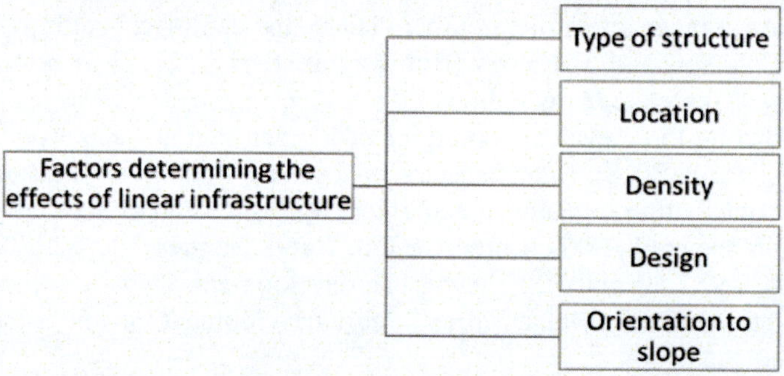

Fig. 7.3 Factors determining the type and degree of impacts of linear infrastructure (Raiter et al., 2018)

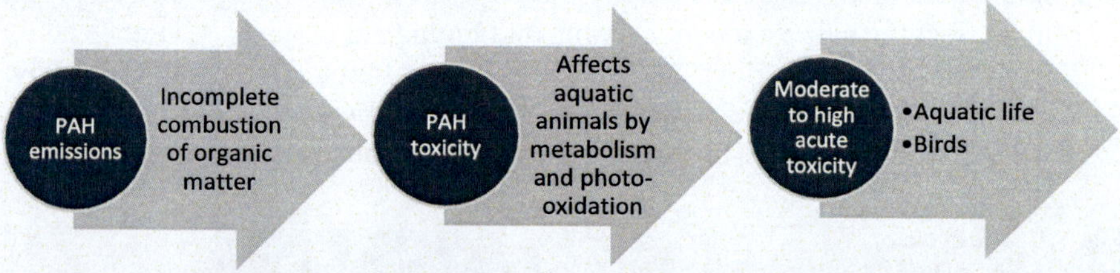

Fig. 7.4 Source and effects of PAHs (Abdel-Shafy & Mansour, 2016)

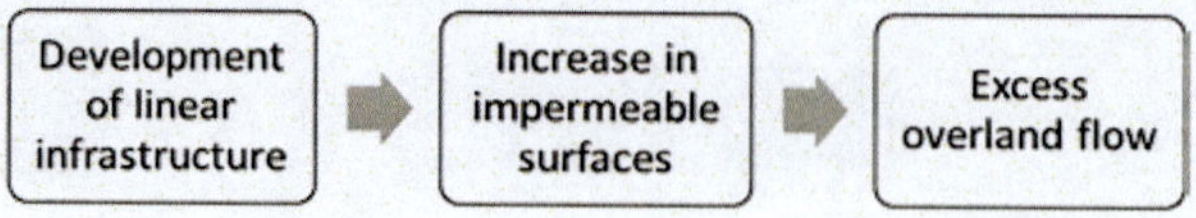

Fig. 7.5 Linear infrastructure and its effect on the surface hydrology (Laurence et al., 2009; Raiter et al., 2018)

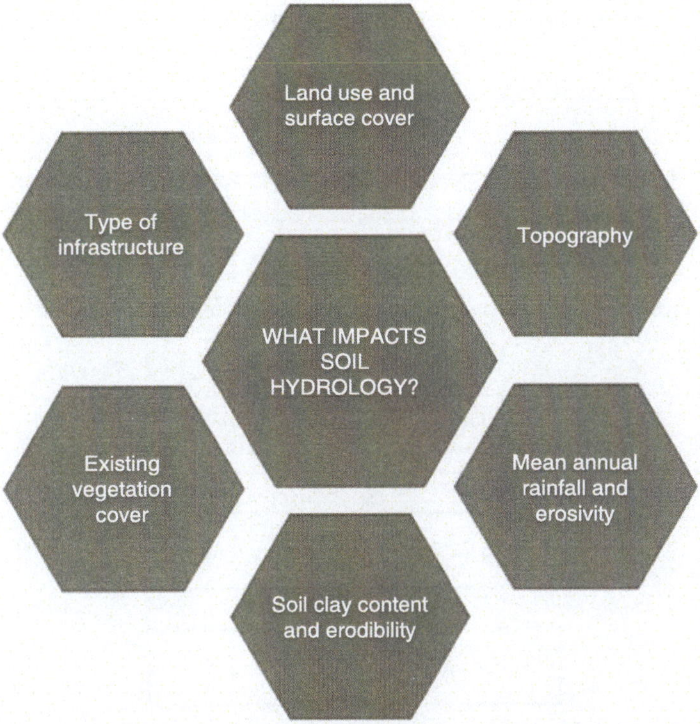

Fig. 7.6 Factors impacting soil hydrology in presence of linear infrastructure (Raiter et al., 2018)

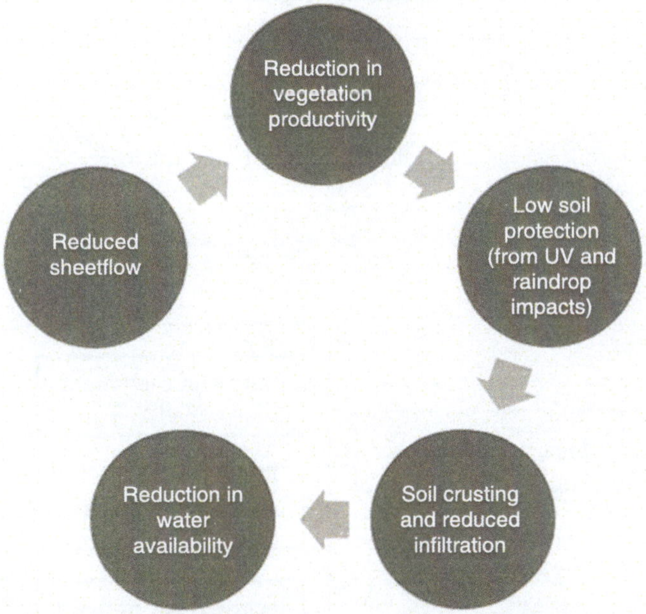

Fig. 7.7 Synergistic interactions leading to major hydrological issues because of impermeable linear infrastructure development (Laurence et al., 2009; Raiter et al., 2018)

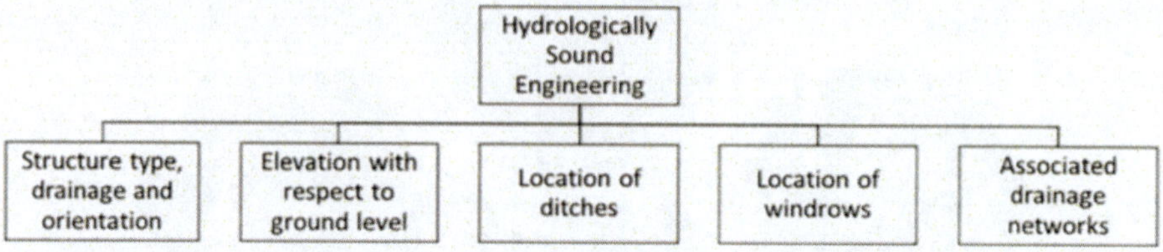

Fig. 7.8 Parameters to be considered for hydrologically sound engineering of linear infrastructure (Raiter et al., 2018)

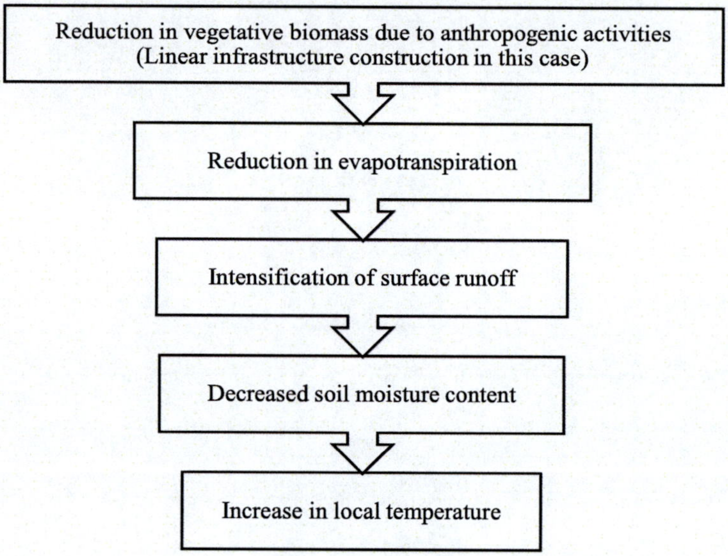

Fig. 7.9 Flowchart depicting cascade effects resulting due to anthropogenic activities (Covey et al., 2021)

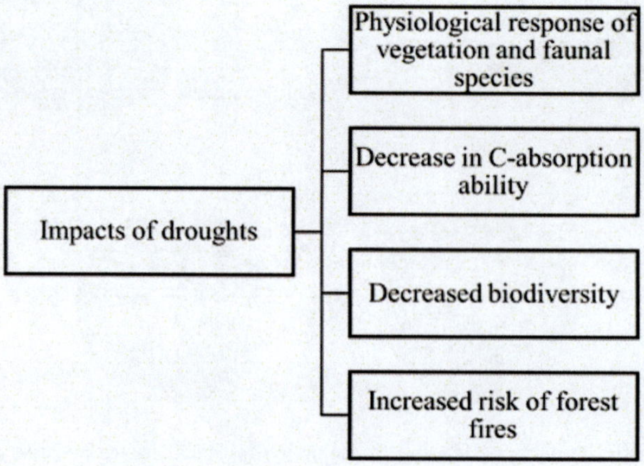

Fig. 7.10 Impacts of droughts on the ecosystem

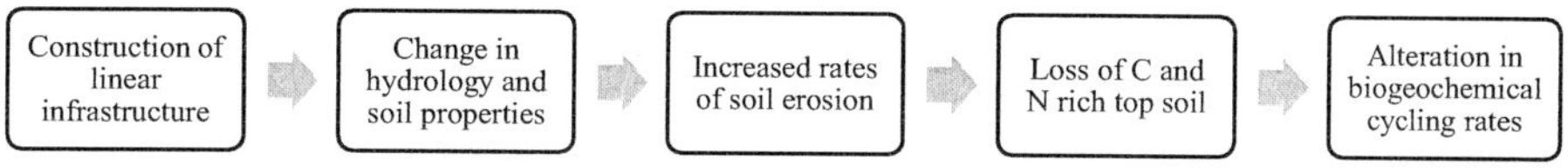

Fig. 7.11 Effects of linear infrastructure on biogeochemical cycling of C and N in soil (Lorenz & Lal, 2009)

roads, or other structures, which depending on the behavioral adaptations of the animals, reduces/conduces animal movement. Some animals become habituated to the disturbance and start using these structures. For example, in one case, the female greater sage grouse (*Centrocercus urophasianus*, Bonaparte, 1827) showed increased utilization of winter breeding grounds with increasing density of natural gas wells and associated roads (Duniway & Herrick, 2011).
- *Indirect effects*, which include process feedbacks and those impacts that cannot be seen in the immediate physical vicinity of the structure. Synergistic interactions and feedback loops may exacerbate these issues. Unlike initial or direct effects, which may be immediately evident, the indirect effects may be present several years after the development of the project. This makes it more difficult to quantify and study these effects (Duniway & Herrick, 2011).
- *Opportunistic effects*, which arise due to certain conditions at a particular time. For example, in arid and semiarid regions, droughts maybe intensified when coupled with climate change. Further, due to changes in soil moisture and hydrology, floods may intensify, increasing runoff and changing chemical cycling (Duniway & Herrick, 2011; Rong et al., 2022).

The introduction of nonnative plant species in an environment is a common issue when it comes to linear infrastructure development (Borda-de-Água et al., 2017). Nitrogen and carbon cycling, primary productivity, and other biogeochemical cycles can be altered because of developmental activities (Kaye et al., 2006; Smith et al., 2017).

Biogeochemical processes and cycling may be altered because of transportation and the transformation of energy and matter (Lorenz & Lal, 2009; Lin et al., 2014). The energy expenditure of an urban area is 10 to 100 times more than the unmanaged, unaltered ecosystems. This energy and material input is solely external. The ecological footprint of a city is the total land area required to meet the needs of its population (considering both waste accumulation and assimilation). This depends on the biophysical setting, urban human population size, and the per capita consumption rate. Human activities, mostly termed developmental activities, may alter the parameters listed as follows (Kaye et al., 2006):

- Hydrological settings, including water flow that may alter chemical cycling.
- Atmospheric chemistry – due to elevated levels of CO_2, NO_x, O_3, organic aerosols and heavy metals which effects plant growth, carbon, and nitrogen cycling.
- Climatic conditions – by Urban Heat Island (UHI) effect and irrigation which alters the evapotranspiration and other biogeochemical cycling rates.

- Nutrient quantities – due to food and fertilizer cycling which further alters plant growth, hydrology, and atmospheric chemistry.
- Vegetation, landuse changes, the plant composition and ultimately, the ecosystem function, and the biogeochemical stores that plants provide.

In any urban environment, carbon sequestration is lower, but there are larger carbon imports. Biomass reduction leads to higher CO_2 contributions. Cities also serve as sinks for elements like nitrogen, phosphorous, and metals (Rindy et al., 2019; Ponette-González et al., 2022). The concentrations of nitrogen, carbon, aerosols, metals, and ozone are higher in urban areas (Grzywa-Celińska et al., 2020; Manisalidis et al., 2020). The alteration of biogeochemical cycles occurs because of several factors, one of which is the reduction of ground-level short-wave radiation (approximately 20%) (Kaye et al., 2006; Kinnunen et al., 2022).

Focus: Pipelines and Their Effects

Pipelines are crucial for the transportation and distribution of goods (frequently crude oil and gas) across vast distances (Hendrick et al., 2016; Weller et al., 2022). Their cargo may range from inert substances like water to harmful chemicals, inflammable substances like crude oil and natural gas. Like most other infrastructure, there may be various immediate environmental impacts during and after construction. Leaks and spillages also occur and are major issues that adversely affect the surrounding environment depending on their intensity (Phillips et al., 2013; Jackson et al., 2014; Chamberlain et al., 2016; Weller et al., 2022). Major aftereffects of spillages/leaks are listed as follows (Eganhouse et al., 1994):

(a) Changes in soil biological composition and processes
(b) Formation of a layer on water surfaces (in case of hydrophobic substances)
(c) Contamination of water sources due to dissolution of substances

Based on the chemical properties of the contaminant substance in question, several geochemical layers may be formed in open water bodies (Eganhouse et al., 1994). In Chennai, Tamil Nadu, India, a defunct pipeline leakage contaminated water and soil in 2022. Remediation work was suggested by the authorities to repair the damage caused to the environment because of the leakage (Special Correspondent, 2022). Pipelines, despite being a very efficient mode of transportation of liquid and gas cargo, especially and harmful and combustible materials, may fail with catastrophic consequences (Dey, 2002; Chamberlain et al., 2016; Weller et al., 2022). These structures require regular maintenance by surveys, for early detection of issues. Early detection of leaks and corrosion and proper maintenance of these structures will help in averting disasters (Medverec Knežević et al., 2011; Ramírez-Camacho et al., 2015; Wang et al. 2021a, b). Like any other structure, pipelines are constructed to be fail-proof, but issues like corrosion, pilferage, and other activities (third-party external interference) can result in significant losses (Dey, 2002). For pipeline construction, important considerations to minimize damage and maximize profits include the following:

1. Route efficiency – Avoid populated areas to minimize the damage in case of any untoward incidences. This can be achieved by tools like Analytical Hierarchy Process (AHP), developed by Saaty, a form of multicriteria decision-making process, integrated in the GIS environment (Dey, 2002).
2. Materials used for the pipeline construction and the substance that needs to be transported must be scientifically chosen to ensure minimum damage (like corrosion and adverse reactions).
3. Avoid hilly/rocky terrains and forests (basically any obstacles that make it difficult for construction, subsequent maintenance and has more potential for causing damage) as much as possible to prevent ecological and environmental damage and to make the pipelines easily accessible for maintenance (Dey, 2002).

Developmental infrastructure construction requires profit maximization; however, insufficient planning has occasioned losses (Ramírez-Camacho et al., 2015). Similarly, for pipeline construction, minimal route distances, approachability, and constructability must be considered, as well as factors like operability, room for augmentation and maintainability (Dey, 2002). Even after their decommissioning, pipelines can be harmful to the environment, as chemical elements like mercury may be present, which may affect organisms (Gissi et al., 2022).

Effects of Linear Infrastructure Development

Pollution

National development generally requires expansion of the transport sector (roads and railways) for economic exchanges, with attendant increases in adverse biotic and xenobiotic impacts (Pataki et al., 2011). Phillips et al. (2021) write that "roads form vast, pervasive and growing networks across the Earth, causing negative environmental impacts that spill out into a 'road-effect zone'," with some pollutants pervading the land surface, and based on their research findings, "whilst the greatest levels of road pollution are relatively localized around the busiest roads, low levels of road pollution (which may be ecologically significant) are pervasive." Pollution is a common result, with increases in the occurrence of xenobiotics, which include those substances which are not naturally found in an animal's body like heavy metals, cosmetics, plant constituents, pesticides, drugs, and environmental pollutants (Patterson et al., 2010; Campbell 2017).

Herbicides are also commonly used in the vicinity of linear infrastructure to maintain connectivity across the structures (Wilkomirski et al. 2011). These compounds may pose serious threats to the soil biota as well as to tertiary consumers, as some artificial compounds may undergo biomagnification and bioaccumulation

(Annett et al., 2014; Gustafson et al., 2015; Yang et al., 2021; Lechinovski et al., 2022). Pollutants are generally distributed to the periphery of these structures or into water bodies through surface runoff. Chemical pollutants, including heavy metals, nutrients, ozone, and organic molecules may have higher concentrations near roads (Van Wijnen & van der Zee, 1998). Lead, mostly due to the use of leaded gasoline, and de-icing salts (mostly in developed countries) harm roadside vegetation and alter soil and water chemistry in the immediate vicinity of roads (Meyer et al., 2008; Angrand et al., 2022). Such settled pollutants may be dispersed during heavy rains, entering water bodies and wetlands (Walaszek et al., 2018). Some of these nutrients may lead to eutrophication in open water bodies, while heavy metals may undergo biomagnification (Laurence et al., 2009; Lorenz & Lal, 2009).

Railway and road transport are associated with the release of both heavy metals and polycyclic aromatic hydrocarbons (PAHs), both being major issues of concern for environmental management (Romagnoli et al., 2017; Gbeddy et al., 2020; Kumpiene et al., 2021). PAHs are a subset of the group Polycyclic Organic Compounds (POCs). PAHs are among the commoner sources of organic pollutants from the road and railway sector, and the increased use of railway infrastructure, has increased the concentration of heavy metals and PAHs (Abdel-Shafy & Mansour, 2016). Road construction makes use of asphalt, which is a source of PAHs (Campo et al., 2006; Mo et al., 2019). Motor vehicle exhaust is also a major cause of emission of PAHs. Incomplete combustion of fossil-derived derived fuel can lead to the release of PAHs and has been identified as the highest anthropogenic contributor to their concentration in the environment. Srogi (2007: 169) notes that "PAHs are widespread environmental contaminants resulting from incomplete combustion of organic materials. The occurrence is largely a result of anthropogenic emissions such as fossil fuel-burning, motor vehicle, waste incinerator, oil refining, coke and asphalt production, and aluminum production, etc." Vehicular exhaust emissions, though not significant, are also point sources of PAH emissions (Wilkomirski et al., 2011; Mętrak et al., 2015; Abdel-Shafy & Mansour, 2016; Miura et al., 2019).

The concentration of PAHs in the atmosphere also increases with urbanization, and therefore, these compounds are commonly detected around urbanized transport network areas, in much higher concentrations than in undisturbed rural areas (Lighty et al., 2000; Bekki et al., 2012; Abdel-Shafy & Mansour, 2016; Miura et al., 2019). Greater amounts of atmospheric dust are positively correlated with the concentration of PAHs in any area. These are then deposited on the soil surface and are then bound to the soil particles and sediments. Storm water runoff and sanitary sewer effluents are also major contributors for the addition of PAHs into the soils and the water bodies. Certain PAHs also have the ability of bioaccumulation. However, they also have a high bio-transformation potential, which makes bioaccumulation rare. Long distance transportation of these PAHs also pollutes areas away from the source. These are then taken up by crop plants and introduced into food chains (Abdel-Shafy & Mansour, 2016). Issues with emissions and toxicity are shown in Fig. 7.4 and Table 7.1.

Roads are also sources of microplastics, which are defined as "particles from tires, road markings and polymer-modified bitumen... all released through the wear

Table 7.1 Major pollutants and their sources in the railway transportation sector

Polycyclic aromatic hydrocarbons (PAHs)		Heavy metals	
Source	PAH	Source	Metals
Petroleum products used for rail tracks, rolling stock conservation and maintenance	Lubricants and condenser fluids, rail ties, coal combustion and from goods like fertilizers, harmful chemicals, and metal ores	Brakes	Fe, Cu, Mn, Cr, Pb
		Rails	Fe, Mn, Cr
		Wheels	Fe, Mn, Ni, V
		Galvanization	Zn, Cd

Sources: Wilkomirski et al. (2011), Mętrak et al. (2015)

and tear of tires and the road surface" (Rødland et al., 2020). Other components are from thermo-plastic paints on roads (Vogelsang et al., 2019; Kim et al., 2020). Wind and water action deposit these microplastics into the topsoil and soil matrix and alter the structure and physical properties of the soil. Wang et al. (2022) note that microplastics "occur widely in all surveyed soil types, such as agricultural soils, industrial soils, urban soils, and unused soils, but show variation in their abundance, type, shape, and size. In most cases, MPs can change soil physical, chemical, and microbiological properties, but the effects vary, and are dependent on polymer type, shape, dose, and size." Urban systems are also sources of nutrient pollution for receiving waters (Kaye et al., 2006). Exposure of tree saplings to higher ground-level ozone concentrations in urban areas may result in a slower growth rate. This occurs due to warmer temperatures and nitrogen enrichment in the core (Kaye et al., 2006). Fossil fuel consumption by vehicles serves as an important source of reactive nitrogen species in the atmosphere (Lorenz & Lal, 2009).

Effect on Hydrology

The hydrological and hence geochemical setting of a landscape is greatly altered due to the construction of linear infrastructure, such as roads (Raiter et al., 2018). Such water flows may alter the chemical movements, impacting on biogeochemical systems based on local ecosystems (Campbell, 2020). For example, in slopes, roads can capture overland sheet flows, and decrease downslope flows, due to surface changes. Downslope plant communities are impacted, due to decreases in water availability, creating water starvation and those upslope due to inundation (Laurence et al., 2009; Duniway & Herrick, 2011; Raiter et al., 2018). Surface hydrology is greatly altered by impermeable linear infrastructures like road surfaces. Pooling and erosion occur 5–6 times more on roads as compared to areas devoid of infrastructure. Simultaneously, groundwater levels are affected by the wider extent of impermeable surfaces and greater flow rates, resulting in lower rates of percolation (Raiter et al., 2018). Changes in sedimentation rates in water bodies may also contribute to changes in the biogeochemical cycling rates associated with the aquatic environment due to an imbalance in the respiration and photosynthesis rates (Covey et al., 2021).

Railway tracks and their supporting ground structures may also contribute to flow impedance, with increased erosion severity. For example, water movements may be impacted (~98% of ephemeral streamlines in one study), especially in arid and semiarid regions (Raiter et al., 2018). Raiter et al. (2018: 446) commented that "erosion severity was greater in the presence of tracks, and 98% of crossings of ephemeral streamlines showed some evidence of impact on water movement (flow impedance (62%); diversion of flows (73%); flow concentration (76%); and/or channel initiation (31%)." Affecting variables included "infrastructure type, pastoral land use, culvert presence, soil clay content and erodibility, mean annual rainfall, rainfall erosivity, topography and bare soil cover" which affected the frequency and severity of the impacts (Raiter et al., 2018: 44). The severity of the effects was directly proportional to the density of the structures. The authors concluded that linear infrastructure affects "ephemeral stream flows and intercepts natural overland and near-surface flow" thereby changing the moisture regimes (some areas becoming wetter, some drier, with increased erosion, soil loss, and degradation in some areas (Raiter et al., 2018: 446). See Figs. 7.5 and 7.6 for some impacts.

Another study (Kaye et al., 2006) found that impervious structures like roads alter surface hydrology, increase surface runoff, and decrease the infiltration by vegetation, resulting in altered infiltration to runoff ratio and the impacts listed as follows:

- Reduction in the influence of soil and plant processes on water chemistry
- Reduction in the stream baseflow
- Increased stream temperatures
- Lowered water tables (especially in riparian areas which help in nutrient removal between aquatic and terrestrial environments)

Major hydrological impacts of linear infrastructure can be broadly grouped as follows:

(a) *Increased runoff*, resulting in chronic erosion along the infrastructure and its upslope and downslope areas. Excess flows may breach road edges and increase the flow downslope due to channel initiation. The roadway depressions cause pooling and water enters the subsoil profiles. This leads to minor instability and landslides (Laurence et al., 2009; Raiter et al., 2018).

(b) *Altered stream function/flow*, leading to alteration of the functional health of stream and its stability and facilitates soil organic matter movement. This organic matter flows into streams downslope and alters the flow regime, as well as the composition of the water (Raiter et al., 2018; Kaye et al., 2006).

(c) *Downslope starvation*, from the impermeable substances used in construction of linear infrastructure and change in topology, which reduce the water flow downslope. This will result in downslope starvation of water bodies, eventually affecting the vegetation communities downstream (Laurence et al., 2009; Raiter et al., 2018).

Water availability may increase, due to linear infrastructure development, which may change the near-road vegetation flowering times and result in species hybridization, herbivory, and in increased mortality due to collisions (Raiter et al., 2018). In slopes, however, there may be downslope starvation and upslope inundation. Both result in reduced viability of the habitat for species dependent on water and diminished stream condition (Laurence et al., 2009; Raiter et al., 2018). Downslope starvation, and climate change effects (global warming), may exacerbate water scarcity over the long term, especially in arid and semiarid regions (Raiter et al., 2018). Increased density of linear infrastructure may lead to increased access to previously inaccessible areas, which may result in reduced vegetation cover due to timber harvesting. This results in higher flow intensity and increased soil loss and erosion (Raiter et al., 2018). See Fig. 7.7 for some impacts.

Better assessment, minimization (wherever possible), monitoring, and mitigation must be practiced to offset the adverse effects of linear infrastructure on hydrology. Mitigation must be targeted only in those areas which have the greatest potential for recovery, while minimization must be done in areas which have the greatest potential to be affected (Duniway & Herrick, 2011). Impact prediction allows the formulation of mitigation strategies. Mitigation and minimization may be practiced at local (the reduction of impacts of existing and new linear infrastructure) and regional scales (limiting the expansion of the infrastructure into ecologically sensitive areas) (Laurence et al., 2009). At the planning stage, risk mapping (i.e., identification of parts of the landscape which potentially have the greatest hydrological impacts) may be attempted. At the engineering stage hydrologically, considerate protocols may be developed to maintain water flow and minimize impedance (Raiter et al., 2018). See Fig. 7.8.

Some more points, listed as follows, may be considered for the mitigation of hydrological impacts (Laurence et al., 2009):

1. Restriction of the construction of roads (and other infrastructure) to flatter slopes and ridgelines, if and wherever possible.
2. Limitation of the width of roads.
3. Limitation of the construction to dry seasons to limit erosion and stream sedimentation.
4. Construction of railroads instead of traditional bitumen roads to minimize most effects. However, in both the cases, it is most important to put all mitigation measures in place.

Effects on Soil Biogeochemistry and Physical Properties

Geochemical changes, especially those of soil temperature and moisture changes (due to sediment deposition/soil inundation), gas diffusivity, and changes in nutrient cycling are important effects of climate variability and disturbances (Nottingham et al., 2019; Wang et al., 2021a; b). Figures 7.9 and 7.10 illustrate some issues.

Nitrogen fluxes are affected by logging and clearing activities in forests, and deforestation is common during the construction of linear infrastructure (Törmänen et al., 2020). Even if the magnitude by which the alteration occurs has not been researched, the direction of alteration is well noted (Cui et al., 2020). Reduced soil moisture in forest environments may result in the reduction of N_2O emissions (Wang et al., 2023). Soil enzyme activity may increase due to warming, leading to more N_2O emissions, especially in humid tropical forests (Holtgrieve et al., 2006). Soils are an important natural CH_4 sink. The erosion of these soils and degradation of their quality due to compaction will result in a significant amount of CH_4 in the atmosphere, resulting in further warming. Biogenic volatile organic compounds (BVOCs) are released by plants into the atmosphere, and these may increase due to increased CO_2 and N_2O concentrations (Loreto et al., 2014; Ren et al., 2017; Portillo-Estrada et al., 2018; Huang et al., 2020; Lun et al., 2020). BVOCs are known to have positive feedback in the greenhouse effect and result in net positive radiative forcing (this refers to the "change in energy flux caused by natural or anthropogenic drivers of climate change" see Shindell, 2013). Landuse change can alter these emissions when vegetation cover is reduced. NO_x, CO, and VOCs are known precursors of O_3. The concentrations of these precursors are increasing in the atmosphere due to anthropogenic activities like deforestation and urbanization (Covey et al., 2021). Zhang et al. (2018: 31) note that "tropospheric ozone (O_3) is a major air pollutant and causes serious injury to vegetation," by inhibiting photosynthesis, reducing tree growth and the survivability of seedlings. It also makes vegetation more susceptible to damage due to stressors like diseases, pests, and adverse weather conditions (Iowa Department of Natural Resources, n.d.).

Transpiration is an important driving factor for regulating the onset of the wet season (Wright et al., 2017). Deforestation may have minimal impacts on evapotranspiration during the wet season, but the impact is significant during the dry season (Longo et al., 2020; Xu et al., 2022; Lapola et al., 2023). An overall decrease in evapotranspiration rates in the dry season results in an alteration of the onset and duration of the dry season. With an increase in the dry season length, secondary forests become more susceptible to reduction in overall growth and become vulnerable to drought stress and fire risks (Hanson & Weltzin, 2000; Clark et al., 2016; Littell et al., 2016; Jump et al., 2017; Buotte et al., 2019; Zou et al., 2020). It has been shown from research that these effects are higher in areas with greater deforestation rates and on the edges where the microclimate is comparatively drier and hotter than the continuous forest (Engelbrecht et al., 2007; Covey et al., 2021; Da Rosa Ferraz Jardim et al., 2022).

Linear infrastructure construction activities disturb soils, changing the soil properties and creating urban soils. This anthropedogenesis (soil creation due to human actions) refers to soils formed by human action: "Anthropedogenic processes are characterised as artificial settings that dramatically alter natural soil formation processes or causes" and "in any given setting, human actions alter one or more of the five soil-forming factors" (Mageshen et al., 2022) as opposed to natural pedogenesis, has a rapid transformation cycle. Carbon and nitrogen cycles are altered through the formation of urban soils, with parent material, microclimatic conditions, soil microbes, relief, and vegetation also changing (Lorenz & Lal, 2009). The surface runoff may increase, with a reduction in the baseflow by the impervious layers. In certain places, de-icing salts causes major soil and plant-related issues in the immediate vicinity of roads (Łuczak et al., 2020). Use of construction machinery alters the soil porosity, with variations depending on the depth of disturbance. Results include the destruction of microhabitats, and exposure of subsoil decomposer communities to harsh, dry conditions (solar radiation, temperature fluctuations, and low moisture) and the homogenization of topsoil and subsoil with possible biogeochemical results (D'Acqui et al., 2020; Shehadeh et al., 2022). Artificial soil fertilization may also alter soil pH, with consequent changes in C and N cycling, resulting in serious biogeochemical consequences. Biogeographical change, brought on by the planting of green spaces and fertilization of soils, may alter the biogeochemistry of the area (Pouyat et al., 2006; Lorenz & Lal, 2009).

When urbanization activities change, the rate of cycling of C also changes (Lorenz & Lal, 2009; Churkina, 2012; Lal, 2012; Hutyra et al., 2014; Kinnunen et al., 2022; Patoine et al., 2022). See Fig. 7.11 for some impacts on chemical cycling. In the long term, urban construction may contribute to soil compaction in the subsoil layers, decreasing space for roots and decomposer organisms to carry out processes like turning and thus affect biogeochemical cycling. The transformation and transfer rates of nutrients between the soil and atmosphere decrease due to soil compaction and impervious layers between them (Le et al., 2020; Rivers et al., 2021; Das et al., 2023). Cascade effects may alter the cycling of nutrients over the long term. These structures also enable nonnative, sometimes invasive species to enter an ecosystem and alter the biogeochemistry, species compositions, food webs, and decomposition rates of the area. Sulfur and nitrogen (especially reactive nitrogen species) deposition gradients may also be found along transportation infrastructure because of combustion in automobile engines (Lorenz & Lal, 2009).

Effects Due to Landuse and Landcover Changes

Ecosystems are very delicate, and the balance can be disturbed even with very minute disturbances (De Laender et al., 2016; Weiskopf et al., 2020). For example, over the last two decades, the road network in India has expanded, leading to various problems, including changes in microclimate and degradation of habitat and animal corridors, with possible changes in local and regional biogeochemistry (Dutta et al., 2019; Anasuya et al., 2020; Jana et al., 2020). Ecological risk is the probability that any ecosystem returns to its earlier state (low-energy equilibrium) after an external disturbance (Fraterrigo & Rusak, 2008). The development of linear infrastructure results in a net change in the landuse and landcover patterns of a given area, which further changes the resource use patterns (Sharma & Raghubanshi, 2009; Dutta et al., 2018). Sustainable road development plans include proximate forest cover, with constant monitoring of the LULC change patterns along any linear infrastructure over time (Mann et al., 2020). Change in landcover can affect the existing water, energy, nutrient, and carbon cycle of any given ecosystem and those connected to it (Lanfri, 2010).

Rapid deforestation and forest degradation due to developmental activities leads to the release of CO_2 and affects the net C balance of the region (Aragão et al., 2014). For example, in the Amazon Basin, landuse and landcover change and forest degradation are the major driving forces of CO_2 emissions (Smith et al., 2020; Covey et al., 2021). Soil C pools are vulnerable to change due to landuse activities and climate change. This influences impacts on soil biodiversity and other biogeochemical effects (Kauffman et al., 2009; Berenguer et al., 2014; Grace et al., 2014; Bustamante et al., 2016; Chazdon et al., 2016). As most C is stored in the top 50 cm of soil, developmental activities which lead to erosion result in further decrease in soil C content. With reduced vegetation to take up CO_2, the increased N_2O and CH_4 emissions due to anthropogenic activities (disturbed ecosystems) are unbalanced and may change biogeochemical cycling (Galford et al., 2010, 2011; Cohn et al., 2014). For example, Galford et al. (2010: 19649) note that "the Brazilian Amazon is one of the most rapidly developing agricultural areas in the world and represents a potentially large future source of greenhouse gases from land clearing and subsequent agricultural management" and conclude that "Both deforestation and future land-use management play important roles in the net greenhouse gas emissions of this frontier, suggesting that both should be considered in emissions policies. We find that avoided deforestation remains the best strategy for minimizing future greenhouse gas emissions from Mato Grosso." Increased CO concentrations in the atmosphere extends the average atmospheric lifetime of CH_4, which are otherwise removed by the OH^- (Covey et al., 2021). Urbanized systems with more impermeable surfaces have a greater chance of Urban Heat Island (UHI) effect. This alters plant growth and flowering cycles, and plant and microbial growth rates, which in turn changes nutrient cycles (Kaye et al., 2006).

Case Study: How Has Urban Development Affected the Himalayas?

Urban biogeochemistry may examine the biogeochemical changes occasioned by urban infrastructural developments in variably rural contexts, with both biogeographical and geochemical relevance (e.g., deforestation, soil, and climate change) (Campbell, 2020). In this, a constructive case study may focus on examining little-researched areas, with fragile, vulnerable ecosystems, where the impacts of human action may be discerned easily, as "the fragility of the environment directly affects the stability of the regional ecosystem and the sustainable development of the ecological environment" (Chen et al., 2018b). Case studies of mountainous regions are instructive in this regard, as "mountains are an essential component of the global life-support system" and they are "exposed to multitudinous threats that range from climate change to direct habitat degradation by human activity" and "these impacts have biological consequences in sensitive mountain ecosystems" (Schmeller et al., 2022).

Biogeographical, mostly dense vegetation, change and geochemical dynamics such as climate change are the key dynamics for environmental stability for such fragile systems. For example, Li et al. (2022) note that "landslides are one disaster that has become more prevalent because of anthropogenic disturbances, such as land-cover changes, land degradation, and expansion of infrastructure." The study was based on 247 mountain ranges (about a quarter of the global mountainous areas) identified as highly vulnerable, with "exceptional biodiversity and landslide risks" and 31 biodiverse mountains which are classified as future vulnerable mountains, with increasing landslide risks linked to climate change and deforestation. The authors argue that of the 278 studied mountain ranges, there were 52 that required immediate conservation actions due to high vulnerability, mostly severe threats from future deforestation and precipitation events, low current protection, and high human population density and actions. It is argued that necessary actions include increased protected areas, forest conservation, and restoration to reduce landslide risks (Li et al., 2022).

Grima et al. (2020), writing on a case study of the South American Andes mountains, argue that "landslides cause billions of dollars (USD) in damage and hundreds of life losses every year in mountainous areas globally, and these effects are exacerbated by climate change and increased human occupation of vulnerable areas... forests deliver slope stability, helping to prevent landslides. However, forests are progressively converted into other land uses in many mountainous regions." Linear infrastructure, namely roads are cited: "Our most conservative estimates indicate it is 16 times more cost-effective to promote forest corridors, via conservation or reforestation along roads by paying farmers and cattle herders their opportunity costs, than for the public to pay the expected value of landslide damage." The authors conclude that the "analysis provides strong evidence that vegetated hillsides can provide a cost-effective ecosystem service approach to mitigate economic losses due to landslides in one of the world's most landslide prone areas" (Grima et al., 2020). This study powerfully underscores the importance of biogeographical patterns and change, in relation to the geochemical factors of soil structure, climate and topography, and human actions contributing to landcover change, and situates urban and other structural changes landcover as key drivers for such biogeochemical changes (Grima et al., 2020).

The selected case study, the Himalayas has a very fragile ecosystem, being earthquake prone (Rosser et al., 2021) and currently undergoing rapid developmental activities, which may create substantial landcover deforestation and even topographical change (Mishra & Mainali, 2017). For example, Sultan et al. (2022) point out that "the Himalayan region is a fragile high mountain landscape where the population experiences acute vulnerability within a complex coupled human-natural system due to environmental, social, and economic linkages. The lack of significant regional and spatial knowledge of multifaceted vulnerabilities hinders any potential recommendations to address these vulnerabilities." Major forest losses and landcover developments such as road construction have resulted in steep and rapidly weathering roadcut cliffs. Forest degradation and deforestation have been shown to be important causes of instability, these amplified by the possibilities for earthquakes, results in massive landslides (Mishra & Mainali, 2017; Rosser et al., 2021). As a result of land denudation, there may also be events such as landslides (Chalise & Kumar, 2020; Dahal et al., 2020; Thapa et al., 2021; Yadav et al., 2021). A Nepalese study by Brunsden et al. (1981) indicated higher mass wasting on steep forested slopes than along the gentle slopes used for agriculture. Landslide debris may also affect downslope tree cover (Haigh et al., 1995).

Abraham et al. (2020) wrote that the Himalayas "are among the most landslide-prone areas in the world, and attempts have been made to determine the rainfall thresholds for possible occurrence of landslides in the region." As per a report by the Geological Survey of India (GSI), almost 0.14 million square kilometers in the North-western region of Himalaya is very prone to landslides (Sur et al., 2020). These landslides can even cause the damming of rivers, with economic and human life losses, and indirect effects including further reduction in the forest cover and decreased slope stability downslope after landslides (Guri et al., 2015; Gagné, 2019; Pandit, 2020). Major drivers of landslides include excess moisture in soil, mass movement acceleration because of deforestation, slope instability and root wedging into expanded and weathered joints, and earthquakes (Rosser et al., 2021; Jahan et al., 2023). Forest cover may be a driver of landslides, but may also prevent landslides, as forest tree roots may wedge the bedrock and anchor slopes, but the loosened soil may accumulate moisture, making the soil heavy and accelerating landslides (Haigh et al., 1995).

Major earthquakes have also caused massive numbers of landslides in Nepal and Indian Himalayas, including along the along the Himalayan Frontal Thrust (HFT) (Jahan et al., 2023). Rosser et al. (2021) describe the 2015 Gorkha Earthquake, arguing that "the 2015 Mw 7.8 Gorkha, Nepal Earthquake triggered in excess of 20,000 landslides across 14 districts of Central and Western Nepal" and noting that their study provides "insight into how landsliding has evolved since the earthquake, and how it has impacted those living in the affected area" by creating a "detailed time-series landslide mapping campaign… to monitor the evolution of co-seismic landslides and the initiation of new post-seismic landslides." The research found that the landslide hazard was higher in November 2019 than immediately after the 2015 earthquake, and "while pre-existing landslides continued to pose the majority of hazard in the aftermath of the earthquake, a significant number of landslides also occurred in new locations" (Rosser et al., 2021). The earth surface cracking resulting from seismic events may increase landslide possibilities post earthquake. Such coseismic landsliding may result from amplified seismic wave impacts on ridgelines and convexities with later possibilities of debris slide from higher elevations, also based on rainfall, deforestation, and slope instability (Rosser et al., 2021).

In one example of landslide, a small town of Joshimath, situated in the North-western Himalayas, in the state of Uttarakhand, India, had to be evacuated in the early weeks of January 2023. In a span of 12 days (December 27, 2022, to January 8, 2023), a rapid land subsidence event occurred in which the town sank 5.4 cm. Reports suggested long term, local land sinking (~10 cm every year since 2018), and the town was located on an ancient landslide complex (Saha, 2023). Contributory factors included water flow and accumulation and how it affects the soil (Gantzer &

Gantzer, 2023). Multiple factors acting together occasioned this disaster: the geological setting of the area and anthropogenic developments (mostly urbanization), which disrupted the drainage, uncontrolled water discharge and slopes' undercutting for construction activities (Bisht & Rautela, 2010; Gantzer & Gantzer, 2023; Saha, 2023). Over concretization – including roads – alters the water percolation and results in rapid runoff in the slopes. Static traffic on the roads may contribute to more pressure on the hillside, which adds to the burden and causes instability. The new tunnel and ropeway project under development in Joshimath was suspended (Gantzer & Gantzer, 2023). Different strategies for development will have to be taken up based on the geographical settings of any area (Saha, 2023).

Conclusion

Urban biogeochemistry is developing into an indispensable, core area of biogeochemistry, based on the increasingly dominant role of urbanization in global environmental change. Current urban developments, based mainly on building and infrastructure developments, create extreme environmental changes, including soil compaction and disturbance, impermeable ground layers, drainage disruption, deforestation, and pollution, which affect chemical cycling, especially elements such as carbon and nitrogen, and pollutants. These, based on the global urban footprints, also impact wider environments that support transport systems. Like its parent discipline, biogeochemistry, urban biogeochemistry is closely related to and supported by related sciences, especially urban-related aspects of biogeography and geochemistry, which in turn are linked to other environmental and earth sciences. Linear infrastructure, including roads, railways, and streets, are vital, as they are the basic structures for all urban movement and connect all urban systems through rural lands, and affect hydrological, pedological, geochemical, ecological and biogeographical systems, to the point of minor to extreme biogeochemical and chemical cycling change.

This chapter has reviewed a few examples of what is evidently a vast topic and attempted to examine the supporting environmental and earth sciences for urban biogeochemistry. The case studies show how important urban biogeochemistry is, especially in fragile environments where the impacts may be more pronounced, and serious action may be required for amelioration. The findings indicate that construction and infrastructure planning must consider environmental issues, especially biogeochemical impacts, for sustainable ecological outcomes. This contributes to the investigation of the biogeochemical bases of linear development projects, and urban biogeochemistry in general. The impacts of urban biogeochemistry are directly and indirectly dependent on human action, and also complemented or even enhanced by natural events like earthquakes and climate change and consequent alterations of geochemical cycling and biogeographical dynamics. Further studies may examine to a greater extent the chemical bases of biogeochemical changes, and also the human structures studied by subjects such as civil engineering and urban geography. The future advancement and utility of biogeochemistry may depend on the support derived from associated technical, natural, and human-centered disciplines, and enduring sensitivity to environmental dynamics. Crucially, development and sustainability must be balanced together.

References

Abdel-Shafy, H. I., & Mansour, M. (2016). A review on polycyclic aromatic hydrocarbons: Source, environmental impact, effect on human health and remediation. *Egyptian Journal of Petroleum, 25*(1), 107–123. https://doi.org/10.1016/j.ejpe.2015.03.011

Abraham, M. T., Satyam, N., Pradhan, B., & Alamri, A. M. (2020). IoT-Based Geotechnical Monitoring of Unstable Slopes for Landslide Early Warning in the Darjeeling Himalayas. *Sensors (Basel, Switzerland), 20*(9), 2611. https://doi.org/10.3390/s20092611

Anasuya, B., Swain, D., & Vinoj, V. (2020). Rapid urbanization and associated impacts on land surface temperature changes over Bhubaneswar Urban District, India. *Environmental Monitoring and Assessment, 191*(Suppl 3), 790. https://doi.org/10.1007/s10661-019-7699-2

Angrand, R. C., Collins, G., Landrigan, P. J., & Thomas, V. M. (2022). Relation of blood lead levels and lead in gasoline: An updated systematic review. *Environmental Health, 21*(1), 138. https://doi.org/10.1186/s12940-022-00936-x

Annett, R., Habibi, H. R., & Hontela, A. (2014). Impact of glyphosate and glyphosate-based herbicides on the freshwater environment. *Journal of Applied Toxicology: JAT, 34*(5), 458–479. https://doi.org/10.1002/jat.2997

Aragão, L. E., Poulter, B., Barlow, J. B., Anderson, L. O., Malhi, Y., Saatchi, S., Phillips, O. L., & Gloor, E. (2014). Environmental change and the carbon balance of Amazonian forests. *Biological Reviews of the Cambridge Philosophical Society, 89*(4), 913–931. https://doi.org/10.1111/brv.12088

Bekki, K., Toriba, A., Tang, N., Kameda, T., Takigami, H., Suzuki, G., & Hayakawa, K. (2012). Yakugaku zasshi. *Journal of the Pharmaceutical Society of Japan, 132*(3), 325–329. https://doi.org/10.1248/yakushi.132.325

Berenguer, E., Ferreira, J., Gardner, T. A., Aragão, L. E., De Camargo, P. B., Cerri, C. E., Durigan, M., Cosme De Oliveira Junior, R., Vieira, I. C., & Barlow, J. (2014). A large-scale field assessment of carbon stocks in human-modified tropical forests. *Global Change Biology, 20*(12), 3713–3726. https://doi.org/10.1111/gcb.12627

BIOGRIP. (2019). *Antarctic legacy archive*. [Online] Retrieved from https://alp.lib.sun.ac.za/handle/123456789/27448

Bisht, M. P. S., & Rautela, P. (2010). Disaster looms over Joshimath. *Current Science, 98*(10), 1271.

Borda-de-Água, L., Barrientos, R., Beja, P., & Pereira, H. M. (2017). *Railway ecology*. Springer International Publishing AG. https://doi.org/10.1007/978-3-319-57496-7

Brunsden, D., Jones, D. K. C., Martin, R. P. & Doornkamp, J. C. (1981). The geomorphological character of part of the Low Himalaya of Eastern Nepal. *Zeitschrift fur Geomorphologie, Supplementband, 37*, 25–72.

Buotte, P. C., Levis, S., Law, B. E., Hudiburg, T. W., Rupp, D. E., & Kent, J. J. (2019). Near-future forest vulnerability to drought and fire varies across the western United States. *Global Change Biology, 25*(1), 290–303. https://doi.org/10.1111/gcb.14490

Bustamante, M. M., Roitman, I., Aide, T. M., Alencar, A., Anderson, L. O., Aragão, L., Asner, G. P., Barlow, J., Berenguer, E., Chambers, J., Costa, M. H., Fanin, T., Ferreira, L. G., Ferreira, J., Keller, M., Magnusson, W. E., Morales-Barquero, L., Morton, D., Ometto, J. P., Palace, M., et al. (2016). Toward an integrated monitoring framework to assess the effects of tropical forest degradation and recovery on carbon stocks and biodiversity. *Global Change Biology, 22*(1), 92–109. https://doi.org/10.1111/gcb.13087

Campbell, M. (2017). *Biological conservation in the 21st century: A conservation biology of large wildlife*. Nova Science Publishers.

Campbell, M. (2020). *The physical geography of El Salvador: An ecological and geophysical approach*. Nova Science Publishers.

Campo, L., Buratti, M., Fustinoni, S., Cirla, P. E., Martinotti, I., Longhi, O., Cavallo, D., & Foà, V. (2006). Evaluation of exposure to PAHs in asphalt workers by environmental and biological monitoring. *Annals of the New York Academy of Sciences, 1076*, 405–420. https://doi.org/10.1196/annals.1371.013

Chalise, D., & Kumar. (2020). Land use change affects water erosion in the Nepal Himalayas. *PLoS One, 15*(4), e0231692. https://doi.org/10.1371/journal.pone.0231692

Chamberlain, S. D., Ingraffea, A. R., & Sparks, J. P. (2016). Sourcing methane and carbon dioxide emissions from a small city: Influence of natural gas leakage and combustion. *Environmental Pollution (Barking, Essex: 1987), 218*, 102–110. https://doi.org/10.1016/j.envpol.2016.08.036

Chambers, L. G., Chin, Y. P., Filippelli, G. M., Gardner, C. B., Herndon, E. M., Long, D. T., Lyons, W. B., Macpherson, G. L., McElmurry, S. P., McLean, C. E., Moore, J., Moyer, R. P., Neumann, K., Nezat, C. A., Soderberg, K., Teutsch, N., & Widom, E. (2016). Developing the scientific framework for urban geochemistry. *Applied Geochemistry, 67*, 1–20. https://doi.org/10.1016/j.apgeochem.2016.01.005

Chazdon, R. L., Broadbent, E. N., Rozendaal, D. M., Bongers, F., Zambrano, A. M., Aide, T. M., Balvanera, P., Becknell, J. M., Boukili, V., Brancalion, P. H., Craven, D., Almeida-Cortez, J. S., Cabral, G. A., de Jong, B., Denslow, J. S., Dent, D. H., DeWalt, S. J., Dupuy, J. M., Durán, S. M., Espírito-Santo, M. M., et al. (2016). Carbon sequestration potential of second-growth forest regeneration in the Latin American tropics. *Science Advances, 2*(5), e1501639. https://doi.org/10.1126/sciadv.1501639

Chen, L., Zhi, X., Shen, Z., Dai, Y., & Aini, G. (2018a). Comparison between snowmelt-runoff and rainfall-runoff nonpoint source pollution in a typical urban catchment in Beijing, China. *Environmental Science and Pollution Research International, 25*(3), 2377–2388. https://doi.org/10.1007/s11356-017-0576-z

Chen, P., Hou, K., Chang, Y., Li, X., & Zhang, Y. (2018b). Study on the progress of ecological fragility assessment in China. *IOP Conference Series: Earth and Environmental Science, 113*, 012088. https://doi.org/10.1088/1755-1315/113/1/012088

Chiari, C. G., & Seeland, K. (2004). Are urban green spaces optimally distributed to act as places for social integration? Results of a geographical information system GIS approach for urban forestry research. *Forest Policy and Economics, 6*(1), 3–13. https://doi.org/10.1016/S1389-9341(02)00067-9

Churkina, G. (2012). Carbon cycle of urban ecosystems. In R. Lal & B. Augustin (Eds.), *Carbon sequestration in urban ecosystems*. Springer. https://doi.org/10.1007/978-94-007-2366-5_16

Claireau, F., Bas, Y., Pauwels, J., Barré, K., Machon, N., Allegrini, B., Puechmaille, S. J., & Kerbiriou, C. (2019). Major roads have important negative effects on insectivorous bat activity. *Biological Conservation, 235*, 53–62. https://doi.org/10.1016/j.biocon.2019.04.002

Clark, J. S., Iverson, L., Woodall, C. W., Allen, C. D., Bell, D. M., Bragg, D. C., D'Amato, A. W., Davis, F. W., Hersh, M. H., Ibanez, I., Jackson, S. T., Matthews, S., Pederson, N., Peters, M., Schwartz, M. W., Waring, K. M., & Zimmermann, N. E. (2016). The impacts of increasing drought on forest dynamics, structure, and biodiversity in the United States. *Global Change Biology, 22*(7), 2329–2352. https://doi.org/10.1111/gcb.13160

CMS. (2020). *Addressing the impacts of linear infrastructure on migratory species*. United Nations Environment Programme.

Cohn, A. S., Mosnier, A., Havlík, P., Valin, H., Herrero, M., Schmid, E., O'Hare, M., & Obersteiner, M. (2014). Cattle ranching intensification in Brazil can reduce global greenhouse gas emissions by sparing land from deforestation. *Proceedings of the National Academy of Sciences of the United States of America, 111*(20), 7236–7241. https://doi.org/10.1073/pnas.1307163111

Covey, K., et al. (2021). Carbon and beyond: The biogeochemistry of climate in a rapidly changing Amazon. *Frontiers in Forests and Global Change, 4*, 1–20. https://doi.org/10.3389/ffgc.2021.618401

Cui, J., Cui, J., Peng, Y., Yao, D., Chan, A., Chen, Z., & Chen, Y. (2020). Sources and trends of oxidized and reduced nitrogen wet deposition in a typical medium-sized city of eastern China during 2010-2016. *The Science of the Total Environment, 744*, 140558. https://doi.org/10.1016/j.scitotenv.2020.140558

Da Rosa Ferraz Jardim, A. M., de Morais, J. E. F., de Souza, L. S. B., & da Silva, T. G. F. (2022). Understanding interactive processes: A review of CO2 flux, evapotranspiration, and energy partitioning under stressful conditions in dry forest and agricultural environments. *Environmental Monitoring and Assessment, 194*(10), 677. https://doi.org/10.1007/s10661-022-10339-7

D'Acqui, L. P., Certini, G., Cambi, M., & Marchi, E. (2020). Machinery's impact on forest soil porosity. *Journal of Terramechanics, 91*, 65–71. https://doi.org/10.1016/j.jterra.2020.05.002

Dahal, P., Shrestha, M. L., Panthi, J., & Pradhananga, D. (2020). Modeling the future impacts of climate change on water availability in the Karnali River basin of Nepal Himalaya. *Environmental Research, 185*, 109430. https://doi.org/10.1016/j.envres.2020.109430

Dansereau, P. (2015). Biogeography. *The Canadian Encyclopaedia*. Retrieved from https://www.thecanadianencyclopedia.ca/en/article/biogeography

Das, T. K., Kabir, A., Zhao, W., Stenstrom, M. K., Dittrich, T. M., & Mohanty, S. K. (2023). A review of compaction effect on subsurface processes in soil: Implications on stormwater treatment in roadside compacted soil. *The Science of the Total Environment, 858*(Pt 3), 160121. https://doi.org/10.1016/j.scitotenv.2022.160121

De Laender, F., Rohr, J. R., Ashauer, R., Baird, D. J., Berger, U., Eisenhauer, N., Grimm, V., Hommen, U., Maltby, L., Meliàn, C. J., Pomati, F., Roessink, I., Radchuk, V., & Van den Brink, P. J. (2016). Reintroducing environmental change drivers in biodiversity-ecosystem functioning research. *Trends in Ecology & Evolution, 31*(12), 905–915. https://doi.org/10.1016/j.tree.2016.09.007

De Magalhães, R. F., Danilevicz, Â. M. F., & Saurin, T. A. (2017). Reducing construction waste: A study of urban infrastructure projects. *Waste Management (New York, N.Y.), 67*, 265–277. https://doi.org/10.1016/j.wasman.2017.05.025

Decker, E. H., Elliott, S., & Smith, F. A. (2002). Megacities and the environment. *The Scientific World Journal, 2*, 374–386. https://doi.org/10.1100/tsw.2002.103

Dey, P. K. (2002). An integrated assessment model for cross country pipelines. *Environmental Impact Assessment Review, 22*, 703–721. https://doi.org/10.1016/S0195-9255(02)00020-3

Douglas, B., & Philip, J. (Eds.). (2014). *Urban ecology: An introduction*. Routledge. https://doi.org/10.4324/9780203108703

Duniway, M. C., & Herrick, J. E. (2011). Disentangling road network impacts: The need for a holistic approach. *Journal of Soil and Water Conservation, 66*(2), 31A–36A. https://doi.org/10.2489/jswc.66.2.31A

Dutta, T., Sharma, S., & DeFries, R. (2018). Targeting restoration sites to improve connectivity in a tiger conservation landscape in India. *PeerJ, 6*, e5587. https://doi.org/10.7717/peerj.5587

Dutta, D., Rahman, A., Paul, S. K., & Kundu, A. (2019). Changing pattern of urban landscape and its effect on land surface temperature in and around Delhi. *Environmental Monitoring and Assessment, 191*(9), 551. https://doi.org/10.1007/s10661-019-7645-3

Eganhouse, R. P., Baedecker, M. J., & Cozzarelli, I. M. (1994). Biogeochemical processes in an aquifer contaminated by crude oil: An overview of the studies at Bemidji, Minnesota, research site. In *Symposium on natural attenuation of ground water* (pp. 89–94). Environmental Protection Agency.

Engelbrecht, B. M., Comita, L. S., Condit, R., Kursar, T. A., Tyree, M. T., Turner, B. L., & Hubbell, S. P. (2007). Drought sensitivity shapes species distribution patterns in tropical forests. *Nature, 447*(7140), 80–82. https://doi.org/10.1038/nature05747

Ferreira, C., Kalantari, Z., Seifollahi-Aghmiuni, S., Ghajarnia, N., Rahmati, O., & Kapovic Solomun, M. (2021). Rainfall-runoff-erosion processes in urban areas. In *Precipitation: Earth surface responses and processes* (pp. 481–498). https://doi.org/10.1016/B978-0-12-822699-5.00018-5

Fraterrigo, J. M., & Rusak, J. A. (2008). Disturbance-driven changes in the variability of ecological patterns and processes. *Ecology Letters, 11*(7), 756–770. https://doi.org/10.1111/j.1461-0248.2008.01191.x

Gagné, K. (2019). Waiting for the flood: Technocratic time and impending disaster in the Himalayas. *Disasters, 43*(4), 840–866. https://doi.org/10.1111/disa.12379

Galford, G. L., Melillo, J. M., Kicklighter, D. W., Cronin, T. W., Cerri, C. E., Mustard, J. F., & Cerri, C. C. (2010). Greenhouse gas emissions from alternative futures of deforestation and agricultural management in the southern Amazon. *Proceedings of the National Academy of Sciences of the United States of America, 107*(46), 19649–19654. https://doi.org/10.1073/pnas.1000780107

Galford, G. L., Melillo, J. M., Kicklighter, D. W., Mustard, J. F., Cronin, T. W., Cerri, C. E., & Cerri, C. C. (2011). Historical carbon emissions and uptake from the agricultural frontier of the Brazilian Amazon. *Ecological Applications, 21*(3), 750–763. https://doi.org/10.1890/09-1957.1

Gantzer, H., & Gantzer, C. (2023). *Joshimath disaster a warning for Mussorie*. The Tribune.

Gao, S., Wang, Z., Wu, Q., Wang, W., Peng, C., Zeng, J., & Wang, Y. (2021). Urban geochemistry and human-impacted imprint of dissolved trace and rare earth elements in a high-tech industrial city, Suzhou. *Elementa: Science of the Anthropocene, 9*(1). https://doi.org/10.1525/elementa.2020.00151

Gbeddy, G., Goonetilleke, A., Ayoko, G. A., & Egodawatta, P. (2020). Transformation and degradation of polycyclic aromatic hydrocarbons (PAHs) in urban road surfaces: Influential factors, implications, and recommendations. *Environmental Pollution (Barking, Essex: 1987), 257*, 113510. https://doi.org/10.1016/j.envpol.2019.113510

Gissi, F., et al. (2022). A review of the potential risks associated with mercury in subsea oil and gas pipelines in Australia. *Environmental Chemistry, 19*(3&4), 210–227. https://doi.org/10.1071/EN22048

Gong, C., Xian, C., Wu, T., & Ouyang, Z. (2023). Role of urban vegetation in air phytoremediation: Differences between scientific research and environmental management perspectives. *npj Urban Sustainability, 3*, 24. https://doi.org/10.1038/s42949-023-00105-0

Grace, J., Mitchard, E., & Gloor, E. (2014). Perturbations in the carbon budget of the tropics. *Global Change Biology, 20*(10), 3238–3255. https://doi.org/10.1111/gcb.12600

Grima, N., Edwards, D., Edwards, F., Petley, D., & Fisher, B. (2020). Landslides in the Andes: Forests can provide cost-effective landslide regulation services. *The Science of the Total Environment, 745*, 141128. https://doi.org/10.1016/j.scitotenv.2020.141128

Grimm, N. B., Faeth, S. H., Golubiewski, N. E., Redman, C. L., Wu, J., Bai, X., & Briggs, J. M. (2008). Global change and the ecology of cities. *Science (New York, N.Y.), 319*(5864), 756–760. https://doi.org/10.1126/science.1150195

Grzywa-Celińska, A., Krusiński, A., & Milanowski, J. (2020). 'Smoging kills' - effects of air pollution on human respiratory system. *Annals of Agricultural and Environmental Medicine: AAEM, 27*(1), 1–5. https://doi.org/10.26444/aaem/110477

Gu, B., Dong, X., Peng, C., Luo, W., Chang, J., & Ge, Y. (2012). The long-term impact of urbanization on nitrogen patterns and dynamics in Shanghai, China. *Environmental Pollution (Barking, Essex: 1987), 171*, 30–37. https://doi.org/10.1016/j.envpol.2012.07.015

Guri, P. K., Ray, P. K., & Patel, R. C. (2015). Spatial prediction of landslide susceptibility in parts of Garhwal Himalaya, India, using the weight of evidence modelling. *Environmental Monitoring and Assessment, 187*(6), 324. https://doi.org/10.1007/s10661-015-4535-1

Gustafson, K. D., Belden, J. B., & Bolek, M. G. (2015). The effects of the herbicide atrazine on freshwater snails. *Ecotoxicology (London, England), 24*(5), 1183–1197. https://doi.org/10.1007/s10646-015-1469-x

Haigh, M. J., et al. (1995). Interactions between forest and landslide activity along new haighways in the Kumaun Himalaya. *Forest Ecology and Management, 78*(1–3), 173–189. https://doi.org/10.1016/0378-1127(95)03584-5

Hanson, P. J., & Weltzin, J. F. (2000). Drought disturbance from climate change: Response of United States forests. *The Science of the Total Environment, 262*(3), 205–220. https://doi.org/10.1016/s0048-9697(00)00523-4

Hendrick, M. F., Ackley, R., Sanaie-Movahed, B., Tang, X., & Phillips, N. G. (2016). Fugitive methane emissions from leak-prone natural gas distribution infrastructure in urban environments.

Environmental Pollution (Barking, Essex: 1987), 213, 710–716. https://doi.org/10.1016/j.envpol.2016.01.094

Holland, H. D., & Turiekian, K. K. (2014). *Treatise on geochemistry*. Elsevier.

Holtgrieve, G. W., Jewett, P. K., & Matson, P. A. (2006). Variations in soil N cycling and trace gas emissions in wet tropical forests. *Oecologia, 146*(4), 584–594. https://doi.org/10.1007/s00442-005-0222-1

Huang, X., Lai, J., Liu, Y., Zheng, L., Fang, X., Song, W., & Yi, Z. (2020). Biogenic volatile organic compound emissions from Pinus massoniana and Schima superba seedlings: Their responses to foliar and soil application of nitrogen. *The Science of the Total Environment, 705*, 135761. https://doi.org/10.1016/j.scitotenv.2019.135761

Hutyra, L. R., Duren, R. M., Gurney, K. R., Grimm, N. B., Kort, E. A., Larson, E. K., & Shrestha, G. (2014). Urbanization and the carbon cycle: Current capabilities and research outlook from the natural sciences perspective. *Earth's Future, 2*. https://doi.org/10.1002/2014EF000255

Iowa Department of Natural Resources. (n.d.). *Effects of ground level ozone*. https://www.iowadnr.gov/Environmental-Protection/Air-Quality/Air-Pollutants/ Effects-Ozone

Jackson, R. B., Down, A., Phillips, N. G., Ackley, R. C., Cook, C. W., Plata, D. L., & Zhao, K. (2014). Natural gas pipeline leaks across Washington, DC. *Environmental Science & Technology, 48*(3), 2051–2058. https://doi.org/10.1021/es404474x

Jahan, N., Rana, Y. P., & Singh, R. J. (2023). Structural evidences of active tectonics along Himalayan Frontal Thrust of northwest Himalaya: A case study along Kumia river section, Nainital, India. *Journal of Earth System Science, 132*, 57. https://doi.org/10.1007/s12040-023-02078-1

Jana, C., Mandal, D., Shrimali, S. S., Alam, N. M., Kumar, R., Sena, D. R., & Kaushal, R. (2020). Assessment of urban growth effects on green space and surface temperature in Doon Valley, Uttarakhand, India. *Environmental Monitoring and Assessment, 192*(4), 257. https://doi.org/10.1007/s10661-020-8184-7

Jump, A. S., Ruiz-Benito, P., Greenwood, S., Allen, C. D., Kitzberger, T., Fensham, R., Martínez-Vilalta, J., & Lloret, F. (2017). Structural overshoot of tree growth with climate variability and the global spectrum of drought-induced forest dieback. *Global Change Biology, 23*(9), 3742–3757. https://doi.org/10.1111/gcb.13636

Kauffman, J. B., Hughes, R. F., & Heider, C. (2009). Carbon pool and biomass dynamics associated with deforestation, land use, and agricultural abandonment in the neotropics. *Ecological Applications, 19*(5), 1211–1222. https://doi.org/10.1890/08-1696.1

Kaushal, S. S., McDowell, W. H., & Wollheim, W. M. (2014). Tracking evolution of urban biogeochemical cycles: Past, present, and future. *Biogeochemistry, 121*, 1–21. https://doi.org/10.1007/s10533-014-0014-y

Kaye, J. P., Groffman, P. M., Grimm, N. B., Baker, L. A., & Pouyat, R. V. (2006). A distinct urban biogeochemistry? *Trends in Ecology and Evolution, 21*(4), 192–199. https://doi.org/10.1016/j.tree.2005.12.006

Kennedy, C. A., Stewart, I., Facchini, A., Cersosimo, I., Mele, R., Chen, B., Uda, M., Kansal, A., Chiu, A., Kim, K. G., Dubeux, C., Lebre La Rovere, E., Cunha, B., Pincetl, S., Keirstead, J., Barles, S., Pusaka, S., Gunawan, J., Adegbile, M., Nazariha, M., et al. (2015). Energy and material flows of megacities. *Proceedings of the National Academy of Sciences of the United States of America, 112*(19), 5985–5990. https://doi.org/10.1073/pnas.1504315112

Kim, Y.-N., Yoon, J.-H., & Kim, K.-H. (2020). Microplastic contamination of soil environment- a review. *Soil Science Annual, 71*(4), 300–308. https://doi.org/10.37501/soilsa/131646

Kinnunen, A., Talvitie, I., Ottelin, J., Heinonen, J., & Junnila, S. (2022). Carbon sequestration and storage potential of urban residential environment – A review. *Sustainable Cities and Society, 84*, 104027. https://doi.org/10.1016/j.scs.2022.104027

Kirschbaum, D., & Stanley, T. (2018). Satellite-based assessment of rainfall-triggered landslide hazard for situational awareness. *Earth's Future, 6*(3), 505–523. https://doi.org/10.1002/2017EF000715

Konijnendijk, C. C. (2017). Cities and nature: Urban forestry for greater biocultural diversity. In M. Campbell (Ed.), *Biological conservation in the 21st century a conservation biology of large wildlife* (pp. 15–30). Nova Science Publishers.

Kowarik, I. (2011). Novel urban ecosystems, biodiversity, and conservation. *Environmental pollution (Barking, Essex: 1987), 159*(8–9), 1974–1983. https://doi.org/10.1016/j.envpol.2011.02.022

Kuhn, M. (2003). Greenbelt and greenheart: Separating and integrating landscapes in European city regions. *Landscape and Urban Planning, 64*(1-2), 19–27. https://doi.org/10.1016/S0169-2046(02)00198-6

Kumpiene, J., Larsson, M. O., Carabante, I., & Arp, H. P. H. (2021). Roads with underlying tar asphalt - spreading, bioavailability and toxicity of their polycyclic aromatic hydrocarbons. *Environmental Pollution (Barking, Essex: 1987), 289*, 117828. https://doi.org/10.1016/j.envpol.2021.117828

Lal, R. (2012). Urban ecosystems and climate change. In R. Lal & B. Augustin (Eds.), *Carbon sequestration in urban ecosystems*. Springer. https://doi.org/10.1007/978-94-007-2366-5_1

Lanfri, S. (2010). *Vegetation analysis using remote sensing. Master in emergency early warning and response space applications*. Argentinian spatial agency (CONAE): Cordoba National University (UNC).

Lapola, D. M., Pinho, P., Barlow, J., Aragão, L. E. O. C., Berenguer, E., Carmenta, R., Liddy, H. M., Seixas, H., Silva, C. V. J., Silva-Junior, C. H. L., Alencar, A. A. C., Anderson, L. O., Armenteras, D., Brovkin, V., Calders, K., Chambers, J., Chini, L., Costa, M. H., Faria, B. L., Fearnside, P. M., et al. (2023). The drivers and impacts of Amazon forest degradation. *Science (New York, N.Y.), 379*(6630), eabp8622. https://doi.org/10.1126/science.abp8622

Laurence, W. F., Goosem, M., & Laurence, G. W. (2009). Impacts of roads and linear clearings on tropical forests. *Trends in Ecology and Evolution, 24*(1), 659–669. https://doi.org/10.1016/j.tree.2009.06.009

Le, H., Valenca, R., Ravi, S., Stenstrom, M. K., & Mohanty, S. K. (2020). Size-dependent biochar breaking under compaction: Implications on clogging and pathogen removal in biofilters. *Environmental Pollution (Barking, Essex: 1987), 266*(Pt 1), 115195. https://doi.org/10.1016/j.envpol.2020.115195

Lechinovski, L., Bados, M., Rosa, J., Moda, D. B., & Bueno Krawczyk, A. C. D. (2022). Ecotoxicological effects of conventional herbicides and a natural herbicide on freshwater fish (Danio rerio). Journal of environmental science and health. *Part B, Pesticides, Food Contaminants, and Agricultural Wastes, 57*(10), 812–820. https://doi.org/10.1080/03601234.2022.2122664

Leng, S., Sun, R., Yang, X., Jin, M., & Chen, L. (2023). Diverse types of coupling trends in urban tree and nontree vegetation associated with urbanization levels. *npj Urban Sustain, 3*, 33. https://doi.org/10.1038/s42949-023-00111-2

Li, B. V., Jenkins, C. N., & Xu, W. (2022). Strategic protection of landslide vulnerable mountains for biodiversity conservation under land-cover and climate change impacts. *Proceedings of the National Academy of Sciences of the United States of America, 119*(2), e2113416118. https://doi.org/10.1073/pnas.2113416118

Lighty, J. S., Veranth, J. M., & Sarofim, A. F. (2000). Combustion aerosols: Factors governing their size and composition and implications to human health. *Journal of the Air & Waste Management Association (1995), 50*(9), 1565–1622. https://doi.org/10.1080/10473289.2000.10464197

Lin, T., Gibson, V., Cui, S., Yu, C. P., Chen, S., Ye, Z., & Zhu, Y. G. (2014). Managing urban nutrient biogeochemistry for sustainable urbanization. *Environmental Pollution (Barking, Essex: 1987), 192*, 244–250. https://doi.org/10.1016/j.envpol.2014.03.038

Littell, J. S., Peterson, D. L., Riley, K. L., Liu, Y., & Luce, C. H. (2016). A review of the relationships between drought and forest fire in the United States. *Global Change Biology, 22*(7), 2353–2369. https://doi.org/10.1111/gcb.13275

Long, D. T., & Berry Lyons, W. (2021). Urban geochemistry. In *Encyclopedia of geology (Second Edition)* (pp. 235–250). https://doi.org/10.1016/B978-0-12-409548-9.12526-6

Longo, M., Saatchi, S., Keller, M., Bowman, K., Ferraz, A., Moorcroft, P. R., Morton, D. C., Bonal, D., Brando, P., Burban, B., Derroire, G., Dos-Santos, M. N., Meyer, V., Saleska, S.,

Trumbore, S., & Vincent, G. (2020). Impacts of Degradation on Water, Energy, and Carbon Cycling of the Amazon Tropical Forests. *Journal of Geophysical Research. Biogeosciences, 125*(8), e2020JG005677. https://doi.org/10.1029/2020JG005677

Lorenz, K., & Lal, R. (2009). Biogeochemical C and N cycles in urban soils. *Environment International, 35*, 1–8. https://doi.org/10.1016/j.envint.2008.05.006

Loreto, F., Dicke, M., Schnitzler, J. P., & Turlings, T. C. (2014). Plant volatiles and the environment. *Plant, Cell & Environment, 37*(8), 1905–1908. https://doi.org/10.1111/pce.12369

Łuczak, K., Czerniawska-Kusza, I., Rosik-Dulewska, C., & Kusza, G. (2020). Effect of NaCl road salt on the ionic composition of soils and Aesculus hippocastanum L. foliage and leaf damage intensity. *Scientific Reports, 11*, 5309. https://doi.org/10.1038/s41598-021-84541-x

Lun, X., Lin, Y., Chai, F., Fan, C., Li, H., & Liu, J. (2020). Reviews of emission of biogenic volatile organic compounds (BVOCs) in Asia. *Journal of Environmental Sciences (China), 95*, 266–277. https://doi.org/10.1016/j.jes.2020.04.043

Mageshen, V. R., Meena, S., Santhy, P., & Latha, M. R. (2022). Transition of natural pedology into man made anthropedogenesis through anthropocene epoch- a review. *YMER Digital, 21*, 464–483. https://doi.org/10.37896/YMER21.01/43

Manisalidis, I., Stavropoulou, E., Stavropoulos, A., & Bezirtzoglou, E. (2020). Environmental and health impacts of air pollution: A review. *Frontiers in Public Health, 8*, 14. https://doi.org/10.3389/fpubh.2020.00014

Mann, D., Anees, M. M., Rankavat, S., & Joshi, P. K. (2020). Spatio-temporal variations in landscape ecological rish related to road network in the central Himalaya. *Human and Ecological Risk Assessment: An International Journal*, 1–18. https://doi.org/10.1080/10807039.2019.1710693

Manning, W. J. (2013). Urban environment: Recognizing that pollution and social factors can create barriers to development of more healthy and liveable cities. *Environmental Pollution, 183*, 1. https://doi.org/10.1016/j.envpol.2013.07.003

Medverec Knežević, Z., Nadih, M., Josipović, R., Grgić, I., & Cvitković, A. (2011). Zagađenje pitke vode mineralnim uljima u Slavonskom Brodu [Mineral oil drinking water pollution accident in Slavonski Brod, Croatia]. *Arhiv za Higijenu Rada i Toksikologiju, 62*(4), 349–356. https://doi.org/10.2478/10004-1254-62-2011-2119

Mętrak, M., et al. (2015). Does the function of railway infrastructure determine qualitative and quantitative composition of contaminants (PAHs, heavy metals) in soil and plant biomass? *Water, Air, & Soil Pollution, 226*, 253. https://doi.org/10.1007/s11270-015-2516-1

Meyer, P. A., Brown, M. J., & Falk, H. (2008). Global approach to reducing lead exposure and poisoning. *Mutation Research, 659*(1-2), 166–175. https://doi.org/10.1016/j.mrrev.2008.03.00

Mishra, N. B., & Mainali, K. P. (2017). Greening and browning of the Himalaya: Spatial patterns and the role of climatic change and human drivers. *The Science of the Total Environment, 587-588*, 326–339. https://doi.org/10.1016/j.scitotenv.2017.02.156

Miura, K., Shimada, K., Sugiyama, T., Sato, K., Takami, A., Chan, C. K., Kim, I. S., Kim, Y. P., Lin, N. H., & Hatakeyama, S. (2019). Seasonal and annual changes in PAH concentrations in a remote site in the Pacific Ocean. *Scientific Reports, 9*(1), 12591. https://doi.org/10.1038/s41598-019-47409-9

Mo, S., Wang, Y., Xiong, F., & Ai, C. (2019). Effects of asphalt source and mixing temperature on the generated asphalt fumes. *Journal of Hazardous Materials, 371*, 342–351. https://doi.org/10.1016/j.jhazmat.2019.03.025

Molina, M. J., & Molina, L. T. (2004). Megacities and atmospheric pollution. *Journal of the Air & Waste Management Association (1995), 54*(6), 644–680. https://doi.org/10.1080/10473289.2004.10470936

Morancho, A. B. (2003). A hedonic valuation of urban green areas. *Landscape and Urban Planning, 66*(1), 35–41. https://doi.org/10.1016/S0169-2046(03)00093-8

Morrow, S., Smolen, M., Stiegler, J., & Cole, J. (2017). *Using vegetation for erosion control on construction sites*. Oklahoma State University.

Nottingham, A. T., Bååth, E., Reischke, S., Salinas, N., & Meir, P. (2019). Adaptation of soil microbial growth to temperature: Using a tropical elevation gradient to predict future changes. *Global Change Biology, 25*(3), 827–838. https://doi.org/10.1111/gcb.14502

Pandit, M. K. (2020). The Himalaya should be a nature reserve. *Nature, 583*(7814), 9. https://doi.org/10.1038/d41586-020-01809-4

Park, C. R., & Lee, W. S. (2000). Relationship between species composition and area in breeding birds of urban woods in Seoul, Korea. *Landscape and Urban Planning, 51*(1), 29–36. https://doi.org/10.1016/S0169-2046(00)00094-3

Pataki, D., Carreiro, M., Cherrier, J., Grulke, N., Jennings, V., Pincetl, S., Pouyat, R., Whitlow, T., & Zipperer, W. (2011). Coupling biogeochemical cycles in urban environments: Ecosystemservices, green solutions, and misconceptions. *Frontiers Ecology Environment, 9*(1), 27–36. https://doi.org/10.1890/090220

Patoine, G., Eisenhauer, N., Cesarz, S., Phillips, H. R. P., Xu, X., Zhang, L., & Guerra, C. A. (2022). Drivers and trends of global soil microbial carbon over two decades. *Nature Communications, 13*(1), 4195. https://doi.org/10.1038/s41467-022-31833-z

Patterson, A. D., Gonzalez, F. J., & Idle, J. R. (2010). Xenobiotic metabolism- a view through the metabolometer. *Chemical Research in Toxicology, 23*(5), 851–860. https://doi.org/10.1021/tx100020p

Phillips, N. G., Ackley, R., Crosson, E. R., Down, A., Hutyra, L. R., Brondfield, M., Karr, J. D., Zhao, K., & Jackson, R. B. (2013). Mapping urban pipeline leaks: Methane leaks across Boston. *Environmental Pollution (Barking, Essex: 1987), 173*, 1–4. https://doi.org/10.1016/j.envpol.2012.11.003

Phillips, B. B., Bullock, J. M., Osborne, J. L., & Gaston, K. J. (2021). Spatial extent of road pollution: A national analysis. *The Science of the Total Environment, 773*, 145589. https://doi.org/10.1016/j.scitotenv.2021.145589

Ponette-González, A. G., Chen, D., Elderbrock, E., Rindy, J. E., Barrett, T. E., Luce, B. W., Lee, J. H., Ko, Y., & Weathers, K. C. (2022). Urban edge trees: Urban form and meteorology drive elemental carbon deposition to canopies and soils. *Environmental pollution (Barking, Essex: 1987), 314*, 120197. https://doi.org/10.1016/j.envpol.2022.120197

Portillo-Estrada, M., Zenone, T., Arriga, N., & Ceulemans, R. (2018). Contribution of volatile organic compound fluxes to the ecosystem carbon budget of a poplar short-rotation plantation. Global Change Biology. *Bioenergy, 10*(6), 405–414. https://doi.org/10.1111/gcbb.12506

Pouyat, R., Groffman, P., Yesilonis, I., & Hernandez, L. (2002). Soil carbon pools and fluxes in urban ecosystems. *Environmental pollution (Barking, Essex: 1987), 116*(Suppl 1), S107–S118. https://doi.org/10.1016/s0269-7491(01)00263-9

Pouyat, R. V., Yesilonis, I. D., & Nowak, D. J. (2006). Carbon storage by urban soils in the United States. *Journal of Environmental Quality, 35*(4), 1566–1575. https://doi.org/10.2134/jeq2005.0215

Raiter, K. G., et al. (2018). Linear infrastructure impacts on landscape hydrology. *Journal of Environmental Management, 206*, 446–457. https://doi.org/10.1016/j.jenvman.2017.10.036

Ramírez-Camacho, J. G., Pastor, E., Casal, J., Amaya-Gómez, R., & Muñoz-Giraldo, F. (2015). Analysis of domino effect in pipelines. *Journal of Hazardous Materials, 298*, 210–220. https://doi.org/10.1016/j.jhazmat.2015.05.033

Reisinger, A. J., Groffman, P. M., & Rosi-Marshall, E. J. (2016). Nitrogen cycling process rates across urban ecosystems. *FEMS Microbiology Ecology, 92*(12), 1–11. https://doi.org/10.1093/femsec/fiw198

Ren, Y., Qu, Z., Du, Y., Xu, R., Ma, D., Yang, G., Shi, Y., Fan, X., Tani, A., Guo, P., Ge, Y., & Chang, J. (2017). Air quality and health effects of biogenic volatile organic compounds emissions from urban green spaces and the mitigation strategies. *Environmental Pollution (Barking, Essex: 1987), 230*, 849–861. https://doi.org/10.1016/j.envpol.2017.06.049

Rindy, J. E., Ponette-González, A. G., Barrett, T. E., Sheesley, R. J., & Weathers, K. C. (2019). Urban trees are sinks for soot: Elemental carbon accumulation by two widespread Oak species. *Environmental Science & Technology, 53*(17), 10092–10101. https://doi.org/10.1021/acs.est.9b02844

Rivers, E. N., Heitman, J. L., McLaughlin, R. A., & Howard, A. M. (2021). Reducing roadside runoff: Tillage and compost improve stormwater mitigation in urban soils. *Journal of Environmental Management, 280*, 111732. https://doi.org/10.1016/j.jenvman.2020.111732

Rødland, E. S., Okoffo, E. D., Rauert, C., Heier, L. S., Lind, O. C., Reid, M., Thomas, K. V., & Meland, S. (2020). Road de-icing salt: Assessment of a potential new source and pathway of microplastics particles from roads. *The Science of the Total Environment, 738*, 139352. https://doi.org/10.1016/j.scitotenv.2020.139352

Romagnoli, P., Balducci, C., Cecinato, A., L'Episcopo, N., Gariazzo, C., Gatto, M. P., Gordiani, A., & Gherardi, M. (2017). Fine particulate-bound polycyclic aromatic hydrocarbons in vehicles in Rome, Italy. *Environmental Science and Pollution Research International, 24*(4), 3493–3505. https://doi.org/10.1007/s11356-016-8098-7

Rong, Y., Bates, P., & Neal, J. (2022). *Quantifying the impact of soil moisture dynamics on UK flood hazard under climate change*. s.l., EGU General Assembly 2022.

Rosser, N. J., Kincey, M. E., Oven, K. J., Densmore, A. L., Robinson, T. R., Pujara, D. S., Shrestha, R., Smutny, J., Gurung, K., Lama, S., & Dhital, M. R. (2021). Changing significance of landslide Hazard and risk after the 2015 Mw 7.8 Gorkha, Nepal earthquake. Progress in Disaster. *Science, 10*, 100159. https://doi.org/10.1016/j.pdisas.2021.100159

Saha, B. (2023). *Joshimath sank 10 cm every year since 2018, advanced satellite image analysis reveals*. India Today.

Savard, J. P. L., Clergeau, P., & Mennechez, G. (2000). Biodiversity concepts and urban ecosystems. *Landscape and Urban Planning, 48*(3–4), 131–142. https://doi.org/10.1016/S0169-2046(00)00037-2

Schmeller, D. S., Urbach, D., Bates, K., Catalan, J., Cogălniceanu, D., Fisher, M. C., Friesen, J., Füreder, L., Gaube, V., Haver, M., Jacobsen, D., Le Roux, G., Lin, Y. P., Loyau, A., Machate, O., Mayer, A., Palomo, I., Plutzar, C., Sentenac, H., Sommaruga, R., et al. (2022). Scientists' warning of threats to mountains. *The Science of the Total Environment, 853*, 158611. https://doi.org/10.1016/j.scitotenv.2022.158611

Schwab, J. (2009). *Planning the urban forest: Ecology, economy, and community development*. American Planning Association.

Sharma, G. P., & Raghubanshi, A. S. (2009). Plant invasions along roads: A case study from central highlands, India. *Environmental Monitoring and Assessment, 157*(1–4), 191–198. https://doi.org/10.1007/s10661-008-0527-8

Shehadeh, A., Alshboul, O., Tatari, O., Alzubaidi, M. A., & Hamed El-Sayed Salama, A. (2022). Selection of heavy machinery for earthwork activities: A multi-objective optimization approach using a genetic algorithm. *Alexandria Engineering Journal, 61*(10), 7555–7569. https://doi.org/10.1016/j.aej.2022.01.010

Shindell, D. (2013). *Radiative forcing in the AR5*. Retrieved from http://climate.envsci.rutgers.edu/climdyn2013/IPCC/IPCC_WGI12-Radiative Forcing.pdf

Siegel, F. R. (2002). *Environmental geochemistry of potentially toxic metals*. Springer.

Smith, R. M., et al. (2017). Influence of infrastructure on water quality and greenhouse gas dynamics in urban streams. *Biogeosciences, 14*(11), 1–35.

Smith, C. C., Espírito-Santo, F. D. B., Healey, J. R., Young, P. J., Lennox, G. D., Ferreira, J., & Barlow, J. (2020). Secondary forests offset less than 10% of deforestation-mediated carbon emissions in the Brazilian Amazon. *Global Change Biology, 26*(12), 7006–7020. https://doi.org/10.1111/gcb.15352

Special Correspondent. (2022). *Oil leaking from defunct pipeline contaminates water, soil in Tondiarpet*. The Hindu.

Srogi, K. (2007). Monitoring of environmental exposure to polycyclic aromatic hydrocarbons: A review. *Environmental Chemistry Letters, 5*(4), 169–195. https://doi.org/10.1007/s10311-007-0095-0

Stewart, I. D., & Kennedy, C. A. (2017). Metabolic heat production by human and animal populations in cities. *International Journal of Biometeorology, 61*(7), 1159–1171. https://doi.org/10.1007/s00484-016-1296-7

Sultan, H., Zhan, J., Rashid, W., Chu, X., & Bohnett, E. (2022). Systematic review of multidimensional vulnerabilities in the Himalayas. *International Journal of Environmental Research and Public Health, 19*(19), 12177. https://doi.org/10.3390/ijerph191912177

Sur, U., Singh, P., & Meena, S. R. (2020). Landslide susceptibility assessment in a lesser Himalayan Road corridor (India) applying fuzzy AHP technique and earth-observation data. *Geomatics, Natural Hazards and Risk, 11*(1), 2176–2209. https://doi.org/10.1080/19475705.2020.1836038

Tarsitano, E. (2006). Interaction between the environment and animals in urban settings: Integrated and participatory planning. *Environmental Management, 38*(5), 799–809. https://doi.org/10.1007/s00267-005-0148-8

Thapa, S., Li, H., Li, B., Fu, D., Shi, X., Yabo, S., Lu, L., Qi, H., & Zhang, W. (2021). Impact of climate change on snowmelt runoff in a Himalayan basin. *Nepal. Environmental Monitoring and Assessment, 193*(7), 393. https://doi.org/10.1007/s10661-021-09197-6

Thornton, I. (1991). Metal contamination of soils in urban areas. In P. Bullock & P. J. Gregory (Eds.), *Soils in the urban environment*. British Society of Soil Science, Blackwell Scientific Publications.

Thornton, I. (2012). Environmental geochemistry: 40 years research at Imperial College, London, UK. *Applied Geochemistry, 27*(5), 939–953.

Törmänen, T., Lindroos, A. J., Kitunen, V., & Smolander, A. (2020). Logging residue piles of Norway spruce, scots pine and silver birch in a clear-cut: Effects on nitrous oxide emissions and soil percolate water nitrogen. *The Science of the Total Environment, 738*, 139743. https://doi.org/10.1016/j.scitotenv.2020.139743

Van der Ree, R., Smith, D., & Grilo, C. (2015). *Handbook of road ecology*. John Wiley and Sons.

Van Wijnen, J. H., & van der Zee, S. C. (1998). Traffic-related air pollutants: Exposure of road users and populations living near busy roads. *Reviews on Environmental Health, 13*(1–2), 1–25.

Vogelsang, C., et al. (2019). *Microplastics in road dust – Characteristics, pathways and measures*. Norway Institute for Water Resources.

Wade, A. M., Richter, D. D., Craft, C. B., Bao, N. Y., Heine, P. R., Osteen, M. C., & Tan, K. G. (2021). Urban-soil pedogenesis drives contrasting legacies of Lead from paint and gasoline in city soil. *Environmental Science & Technology, 55*(12), 7981–7989. https://doi.org/10.1021/acs.est.1c00546

Walaszek, M., Bois, P., Laurent, J., Lenormand, E., & Wanko, A. (2018). Urban stormwater treatment by a constructed wetland: Seasonality impacts on hydraulic efficiency, physico-chemical behavior and heavy metal occurrence. *The Science of the Total Environment, 637-638*, 443–454. https://doi.org/10.1016/j.scitotenv.2018.04.325

Wang, J., Liu, G. H., Wang, J., Xu, X., Shao, Y., Zhang, Q., Liu, Y., Qi, L., & Wang, H. (2021a). Current status, existent problems, and coping strategy of urban drainage pipeline network in China. *Environmental Science and Pollution Research International, 28*(32), 43035–43049. https://doi.org/10.1007/s11356-021-14802-9

Wang, C., Morrissey, E. M., Mau, R. L., Hayer, M., Piñeiro, J., Mack, M. C., Marks, J. C., Bell, S. L., Miller, S. N., Schwartz, E., Dijkstra, P., Koch, B. J., Stone, B. W., Purcell, A. M., Blazewicz, S. J., Hofmockel, K. S., Pett-Ridge, J., & Hungate, B. A. (2021b). The temperature sensitivity of soil: Microbial biodiversity, growth, and carbon mineralization. *The ISME Journal, 15*(9), 2738–2747. https://doi.org/10.1038/s41396-021-00959-1

Wang, F., Wang, Q., Adams, C. A., Sun, Y., & Zhang, S. (2022). Effects of microplastics on soil properties: Current knowledge and future perspectives. *Journal of Hazardous Materials, 424*(Pt C), 127531. https://doi.org/10.1016/j.jhazmat.2021.127531

Wang, H., Yan, Z., Ju, X., Song, X., Zhang, J., Li, S., & Zhu-Barker, X. (2023). Quantifying nitrous oxide production rates from nitrification and denitrification under various moisture conditions in agricultural soils: Laboratory study and literature synthesis. *Frontiers in Microbiology, 13*, 1110151. https://doi.org/10.3389/fmicb.2022.1110151

Waters, E. R., Morse, J. L., Bettez, N. D., & Groffman, P. M. (2014). Differential carbon and nitrogen controls of denitrification in riparian zones and streams along an urban to exurban gradient. *Journal of Environmental Quality, 43*(3), 955–963. https://doi.org/10.2134/jeq2013.12.0504

Weeks, J. R. (2010). Defining urban areas. In T. Rashed & C. Jürgens (Eds.), *Remote sensing of urban and suburban areas. Remote sensing and digital image processing 10*. Springer. https://doi.org/10.1007/978-1-4020-4385-7_3

Weiskopf, S. R., Rubenstein, M. A., Crozier, L. G., Gaichas, S., Griffis, R., Halofsky, J. E., Hyde, K. J. W., Morelli, T. L., Morisette, J. T., Muñoz, R. C., Pershing, A. J., Peterson, D. L., Poudel, R., Staudinger, M. D., Sutton-Grier, A. E., Thompson, L., Vose, J., Weltzin, J. F., & Whyte, K. P. (2020). Climate change effects on biodiversity, ecosystems, ecosystem services, and natural resource management in the United States. *The Science of the Total Environment, 733*, 137782. https://doi.org/10.1016/j.scitotenv.2020.137782

Weller, Z. D., Im, S., Palacios, V., Stuchiner, E., & von Fischer, J. C. (2022). Environmental injustices of leaks from urban natural gas distribution systems: Patterns among and within 13 U.S. metro areas. *Environmental Science & Technology, 56*(12), 8599–8609. https://doi.org/10.1021/acs.est.2c00097

White, J. G., Antos, M. J., Fitzsimons, J. A., & Palmer, G. C. (2005). Non-uniform bird assemblages in urban environments: The influence of street vegetation. *Landscape and Urban Planning, 71*(2–4), 123–135. https://doi.org/10.1016/j.landurbplan.2004.02.006

Wiłkomirski, B., Sudnik-Wójcikowska, B., Galera, H., Wierzbicka, M., & Malawska, M. (2011). Railway transportation as a serious source of organic and inorganic pollution. *Water, Air, and Soil Pollution, 218*(1–4), 333–345. https://doi.org/10.1007/s11270-010-0645-0

Wong, C. S., Li, X., & Thornton, I. (2006). Urban environmental geochemistry of trace metals. *Environmental Pollution (Barking, Essex: 1987), 142*(1), 1–16. https://doi.org/10.1016/j.envpol.2005.09.004

Wright, J. S., Fu, R., Worden, J. R., Chakraborty, S., Clinton, N. E., Risi, C., Sun, Y., & Yin, L. (2017). Rainforest-initiated wet season onset over the southern Amazon. *Proceedings of the National Academy of Sciences of the United States of America, 114*(32), 8481–8486. https://doi.org/10.1073/pnas.1621516114

Wu, S., & Li, B. (2022). Sustainable linear infrastructure route planning model to balance conservation and socioeconomic development. *Biological Conservation, 266*, 109449. https://doi.org/10.1016/j.biocon.2022.109449

Xu, H., Lian, X., Slette, I. J., Yang, H., Zhang, Y., Chen, A., & Piao, S. (2022). Rising ecosystem water demand exacerbates the lengthening of tropical dry seasons. *Nature Communications, 13*(1), 4093. https://doi.org/10.1038/s41467-022-31826-y

Yadav, J. S., Tiwari, S. K., Misra, A., Rai, S. K., & Yadav, R. K. (2021). High-altitude meteorology of Indian Himalayan region: Complexities, effects, and resolutions. *Environmental Monitoring and Assessment, 193*(10), 654. https://doi.org/10.1007/s10661-021-09418-y

Yang, C., Lim, W., & Song, G. (2021). Reproductive toxicity due to herbicide exposure in freshwater organisms. Comparative biochemistry and physiology. *Toxicology & Pharmacology: CBP, 248*, 109103. https://doi.org/10.1016/j.cbpc.2021.109103

Zhang, L., Hoshika, Y., Carrari, E., Burkey, K. O., & Paoletti, E. (2018). Protecting the photosynthetic performance of snap bean under free air ozone exposure. *Journal of Environmental Sciences (China), 66*, 31–40. https://doi.org/10.1016/j.jes.2017.05.009

Zhang, Z., Gao, X., Zhang, S., Gao, H., Huang, J., Sun, S., Song, X., Fry, E., Tian, H., & Xia, X. (2022). Urban development enhances soil organic carbon storage through increasing urban vegetation. *Journal of Environmental Management, 312*, 114922. https://doi.org/10.1016/j.jenvman.2022.114922

Zhi, X., Chen, L., & Shen, Z. (2018). Impacts of urbanization on regional nonpoint source pollution: Case study for Beijing, China. *Environmental Science and Pollution Research International, 25*(10), 9849–9860. https://doi.org/10.1007/s11356-017-1153-1

Zou, L., Cao, S., & Sanchez-Azofeifa, A. (2020). Evaluating the utility of various drought indices to monitor meteorological drought in Tropical Dry Forests. *International Journal of Biometeorology, 64*(4), 701–711. https://doi.org/10.1007/s00484-019-01858-z

Chapter 8
The Future Developments in Biogeochemistry

Michael O'Neal Campbell

Abstract The future of biogeochemistry and its relationship with environmental change is largely based on how the discipline and associated sciences react to global events and the disciplinary and methodological developments that occur as a result. It may be hypothesized that future developments in such studies will be dominated by models, some analytical, others predictive and retrospective, considering the increasingly integrated nature of the relevant topics of biogeochemical analyses, technological developments, and the increasingly complex and trending nature of environmental change. Current evidence indicates that biogeochemistry has a key, even indispensable, role in the future of the environmental sciences, including biogeography, oceanography, Earth Systems Science, and even geomatics-based applications, but the strength of the role will depend on the structural organization of the discipline, in terms of its links with other related disciplines, its flexibility in reaction to academic, research, and human–environmental changes, and the deeper understanding of the factors (natural and socioeconomic) that contribute to environmental change. Research methodologies are advancing with technology, especially computer-generated developments such as geomatics, and the understanding of such applications may dominate the future as high-tech methods overtaking the older, less sophisticated research methods of the late twentieth century. This assessment contributes to a critical look at the future of biogeochemical applications.

Introduction

Current scientific assessments cite biogeochemistry as one of the most important benchmarks for understanding the future of the relevant environmental sciences (Geodynamics Program, 2023). The most important issues concern atmospheric

M. O. Campbell (✉)
Simon Fraser University, Burnaby, BC, Canada
e-mail: evaz@torontomu.ca

M. O. Campbell, *Biogeochemistry and the Environment*,
https://doi.org/10.1007/978-3-031-47017-2_8

and Earth temperature change, biosphere changes relevant to biogeochemical cycling, and the role of complex variations of oceanic changes (Bianchi, 2020). For example, Bianchi et al. (2021) argue that "biogeochemistry has an important role to play in many environmental issues of current concern related to global change and air, water, and soil quality. However, reliable predictions and tangible implementation of solutions, offered by biogeochemistry, will need further integration of disciplines." The Geodynamics Program (2023) notes the importance of global biogeochemistry: "The biogeochemistry of Earth's Ocean and atmosphere makes it habitable to life, a characteristic likely unique in our solar system. Greenhouse gases in the atmosphere (mainly, CO_2) sustain temperatures that allow liquid water to be present at the surface. Additionally, Earth has supported an oxygen-rich atmosphere over the past couple billion years." It is argued that these environmental variables have created and modified the complex feature of the biosphere, and "thus, understanding the collective influence of the processes affecting the biogeochemical variations of Earth over various timescales is necessary to understand its past history as well as how it will evolve into the future" (Geodynamics Program, 2023).

The future of biogeochemistry is generally deemed complicated, and vital, but even problematic and critical. For example, Likens (2004: 5) poses some "big, important questions to drive biogeochemical research in the future." These listed as questions on the loss and retention of nitrogen watershed-ecosystems, the rates and controls on nitrogen fixation and denitrification in variable ecosystems, the role of spatial and temporal scale in biogeochemical flux and cycling, the regional to global scales of biogeochemical functions, the factors for weathering rates in terrestrial ecosystems and the destiny of weathered products, and the quantitative relations between hydrologic and biogeochemical cycles. Likens (2004: 5) argues that there are some answers, which concern the possibilities of multidisciplinary foci and personnel training, and the use of sophisticated tools such as stable isotopes, spatial statistics, and remote sensing.

Doney et al. (2014: 108) pose a broader approach including anthropogenic issues: "Changing atmospheric composition due to human activities, primarily carbon dioxide (CO_2) emissions from fossil fuel burning, is already impacting ocean circulation, biogeochemistry, and ecology, and model projections indicate that observed trends will continue or even accelerate over this century." Additionally, higher atmospheric CO_2 may change the Earth's radiative balance, causing global-scale warming. Most of the resulting anomalous heat is in the ocean, which also drives physical, chemical, and biological impacts. It is further argued that there may be changes in oceanic chemistry (ocean acidification and declining pH) as increased ocean surface warming and vertical stratification are predicted to reduce global-integrated primary production and export flux, and lower subsurface dissolved oxygen concentrations. Upper trophic levels may be affected by warming and productivity changes, and larger low oxygen zones. It is noted that the ocean currently absorbs about one-quarter of present-day anthropogenic CO_2 emissions, and that future climate warming may slow the ocean CO_2 uptake, but this may not significantly reduce upper ocean acidification. The important development for the understanding of such dynamics and the accuracy of any future model projections may need "better observational constraints on current rates of ocean change and a better understanding of the mechanisms controlling key physical and biogeochemical processes" (Doney et al. (2014: 108).

The rest of this chapter will examine the developments in biogeochemical studies that can assess such future trends and the predictive modeling that may be necessary to understand, measure, hypothesize, or even guess the processes leading to the future. Recent studies have become focused on the required structural changes in biogeochemistry and the associated sciences, and the possible changes in the actual factors for biogeochemical changes, such as possible temperature changes, sea level rises, industrial expansion, and the possibilities of inclusive modeling that may analyze these related and complex dynamics (Campbell, 2017, 2018). Such multidisciplinary modeling concerns the developments in technical sciences, mostly computing, and environmentally aware industrial modeling, and the interfaces between the biological, geological, and chemical sciences. Such analyses may open new avenues for predictive modeling but will require constant reanalysis and monitoring for decisive impacts.

Disciplinary Developments

A major development in biogeochemistry concerns the increased promotion of biogeochemical predictive modeling, a necessity considering the increased attention to current trends and future environmental possibilities in many current studies, which may link strongly with other related disciplines (Campbell, 2018). This is largely fueled by rapid environmental change, mostly from anthropogenic sources and impacting climatic and carbon cycling changes (see for example Fay et al., 2018; Lovenduski et al., 2016; McKinley et al., 2017; Peters et al., 2017; Ridge & McKinley, 2020). The future of the discipline of biogeochemistry may be at least partly dependent on the successful application of such models, as with rapid global environmental change, the scientific analysis may only be successful if it keeps pace with the environmental changes, both current and in the near future. Fransner et al. (2020) argue that "predictions of ocean biogeochemistry, such as primary productivity and CO_2 uptake, would help to understand the changing marine environment and the global climate." It is noted that recent studies in ocean physics are increasingly examining possible future trends, which have achieved successful predictions of ocean biogeochemistry. The authors however express caution at the strength if these studies, as "it is, however, unclear how much these predictions could be improved by also assimilating biogeochemical data to reduce uncertainties of the initial conditions" and "the mechanisms that lead to biogeochemical predictability are poorly understood" (Fransner et al., 2020). Gehlen et al. (2015) concur and posit that "building the capacity for monitoring and forecasting marine biogeochemistry and ecosystem dynamics is a scientific challenge of strategic importance in the context of rapid environmental change and growing public awareness of its potential impacts on marine ecosystems and resources" and there are current efforts in this direction, such as the National Operational Oceanography centers, which are integrating biogeochemical methods into in operational systems (Gehlen et al., 2015).

Biogeography and associated spatial sciences are also forecast to have a useful future, especially with links to conservation biology, human dimensions of the environment, and as a background of global and regional, human-altered biogeochemistry (Campbell, 2018). Some of these trends may be manifested in the leading biogeography journals (Diversity and Distributions, Ecography, Frontiers of Biogeography, Global Ecology and Biogeography, and Journal of Biogeography) (McGill et al. 2018). Due to the impact of human action on the environment, the global ecosystem is changing, and this means that the scientific disciplines that study the environment must adjust accordingly to stay relevant (Campbell, 2017, 2018). Arnillas et al. (2019: 550) hold that "humans have now altered essentially every natural ecosystem in the world, and among the numerous consequences of anthropogenic global change, many of the Earth's species are currently living under drastically different environmental and ecological conditions." The consequences of this massive change in ecosystems are that several species are reaching local, regional, and even global extinction levels, while some more adaptive species are currently becoming invasive in human-dominated spaces. Arnillas et al. (2019: 550) contend that "to address this major challenge, it is critical that conservation practitioners understand the multiple short- and long-term climatological, geological, and evolutionary mechanisms that have resulted in the current distribution of species; understanding how these same mechanisms interact is also key in predicting species distributions – and possible extinctions – into the future."

The emergence of the subdiscipline of conservation biogeography reflects the growing link between biogeography as a spatial discipline and conservation biology, with a backbone of biogeochemistry, and illustrates the disciplinary structural change in reaction to changes in the objects of study (Campbell, 2018). Arnillas et al. (2019: 56) argue that "while the principles of biogeography tend to relate to 'bigger picture' changes in the Earth systems, its conceptual tools and methods are widely applied to address real-world conservation problems." Recent developments on the trend of increasing links between biogeography and conservation biology, and possible consequent developments within conservation biogeography, include global databases such as the Global Biodiversity Information Facility (GBIF), relevant software packages and computer systems, and internet access, which "increase our capacity to analyze and interpret current biogeographic trends and use this information to inform conservation policy and practice" (Arnillas et al. (2019: 56; see also Moritz et al. 2011).

Technological Developments

Technological developments are crucial for the future of biogeochemical studies and applications, which are based mainly on the tools of geomatics, such as satellites and satellite-mounted sensing and survey equipment, and also come with developments within the chemical and computing sciences (Campbell, 2018). Li and Yang (2015) note that "Global Change studies are increasingly considered a vital source of information to understand the Earth Environment, especially in the framework of human-induced, climate change and land use transformation. Satellite Earth Observing systems and geomatics technologies provide a unique tool to monitor and model the changes, respectively." It is also contended that the range of applications and innovative techniques is increasing, and satellite-derived image data offers more critical and detailed information, which underscores the usefulness of geomatics technologies and global environmental issues. Therefore, geomatics technologies provide powerful tools to model and analyze the effects of those global environmental changes towards minimizing their adverse impacts on human health and the environment" (Li & Yang, 2015).

"Big Data" is an important development that is increasingly important for the environmental sciences (Campbell, 2022). For example, Farley et al. (2018: 563) point out that "Ecology has joined a world of big data. Two complementary frameworks define big data: data that exceed the analytical capacities of individuals or disciplines or the 'Four Vs' axes of volume, variety, veracity, and velocity," and "rates of ecological data generation, aggregation, and interpretation are increasing on many fronts, with rapid growth in data volumes, methods of data collection, and new analytical and computational approaches." Recent developments include the collection of petabytes (one petabyte is equivalent to one million gigabytes) of data by the remote sensing Earth observation technology, the combination of limited, small-scale scientific observations and experiments into large, community data resources (with the examples cited of AmeriFlux, Global Biodiversity Information Facility, NutNet, and Neotoma), the development of long-term ecological monitoring networks, ranging from national to continental scales (examples being LTER, NEON, and CZO), and increased use of smaller automated, data-gathering sensor networks (including phenology cameras, wildlife camera traps, and temperature loggers) (Farley et al., 2018: 563). Looking to the future scenarios, however, Farley et al. (2018: 563) caution that "big data present an array of challenges related to large data volumes, high data heterogeneity (variety), varying quality and uncertainty (veracity), and a need for timely information" (velocity) (see also Yang & Huang, 2013).

Wang and Menenti (2021) write about the developments of LiDAR that may have future relevance: "Measurement concepts and technology are evolving simultaneously in different directions" and "the role of LiDAR will be increasingly important in the future. Although there are many potentials for new lidar technology advancements, LiDAR activities are gradually shifting from technology developments to applications." An important issue for applications to vast earth systems such as atmospheric dynamics is the limitation of single LiDAR measurements in their ability in broad application to global-scale systems. Wang and Menenti (2021) mention that many LiDAR networks, including the European Aerosol Research Lidar Network (EARLINET), the Asian Dust Network (AD-Net), the National Aeronautics and Space Administration Micropulse LiDAR network (MPLNET), and the Network for the Detection of Atmospheric Composition Change (NDACC) have been developed with variable successes in applications. A system of network LiDAR operations, at different scales for varying scaled applications, including regional and global levels is vital (Wang & Menenti, 2021: see also Welton et al., 2001; Pappalardo et al., 2014; Wulfmeyer et al., 2015; Sugimoto et al., 2014; Chu & Yu, 2017; De Mazière et al., 2018; Langford et al., 2018).

Van Rens (2020) notes that LiDAR technology, open data, and artificial intelligence/machine learning (AI/ML) are "developing rapidly," the evidence being the increased number of organizations entering the LiDAR market, strong improvements in pulse repetition rates (PRRs) and miniaturization (smaller LiDAR systems on smaller more flexible platforms such as unmanned aerial vehicles or drones), the development of optimized sensors for corridor mapping enabled by vertical takeoff and landing (VTOL) lightweight platforms, which have higher measurement accuracy, speed, and multitarget capability, with these developments making critical contributions to the progress of the geospatial market and its applications. The developments of LiDAR technology and artificial intelligence/machine learning fulfill the demands for more accurate data, which can have superior environmental applications. Van Rens (2020) notes that "the promise of LiDAR is in the accuracy and completeness of the information and the integration of the relevant additional sensor data to provide the high-fidelity visualization that is needed." Examples are the NOAA Shoreline Data Explorer, which can measure and monitor storms and sea rise using bathymetric LiDAR data, as the "impact of smaller UAV-borne bathymetric systems will highlight the ability of local districts to monitor critical wetlands and calculate water loads in rivers and streams to protect us from recurring flooding." The 3DEP program of the United States Geological Survey (USGS) is an example, which has wide applications, where "the massive amount of geospatial data is free to all and it has been downloaded by many in Silicon Valley to enable them to check their maps and correct their products" (Van Rens, 2020).

Anthropogenic Developments

The changes in global biogeochemistry and associated biogeographic impacts are due to human actions, and largely concern carbon and organic chemical level changes (Campbell, 2018). Silva (2022: 19) argues that "the unfolding climate crisis is in many respects a human issue, one caused by anthropogenic emissions of CO_2 to the atmosphere, and that can only be solved through a concerted effort across all sectors of society." Silva (2022: 19) explains that increasing the scope of biogeochemical research may enable a stronger more impactful climate change mitigation and adaptation agenda. Actions for the future include land conservation, management, and restoration. It is contended that there is a spatial correlation between the abundance of threatened plant and animal species and the distribution of carbon- and water-rich habitats in key regions, and these regions can be prioritized for biodiversity protection, which may yield major climatic benefits. It is also argued that "long-term records of socioecological change warrant a revision of current models for sustainable forestry and agriculture in favor of decentralized system-specific prescriptions, including interventions where organic or inorganic carbon capture may improve wood and food production under future climate scenarios" (Silva, 2022: 19). Scientific research on degraded landscapes has shown evidence of carbon stabilization, including changes in organic complexes, which also document the positive outcomes of ecological succession and possibilities of carbon accumulation above thresholds predicted for undisturbed ecosystems. The author concludes that the crucial factor is "the potential for innovation at the intersection of basic and applied biogeoscience, which could accelerate atmospheric carbon capture through new discoveries and collective action" (Silva, 2022: 19).

Galloway et al. (2014: 351) affirm that global-scale alterations of biogeochemical cycles are a current trend and the impacts and implications trend from now into the future. Global carbon dioxide emissions are asserted to be the most significant driver of human-caused climate change, but there are other human-accelerated cycles of other elements, including those of nitrogen, phosphorus, and sulfur, which may be relevant to climate and biogeochemical change. Climate change impacts biogeochemical changes, trending to future impacts on biogeochemical cycles, which may change future impacts on climate and affect our capacity to cope with coupled changes in climate, biogeochemistry, and other factors. Galloway et al. (2014: 351) further maintain that "carbon dioxide (CO_2) is the most abundant of the heat-trapping greenhouse gases that are increasing due to human activities" but "methane (CH_4) and nitrous oxide (N_2O) have higher greenhouse-warming potential per molecule than CO_2, and both are also increasing in the atmosphere" (Galloway et al., 2014: 351). It is also argued that "changes in biogeochemical cycles of carbon, nitrogen, phosphorus, and other elements – and the coupling of those cycles – can influence climate. In turn, this can change atmospheric composition in other ways that affect how the planet absorbs and reflects sunlight (for example, by creating small particles known as aerosols that can reflect sunlight)" (Galloway et al., 2014: 351).

Other biogeochemical problems concern the reduction of biodiversity in water- and land-based ecosystems due to climate change and nitrogen deposition, freshwater and marine eutrophication due to human actions increasing the nitrogen and phosphorus cycles (possibly progressively more dangerous with warmer future climates), increased ozone in the troposphere resulting from methane, organic compounds, and nitrogen oxide emissions, contributing to problems for biological productivity, and "rates of ozone formation are accelerated by higher temperatures, creating a reinforcing cycle between rising temperatures and continued human alteration of the nitrogen and carbon cycles," with possible actions including possibilities that "lowered NOx, NHx, and SOx emissions remove cooling agents from the atmosphere, but improve air quality" and "reductions in excess nitrogen and phosphorus from agricultural and industrial activities can potentially reduce the rate and impacts of climate change, while simultaneously addressing concerns in biodiversity, water quality, food security, and human health" (Galloway et al., 2014: 351: see also Chameides et al., 1994; Baron, 2006; Carpenter, 2008; Jacob & Winner, 2009; Rabalais et al., 2009; Smith & Schindler, 2009; Bobbink et al., 2010; Jeppesen et al., 2010; Howarth et al., 2011; Pardo et al., 2011; Peel et al., 2012; Townsend & Porder, 2012; Porter et al., 2013).

The future development of biogeochemistry and its study topics is also related to biogeography and its potential future developments, with an important link between the two disciplinary clusters being the shared development of a globally relevant, sometimes human relevant, perspective (Campbell, 2017, 2018). For example, concerning the status of biogeography, Parenti and Ebach (2009) point out that there is a possibility of a "global biogeography… that biogeography is a Big Science which deserves the attention and resources given to other large-scale, global scientific efforts. Global projects can accelerate species discovery, taxon, and area descriptions, and cyber taxonomy, and can set a standard for collaborative scientific research in the twenty-first century." Large-scale human actions have influenced global biogeography as they have also influenced global-scale human impacts in biogeochemistry. Hoffmann (2022: 325) also note the development of the spatial issue in biodiversity, in that "area-based conservation is essential to safeguard nature's diversity. In view of expanding human land use, increasing climate change and unmet conservation targets, area-based conservation requires efficiency and effectiveness more than ever" and "protected area management is particularly challenged by human land use, climate change, invasive species, and social, political and economic limitations." Future developments may show awareness of the requirement for reliable data on environmental trends, biocultural conservation, "climate-smart management," biosecurity approaches, effective economic policies, technologies such as advanced monitoring, remote sensing and data infrastructures, adaptive management, more effective protected areas, and "a global information system that is to support area-based conservation by synthesizing challenges and opportunities for protected area management effectiveness and efficiency at the local to global level" (Hoffmann (2022: 325).

Serra-Diaz and Franklin (2019: 492–493) also assert in an article entitled as "What's hot in conservation biogeography in a changing climate? Going beyond species range dynamics" that in the current period biogeochemical dynamics are coevolving with biogeographic change: "In recent decades Earth's rapidly changing climate, driven by anthropogenic greenhouse gas emissions, has affected species distributions and phenology, ecological communities and ecosystem processes, effects that are increasingly being observed globally" (see also Walther et al., 2002; Parmesan, 2006; Allen et al., 2010; Doney et al., 2012; Franklin et al., 2016). The important issue in the period concerns the rapidity of environmental change: compared with Pleistocene shifts in species ranges during glacial–interglacial transitions, "the pace of current anthropogenic warming outstrips past changes in the Earth system and climate, however, leading to new climate novelties and ecological communities" (Serra-Diaz & Franklin (2019: 492–493: see also Williams & Jackson, 2007; MacDonald et al., 2008; Ordonez et al., 2016; Nolan et al., 2018).

Conservation biogeography is presented as the discipline that "addresses the impacts of global change on the distribution of species, communities, and ecosystems with implications for large-scale conservation assessment and planning" and "species distribution models -statistical models associating the spatial distribution of species with climate and other environmental factors… have been used to project the impacts of climate change for large numbers of species, across taxonomic groups, at large spatial scales" (Serra-Diaz & Franklin (2019: 492–493: see also Franklin, 2010a, b; Richardson & Whittaker, 2010; Franklin, 2016). Species distribution models (SDMs) are therefore increasingly posited as relevant to the study of the trends and factors for conservation biogeography, with issues arising including more comprehensive approaches to risk assessments and projections (including disturbances), dispersals, biotic interactions, adaptive species ecologies, ecoevolutionary dynamics, and predictions of climate change (Serra-Diaz & Franklin (2019: 492–493; see also Urban et al., 2016). Serra-Diaz and Franklin (2019: 492–493) argue that "in the context of the persistent lack of widespread biodiversity data, with shortfalls ranging from species discoveries, lack of known ranges and shortage of abundance data, it is not surprising that SDMs are still a dominant tool in conservation biogeography, but we argue that their use under climate change needs to explore further dimensions of conservation biogeography, beyond range change projections."

The study by Serra-Diaz and Franklin (2019) explores methods for projections for climate change impacts on species distributions related to conservation biogeography, which move beyond the simpler, suitability-derived SDM outputs for species. The studies include links between climate and distribution modeling of parapatric species (species with large ranges, with possibilities form speciation), ecological sensitivities to climate change, interfaces between traditional species-cent red conservation biogeography and functional biogeography under climate change, physiological constraints for climate change projections, impacts of temporal dynamics on population trends and ranges, and the importance of examining multiple global change drivers spatially for predictive modeling for conservation biogeography, especially climate change and land-use change (Serra-Diaz & Franklin, 2019; see also Becker et al., 2019; Benedetti et al., 2019; Braz et al., 2019; Caddy-Retalic et al., 2019; Di Febbraro et al., 2019; Casties et al., 2019; Maxwell et al., 2019; Velazco et al., 2019; Wilson et al., 2019). Serra-Diaz and Franklin (2019: 494–495) conclude that "following calls to move beyond static species distribution modelling in order to forecast global change threats to global biodiversity…, researchers, including those featured in the Special Issue, have worked to fill the gap, applying new methods, data and experiments," "new approaches in conservation biogeography are still needed to forecast range shifts, and resulting conservation and restoration decisions," and "new avenues of research are needed to guide ecosystem management, inform conservation policy and design nature-based solutions to global change."

The technological developments within the field of geochemistry may also be important for biogeochemical research in the future (Brown & Calas, 2012). For example, Smith et al. (2018) point to the importance of global-scale, or continental-scale, geochemical surveys, which are used for broad scales of sampling and examine processes including "tectonics, climate, weathering, geochemical and mineralogical composition of the original soil parent material, continental-scale glaciation, topography, regional-scale alteration and mineralization, and in some cases, human activity." These authors note that several global-scale geochemical surveys have covered Australia, China, Europe, India, Mexico, and the United States during the twenty-first century, and there is a "hopeful future for continuing global-scale geochemical mapping in other parts of the world, with the ultimate product being a global geochemical database and atlas derived from this database." Arndt et al. (2017: 1) also point to future technological possibilities for geochemistry, focusing on mineral exploration and the industrial and environmental consequences. The "he critical roles that geoscientists will play in assuring continued supplies of minerals… include the development of new concepts and techniques that will assist the discovery, mining, processing, remediation, and management of mineral resources" (Arndt et al., 2017: 1).

Conclusions

This concluding chapter has examined current directions for the future of biogeochemistry and its links with developments of environmental research. The key components of these developments were identified as changes in the relevant disciplines, the technological advancements that enable the data and analysis of the changing disciplinary environment, and related anthropogenic changes that facilitate developments in the disciplines of biogeochemistry and related environmental sciences. The disciplinary developments were primarily concerned with predictive modeling, technological supports, and human dimensions of environmental change, which enable the understanding and management of related environmental changes and networks. It is evident that the breadth and complexity of the concerned disciplines make predictive modeling extremely difficult. This leads to the link with technological developments, mostly based on Big Data, geomatics, and chemistry techniques, which are required to measure the environmental changes behind biogeochemistry. The geomatics methods require technological advancements such as LiDAR, satellite imagery, a geodetic-background, computer-based analyses, and biochemical laboratory-based studies. Anthropogenic factors for environmental change are also increasingly emerging as important or even dominant factors for biogeochemical and biogeographical changes, and the inclusion of social science research methods may increasingly feature in such studies of environmental change. Future understandings of biogeochemistry and related environmental changes may thus hinge on the integration of these developments within a flexible disciplinary and practical research format, which may determine the utility of research findings for scientific understanding, policy actions and management strategies.

References

Allen, C. D., Macalady, A. K., Chenchouni, H., Bachelet, D., McDowell, N., Vennetier, M., et al. (2010). A global overview of drought and heat-induced tree mortality reveals emerging climate change risks for forests. *Forest Ecology and Management, 259*(4), 660–684. https://doi.org/10.1016/j.foreco.2009.09.001

Arndt, N. T., Fontboté, L., Hedenquist, J. W., Kesler, S. E., Thompson, J. F. H., & Wood, D. G. (2017). Future global mineral resources. *Geochemical Perspectives, 6*(1), 1–171. https://doi.org/10.7185/geochempersp.6.1

Arnillas, C., Smith, S., Ni, F., & Martin, A. (2019). Biogeography in conservation: Tools to explore the past and future of species in a changing world. *Lessons in Conservation, 9*, 55–94.

Baron, J. S. (2006). Hindcasting nitrogen deposition to determine an ecological critical load. *Ecological Applications, 16*, 433–439. https://doi.org/10.1890/1051-0761(2006)016[0433:HNDTDA]2.0.CO;2

Becker, E. A., Forney, K. A., Redfern, J. V., Barlow, J., Jacox, M. G., Roberts, J. J., & Palacios, D. M. (2019). Predicting cetacean abundance and distribution in a changing climate. *Diversity and Distributions, 25*(4), 626–643. https://doi.org/10.1111/ddi.12867

Benedetti, F., Ayata, S.-D., Irisson, J.-O., Adloff, F., & Guilhaumon, F. (2019). Climate change may have minor impact on zooplankton functional diversity in the Mediterranean Sea. *Diversity and Distributions, 25*(4), 568–581. https://doi.org/10.1111/ddi.12857

Bianchi, T. S. (2020). The evolution of biogeochemistry, revisited. *Biogeochemistry, 13*, 199–239. https://doi.org/10.1007/s10533-020-00708-0

Bianchi, T. S., Anand, M., Bauch, C. T., Canfield, D. E., De Meester, L., Fennel, K., Groffman, P. M., Pace, M. L., Saito, M., & Simpson, M. J. (2021). Ideas and perspectives: Biogeochemistry – Some key foci for the future. *Biogeosciences, 18*, 3005–3013. https://doi.org/10.5194/bg-18-3005-2021

Bobbink, R., Hicks, K., Galloway, J., Spranger, T., Alkemade, R., Ashmore, M., Bustamante, M., Cinderby, S., Davidson, E., Dentener, F., Emmett, B., Erisman, J. W., Fenn, M., Gilliam, F., Nordin, A., Pardo, L., & De Vries, W. (2010). Global assessment of nitrogen deposition effects on terrestrial plant diversity: A synthesis. *Ecological Applications, 20*(1), 30–59. https://doi.org/10.1890/08-1140.1

Braz, A. G., Lorini, M. L., & Vale, M. M. (2019). Climate change is likely to affect the distribution but not parapatry of the Brazilian marmoset monkeys (Callithrix spp.). *Diversity and Distributions, 25*(4), 536–550. https://doi.org/10.1111/ddi.12872

Brown, G. E., Jr., & Calas, G. (2012). Mineral-aqueous solution interfaces and their impact on the environment. *Geochemical Perspectives, 1*(4–5), 483–484. https://doi.org/10.7185/geochempersp.1.4

Caddy-Retalic, S., Hoffmann, B. D., Guerin, G. R., Andersen, A. N., Wardle, G. M., McInerney, F. A., & Lowe, A. J. (2019). Plant and ant assemblages predicted to decouple under climate change. *Diversity and Distributions, 25*(4), 551–567. https://doi.org/10.1111/ddi.12858

Campbell, M. (2017). *Biological conservation in the 20th century: A conservation biology of large wildlife*. Nova Science Publishers.

Campbell, M. (2018). *Geomatics and conservation biology*. Nova Science Publishers.

Campbell, M. (2022). *The great eagles their evolution, ecology and conservation*. Taylor and Francis (CRC Press).

Carpenter, S. R. (2008). Phosphorus control is critical to mitigating eutrophication. *Proceedings of the National Academy of Sciences of the United States of America, 105*(32), 11039–11040. https://doi.org/10.1073/pnas.0806112105

Casties, I., Clemmesen, C., & Briski, E. (2019). Environmental tolerance of three gammarid species with and without invasion record under current and future global warming scenarios. *Diversity and Distributions, 25*(4), 603–612. https://doi.org/10.1111/ddi.12856

Chameides, W.L., Kasibhatla, P.S., Yienger, J. & Levy, H. 1994. Growth of continental-scale metro-agro-plexes, regional ozone pollution, and world food production. Science (.) 264(5155): 74–77. Doi:https://doi.org/10.1126/science.264.5155.74.

Chu, X., & Yu, Z. (2017). Formation mechanisms of neutral Fe layers in the thermosphere at Antarctica studied with a thermosphere-ionosphere Fe/Fe+ (TIFe) model. *Journal of Geophysical Research: Space Physics, 122*(6), 6812–6848. https://doi.org/10.1002/2016JA023773

De Mazière, M., Thompson, A. M., Kurylo, M. J., Wild, J. D., Bernhard, G., Blumenstock, T., Braathen, G. O., Hannigan, J. W., Lambert, J.-C., Leblanc, T., McGee, T. J., Nedoluha, G., Petropavlovskikh, I., Seckmeyer, G., Simon, P. C., Steinbrecht, W., & Strahan, S. E. (2018).

The network for the detection of atmospheric composition change (NDACC): History, status, and perspectives. *Atmospheric Chemistry and Physics, 18*(7), 4935–4964. https://doi.org/10.5194/acp-18-4935-2018

Di Febbraro, M., Menchetti, M., Russo, D., Ancillotto, L., Aloise, G., Roscioni, F., et al. (2019). Integrating climate and land-use change scenarios in modelling the future spread of invasive squirrels in Italy. *Diversity and Distributions, 25*(4), 644–659. https://doi.org/10.1111/ddi.12890

Doney, S. C., Ruckelshaus, M., Emmett Duffy, J., Barry, J. P., Chan, F., English, C. A., et al. (2012). Climate change impacts on marine ecosystems. *Annual Review of Marine Science, 4*(1), 11–37. https://doi.org/10.1146/annurev-marine-041911-111611

Doney, S. C., Bopp, L., & Long, M. C. (2014). Historical and future trends in ocean climate and biogeochemistry. *Oceanography, 27*(1), 108–119. http://www.jstor.org/stable/24862126. https://doi.org/10.5670/oceanog.2014.14

Farley, S., Dawson, A., Goring, S. J., & Williams, J. W. (2018). Situating ecology as a big-data science: Current advances, challenges, and solutions. *BioScience, 68*(8), 563–576. https://doi.org/10.1093/biosci/biy068

Fay, A. R., Lovenduski, N. S., McKinley, G. A., Munro, D. R., Sweeney, C., Gray, A. R., Landschützer, P., Stephens, B. B., Takahashi, T., & Williams, N. (2018). Utilizing the Drake Passage time-series to understand variability and change in subpolar Southern Ocean pCO2. *Biogeosciences, 15*, 3841–3855. https://doi.org/10.5194/bg-15-3841-2018

Franklin, J. (2010a). *Mapping species distributions: Spatial inference and prediction*. Cambridge University Press.

Franklin, J. (2010b). Moving beyond static species distribution models in support of conservation biogeography: Moving beyond static species distribution models. *Diversity and Distributions, 16*(3), 321–330. https://doi.org/10.1111/j.1472-4642.2010.00641.x

Franklin, J. (2016). Diversity and distributions is (still) a journal of conservation biogeography. *Diversity and Distributions, 22*(1), 1–2. https://doi.org/10.1111/ddi.12402

Fransner, F., Counillon, F., Bethke, I., Tjiputra, J., Samuelsen, A., Nummelin, A., & Olsen, A. (2020). Ocean biogeochemical predictions—Initialization and limits of predictability. *Frontiers in Marine Science., 7*, 386. https://doi.org/10.3389/fmars.2020.00386

Galloway, J. N., Schlesinger, W. H., Clark, C. M., Grimm, N. B., Jackson, R. B., Law, B. E., Thornton, P. E., Townsend, A. R., & Martin, R. (2014). Ch. 15: Biogeochemical cycles. In J. M. Melillo, T. Richmond, & G. W. Yohe (Eds.), *Climate change impacts in the United States: The third National Climate Assessment* (pp. 350–368). U.S. Global Change Research Program. http://nca2014.globalchange.gov/report/sectors/biogeochemical-cycles

Gehlen, M., Barciela, R., Bertino, L., Brasseur, P., Butenschön, M., Chai, F., Crise, A., Drillet, Y., Ford, D., Lavoie, D., Lehodey, P., Perruche, C., Samuelsen, A., & Simon, E. (2015). Building the capacity for forecasting marine biogeochemistry and ecosystems: Recent advances and future developments. *Journal of Operational Oceanography, 8*(sup1), s168–s187. https://doi.org/10.1080/1755876X.2015.1022350

Geodynamics Program. (2023). *Biogeochemical Cycles: The Past, Present and Future Habitability of the Earth*. Retrieved from https://geodynamicsprogram.whoi.edu/2020-program/

Hoffmann, S. (2022). Challenges and opportunities of area-based conservation in reaching biodiversity and sustainability goals. *Biodiversity and Conservation, 31*, 325–352. https://doi.org/10.1007/s10531-021-02340-2

Howarth, R., Chan, F., Conley, D. J., Garnier, J., Doney, S. C., Marino, R., & Billen, G. (2011). Coupled biogeochemical cycles: Eutrophication and hypoxia in temperate estuaries and coastal marine ecosystems. *Frontiers in Ecology and the Environment, 9*(1), 18–26. https://doi.org/10.1890/100008

Jacob, D. J., & Winner, D. A. (2009). Effect of climate change on air quality. *Atmospheric Environment, 43*(1), 51–63. https://doi.org/10.1016/j.atmosenv.2008.09.051

Jeppesen, E., Meerhoff, M., Holmgren, K., González-Bergonzoni, I., Teixeira de Mello, F., Declerck, S., De Meester, L., Søndergaard, M., Lauridsen, T., Bjerring, R., Conde-Porcuna, J., Mazzeo, N., Iglesias, C., Reizenstein, M., Malmquist, H., Liu, Z., Balayla, D., & Lazzaro,

X. (2010). Impacts of climate warming on lake fish community structure and potential effects on ecosystem function. *Hydrobiologia, 646*, 73–90. https://doi.org/10.1007/s10750-010-0171-5

Langford, A. O., Alvarez, R., Brioude, J., Evan, S., Iraci, L. T., Kirgis, G., Kuang, S., Leblanc, T., Newchurch, M. J., Pierce, R. B., Senff, C., & Yates, E. L. (2018). Coordinated profiling of stratospheric intrusions and transported pollution by the tropospheric ozone Lidar network (TOLNet) and NASA alpha jet experiment (AJAX): Observations and comparison to HYSPLIT, RAQMS, and FLEXPART. *Atmospheric Environment, 174*, 1–14. https://doi.org/10.1016/j.atmosenv.2017.11.031

Li, J., & Yang, X. (2015). *Monitoring and Modeling of global changes: A geomatics perspective*. Springer. https://doi.org/10.1007/978-94-017-9813-6

Likens, G. E. (2004). Biogeochemistry: Some opportunities and challenges for the future. *Water, Air, & Soil Pollution: Focus, 4*, 5–24. https://doi.org/10.1023/B:WAFO.0000028341.75842.08

Lovenduski, N., McKinley, G. A., Fay, A. R., Lindsay, K., & Long, M. C. (2016). Partitioning uncertainty in ocean carbon uptake projections. *Global Biogeochemical Cycles, 30*(9), 1276–1287. https://doi.org/10.1002/2016GB005426

MacDonald, G. M., Bennett, K. D., Jackson, S. T., Parducci, L., Smith, F. A., Smol, J. P., & Willis, K. J. (2008). Impacts of climate change on species, populations and communities: Palaeobiogeographical insights and frontiers. *Progress in Physical Geography: Earth and Environment, 32*(2), 139–172. https://doi.org/10.1177/0309133308094081

Maxwell, S. L., Butt, N., Maron, M., McAlpine, C. A., Chapman, S., Ullmann, A., et al. (2019). Conservation implications of ecological responses to extreme weather and climate events. *Diversity and Distributions, 25*(4), 613–625. https://doi.org/10.1111/ddi.12878

McGill, B., Araújo, M., Franklin, J., Linder, H. P., & Dawson, M. N. (2018). Writing the future of biogeography. *Frontiers of Biogeography, 10*(3–4), 1–8. https://doi.org/10.21425/F5FBG41964

McKinley, G. A., Fay, A. R., Lovenduski, N. S., & Pilcher, D. J. (2017). Natural variability and anthropogenic trends in the ocean carbon sink. *Annual Review of Marine Science, 9*, 125–150. https://doi.org/10.1146/annurev-marine-010816-060529

Moritz, T., Krishnan, S., Roberts, D., Ingwersen, P., Agosti, D., Penev, L., Cockerill, M., & Chavan, V. (2011). Towards mainstreaming of biodiversity data publishing: Recommendations of the GBIF data publishing framework task group. *BMC Bioinformatics, 12*(Suppl 15), S1.

Nolan, C., Overpeck, J. T., Allen, J. R. M., Anderson, P. M., Betancourt, J. L., Binney, H. A., Brewer, S., Bush, M. B., Chase, B. M., Cheddadi, R., Djamali, M., Dodson, J., Edwards, M. E., Gosling, W. D., Haberle, S., Hotchkiss, S. C., Huntley, B., Ivory, S. J., Kershaw, A. P., Kim, S. H., et al. (2018). Past and future global transformation of terrestrial ecosystems under climate change. *Science (New York), 361*(6405), 920–923. https://doi.org/10.1126/science.aan5360

Ordonez, A., Williams, J. W., & Svenning, J.-C. (2016). Mapping climatic mechanisms likely to favour the emergence of novel communities. *Nature Climate Change, 6*(12), 1104–1109. https://doi.org/10.1038/nclimate3127

Pappalardo, G., Amodeo, A., Apituley, A., Comeron, A., Freudenthaler, V., Linné, H., Ansmann, A., Bösenberg, J., D'Amico, G., Mattis, I., Mona, L., Wandinger, U., Amiridis, V., Alados-Arboledas, L., Nicolae, D., & Wiegner, M. (2014). EARLINET: Towards an advanced sustainable European aerosol lidar network. *Atmospheric Measurement Techniques, 7*(8), 2389–2409. https://doi.org/10.5194/amt-7-2389-2014

Pardo, L. H., Fenn, M. E., Goodale, C. L., Geiser, L. H., Driscoll, C. T., Allen, E. B., Baron, J. S., Bobbink, R., Bowman, W. D., Clark, C. M., Emmett, B., Gilliam, F. S., Greaver, T. L., Hall, S. J., Lilleskov, E. A., Liu, L., Lynch, J. A., Nadelhoffer, K. J., Perakis, S. S., Robin-Abbott, M. J., Stoddard, J. L., Weathers, K. C., & Dennis, R. L. (2011). Effects of nitrogen deposition and empirical nitrogen critical loads for ecoregions of the United States. *Ecological Applications, 21*(8), 3049–3082. https://doi.org/10.1890/10-2341.1

Parenti, L. R., & Ebach, M. C. (2009). *The future of biogeography. Comparative biogeography: Discovering and classifying biogeographical patterns of a dynamic earth*. https://doi.org/10.1525/california/9780520259454.003.0010

Parmesan, C. (2006). Ecological and evolutionary responses to recent climate change. *Annual Review of Ecology, Evolution, and Systematics, 37*(1), 637–669. https://doi.org/10.1146/annurev.ecolsys.37.091305.110100

Peel, J. L., Haeuber, R., Garcia, V., Russell, A. G., & Neas, L. (2012). Impact of nitrogen and climate change interactions on ambient air pollution and human health. *Biogeochemistry, 114*, 121–134. https://doi.org/10.1007/s10533-012-9782-4

Peters, G. P., Le Quere, C., Andrew, R., Canadell, J. G., Friedlingstein, P., Ilyina, T., Jackson, R. B., Joos, F., Korsbakken, J. I., McKinley, G. A., Sitch, S., & Tans, P. (2017). Towards real-time verification of CO_2 emissions. *Nature Climate Change, 7*, 848–852. https://doi.org/10.1038/s41558-017-0013-9

Porter, E. M., Bowman, W. D., Clark, C. M., Compton, J. E., Pardo, L. H., & Soong, J. L. (2013). Interactive effects of anthropogenic nitrogen enrichment and climate change on terrestrial and aquatic biodiversity. *Biogeochemistry, 114*, 93–120. https://doi.org/10.1007/s10533-012-9803-3

Rabalais, N. N., Turner, R. E., Díaz, R. J., & Justić, D. (2009). Global change and eutrophication of coastal waters. *ICES Journal of Marine Science, 66*(7), 1528–1537. https://doi.org/10.1093/icesjms/fsp047

Richardson, D. M., & Whittaker, R. J. (2010). Conservation biogeography – foundations, concepts and challenges: Conservation biogeography: Foundations, concepts and challenges. *Diversity and Distributions, 16*(3), 313–320. https://doi.org/10.1111/j.1472-4642.2010.00660.x

Ridge, S. M., & McKinley, G. A. (2020). Advective controls on the North Atlantic anthropogenic carbon sink. *Global Biogeochemical Cycles, 34*, e2019GB006457. https://doi.org/10.1029/2019GB006457

Serra-Diaz, J. M., & Franklin, J. (2019). What's hot in conservation biogeography in a changing climate? Going beyond species range dynamics. *Diversity and Distributions, 25*(4), 492–498. https://doi.org/10.1111/ddi.12917

Silva, L. C. R. (2022). Expanding the scope of biogeochemical research to accelerate atmospheric carbon capture. *Biogeochemistry, 161*, 19–40. https://doi.org/10.1007/s10533-022-00957-1

Smith, V. H., & Schindler, D. W. (2009). Eutrophication science: Where do we go from here? *Trends in Ecology & Evolution, 24*(4), 201–207. https://doi.org/10.1016/j.tree.2008.11.009

Smith, D. B., Demetriade, A., de Carita, P., & Wang, X. (2018). The history, progress, and future of global-scale geochemical mapping. *Geochimica Brasiliensis, 32*(2), 115–135. https://doi.org/10.21715/GB2358-2812.2018322115

Sugimoto, N., Nishizawa, T., Shimizu, A., Matsui, I., & Jin, Y. (2014). Characterization of aerosols in East Asia with the Asian dust and aerosol LiDAR observation network (AD-net). In *Lidar remote sensing for environmental monitoring XIV* (Vol. 9262, pp. 74–82). SPIE. https://doi.org/10.1117/12.2069892

Townsend, A. R., & Porder, S. (2012). Agricultural legacies, food production and its environmental consequences. *Proceedings of the National Academy of Sciences, 109*, 5917–5918. https://doi.org/10.1073/pnas.1203766109

Urban, M. C., Bocedi, G., Hendry, A. P., Mihoub, J.-B., Peer, G., Singer, A., et al. (2016). Improving the forecast for biodiversity under climate change. *Science, 353*(6304), aad8466. https://doi.org/10.1126/science.aad8466

Van Rens, J. (2020). *The future of Lidar is critical to the future of our world*. https://www.gim-international.com/content/article/the-future-of-lidar-is-critical-to-the-future-of-our-world

Velazco, S. J. E., Villalobos, F., Galvão, F., & De Marco Júnior, P. (2019). A dark scenario for Cerrado plant species: Effects of future climate, land use and protected areas ineffectiveness. *Diversity and Distributions, 25*(4), 660–673. https://doi.org/10.1111/ddi.12886

Walther, G.-R., Post, E., Convey, P., Menzel, A., Parmesan, C., Beebee, T. J. C., et al. (2002). Ecological responses to recent climate change. *Nature, 416*(6879), 389–395. https://doi.org/10.1038/416389a

Wang, Z., & Menenti, M. (2021). Challenges and opportunities in Lidar remote sensing. *Frontiers in Remote Sensing, 2*, 641723. https://doi.org/10.3389/frsen.2021.641723

Welton, E. J., Campbell, J. R., Spinhirne, J. D., & Scott, V. S., III. (2001). Global monitoring of clouds and aerosols using a network of micropulse lidar systems. In *Lidar remote sensing for industry and environment monitoring* (Vol. 4153, pp. 151–158). SPIE. https://doi.org/10.1117/12.417040

Williams, J. W., & Jackson, S. T. (2007). Novel climates, no-analog communities, and ecological surprises. *Frontiers in Ecology and the Environment, 5*(9), 475–482. https://doi.org/10.1890/070037

Wilson, K. L., Skinner, M. A., & Lotze, H. K. (2019). Projected 21st-century distribution of canopy-forming seaweeds in the Northwest Atlantic with climate change. *Diversity and Distributions, 25*(4), 582–602. https://doi.org/10.1111/ddi.12897

Wulfmeyer, V., Hardesty, R. M., Turner, D. D., Behrendt, A., Cadeddu, M. P., Di Girolamo, P., Schlüssel, P., Van Baelen, J., & Zus, F. (2015). A review of the remote sensing of lower tropospheric thermodynamic profiles and its indispensable role for the understanding and the simulation of water and energy cycles. *Reviews of Geophysics, 53*(3), 819–895. https://doi.org/10.1002/2014RG000476

Yang, C., & Huang, Q. (2013). *Spatial cloud computing: A practical approach*. Taylor and Francis (CRC Press).

Correction to: Urban Biogeochemistry and Development: The Biogeochemical Impacts of Linear Infrastructure

Correction to:
Chapter 7 in: M. O. Campbell, *Biogeochemistry and the Environment*, https://doi.org/10.1007/978-3-031-47017-2_7

The original version of the chapter "Urban Biogeochemistry and Development: The Biogeochemical Impacts of Linear Infrastructure" was inadvertently published with incorrect author details. The correct details have been updated in the chapter.

The updated version of this chapter can be found at
https://doi.org/10.1007/978-3-031-47017-2_7

M. O. Campbell, *Biogeochemistry and the Environment*,
https://doi.org/10.1007/978-3-031-47017-2_9

Correction to: Biogeochemistry and the Environment

Michael O'Neal Campbell

Correction to:
M. O. Campbell, *Biogeochemistry and the Environment*, https://doi.org/10.1007/978-3-031-47017-2

The original version of the book was inadvertently published with incorrect author details. The initially published version has been corrected.

The updated version of this book can be found at
https://doi.org/10.1007/978-3-031-47017-2

M. O. Campbell, *Biogeochemistry and the Environment*,
https://doi.org/10.1007/978-3-031-47017-2_10

Index

M. O. Campbell, *Biogeochemistry and the Environment*,
https://doi.org/10.1007/978-3-031-47017-2